T0156287

Communications in Computer and Information Science 1613

More information about this series at https://link.springer.com/bookseries/7899

Mayank Singh · Vipin Tyagi · P. K. Gupta ·
Jan Flusser · Tuncer Ören (Eds.)

Advances in Computing and Data Sciences

6th International Conference, ICACDS 2022
Kurnool, India, April 22–23, 2022
Revised Selected Papers, Part I

Editors
Mayank Singh
University of KwaZulu-Natal
Durban, South Africa

Vipin Tyagi
Jaypee University of Engineering
and Technology
Guna, India

P. K. Gupta
Jaypee University of Information Technology
Waknaghat, India

Jan Flusser
Institute of Information Theory
and Automation
Prague, Czech Republic

Tuncer Ören
University of Ottawa
Ottawa, ON, Canada

ISSN 1865-0929 ISSN 1865-0937 (electronic)
Communications in Computer and Information Science
ISBN 978-3-031-12637-6 ISBN 978-3-031-12638-3 (eBook)
https://doi.org/10.1007/978-3-031-12638-3

This Springer imprint is published by the registered company Springer Nature Switzerland AG
The registered company address is: Gewerbestrasse 11, 6330 Cham, Switzerland

Preface

Computing techniques like big data, cloud computing, machine learning, the Internet of Things (IoT), etc. are playing a key role in the processing of data and retrieval of advanced information. Several state-of-art techniques and computing paradigms have been proposed based on these techniques. This volume contains papers presented at 6th International Conference on Advances in Computing and Data Sciences (ICACDS 2022) held during April 22–23, 2022, at G. Pullaiah College of Engineering and Technology (GPCET), Kurnool, Andhra Pradesh, India. The conference was organized specifically to help bring together researchers, academicians, scientists, and industry experts and to derive benefits from the advances of next-generation computing technologies in the areas of advanced computing and data sciences.

The Program Committee of ICACDS 2022 is extremely grateful to the authors who showed an overwhelming response to the call for papers, with over 411 papers submitted. All submitted papers went through a double-blind peer-review process, and finally 69 papers were accepted for publication in the Springer CCIS series. We are thankful to the reviewers for their efforts in finalizing the high-quality papers.

The conference featured many distinguished personalities such as Buddha Chandrashekhar, Advisor and CCO, All India Council for Technical Education, India; D V L N Somayajulu, Director, Indian Institute of Information Technology Design and Manufacturing, Kurnool, India; Shailendra Mishra, Majmaah University, Saudi Arabia; Dimitrios A. Karras, National and Kapodistrian University of Athens, Greece; Athanasios (Thanos) Kakarountas, University of Thessaly, Greece; Huiyu Zhou, University of Leicester, UK; Antonino Galletta, University of Messina, Italy; Arun Sharma, Indira Gandhi Delhi Technical University for Women, India; and Sarika Sharma, Symbiosis International (Deemed University), India; among many others. We are very grateful for the participation of all speakers in making this conference a memorable event.

The Organizing Committee of ICACDS 2022 is indebted to Sri. G. V. M. Mohan Kumar, Chairman, GPCET, India, for the confidence that he gave to us during organization of this international conference, and all faculty members and staff of GPCET, India, for their support in organizing the conference and for making it a grand success.

We would also like to thank Vamshi Krishna, CDAC, India; Sandip Swarnakar, GPCET, India; Sameer Kumar Jasra, University of Malta, Malta; Hemant Gupta, Carleton University, Canada; Nishant Gupta, Sharda University, India; Archana Sar, MGM CoET, India; Arun Agarwal, University of Delhi, India; Mahesh Kumar, Divya Jain, Kunj Bihari Meena, Neelesh Jain, Nilesh Patel, and Kriti Tyagi, JUET Guna, India; Vibhash Yadav, REC Banda, India; Sandhya Tarar, Gautam Buddha University, India; Vimal Dwivedi, Abhishek Dixit, and Vipin Deval, Tallinn University of Technology, Estonia; Sumit Chaudhary, Indrashil University, India; Supraja P, SRM Institute of Science and Technology, India; Lavanya Sharma, Amity University, Noida, India; Poonam Tanwar and Rashmi Agarwal, MRIIRS, India; Rohit Kapoor, SK Info Techies,

Noida, India; and Akshay Chaudhary and Tarun Pathak, Consilio Intelligence Research Lab, India, for their support.

Our sincere thanks to Consilio Intelligence Research Lab, India; the GISR Foundation, India; SK Info Techies, India; Adimaginz Marketing Services, India; and Print Canvas, India, for sponsoring the event.

May 2022

Mayank Singh
Vipin Tyagi
P. K. Gupta
Jan Flusser
Tuncer Ören

Organization

Steering Committee

Alexandre Carlos Brandão Ramos — UNIFEI, Brazil
Mohit Singh — Georgia Institute of Technology, USA
H. M. Pandey — Edge Hill University, UK
M. N. Hooda — BVICAM, India
S. K. Singh — IIT BHU, India
Jyotsna Kumar Mandal — University of Kalyani, India
Ram Bilas Pachori — IIT Indore, India
Alex Norta — Tallinn University of Technology, Estonia

Chief Patrons

G. V. M. Mohan Kumar — GPCET, India
G. Pullaiah — GPCET, India

Patrons

C. Srinivasa Rao — GPCET, India
K. E. Sreenivasa Murthy — RCEW, India

Honorary Chairs

Shailendra Mishra — Majmaah University, Saudi Arabia
M. Giridhar Kumar — GPCET, India
S. Prem Kumar — GPCET, India

General Chairs

Jan Flusser — Institute of Information Theory and Automation, Czech Republic
Mayank Singh — Consilio Research Lab, Estonia

Advisory Board Chairs

P. K. Gupta — JUIT, India
Vipin Tyagi — JUET, India

Technical Program Committee Chairs

Tuncer Ören — University of Ottawa, Canada
Viranjay M. Srivastava — University of KwaZulu-Natal, South Africa
Ling Tok Wang — National University of Singapore, Singapore
Ulrich Klauck — Aalen University, Germany
Anup Girdhar — Sedulity Group, India
Arun Sharma — IGDTUW, India
Mahesh Kumar — JUET, India

Conference Chair

N. Ramamurthy — GPCET, India

Conference Co-chairs

M. Rama Prasad Reddy — GPCET, India
G. Ramachandra Reddy — Ravindra College of Engineering for Women, India

Conveners

Sandip Swarnakar — GPCET, India
Sameer Kumar Jasra — University of Malta, Malta
Hemant Gupta — Carleton University, Canada

Co-conveners

Arun Agarwal — Delhi University, India
Suprativ Saha — Brainware University, India
B. Madhusudan Reddy — Ravindra College of Engineering for Women, India
Gaurav Agarwal — IPEC, India
Ghanshyam Raghuwanshi — Manipal University, India
Prathamesh Churi — NMIMS, India
Lavanya Sharma — Amity University, India

Organizing Chairs

Shashi Kant Dargar — KARE, India
K. C. T. Swamy — GPCET, India

Organizing Co-chairs

Abhishek Dixit	Tallinn University of Technology, Estonia
Vibhash Yadav	REC Banda, India
Nishant Gupta	MGM CoET, India
Nileshkumar Patel	JUET, India
Neelesh Kumar Jain	JUET, India

Organizing Secretaries

Akshay Kumar	CIRL, India
Rohit Kapoor	SKIT, India
Syed Afzal Basha	GPCET, India

Creative Head

Tarun Pathak	Consilio Intelligence Research Lab, India

Program Committee

A. K. Nayak	Computer Society of India, India
A. J. Nor'aini	Universiti Teknologi MARA, Malaysia
Aaradhana Deshmukh	Aalborg University, Denmark
Abdel Badeeh Salem	Ain Shams University, Egypt
Abdelhalim Zekry	Ain Shams University, Egypt
Abdul Jalil Manshad Khalaf	University of Kufa, Iraq
Abhhishek Verma	IIITM Gwalior, India
Abhinav Vishnu	Pacific Northwest National Laboratory, USA
Abhishek Gangwar	Center for Development of Advanced Computing, India
Aditi Gangopadhyay	IIT Roorkee, India
Adrian Munguia	AI MEXICO, Mexico
Amit K. Awasthi	Gautam Buddha University, India
Antonina Dattolo	University of Udine, Italy
Arshin Rezazadeh	University of Western Ontario, Canada
Arun Chandrasekaran	National Institute of Technology Karnataka, India
Arun Kumar Yadav	National Institute of Technology Hamirpur, India
Asma H. Sbeih	Palestine Ahliya University, Palestine
Brahim Lejdel	University of El-Oued, Algeria
Chandrabhan Sharma	University of the West Indies, West Indies
Ching-Min Lee	I-Shou University, Taiwan
Deepanwita Das	National Institute of Technology Durgapur, India
Devpriya Soni	Jaypee Institute of Information Technology, India

Donghyun Kim	Georgia State University, USA
Eloi Pereira	University of California, Berkeley, USA
Felix J. Garcia Clemente	Universidad de Murcia, Spain
Gangadhar Reddy Ramireddy	RajaRajeswari College of Engineering, India
Hadi Erfani	Islamic Azad University, Iran
Harpreet Singh	Alberta Emergency Management Agency, Canada
Hussain Saleem	University of Karachi, Pakistan
Jai Gopal Pandey	Central Electronics Engineering Research Institute, India
Joshua Booth	University of Alabama in Huntsville, Alabama
Khattab Ali	University of Anbar, Iraq
Lokesh Jain	Delhi Technological University, India
Manuel Filipe Santos	University of Minho, Portugal
Mario José Diván	National University of La Pampa, Argentina
Megat Farez Azril Zuhairi	Universiti Kuala Lumpur, Malaysia
Mitsunori Makino	Chuo University, Japan
Moulay Akhloufi	Université de Moncton, Canada
Naveen Aggarwal	Panjab University, India
Nawaz Mohamudally	University of Technology, Mauritius
Nileshkumar R. Patel	Jaypee University of Engineering and Technology, India
Nirmalya Kar	National Institute of Technology Agartala, India
Nitish Kumar Ojha	IIT Allahabad, India
Paolo Crippa	Università Politecnica delle Marche, Italy
Parameshachari B. D.	GSSS Institute of Engineering and Technology for Women, India
Patrick Perrot	Gendarmerie Nationale, France
Prathamesh Churi	NMIMS Mukesh Patel School of Technology Management and Engineering, India
Pritee Khanna	IITDM Jabalpur, India
Purnendu Shekhar Pandey	IIT (ISM) Dhanbad, India
Quoc-Tuan Vien	Middlesex University London, UK
Rubina Parveen	Canadian All Care College, Canada
Saber Abd-Allah	Beni-Suef University, Egypt
Sahadeo Padhye	Motilal Nehru National Institute of Technology, India
Sarhan M. Musa	Prairie View A&M University, USA
Shamimul Qamar	King Khalid University, Saudi Arabia
Shashi Poddar	University at Buffalo, USA
Shefali Singhal	Madhuben & Bhanubhai Patel Institute of Technology Engineering College, India
Siddeeq Ameen	University of Mosul, Iraq

Sotiris Kotsiantis	University of Patras, Greece
Subhasish Mazumdar	New Mexico Tech, USA
Sudhanshu Gonge	Symbiosis International University, India
Tomasz Rak	Rzeszow University of Technology, Poland
Vigneshwar Manoharan	Bharath Corporation, India
Xiangguo Li	Henan University of Technology, China
Youssef Ouassit	Hassan II University of Casablanca, Morocco

Sponsor

Consilio Intelligence Research Lab, India

Co-sponsors

GISR Foundation, India
Print Canvas, India
SK Info Techies, India
Adimaginz Marketing Services, India

Contents – Part I

Contents – Part II

Hardware Description Language Enhancements for High Level Synthesis of Hardware Accelerators

Gurusankar Kasivinayagam, Romaanchan Skanda, Aditya G. Burli, Shruti Jadon(✉), and Reetinder Sidhu

PES University, Bengaluru, India
shrutijadon@pes.edu

Abstract. High level synthesis of hardware accelerators is one of the many complex hardware operations that unfortunately cannot be efficiently performed with languages like Verilog or VHDL. Hardware designers, in order to bridge these gaps present in traditional HDLs, have taken to implementing such high-level hardware operations using functional programming (FP) languages or languages derived from FP languages. This is because FP languages (or their derivatives) have many important features like MapReduce, Immutable variables and Lazy evaluation. Today, only languages like Chisel, MyHDL or Haskell are used to perform high-level hardware operations. This obviously presents itself as a learning curve that hardware designers and experts have to go through in addition to learning Verilog or VHDL. This paper presents a novel approach that aims to take the standard syntax of Verilog and provide necessary enhancements for it to support basic functional programming constructs like Chain and Tree. The main component of this approach is the translation of the enhanced Verilog syntax to standard Verilog. This will be achieved using relevant Python libraries, methods and syntactic macros.

Keywords: Functional programming · Hardware accelerators · High-level synthesis · Enhanced verilog

1 Introduction

Programming languages on the software side have seen continuous change, variety and diversification based on the needs and wants of the time. Languages such as Java and Python have been updated multiple times to improve and expand their respective functionalities and uses. This however, has not been the case with respect to hardware description languages like Verilog and VHDL. Even as hardware operations and designing has reached new heights in terms of complexity, volume and variety, the scope and usage of these languages have remained largely stagnant. The fuzzy logic implementation in [5] allows us to know the limitation of Verilog.

The lack of support provided by these traditional HDL languages for high-level hardware operations and designing of advanced hardware circuits, has led to the growing

M. Singh et al. (Eds.): ICACDS 2022, CCIS 1613, pp. 1–12, 2022.
https://doi.org/10.1007/978-3-031-12638-3_1

community of hardware designers and experts to look for other suitable alternatives to achieve the above tasks, owing largely to the meticulous nature of their development. The suitable alternative turned out to be functional programming.

Functional programming was very useful in performing such high-level hardware operations due to many of its unique features. [8] goes into a lot more detail about the features. The following are some of the features:

1. MapReduce - functionalities like map and reduce allow programmers to transform and deduce values and results from big data clusters with efficient parallelism and modularity.
2. Immutable variables - in functional programming the variables created upon transformation and actions are unique and not modifiable.
3. Lazy evaluation - or call-by-need, is an evaluation strategy which delays the evaluation of an expression until its value is needed.

We learn from [4] and [6] that languages like Chisel and MyHDL are either derived from FP language Scala or work within languages support that support functional programming like Python. This paper will go into the details of these languages in the literature review section. Although the creation of such languages has allowed hardware designers to perform different high-level operations, it cannot be denied that having already learnt Verilog, the process of learning another language is always going to be a substantial learning curve.

Through this paper, we aim to document our effort in developing a tool that will introduce certain elements of functional programming into the Verilog syntax. This will be achieved using a syntax translation tool which will efficiently convert the enhanced Verilog syntax into the standard Verilog syntax. This tool, which is the main core of this development will be created using the Python libraries, methods and syntactic macros. Ultimately, the objective is to provide basic functional programming features like higher order functions and transformers within the most common hardware description language that many hardware designers already know, which in this case is the Verilog HDL. The Chain and Tree constructs will be the specific focus of this paper.

2 Literature Review

We performed a significant survey in this area of research and the most significant work are reviewed here. The work proposed by Mathew Pole [1] briefly introduces some interesting aspects of Haskell-a functional programming language. The first of these is that all functions in Haskell are considered to be single-argument functions and that nested functions are curried together. Essentially, curried functions are functions receiving other functions as input. Mathematically, if we have two function f(x) and g(x), currying would be similar to f(g(x)). With partial application we would substitute g(x) in f(x), resulting in a function in terms of g(x). The output of such a function is a partially applied function. Type Classes in Haskell can be described as a set of types that share certain operations, with each type being supported by different functions, hence supporting ad-hoc polymorphism (overloading). Another interesting idea in Haskell was parameterized types with a brief discussion on descriptors like Maybe, Just and Either.

Another work was proposed by Lenon and Gahan [2] where a comparative study of Chisel with Verilog for FPGA Design is discussed. Chisel is a hardware description language developed on Scala to introduce OOP elements into hardware design. Their paper quantitatively compares the differences observed over 4 different parameters when implementing an N-bit FIFO, a round robin arbiter and a complex, scalable arbiter. The four parameters that were observed and recorded were: Code Density and Maintainability, Design Flow Run Time, Silicon Area, Performance (maximum operating frequency). It was observed that for each, the design flow run time, silicon area and performance were comparable between Chisel and Verilog. In terms of code density and maintainability, although both were comparable in number of lines of code, the complexity of writing Verilog code continued to increase with increasingly difficult components, and here is where Chisel seems to make a better impression. Additionally, Chisel's rapid C++ simulation support makes it very attractive as a tool for hardware design.

The work proposed by Khanfor and Yang [3] takes a deep dive into the ocean of functional programming and tries to discover or analyse the various practical impacts functional programming has had on the programming world. To understand the impact of functional programming, the authors decided to use 'functional programming' as a search term to their selected online libraries. The figure below shows the number of papers published in the domain of functional programming over the past few decades. As shown in Fig. 1, in the year 2017, there were slightly more than 180 papers published relating to functional programming which is a significant increase from slightly over 10 papers published in 1987. This trend reflects the growing interest in functional programming and its usage.

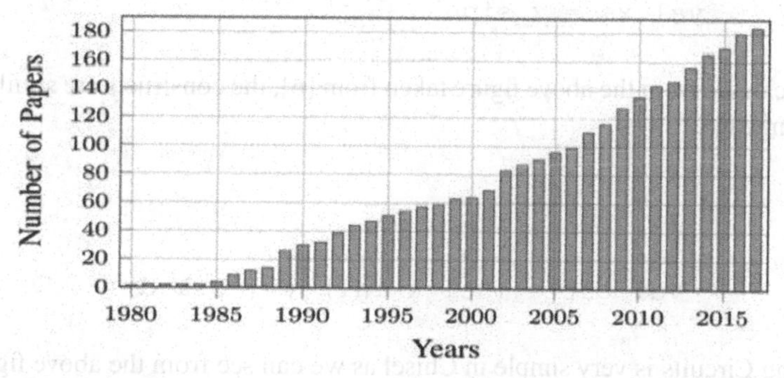

Fig. 1. Increase in the research of functional programming with time

The paper [3] poses research questions upon these papers as follows- RQ1: What is the research outlook in FP paradigm? RQ2: In which software engineering development practice FP paradigm had been studied? The paper [3] moves on to analyze the practical impact of functional programming languages on the criterions such as- software design, security, software reuse, cost estimation etc. The paper asserts that functional programming is a well-established research topic, which will eventually see more extensive studies and analysis into its workings, benefits and drawbacks.

In 2015, Jaic and Smith [4] introduced the features of MyHDL. MyHDL is a hardware development environment based on Python. It makes good use of the dynamic nature of Python for hardware development. Some of the major functions served by MyHDL within hardware development are: modelling, simulation and verification. Rapid prototyping of FPGAs is also possible in MyHDL. MyHDL achieves model concurrency by using generators and utilizes decorators for facilitating hardware description.

MyHDL designs hardware by supporting RTL (Register Transfer Level). It provides two decorators- @always_comb and @always_seq. The authors provide sample code to illustrate combinatorial multiplexer and sequential incrementor with enable signal. MyHDL also can convert its design to Verilog or VHDL. It produces a testbench which can be co-simulated with external simulators like iVerilog. MyHDL also performs attribute conversion by analyzing abstract syntax tree (AST) of the generators. It works by assuming all objects to either be signals or integrated bit vectors. With this added feature of attribute conversion MyHDL can raise its abstraction levels. It can use Python classes to create parameterized interfaces.

Chisel is more or less like a platform which provides modern programming language features for accurately specifying low-level hardware blocks. One of the other advantages of using Chisel is the presence of Bundle classes. It is defined as follows:

```
class MyFloat extends Bundle {
  val sign = Bool()
  val exponent = UFix(width = 8)
val significand = UFix(width = 23)
}
val x = new MyFloat()
val xs = x.sign
```

As we can see from the above figure taken from [6], the constructs are similar to that of Scala language.

```
(a & b) | (~c & d)
val sel = a | b
val out = (sel & in1) | (~c & d)
```

Defining Circuits is very simple in Chisel as we can see from the above figure. Had it been the traditional Verilog we would have to define and instantiate modules for usage and look for the proper connections within the circuits. All of these expressions are stored in the form of nodes of a graph.

To define concisely, hygienic macros are macros that preserve referential transparency. In simpler words, hygienic macros automatically preserve the lexical scoping of all identifiers within a macro definition thereby preventing confusion between values when there are multiple variables with the same identifier, but in different scopes.

Hygienic macros ensure this by creating hygiene conditions by following two steps described in [7]:

1. Identifying situations where the condition of hygiene does not hold.
2. Applying an appropriate number of alpha-conversions (as defined in lambda calculus) to restore hygiene.

With each expansion, the clock value is increased and this is used to determine the origin of the identifier. Tokens with different time-stamps come from different scopes, and upon coming across more than one identifier, the difference in scope is preserved by applying an appropriate number of alpha conversions. Free-identifiers are identifiers which are not nested within a macro or a function definition. These will not require any modification.

Finally, the time-stamped syntax tree is converted back into a standard syntax tree, resulting in a hygienic macro expansion. With the introduction of hygienic macros, developers can now focus on the functional aspects of the macro expansion without having to deal with scope issues.

3 Methodology

As we have noticed in the literature review section of this paper, that over the last few years, various solutions with different approaches have been tried and implemented to get around the problem that hardware description languages like Verilog and VHDL face of simply being incapable and inefficient of handling complex hardware operations and circuit simulations.

The two key functionalities addressed in this paper are the Chain and Tree functionalities. The detailed description and use of these functions will be elucidated in the coming section. The successful implementation of these two functions within the Verilog construct will signify the enhancement that can be made to Verilog for it to be compatible with complex hardware operations like high-level synthesis.

The overall methodology being employed in the enhancement of the Verilog syntax is divided into three different phases.

3.1 Syntax Identification and Code Extraction

The Verilog file written in the non-standard (enhanced) syntax is taken as the input file of the process. The first phase of the process involves the identification of the non- standard syntax of Verilog. The identification is important in distinguishing between the standard and the enhanced Verilog syntax that are present in the input Verilog code. The method used to perform the identification involves the use of string pattern matching to match the enhanced syntax. The time complexity of this process is O(n) where n is the number of lines with extended syntax. This holds true because the source code is parsed only once to identify the lines containing extended syntax.

The identification is followed up with a targeted extraction of the subset of the code present in the file that has been written in the enhanced syntax. The extraction of the code subset allows in the further processing of the given code to produce discernible and transformable forms of the code. Any comments in the extracted modules are removed.

What this essentially means, is that high-level/abstract functions implemented with a concise format in the enhanced Verilog syntax can be expanded to simpler modules, functions, variables, and wires for further processing upon successful extraction.

3.2 Syntax Transformation

The identification and the extraction phase can be compared to the process of removing the gift wrapper to reveal the actual gift(s). Once that is done, the precisely extracted code can now be processed to initiate the decomposition procedure. This is done to come a step closer to the eventual objective of transforming the fancy and enhanced Verilog code to the standard syntax.

In order to perform the transformation of the non-standard syntax, higher-order functions are used. From the extraction process, we have with us the individual statements that are to be transformed.

The higher-order function performs the task of converting the extracted higher-order statements, which have a higher degree of abstraction (a tree or chain statement) into much simpler and smaller modules that are typically implemented in standard Verilog with relative ease. This is the reason why they are referred to as higher-order functions, as they take functions as inputs and return functions as outputs.

3.3 Code Replacement

Once we are done with the task of transforming the extracted non-standard code to the standard syntax of Verilog, the final and the most practical task that is left on the table is a successful replacement of the code back into the parent code, in place of the enhanced syntax statements. This final output is stored as a new file in the same directory.

4 Implementation

The framework consists of 3 modules, viz., the Keyword Extractor which extracts lines containing extended syntax, the Argument Extractor which extracts the arguments passed into the higher-order functions and the Syntax Transformation of Chain and Tree constructs.

4.1 Keyword Extractor

The keyword identifier makes use of simple string functions to determine the beginning and the end of the line containing the extended syntax. The two syntax constructs it looks for are 'tree' and 'chain'.

The pattern for the construct is searched for throughout the code. If such a construct does not exist, it fails silently and returns an empty string. If it does exist, a substring from the input Verilog file containing the entire line is returned.

The pseudocode is as follows:

```
text <- input verilog file
list_of_statements <- find_enhanced_statements(text)
for each statement in list_of_statements:
  op_file += transform(statement)
```

4.2 Argument Extractor

The argument Extractor runs following the Keyword Identifier. Using string operations, the various arguments as part of the extended syntax statement are identified as part of the syntax definition and passed onto the syntax transformer.

For tree statements, the syntax is as follows. The syntax for chain statements is similar in structure.

```
tree (base_module_name_prefix,
      output_module_name,
      output_module_number_of_outputs)

# with common input wire
tree (base_module_name_prefix,
      output_module_name,
      output_module_number_of_outputs,
      common_input_wire)
```

4.3 Syntax Transformation

Tree Construct

This construct will help us to create an optimal tree for the given higher order gates. In this way the internal mapping done will take care of the input order and passing. There are 4 different cases present for construction of the trees based on powers of 2: 2n, 2n + 1, (2n + 2m + …), (2n + 2m + … + 1). Each case is handled separately, but ultimately the only gates used for implementing the so-called module will be the gates taking 2^n inputs. The order of the units is degraded to 2 inputs finally. This is done for all gates such as and, or, xor, xnor, etc. The declaration of the construct is as follows:

```
tree (or2, 2, myor32, 32);
```

These follow the functionality similar to that of Higher order functions in which a function is passed as a parameter to another function. In the above given declaration, there are 4 parameters. The first parameter is the basic module used. This module is supposed to be written in the input file. The second parameter tells us the number of inputs for the basic module. The third parameter tells us the name of the newly created module in the output file. This module is the one which is generated. The fourth parameter tells us the number of inputs to the corresponding newly module definition.

In order to find out how many module instantiations are of each kind are needed we run a recursive degradation algorithm which will keep on reducing the modules to the size of 2 inputs which is more convenient to manipulate and manage. Many helper functions have been identified and created so that we can create the tree in the most efficient way possible.

These functions have individual functionality. The first function is to create the declaration of the new function. It extracts the required input and output wires based on these constraints the module declaration with the right format.

A second function is defined which is used to find the number of internal wire mappings necessary for the module and takes care of its declaration for the module. This is necessary as many inputs will be mapped to a particular unit. This way nothing goes wrong in the further steps.

A third function is responsible for the instantiation of the base module the necessary number of times and in the right order. This also includes the use of the temporary wires present inside the module definition. Each of the module is defined the necessary number of times.

A fourth function combines all the outputs generated from the previous function to generate a final output thereby completing the entire module definition. Essentially using the exact structure of a tree to create the module definition. This is depicted as follows in Fig. 2.

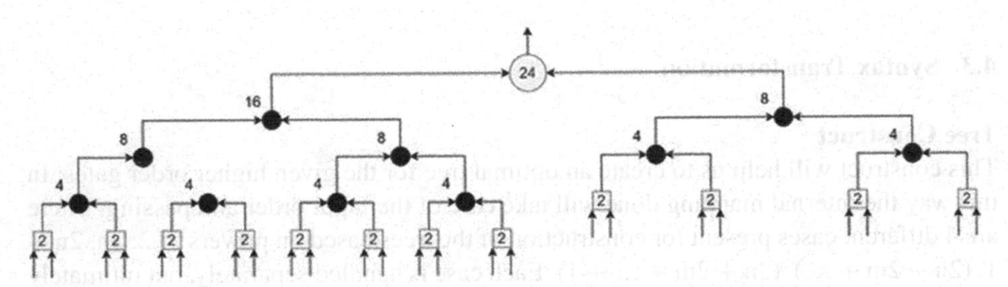

Fig. 2. Tree structure for an 'n' input AND gate

Chain Construct

The chain constructs handle two cases. The first case involves instantiating a series of modules where there are two classes of inputs. The first set of inputs are provided by the user and the second input is mapped from the previous modules output. This in turn creates a chain of modules connected because the outputs are passed from one module to the next. A prime example of this would be a ripple carry adder as shown in Fig. 3.

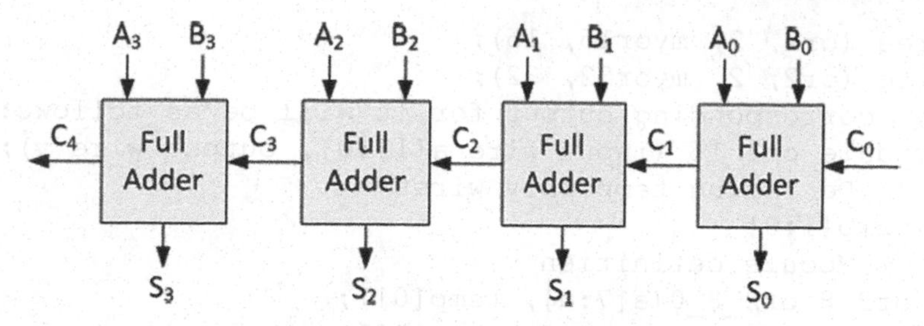

Fig. 3. Ripple Carry Adder as an example for chain

The second case involves instantiating a series of modules with a common input. This would be better explained using an example as shown in Fig. 4. Let's assume you want to create a switch that chooses between 4 sixteen-bit inputs. To achieve this, you would create sixteen 4:1 Multiplexers and attach them to the same set of select lines (common inputs). This way, you can chain multiple 4:1 Multiplexers with the same set of select lines.

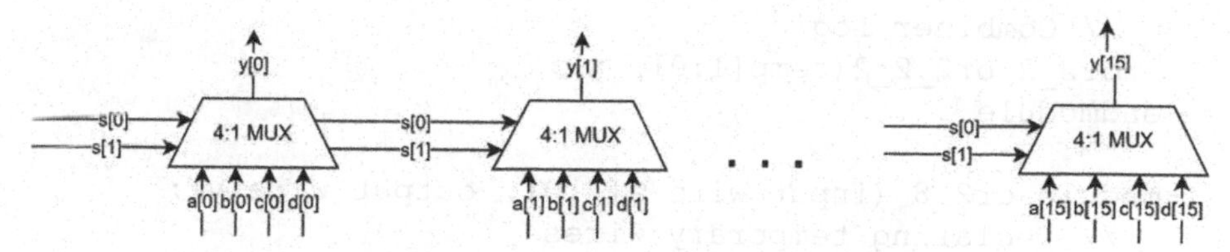

Fig. 4. Chain structure for 4 16-bit input switch

5 Results

Unit testing was performed for all individual components created. This includes the independent helper functions for the trees and chains. This guaranteed that all the functional components are working as intended. This also helped in debugging the edge test cases. Integration testing was performed on the overall framework as well. It can be shown that it works on all the test cases which are used by any and every user. The outputs generated for one such test case is shown below. Consider that a tree statement is declared in the source file.

```
tree (or2, 2, myor15, 15);
tree (or2, 2, myor32, 32);
The corresponding output for it will be as follows:
 module or2_16 (input wire a[15:0], output wire y);
   // Declaring temporary wires
   temp[1:0]
   // Module definition
   or2_8 or2_8_0(a[7:0], temp[0]);
   or2_8 or2_8_1(a[15:8], temp[0]);
   // Combiner Logic
   or2_2 or2_2_2(temp[1:0], y);
endmodule

 // Declaring temporary wires temp [1:0];
 module or2_4 (input wire a[3:0], output wire y);
   or2_2 or2_2_0(a[1:0], temp[0]);
   or2_2 or2_2_1(a[3:2], temp[0]);
   // Module definition
   // Combiner Logic
   or2_2 or2_2_2(temp[1:0], y);
 endmodule

 module or2_8 (input wire a[7:0], output wire y);
   // Declaring temporary wires
   temp[1:0];
   // Module definition
   or2_4 or2_4_0(a[3:0], temp[0]);
   or2_4 or2_4_0(a[7:4], temp[0]);
   // Combiner Logic
   or2_2 or2_2_2(temp[1:0],y);
 endmodule

 module myor15 (input wire a[14:0], output wire y);
   // Declaring temporary wires
   temp[3:0];
   // Module definition
   or2_8 or2_8_0(a[7:0], temp[0]);
   or2_4 or2_4_1(a[11:8], temp [1]);
   or2_2 or2_2_2(a[13:12], temp[2]);
   assign temp[3] = a[14];
  // Combiner Logic
   or2_4 or2_4_4(temp [3:0], y);
  endmodule
```

```
module myor32 (input wire a[31:0], output wire y);
    // Declaring temporary wires
    temp[1:0];
    // Module definition
    or2 16 or2_16_0(a[15:0], temp[0]);
    or2_16 or2_16_0(a[31:16], temp [0]);
    // Combiner Logic
    or2_2 or2_2_2(temp [1:0], y);
endmodule
```

The output is generated when both the trees are present in the input file. The first declaration is an or gate with 15 inputs. This can be degraded into an 8 input or gate, a 4 input or gate, a 2 input or gate and a single assignment statement. Therefore, individual definitions are created for each of these gates. As we can see from the output, we have a definition for 8 and 4 input or gates. The 2 input or gate definition is given by the user in the input file without any kind of changes.

When the second tree statement is parsed, this results in the creation of a tree with 32 inputs. This 32-input gate is further divided into 2 gates as we can see in the definition. The combiner takes care of combining the output generated from the 2 16-input or gates. The 16 input or gates are further divided into 8 & 8 and so on. Now since we have already defined the 8 input or gate already in the file, we don't have to redefine it. Hence its only printed once. Therefore, every definition is declared exactly once as necessary. Every definition will have a combiner logic which will combine the outputs from the instantiated gates up the tree to give out a final output.

For chain statements, we find similar outputs where a chain statement is identified, its arguments extracted and then the code generated according to the parameters passed.

6 Conclusion and Future Work

The framework implementation is done with regards to the usage of tree and chain constructs. This handles a majority of the use cases for which this was intended to be developed. By dividing the implementation into suitable number of components similar to that of microservices architecture, we achieve modularity and the necessary speed for translation. We divided the entire framework into 4 components and regulating each component independently, we were able to achieve the necessary goal.

When tree and chain work together, it's possible to create any large hardware definition because these are the two most essential procedures created to scale hardware components using the smallest logical unit. Despite being limited in functionality; the idea of trees and chains has shown great potential.

The future goals include, addition of more functional features into the framework. Since this will be open sourced, the user can customize the framework based on his/her needs and add compatibility for their own developed functions. There is also an intention to create mapping files which will be responsible to show the mapping from input to out- put based on the testbench values. This feature can be enabled or disabled with the help of flags given in the command line.

In addition to that, the tree and chain function are currently limited and thus, can only accept simple modules with one set of inputs and outputs with an optional common input line. The sophistication of the modules handled by tree and chain could be improved and thereby improve its utility and the flexibility with which it can be used.

And finally, the tree and chain functions could be further abstracted and the different chain functions could be combined creating more complex use cases.

References

1. Mathew P.: A block design for introductory functional programming in Haskell. In: 2019 IEEE Blocks and Beyond Workshop (B&B), pp. 31–35 (2019)
2. Lenon, P., Gahan, R.: A comparative study of chisel for FPGA design. In: 2018 29th Irish Signals and Systems Conference (ISSC), pp. 1–6 (2018)
3. Khanfor, A., Yang, Y.: An overview of practical impacts of functional programming. In: 24th Asia-Pacific Software Engineering Conference Workshops (APSECW), pp. 50–54 (2017)
4. Jaic, K., Smith, M.: Enhancing hardware flows with MyHDL. In: ACM/SIGDA International Symposium on FPGA, pp. 28–31 (2015)
5. Baldania, M.D., Patki, A.B., Sapkal, A.M.: Verilog - HDL based implementation of a fuzzy logic controller for embedded systems. In: IEEE International Conference on Computational Intelligence and Research, pp. 1–4 (2013)
6. Bachrach, J., et al.: Chisel-constructing hardware in a scala embedded language. In:DAC Design automation conference 2012, pp. 1212–1221 (2012)
7. Kohlbecker, E., Friedman, D.P., Felleisen, M., Duba, B.: Hygienic macro expansion. In: ACM conference on LISP and functional programming, pp. 151–161 (1986)
8. Hinsen, K.: The promises of functional programming. Comput. Sci. Eng. J. **11**(4), 86–90 (2009)
9. Villar, J.I., Juan, J., Bellido, M.J., Viejo, J., Guerrero, D., Decaluwe, J.: Python as a hardware description language: a case study. In: 2011 VII Southern Conference on Programmable Logic (SPL), pp. 117–122 (2011)
10. IEEE Standard Verilog Hardware Description Language. In: IEEE Std 1364–2001, pp.1–792, 28 (2001)
11. Ebeling, C., French, B.: Abstract verilog: a hardware description language for novice students. In: 2007 IEEE International Conference on Microelectronic Systems Education (MSE'07), pp. 105–106 (2007)
12. Bove, A., Arbilla, L.: A confluent calculus of Macro expansion and evaluation. In: ACM Conference on LISP and Functional Programming, pp. 278–287 (1992)
13. Bartley, D., Hanson, C., Miller, J.: IEEE Standard for the Scheme Programming Language. IEEE (1991)
14. Herman, D., Wand, M.A.: Theory of hygienic macros. In: Drossopoulou, S. (eds.) Programming Languages and Systems. ESOP 2008. LNCS, vol. 4960. Springer, Heidelberg (2008). https://doi.org/10.1007/978-3-540-78739-6_4
15. Keeney, J., Cahill, V.: Chisel: a policy-driven, context-aware, dynamic adaptation framework. In: IEEE 4th International Workshop on Policies for Distributed Systems and Networks, pp. 3–14 (2003)

CrDrcnn: Design and Development of Crow Optimization-Based Deep Recurrent Neural Network for Software Defect Prediction

S. Sai Satyanarayana Reddy[1], Ashwani Kumar[2(✉)], N Mounica[3], and Donakanti Geetha[3]

[1] Sreyas Institute of Engineering and Technology, Hyderabad 500068, India
[2] Department of Computer Science and Engineering (AIML), Sreyas Institute of Engineering and Technology, Hyderabad 500068, India
ashwani.kumarcse@gmail.com
[3] Department of Computer Science and Engineering, Sreyas Institute of Engineering and Technology, Hyderabad 500068, India

Abstract. Software defect prediction (SDP) is the emerging research direction, which attains significant consideration among the advanced research community. The SDP is utilized to evaluate the reliability and quality of the software together with better allocation of limited testing resources. Numerous data mining methods are available in SDP for analyzing the source code, and developmental process and also for extracting critical metrics. The Deep Recurrent Neural Network (DNN) is utilized in this research for the prediction purpose, in which the hyperparameters of the classifier are effectively tuned by the crow search optimization. Further, to minimize the computational time most relevant features are selected with the help of the wrapper selection technique. The analysis is done using the PROMISE data set, based on the performance metrics such as specificity, sensitivity, and accuracy. The specificity, sensitivity, and accuracy of the proposed method are found to be 98.6248%, 93.5694%, and 92.0506% in accordance with the training percentage.

Keywords: Software defect prediction · Feature selection · Deep learning · Optimization · Software development

1 Introduction

Software system dominates the entire world due to their applicability in all events of human life, which makes them indispensable [1, 2]. Software testing is an essential part of the entire software evolution process as it guarantees the reliability and the quality of the software, which greatly influences the technical advancement of the software evolution process [3, 4]. The advanced software is proved to be more productive than ever, which makes them more complex and entangled [5]. The complexity of the advanced software system degrades the reliability and the quality of the f the software. Hence, a separate team was employed by the software companies to ensure the quality of the software program and it is proven to be a more expensive and intensive process. Furthermore,

M. Singh et al. (Eds.): ICACDS 2022, CCIS 1613, pp. 13–25, 2022.
https://doi.org/10.1007/978-3-031-12638-3_2

it is a time-consuming process [6]. The prior determination of the defective module enables the software programmers (developer) to concentrate on the particular defective module to enhance the quality of the software. Analyzing the particular defective model significantly increases the quality and minimizes the time [4]. However, software testing often consumes more time to determine all the defects in the entire test case. It is impossible to conduct tests for an entire project with a tight schedule and low budgets [7]. Hence, an automated system known as software defect prediction is presented to detect the defects in the software with low budgets and less time [6]. SDP is utilized to identify the defect in the software framework, which makes use of the historical data to predict the location of defective code in the new program. Both deep learning and machine learning techniques are used to construct the productive defect forecasting framework [6, 8]. The SDP procedures consist of three significant steps such as a) gathering of historical data, b) training of the classification or the regression model using the Deep learning or the ML framework, and c) utilizing the trained DL framework to forecast the possibility and the total defect in the given source code [9]. Some of the defect prediction frameworks such as McCabe's cyclomatic complexity metrics and Halstead's software volume metrics utilize the existing features to forecast the defects. Some of the approaches are based on machine learning and statistical theories. Support vector regression (SVR) [10], neural Network [11], Random Forest (RF) [12], and fuzzy support vector regression (FSVR) [13] are some of the approaches that depend on machine learning theories. These models make an attempt to construct the classification or the regression frameworks to forecast the possibility of faults in the particular software [14]. These methods are found to be effective in predicting the defects in the software. Yet, their performance, such as Squared correlation coefficient and mean square error needs considerable improvement [9, 15].

Motivation

The software defect prediction (SDP) helps to identify the defect-prone module that indicates further testing. Thus, the SDP is helpful in assisting the testing phase of software development. However, the detection of the defect is not an easy task and is difficult to predict. Several methods were developed for the prediction of the software defect, still, the inaccurate prediction, requirement of several parameters for the prediction, and failure to consider the general ability are considered as the challenging task. Hence, this research introduces an efficient technique for the prediction of software defects using the deep learning technique with the optimization approach. Here, the most relevant features are selected through the wrapper method so as to reduce the dimensionality of the data, which in turn reduces the computational complexity and the computation time of the data. Further, the selected features are fed forward to the optimized DNN for the effective prediction of the software defects. For accurate prediction, the hyperparameters of the DNN are optimally tuned by the crow search optimization. The organization of the research article follows: The review of the literature with the need for the software defect prediction model is enumerated in Sect. 2. The proposed SDP model is elucidated in Sect. 3 and the result analysis is deliberated in Sect. 4. Finally, the article is concluded in Sect. 5.

2 Related Works

This section provides deep insight into the existing methods utilized for software defect prediction, which further motivates to development of the dynamic model that handles the limitations of the existing methods. Lei Qiao et al. [9] presented the SDP model based on the deep learning techniques, which effectively predicts the total defects in the software. The reduced mean square value (MSE) is the main advantage of this SDP model. Yet, the model is not applicable to large commercial projects. Yuanxun Shao et al.[16] utilize the correlation weighted class association rule mining (CWCAR). This method consists of pruning, ranking, and prediction stages that are established based on the weight support. The similarities and variations of the association rules are not yet explored in this research. Zhongbin Sun et al.[17] presented the recommendation algorithm so-called collaborative filtering-based sampling recommendation algorithm for the accurate prediction of software defects. Kumar et al. given many solutions for detecting the object from the images using machine learning algorithms [18–21]. Zhiguo Ding [4] presented an advanced SDP framework with the help of Pruned Histogram-based isolation forest. This method effectively restrains the issues related to the imbalanced training dataset. However, the model requires further enhancement with a clustering framework to demonstrate the performance and applicability of the system. Kechao Wang et al. [2] utilize machine learning to forecast the software programming defects in the software.

2.1 Challenges

The major challenges addressed by this research are given below.

- The conventional SDP model is not suitable for high dimensional data set, which deteriorates the performance of the system by increasing the training time and computational perplexity.
- The contrasting nature of the software defect is the major issue that hinders the prediction of the conventional methods.

Thus the challenges faced by the conventional SDP methods are higher computational complexity, training time, inaccurate detection, slow convergence rate, and so on. Hence to overcome these challenges optimization based deep learning method is introduced by the proposed method, in which the deep NN is trained with the crow optimization algorithm, which enhances the convergence rate and training time through optimal tuning of weights, improves the accuracy of prediction through the selection of appropriate features, which may also reduce the computational complexity.

2.2 Feature Selection Using Wrapper Algorithm

The wrapper method is one of the leading techniques, which depicts better results than the conventional filter method. The wrapper algorithm obtains the best feature subsets with respect to the predictive power of the features. The wrapper method known as Recursive Feature Elimination (RFE) is categorized as the feature selection algorithm that consists of the kernel function. The prime intention is to determine the subset of size S within V number of variables ($S < V$) with the help of the SVM classifier that enhances the performance of the predictor. The method aims to determine the optimal subset of S feature. This only permits the optimal features to attain the final destination. The wrapper algorithm attempts to find out the subsets from the S features. It attempts to select the S features that result in the largest extremity of class categorization. The backward selection process is utilized to eliminate one feature from the V number of features. The elimination of the feature is continued until it attains the S number of features from the entire features in the input data. The feature that illustrates maximum variation from the other features is eliminated so as to reduce the variation of predictive ability $P(a)$. The predictive ability $P(a)$ is estimated by the following equation

$$P^2(a) = \sum_{i,j=1}^{m} a_i a_j r_i r_j \mathrm{K}\left(t_i, t_j\right) \tag{1}$$

where, a defines the vector of Lagrange multipliers, K represents the kernel function, $r \in [-1, 1]$. Estimate the prediction ability for each feature F:

$$P^2_{-F}(a) = \sum_{i,j=1}^{m} a_i a_j r_i r_j \mathrm{K}\left(t_i^{-F} t_j^{-F}\right) \tag{2}$$

where, t_i^{-F} demonstrate the training object i, and t_j^{-F} demonstrate the training object j, in which the feature F is removed.

3 Proposed Software Defect Prediction Model Using the Optimized Deep NN Classifier:

This research concentrates to develop the productive SDP model, which accurately forecasts the defects in the software. In the proposed methodology the most relevant features are selected through the wrapper methods. These highly relevant features are now fed forward to the optimized deep neural network for the accurate prediction of software defects. The hyperparameters of the deep neural network are effectively tuned by a meta-heuristic nature-inspired algorithm known as the crow search optimization algorithm. This optimization algorithm enhances the prediction accuracy of the Deep Neural Network by tuning the hyperparameters. The diagrammatic representation of the proposed SDP framework is illustrated in Fig. 1.

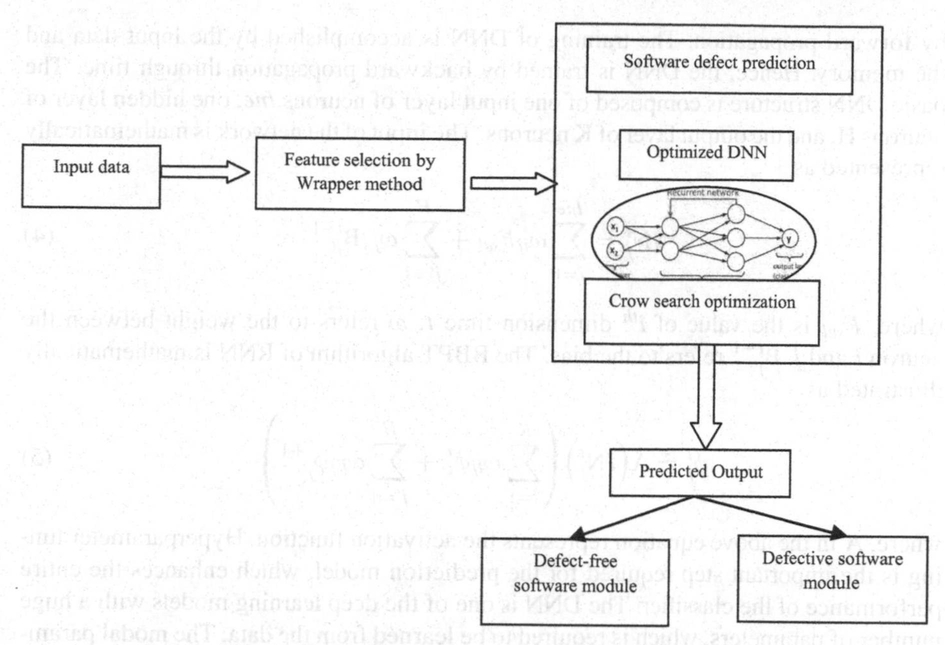

Fig. 1. Block diagram of the proposed methodology

At first, the input data is collected from the PROMISE dataset, which is mathematically symbolized as,

$$H = \{K_i\}; \ (1 \leq i \leq m) \tag{3}$$

where, m in the above equation refers to the total data in the PROMISE dataset and K_i refers to the ith data acquired from the dataset. For the selected input, the most significant features are extracted using the Wrapper method, which is detailed in Sect. 2.2., which is fed as input to the classifier for the prediction of the defect in the software.

3.1 Software Defect Prediction Using Proposed Crow Optimization-Based DNN for Software Defect Prediction

The software defects are predicted using the proposed crow-based DNN classifier, which is detailed in the below section.

3.1.1 DNN Classifier for Defect Prediction

The selected features are now fed forward to the DNN for the effective prediction of software defects. The main advantage of the DNN is that it introduces the recurrent structure by the self-controlled neurons. By using these neurons the historical data is memorized and these data will influence the network output. The DNN outperforms all other neural networks because of the memory posses by them. The internal process of the DNN is similar to that of the Feed forward Neural Network (FNN), which is completed by forward propagation. The training of DNN is accomplished by the input data and the memory. Hence, the DNN is trained by backward propagation through time. The basic DNN structure is composed of one input layer of neurons *Ine*, one hidden layer of neurons H, and the output layer of K neurons. The input of the network is mathematically represented as

$$IN_j^t = \sum_{i=1}^{Ine} \omega_{ij} F_{sel} + \sum_{j'=1}^{H} \omega_{j'j} B_{j'}^{t-1} \tag{4}$$

where, F_{sel} is the value of ith dimension time t, ω refers to the weight between the neuron i and j, $B_{j'}^{t-1}$ refers to the bias. The BBPT algorithm of RNN is mathematically illustrated as

$$\phi_j^t = A\left(IN_j^t\right)\left(\sum_{i=1}^{K} \omega_{ij}\phi_k^t + \sum_{j'=1}^{H} \omega_{j'j}\phi_{j'}^{t+1}\right) \tag{5}$$

where, A in the above equation represents the activation function. Hyperparameter tuning is the important step required for the prediction model, which enhances the entire performance of the classifier. The DNN is one of the deep learning models with a huge number of parameters, which is required to be learned from the data. The modal parameters can be fixed by training the modal with the available data. Yet, there is other categories of parameter so-called hyper-parameter that cannot be fixed during the training process. The hyperparameters are needed to be fixed before the training process. Hence, an advanced algorithm is required to fix or tune the hyperparameters prior to the training process. The nature-inspired algorithms, such as Grey wolf optimization (GWO), particle swarm optimization (PSO), cuckoo search optimization, and genetic algorithm prove to be effective in tuning the hyperparameters of the classifiers. Yet, the slow convergence rate, low precision is some of the drawbacks experienced in the aforementioned algorithm. Hence, the crow search optimization, which restrains the limitations of the aforementioned algorithm, is utilized in this research to optimally tune the hyperparameters of the classifier. The motivation and the mathematical model of the algorithm are deliberated in the following section. The main assumptions of this algorithm are listed as follows.

- Corvids prefer to reside in group
- The corvids effectively memorize the location of the hiding place.
- The corvids observe each other to steal the food.
- The corvids protect their food sources from thievery by probability.

Let us assume that there is X dimensional including the number of crows. Let n number of corvids is in the groups and the position of corvids in the search space is demonstrated by the vector $V^{i,I}$, where $i = 1, 2, \ldots\ldots n$ and $I = 1, 2\ldots\ldots, I_{\max}$, where the $I_{\max}$ is the maximum iterations. Hence the vector is given as $V^{i,I} = \left[V_1^{i,I}, V_2^{i,I}, \ldots\ldots V_X^{i,I}\right]$. Each corvid possesses the memory matrix, where the hiding place location is memorized. The hiding place of the corvid at iteration I is stored in $Me^{i,I}$. The best position attained obtained by the ith corvid is also stored in the memory matrix.

Let us assume an iteration I that the corvid j needs to visit the hiding position $Me^{i,I}$. The corvid i now follows the corvid j to find out the hiding locality of the corvid j. This circumstance results in two possibilities.

Possibilities -1: Corvid j not aware of the corvid i follows it. Hence, the crow i obtain the hiding region of j and they can update its position as represented

$$V^{i,I+1} = V^{i,iter} + ran_i \times flen^{i,I} \times \left(M^{j,I} - V^{i,iter}\right) \tag{6}$$

Where ran_i represents the random number that possesses the uniform distribution between 0 and 1 and $flen^{i,I}$ represents the flight length of the corvid at the iteration i. The small values of *flen* depicts that it leads to local search, whereas the large *flen* directs to the global search.

Possibilities-2: Corvid j aware of the corvid i follows it. Hence, to protect the food the corvid j fools the corvid i. It moves towards another position of search space.

$$V^{i,I+1} = \begin{cases} V^{i,iter} + ran_i \times flen^{i,I} \times \left(M^{j,I} - V^{i,iter}\right) \; ran_j \geq Awp^{i,iter} \\ random\,position; \;\; otherwise \end{cases} \tag{7}$$

The r_j is the random number within the uniform distribution between 0 and 1.

Implementation of CSA: The stepwise procedure for implementation is given in this section

Step-1: Initialize the adjustable hyperparameters: The hyperparameters like flock size n, Maximum iteration $I_{\max}$, *flen* and *Awp* are initialized.
Step-2: Initialize position and memory of crow.

$$Crovid = \begin{bmatrix} V_1^1 \; V_2^1 \;\; V_X^1 \\ V_1^2 \, V_2^2 \; V_X^2 \\ . \\ . \\ . \\ V_1^n \; V_2^n \;\; V_X^n \end{bmatrix} \tag{8}$$

$$Memory = \begin{bmatrix} M_1^1 \; M_2^1 \;\; M_X^1 \\ M_1^2 \, M_2^2 \;\; M_X^2 \\ . \\ . \\ . \\ M_1^n \, M_2^n \;\; M_X^n \end{bmatrix} \tag{9}$$

Step 3: Estimate fitness function: The quality of the location is estimated by using the decision variables in the objective function.
Step 4: Obtain the new position: The corvids need to update their new position to safeguard their food from the other corvids. Hence the locality of the hidden is obtained from the Eq. (6).

Step 5: Evaluation of fitness parameters of a new position: The position of the corvids is updated based on the feasibility of the position. If it is not feasible the corvids stay in the current position and fail to generate a new position.
Step 6: Evaluate fitness function of a new position: Estimate the fitness function of the new position of crow.
Step 7: Update memory: The memory of the crow is updated as

$$Me^{i,iter+1} = \begin{cases} V^{i,iter+1} & if\ f\left(V^{i,iter+1}\right) is\ better\ than\ \left(Me^{i,\ iter}\right) \\ Me^{i,iter+1}, & Otherwise \end{cases} \quad (10)$$

where, the objective value is demonstrated as,$f()$.
Step 8: Check termination criteria.

Step-4 to 7 are repeated until $I_{\max}$ is reached.

Thus, the hyperparameter of the DNN classifier is effectively tuned by using the crow search optimization algorithm so as to boost the prediction accuracy of the classifier.

4 Results and Discussion

The defects in the software are analyzed in order to predict the polarity of the model so that the promotion of a product or growth of an organization can be achieved. The proposed Cr-DNN is implemented in two scenarios 1) with respect to training percentage and 2) with respect to k-fold value by using PROMISE dataset [22]. The experimental results are enumerated in the following section in order to prove their efficiency.

4.1 Experimental Setup:

The experiment is implemented in Python running in Windows 10 operating system with 8 GB RAM memory. In the proposed method, the step size used for the evaluation of the performance is 0.001, the drop out ratio is 0.05, the number of layers used is 2 and the number of neurons used is 100.

4.2 Performance Metrics

Performance metrics is used to analyze the performance of research where the improvement attained can be determined.

Accuracy: Accuracy can be defined as the state of being correct in the case of classification which can be expressed as

$$Accuracy = \frac{Trpos + Trneg}{Trpos + Trneg + Fapos + Faneg} \quad (11)$$

T*rpos* is the true positive value,T*rneg* is the true negative,*Fapos* is the false positive, and *Faneg* is the false negative value.

Sensitivity: The total number of true positive values accurately identified by the system is known as sensitivity. The sensitivity of the system is mathematically represented as

$$Sensitivity = \frac{\mathrm{T}rpos}{\mathrm{T}rpos + Faneg} \tag{12}$$

Specificity: The total number of true negative values accurately identified by the system is known as specificity. The specificity of the system is mathematically represented as

$$Specificity = \frac{\mathrm{T}rneg}{\mathrm{T}rneg + Fapos} \tag{13}$$

4.3 Comparative Analysis

In this analysis, the comparison is made in two scenarios such as a) training percentage and b) K-fold analysis. The parameters accuracy, sensitivity, and specificity are taken into account to reveal the importance of the proposed method.

4.3.1 Comparative Analysis with Respect to the Training Percentage

Figure 2 depicts the comparative analysis of the software defect prediction methods corresponding to training percentage. The accuracy, specificity, and sensitivity are depicted in Figs. 2 a), 2 b), and 2 c), respectively [18, 23–26]. The accuracy of methods, such as Decision tree, KNN, SVM, Neural Network, and proposed Cr-DNN is 84.6335%, 87.2529%, 88.5700%,89.1767%, and 92.0506% respectively at 90% of training. Similarly for the sensitivity, the percentage of the methods Decision tree, KNN, SVM, Neural Network and proposed Cr-DNN is 84.4389%, 86.5740%, 88.0195%, 88.9525%, and 93.5694% respectively at 80% of training. Furthermore, the sensitivity of the methods Decision tree, KNN, SVM, Neural Network, and proposed Cr-DNN is 95.0125%, 95.9155%, 96.8186%, 97.7217%, and 98.6248% for 90% of training.

4.3.2 Comparative Analysis with Respect to K-fold Value

Figure (3) depicts the comparative evaluation of the software defect prediction methods with respect to the K-fold value. The accuracy rate of the methods is given as 81.9983%, 82.4637%, 83.6880%, 84.0157%, and 84.7350% when the k-fold value is 8. Correspondingly the sensitivity of the Decision tree, KNN, SVM, Neural Network, and proposed Cr-DNN for a k-fold value of 9 is given by 84.7409%, 86.7811%, 87.8488%, 88.3588%, and 90.0326% respectively. Comparably the recall rate of the methods Decision tree, KNN, SVM, Neural Network, and proposed Cr-DNN is given by 93.4884%, 94.1237%, 95.0125%, 97.7217%, and 98.6248% respectively.

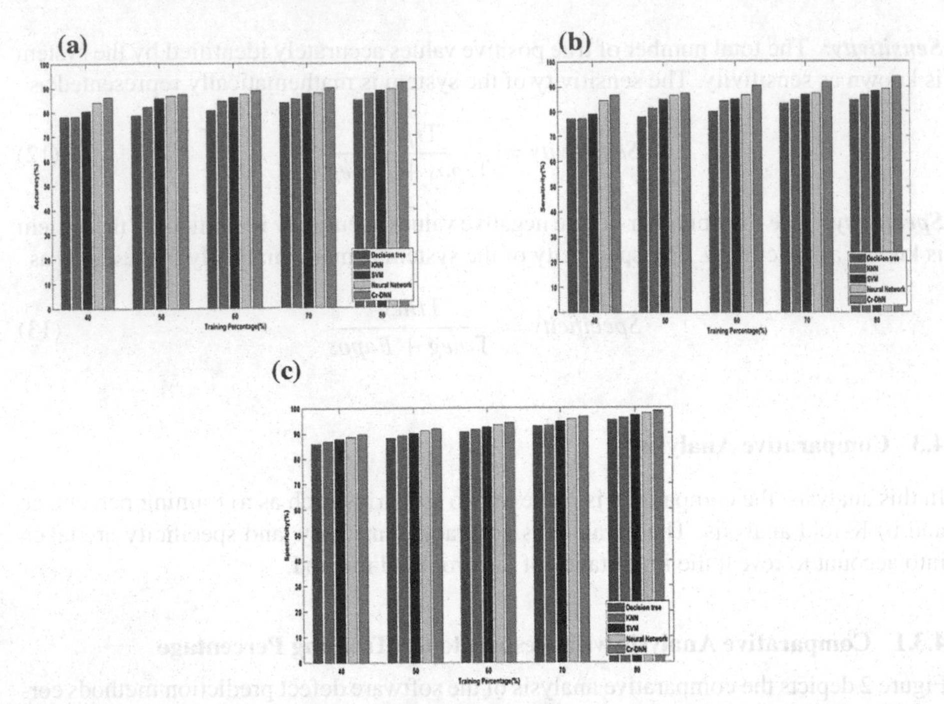

Fig. 2. Comparative analysis by varying training percentage in terms of a) Accuracy b) Sensitivity and c) Specificity

4.4 Comparative Discussion

This section deliberates the methods employed for the prediction of software defects. The methods utilized for the comparison are Decision tree, KNN, SVM, Neural Network, and proposed Cr-DNN which are analyzed from the basic level of the defect prediction. The performance metrics attained by the software defect prediction model with respect to the training percentage of 90% and the K-fold value of 10 are shown in Table 1. Here, the maximal accuracy obtained by the proposed Cr-DNN is 92.05%, which is 8.06%, 5.21%, 3.78%, and 3.12% better compared to the traditional Decision tree, KNN, SVM, and Neural Network. Likewise, the maximal sensitivity and specificity obtained by the proposed method are 93.57% and 98.62% respectively. From the table, it is demonstrated that the proposed Cr-DNN outperforms all the competent models in terms of accuracy, sensitivity, and specificity.

Thus, from the comparative analysis, it is concluded that the proposed Cr-DNN software defect prediction technique obtained elevated performance compared to the existing techniques in terms of all the performance metrics. The challenges faced by the existing methods for the low accuracy in prediction are: the decision tree-based software prediction technique suffers from time complexity, the KNN and SVM are not appropriate for the application of large data, and the Neural Network suffers from vanishing gradient issues. In the proposed method, the most informative feature selection technique using the Wrapper technique helps to reduce the computational complexity by removing the insignificant feature for the prediction. Then, the optimized DNN enhances

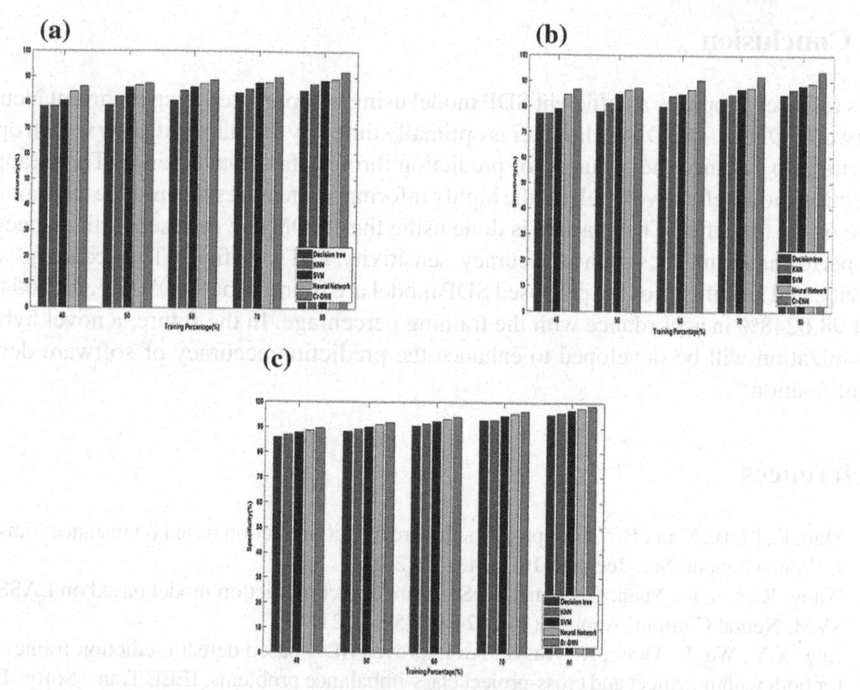

Fig. 3. Comparative analysis by varying K-fold value in terms of a) Accuracy b) Sensitivity and c) Specificity

Table 1. Comparative Analysis

Methods	Training percentage			K-Fold		
	Accuracy (%)	Sensitivity (%)	Specificity (%)	Accuracy (%)	Sensitivity (%)	Specificity (%)
Decision tree	84.63	84.44	95.01	87.43	87.42	93.49
KNN	87.25	86.57	95.92	89.50	88.93	94.12
SVM	88.57	88.02	96.82	89.57	89.73	95.01
Neural Network	89.18	88.95	97.72	90.31	90.33	97.72
Proposed Cr-DNN	92.05	93.57	98.62	90.75	93.30	98.62

the convergence rate and helps to predict the software defect more accurately through the global best optimal solution. Hence, the proposed method obtained enhanced prediction accuracy which refers to the closeness of the detection towards the identification of the software defect. The sensitivity refers to the correctly predicted positive cases and the specificity refers to the correctly predicted negative cases by the proposed method.

5 Conclusion

This research proposes an efficient SDP model using an optimized deep recurrent Neural Network (DNN). The DNN classifier is optimally tuned by the standard crow search optimization to enhance the accuracy of prediction through fast convergence. The wrapper selection model effectively selects the highly informative features that reduce the training time of the classifier. The analysis is done using the PROMISE data set, with respect to the performance metrics such as accuracy, sensitivity, and specificity. The accuracy, sensitivity, and specificity of the proposed SDP model are found to be 92.0506%, 93.5694%, and 98.6248% in accordance with the training percentage. In the future, a novel hybrid optimization will be developed to enhance the prediction accuracy of software defect identification.

References

1. Mao, F., Li, B., Shen, B.: Cross-project software defect prediction based on instance transfer. J. Front. Comput. Sci. Technol. **10**(1), 43–55 (2016)
2. Wang, K., Liu, L., Yuan, C., Wang, Z.: Software defect prediction model based on LASSO–SVM. Neural Comput. Appl. **33**(14), 8249–8259 (2021)
3. Jing, X.Y., Wu, F., Dong, X., Xu, B.: An improved SDA based defect prediction framework for both within-project and cross-project class-imbalance problems. IEEE Trans. Softw. Eng. **43**(4), 321–339 (2016)
4. Ding, Z., Xing, L.: Improved software defect prediction using pruned histogram-based isolation forest. Reliab. Eng. Syst. Saf. **204**, 107170 (2020)
5. Malhotra, R., Jain, J.: Handling imbalanced data using ensemble learning in software defect prediction. In: 2020 10th International Conference on Cloud Computing, Data Science \& Engineering (Confluence), pp. 300–304 (2020)
6. Deng, J., Lu, L., Qiu, S.: Software defect prediction via LSTM. IET Softw. **14**(4), 443–450 (2020)
7. NezhadShokouhi, M.M., Majidi, M.A., Rasoolzadegan, A.: Software defect prediction using over-sampling and feature extraction based on Mahalanobis distance. J. Supercomput. **76**(1), 602–635 (2019). https://doi.org/10.1007/s11227-019-03051-w
8. Chakraborty, T., Chakraborty, A.K.: Hellinger net: a hybrid imbalance learning model to improve software defect prediction. IEEE Trans. Reliab. **70**(2), 481–494 (2020)
9. Qiao, L., Li, X., Umer, Q., Guo, P.: Deep learning based software defect prediction. Neurocomputing **385**, 100–110 (2020)
10. Xing, F., Guo, P.: Support vector regression for software reliability growth modeling and prediction. In: Wang, J., Liao, X., Yi, Z. (eds.) Advances in Neural Networks – ISNN 2005. ISNN 2005. Lecture Notes in Computer Science, vol. 3496, pp. 925–930. Springer, Berlin, Heidelberg (2005). https://doi.org/10.1007/11427391_148
11. Patel, S.K.: Attack detection and mitigation scheme through novel authentication model enabled optimized neural network in smart healthcare. Comput. Meth. Biomech. Biomed. Eng. 1-27. Printing ahead (2022)
12. Bennin, K.E., Toda, K., Kamei, Y., Keung, J., Monden, A., Ubayashi, N.: Empirical evaluation of cross-release effort-aware defect prediction models. In: proceedings of IEEE International Conference on Software Quality, Reliability and Security (QRS), pp. 214–221 (2016)

13. Yan, Z., Chen, X., Guo, P.: Software defect prediction using fuzzy support vector regression. In: Zhang, L., Lu, BL., Kwok, J. (eds.) Advances in Neural Networks - ISNN 2010. ISNN 2010. LNCS, vol. 6064, pp. 17–24. Springer, Berlin, Heidelberg (2010). https://doi.org/10.1007/978-3-642-13318-3_3
14. Tan, M., Tan, L., Dara, S., Mayeux, C.: Online defect prediction for imbalanced data. In: proceedings of IEEE/ACM 37th IEEE International Conference on Software Engineering, vol. 2, pp. 99–108 (2015)
15. Amasaki, S.: Cross-version defect prediction using cross-project defect prediction approaches: does it work?. In: Proceedings of the 14th International Conference on Predictive Models and Data Analytics in Software Engineering, pp. 32–41 (2018)
16. Shao, Y., Liu, B., Wang, S., Li, G.: Software defect prediction based on correlation weighted class association rule mining. Knowl.-Based Syst. **196**, 105742 (2020)
17. Sun, Z., Zhang, J., Sun, H., Zhu, X.: Collaborative filtering based recommendation of sampling methods for software defect prediction. Appl. Soft Comput. **90**, 106163 (2020)
18. Kumar, A.: A cloud-based buyer-seller watermarking protocol (CB-BSWP) using semi-trusted third party for copy deterrence and privacy preserving. Multi. Tools Appl. 1–32 (2022)
19. Kumar, A.: Design of secure image fusion technique using cloud for privacy-preserving and copyright protection. Int. J. Cloud Appl. Comput. (IJCAC) **9**(3), 22–36 (2019)
20. Kumar, A.: A review on implementation of digital image watermarking techniques using LSB and DWT. In: The Third International Conference on Information and Communication Technology for Sustainable Development (ICT4SD 2018), held during 30–31 August (2018)
21. Kumar, A., Zhang, Z.J., Lyu, H.: Object detection in real time based on improved single shot multi-box detector algorithm. EURASIP J. Wirel. Commun. Netw. **2020**(1), 1–18 (2020). https://doi.org/10.1186/s13638-020-01826-x
22. PROMISE dataset http://promise.site.uottawa.ca/SERepository/datasets/cm1.arff. Accessed Mar 2022
23. Pelayo, L., Dick, S.: Applying novel resampling strategies to software defect prediction. In: Proceedings of NAFIPS North American Fuzzy Information Processing Society 2007–2007 Annual Meeting of the North American Fuzzy Information Processing Society, pp. 69–72 (2007)
24. Goyal, S.: Handling class-imbalance with KNN (Neighbourhood) under-sampling for software defect prediction. Artif. Intell. Rev. 1–42 (2021)
25. Wei, H., Hu, C., Chen, S., Xue, Y., Zhang, Q.: Establishing a software defect prediction model via effective dimension reduction. Inf. Sci. **477**, 399–409 (2019)
26. Zheng, J.: Cost-sensitive boosting neural networks for software defect prediction. Expert Syst. Appl. **37**(6), 4537–4543 (2010)

Text Sentiment Analysis Using the Bald Eagle-Based Bidirectional Long Short-Term Memory

Garadasu Anil Kumar[1(✉)], S. Sai Satyanarayana Reddy[1], Punna Sripallavi[2], Bollam Parashuramulu[2], and B. Suresh Banu[3]

[1] Department of CSE, Sreyas Institute of Engineering and Technology, Hyderabad 500068, India
saisn90@gmail.com

[2] Department of ECE, Sreyas Institute of Engineering and Technology, Hyderabad, India

[3] Department of Humanities and Sciences, Sreyas Institute of Engineering and Technology, Hyderabad, India

Abstract. Extraction of emotions that are conveyed through text is termed as sentimental analysis, which is also called as opinion mining. The sentimental analysis is enabled to extract the information about particular objects or an event or various subjects, which helps to analyze and detect the problems associated with them. The detection of this opinion helps in resolving the problems and it is embedded with the growth of the organization. In this research, a bald eagle based BiLSTM classifier is proposed to extricate the sentiment in the text. Initially, the data are collected and preprocessing is performed using the techniques tokenization, stopword removal, stemming and POS tagging. Features are extracted from the preprocessed data and classification is performed using the bald eagle optimization enabled BiLSTM classifier. The experiment shows that the proposed method works more efficient with a precision of 92.65%, and recall of 93.82%.

Keywords: Sentiments · Bald eagle · Tokenization · BiLSTM · Text sentiment analysis

1 Introduction

Human emotions are intricate, multidimensional attributes which contemplates the behavior and personality of individuals. The emotions of humans are expressed in different types such as facial expression, speech, communication with other individuals and certain other ways. The emerging technologies allow the individuals to express their emotions through social media [4]. The increased usage of the social media platform provides numerous numbers of comments in the form of texts which is difficult to classify them manually. There is a need for the automatic identification or recognition of the emotions present in the texts which helps in making decisions, gathering information about a particular event or a subject and the collected information will be greatly useful in various domains such as business, market, government intelligence and so on [7]. In a precise manner, sentimental analysis is defined as the analogy of extracting the opinion,

M. Singh et al. (Eds.): ICACDS 2022, CCIS 1613, pp. 26–36, 2022.
https://doi.org/10.1007/978-3-031-12638-3_3

sentiment and attitude of the individuals. The sentimental analysis is a natural language processing (NLP) problem that deals with enormous amount of data from various social platforms [3].

One of the main purposes of sentimental analysis is to extract and classify the polarity as positive, negative or neutral relying upon the perspective of the text. For identifying the sentiment it is necessary to learn the word vector representation of a text which contains low dimensional, non sparse representation of word vectors [2, 8]. Sentimental analysis is classified into news comment analysis [9], product comment analysis [10], film comment analysis [11] which is useful in projecting the perspective of the users to the providers [2].The opinion of the people are extracted using the sentimental analysis by availing the techniques machine learning, lexicon based, hybrid based techniques, deep learning and so on [3, 12]. There are multiple languages available on the internet such as Spanish, Arabic, hindi, Chinese, French but there is not available dictionary resources for every language [13] which is a strenuous task to be handled. When there is a lack in the availability of the dictionary resources the information present in the text couldn't utilized properly. To effectively recognize the emotions all the information from different languages should be extracted in a efficient manner.

The main intent of the research is to develop a sentiment prediction model that predicts the polarities present in the text. At the beginning the input data are collected from the sentiment 140 dataset and twitter sentiment analysis dataset, and the preprocessing of this data is performed. The different process involved in the preprocessing is tokenization, stemming, stopwords removal and POS tagging. After preprocessing the features is extracted using the TF-IDF vector and then the collected features are trained using the BiLSTM classifier optimized using the Bald-Eagle optimization. Finally the sentiments are predicted as positive or negative and the main contribution of this experimentation is as follows

- The proposed optimization enabled network improves the speed of the prediction by analyzing the texts deliberately.

The section is organized as: Sect. 2 discusses the motivation of the research, Sect. 3 demonstrates the proposed method with the detailed result section in Sect. 4. Finally Sect. 5 concludes the paper.

2 Motivation

In the following section, the need for the proposed sentimental analysis model is explained and the cons and pros of the previous studies are also enlisted in order to emphasis the importance of proposed method.

Literature Review

Feiran Huang *et al.* [1] established a novel image-text sentiment analysis model that effectively focus and recognizes the features and works well in both weakly and manually labeled dataset but the pitfall is that further enhancement are needed for effective analysis. Guixian Xu *et al.* [2] developed an improved word representation method that avails all the information present in the text but it consumes more time. Akshi Kumar *et al.* [3] initiated a system to capture the expressiveness of the image, and the performance of the system shows that it has improved efficiency in generic sentiment analysis but the drawback is that the text recognition is limited by the computer vision API. Kashfia Sailunaz and Reda Alhajj [4] detected and analyzed the sentiment expressed by the people and utilized those information for generating suggestions. Although it is a innovated detection it is limited by time, data as well as missing of information are also takes place. Bing Liu *et al.* [5] introduced a text sentimental analysis method by using the deep learning model along with bag of words which effectively reduce the over fitting but it focuses on few languages

Challenges

- The image will be more conceptual and personalized which is difficult to classify. It is also a challenging task to deal with multimodal data while performing sentimental analysis [1].
- The social media consists of numerous images that express different conceptualities but when it comes to opinion extraction it is more challenging [3].
- The emotions are conveyed through small texts which makes nearly impossible to detect the emotions due to the implicit information and the presence of sarcastic texts makes the analysis more complex [4].

3 Proposed Model for Sentiment Analysis Using the Textual Data

Sentimental analysis of data helps in gaining information about the events or objects that proportionally influence the growth of the organization. Moreover, the continuous monitoring of the feedback of an individual could also be helpful in recognizing the obstacles, employee concerns and various other factors. The process of the sentiment analysis is explained as follows: Initially, the data is collected as a text from the twitter sentiment analysis dataset and then the data are preprocessed for the effective functioning without distortions. The preprocessing is performed using the methods tokenization, stop words removal, stemming, and POS tagging. From the preprocessed data, the important features are extracted using TF-IDF features. Finally, the emotion from the texts is classified into positive and negative using the BiLSTM classifier where the efficiency is improved by using the bald eagle optimization (Fig. 1).

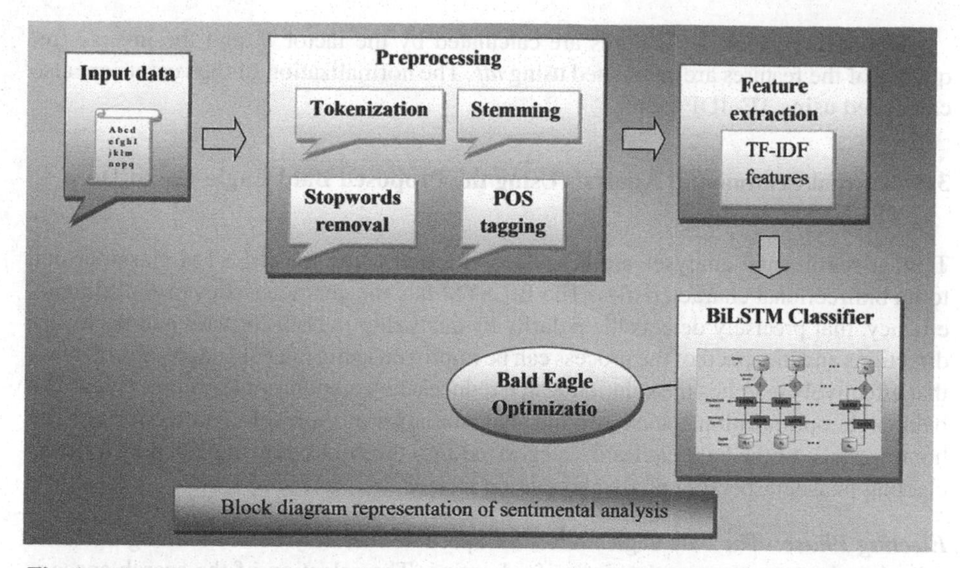

Fig. 1. Block diagram representation of sentimental analysis using bald eagle-based BiLSTM

3.1 Data Pre-processing:

The raw data obtained from the dataset are preprocessed using tokenization, stopword removal, stemming, POS tagging.

Tokenization: The initial process performed in preprocessing are the tokenization where the long texts are converted into smaller texts. Token refers to the building blocks of the sentences and furthermore paragraphs. It is classified into sentence tokenization and word tokenization. Sentence tokenization splits the paragraph into sentences. Similarly word tokenization performs the splitting of paragraph into word.

Stop Words Removal: The supporting words such as is, the, as and so on which carries little information in the sentence is removed using the stop words removal method.

Stemming: Stemming refers to the conversion of the common word into their root words in order to reduce the complexity. For example the word connecting is converted into connect so that the analysis can be performed easily.

POS Tagging: The POS refers to the part of speech that identifies the grammatical group of the sentence as verb, noun or adjective or any other form and assigns a tag corresponding to their identification.

3.2 Feature Extraction

The weight of the word and extraction of the important word is performed in the feature extraction step using TF-IDF vector, which is calculated as,

$$TF - IDF = tf * log(idf) \tag{1}$$

The frequency of the features are calculated by the factor *tf* and the inverse frequency of the features are calculated using *idf*. The normalization of the vectors are also calculated using TF- IDF factor.

3.3 Textual Sentimental Analysis Using the Proposed Bald Eagle-Based Deep BiLSTM Classifier

Textual sentimental analyses are broadly performed using the BiLSTM classifier due to its bidirectional characteristics. The BiLSTM has the characteristics of multilingual efficacy, that precisely detects the polarity by analyzing in both forward and backward directions and the speed of the process can be improved using the bald eagle optimization that effectively searches the solution which reduces the computational complexities. The bald eagle optimization focuses on the characteristics of the eagle while hunting. The hunting stage of the bald eagles is categorized into three and they are listed as follows: electing phase, inspecting phase and raiding phase.

Electing Phase: The bald eagles select the space deliberately before hunting and after selecting the search space they starts for hunting. The selection of the search space is mathematically represented by the equation is as follows

$$M_{new,a} = M_{best} + \beta * b(M_{mean} - M_a) \tag{2}$$

The term M_{best} represents the best space selected for the searching of prey depending upon the previous searches denotes the parameter that controls changes in the position of eagle and it has a random value assigned between 1.5 and 2 . b denotes the random number in the range 0 and 1. M_{mean} indicates that all the information collected from the previous searches are utilized. The search of the bald eagle will be mostly depends on the information gathered from the previous searches. The current search position of the eagle provides the prior information. M_a depends on the random factor which ***Inspecting phase:*** After selecting the search space, the bald eagles start searching for the prey by moving in different directions. The moving of the eagle forms a spiral shape to select the correct position of the prey. The best solution is selected during this inspecting phase and is mathematically expressed as

$$M_{a,new} = M_a + p(a) * (M_a - M_{a+1}) + q(a) * (M_a - M_{mean}) \tag{3}$$

The searching of the bald eagle in a spiral manner is designated by the polar plot and the values in the polar plot ranges from –1 to 1 and are mathematically expressed as

$$q(a) = \frac{qb(a)}{\max(|qb|)} \tag{4}$$

$$p(a) = \frac{pb(a)}{\max(|pb|)} \tag{5}$$

$$qb(a) = b(a) * \sin(\theta(a)) \tag{6}$$

$$pb(a) = b(a) * \cos(\theta(a)) \tag{7}$$

$$\theta(a) = c * \pi * rand \tag{8}$$

$$b(a) = \theta(a) + A * rand \tag{9}$$

Here c is the parameter that consists of the values ranges from 5 and 10 and this parameter helps the bald eagle to determine the distance between the starting and the center point. A assigns the value between 0.5 and 2 and gives the number of cycles taken by the bald eagle to consume the prey. The change in the position of the bald eagle from its spiral behavior is also represented by the factors A and c which notifies that diversification of the eagles.

Raiding Phase: The bald eagle selects a best position and swoop towards the prey for raiding. The movement towards the best position of the prey is mathematically expressed as

$$M_{a,new} = rand * M_{best} + q_1(a) * (M_a - i_1 * M_{mean}) \\ + P_1(a) * (M_a - i_2 * M_{best}) \tag{10}$$

$$q(a) = \frac{qb(a)}{\max(|qb|)} \tag{11}$$

$$p(a) = \frac{pb(a)}{\max(|pb|)} \tag{12}$$

$$qb(a) = b(a) * \sinh[\theta(a)] \tag{13}$$

$$pb(a) = b(a) * \cosh[\theta(a)] \tag{14}$$

$$\theta(a) = c * \pi * rand \tag{15}$$

$$b(a) = \theta(a) \tag{16}$$

The parameters i_1 and i_2 represents the increase in the intensity of the movement of bald eagles towards the best and centre points and the values of these parameters ranges from [$i_1, i_2 \in 1,2$]

3.4 Architecture of BiLSTM Classifier

The architecture of the BiLSTM classifier are shown in Fig. 2. The BiLSTM classifier effectively resolves the gradient vanishing problems which consist of input, output, hidden layers and activation function each performs a specific task. The sum of the weighted input is transformed to the output nodes by the activation layer. The output will be based on both past information and future prediction which works bidirectional. The forwarding and backward characteristics are useful in gathering more information and produce an efficient output.

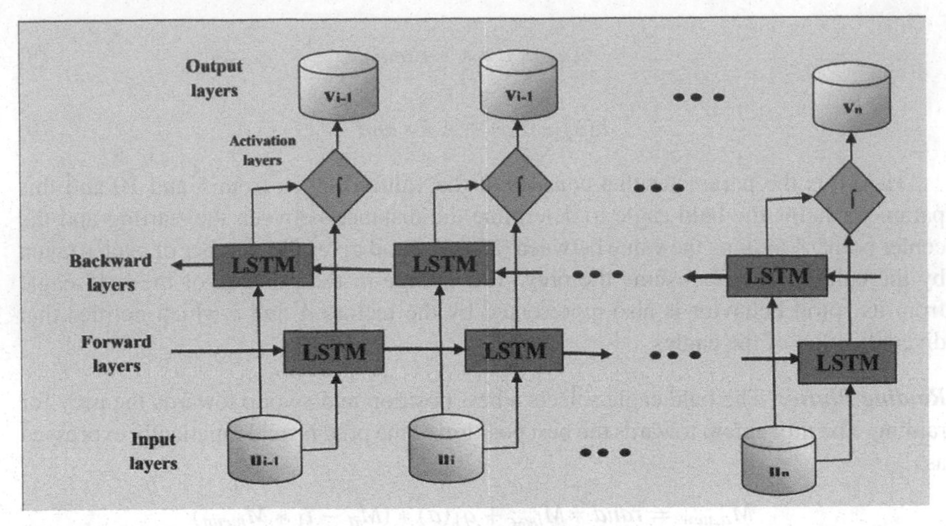

Fig. 2. Architecture of BiLSTM classifier

4 Results and Discussion

4.1 Dataset Description

The dataset used for the detection of emotion in the text is performed using sentiment 140 dataset [14] and twitter sentiment analysis dataset [15].

Sentiment 140 Dataset
The sentiment 140 dataset consists of 1,600,000 tweets extracted from the tweets utilizing twitter API. The sentiment present in the tweets is classified into positive and negative.

Twitter Sentiment Analysis Dataset
It is a natural language processing problem that recognizes the positive and negative tweets by using various machine learning models that has the capability to train thousands of documents

4.2 Parameter Metrics

The parameter metrics are used to explicit the significance of the proposed model and the parameters used in our experimentations are precision, recall and F1 measure.

Precision: Precision is defined as the number of instances that correctly predicts the sentiment from the total extracted emotions and is given by

$$F_{precision} = \frac{F_{TP}}{F_{TP} + F_{FN}} \tag{17}$$

Recall: Recall is defined as the fraction of the number of correctly predicted instances to the total extracted emotions and is given by

$$F_{recall} = \frac{F_{TP}}{F_{TP} + F_{FN}} \tag{18}$$

F1-measure: F1-measure demonstrate the performance of individual texts relying upon the precision and recall functions and is given by

$$F_{F1-measure} = \frac{F_{precision} * F_{recall}}{F_{precision + F_{recall}}} \tag{19}$$

4.3 Experimental Setup

The experiment of text sentimental analysis are implemented using the software python and the system configuration is enumerated as follows: Windows 10 with 8 gb RAM carried out in Python 3.7.6 (64 bit) in the IDE Pycharm 2020- Community Edition.

4.4 Comparative Methods

The methods used for the comparison of the proposed Bald eagle-based BiLSTM are Naïve Bayes [16], K-Nearest neighbor [17], Random forest [18], Bidirectional Long short term memory (BiLSTM) [2]. The comparative analysis shows that the proposed method works more efficient compared with the previous methods.

4.5 Comparative Analysis of Bald Eagle Based BiLSTM Classifier

See (Fig. 3).

Comparative Analysis Using the Sentimental Analysis Dataset
The comparison is performed using this sentimental analysis dataset and the parameters precision, recall and f1-measure are measured and are shown in Fig. 4. Initially the precision rate of the methods Naive Bayes, K-Nearest neighbor, Random forest, BiLSTM and the proposed Bald eagle-based BiLSTM for 90% of training data are measured and the values are depicted as 0.845, 0.889, 0.902, 0.915, 0.927 respectively for 90% of data. Secondly the recall rate of the methods Naive Bayes, K-Nearest neighbor, Random forest, BiLSTM and the proposed Bald eagle-based BiLSTM are measured and enumerated as 0.777, 0.885, 0.918, 0.924, 0.938 for 90% of training data. Finally the f1-measure is measured for the methods Naive Bayes, K-Nearest neighbor, Random forest, BiLSTM and the proposed Bald eagle-based BiLSTM and enlisted as 0.869, 0.909, 0.918, 0.922, 0.938 respectively for 90% of the training data.

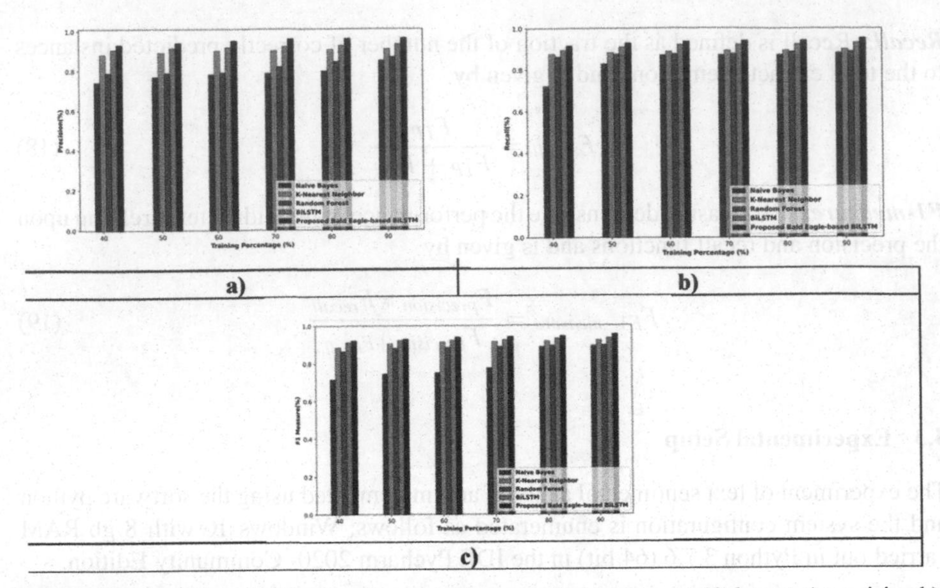

Fig. 3. Comparative analysis of sentiment prediction using sentiment 140 dataset a) precision b) recall c) f1-measure

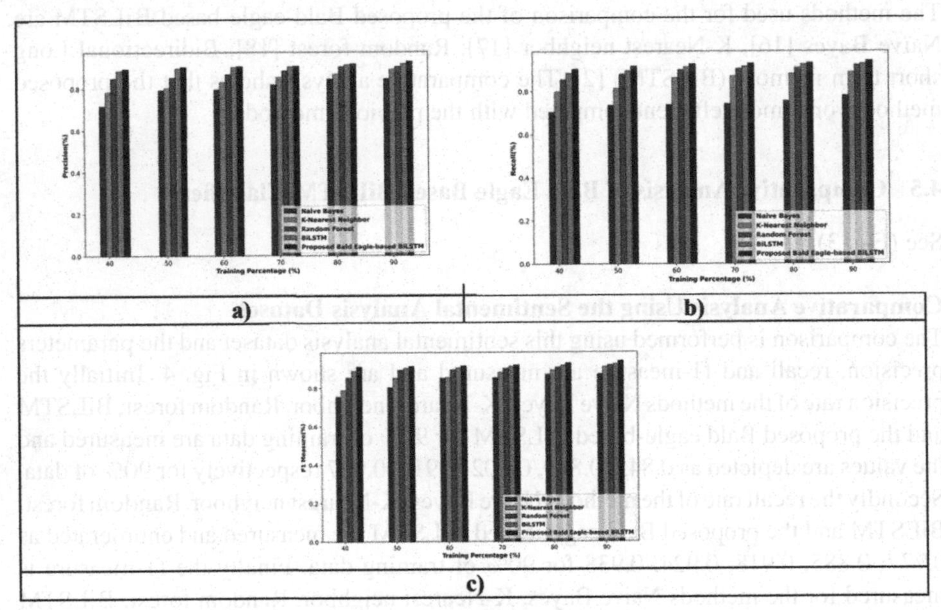

Fig. 4. Comparative analysis of sentiment prediction using sentiment analysis dataset a) precision b) recall c) f1-measure

5 Conclusion

Sentimental analysis plays a major role in the opinion mining of individuals. The proposed bald eagle-based BiLSTM classifier also predicts the sentiment effectively which has a direct impact in the growth of their respective fields. Initially the data are collected and pre-processed using various techniques. Tokenization, stopword removal, stemming and POS tagging, which effectively reduces the complexity and then the features are selected for the classification. In this experimentation, the TF-IDF vector is considered into account and the sentiments are predicted and classified using the proposed bald eagle BiLSTM classifier. The experimental results shows that it performs well in terms of precision by 0.927, recall by 0.938 and F1 measure by 0.938 and overcome the drawbacks of state-of-art methods. In future, the sentimental analysis can be improved by using various optimizations concentrating on single languages rather than English.

References

1. Huang, F., Zhang, X., Zhao, Z., Xu, J., Li, Z.: Image–text sentiment analysis via deep multimodal attentive fusion. Knowl.-Based Syst. **167**, 26–37 (2019)
2. Xu, G., Meng, Y., Qiu, X., Yu, Z., Wu, X.: Sentiment analysis of comment texts based on BiLSTM. Ieee Access **7**, 51522–51532 (2019)
3. Kumar, A., Garg, G.: Sentiment analysis of multimodal twitter data. Multi. Tools Appl. **78**(17), 24103–24119 (2019). https://doi.org/10.1007/s11042-019-7390-1
4. Sailunaz, K., Alhajj, R.: Emotion and sentiment analysis from twitter text. J. Comput. Sci. **36**, 101003 (2019)
5. Liu, B.: Text sentiment analysis based on CBOW model and deep learning in big data environment. J. Ambient. Intell. Humaniz. Comput. **11**(2), 451–458 (2018). https://doi.org/10.1007/s12652-018-1095-6
6. Bhargava, R., Sharma, Y.: MSATS: multilingual sentiment analysis via text summarization. In: 2017 7th International Conference on Cloud Computing, Data Science & Engineering-Confluence, pp. 71–76. IEEE (2017)
7. Kumar, A., Sharma, A.: Systematic literature review on opinion mining of big data for government intelligence. Webology **14**(2) (2017)
8. Zhang, Q., Zhang, S., Lei, Z.: Chinese sentiment classification based on improved convolutional neural network. Comput. Eng. Appl. **53**(22), 116–120 (2017)
9. Krishnamoorthy, S.: Sentiment analysis of financial news articles using performance indicators. Knowl. Inf. Syst. **56**(2), 373–394 (2017). https://doi.org/10.1007/s10115-017-1134-1
10. Shelke, N., Deshpande, S., Hakare, V.T.: Domain independent approach for aspect oriented sentiment analysis for product reviews. In: Proceedings of the 5th International Conference on Frontiers in Intelligent Computing: Theory and Applications, pp. 651–659 (2017)
11. Sharma, P., Mishra, N.: Feature level sentiment analysis on movie reviews. In: 2016 2nd International Conference on Next Generation Computing Technologies (NGCT), pp. 306–311. IEEE, Dehradun, India (2016)
12. Kumar, A., Sharma, A.: Socio-Sentic framework for sustainable agricultural governance. Sustain. Comput. Inform. Syst. **28**, 100274 (2020)
13. Lo, S.L., Cambria, E., Chiong, R., Cornforth, D.: Multilingual sentiment analysis: from formal to informal and scarce resource languages. Artif. Intell. Rev. **48**(4), 499–527 (2017)
14. https://www.kaggle.com/kazanova/sentiment140

15. https://github.com/sharmaroshan/Twitter-Sentiment-Analysis
16. Webb, G.I., Keogh, E., Miikkulainen, R.: Naïve Bayes. Encycl. Mach. Learn. **15**, 713–714 (2010)
17. Ni, K.S., Nguyen, T.Q.: An adaptable k -nearest neighbors algorithm for MMSE image interpolation. IEEE Trans. Image Process. **18**(9), 1976–1987 (2009)
18. Pal, M.: Random forest classifier for remote sensing classification. Int. J. Remote Sens. **26**, 217–222 (2005)

Comparison of Multiple Machine Learning Approaches and Sentiment Analysis in Detection of Spam

A. N. M. Sajedul Alam(✉), Shifat Zaman, Arnob Kumar Dey, Junaid Bin Kibria, Zawad Alam, Mohammed Julfikar Ali Mahbub, Md. Motahar Mahtab, and Annajiat Alim Rasel

Department of Computer Science and Engineering, Brac University, 66 Mohakhali, 1212 Dhaka, Bangladesh
{a.n.m.sajedul.alam,shifat.zaman,arnob.kumar.dey,junaid.bin.kibria, zawad.alam,mohammed.julfikar.ali.mahbub,md.motahar.mahtab}@g.bracu.ac.bd
annajiat@bracu.ac.bd

Abstract. Currently, all of our communications are made through various electronic communication mediums. While it has made communication between people from different parts of the world very easy, things like spamming have made life difficult for many people. Spammers are found in almost every electronic communication platform like email, mobile SMS, social networking sites, etc. So, with time detecting spam messages and filtering them out from our important messages have become more and more important. For many years, Natural Language Processing (NLP) researchers have proposed different techniques to detect spam messages. In this paper, our objective is to detect spam messages in a dataset using vectorization along with various machine learning algorithms and compare their results to find out the best classifier for detecting spam messages.

Keywords: Machine learning · Natural Language Processing · Spam detection

1 Introduction

In the modern world, where electronic communication is a part of our day-to-day activities, spamming has become a major problem. It affects users of various major platforms like email, mobile SMS, social networking sites, etc. Nowadays, most email clients have their own spam filters, which also block out specific spam email addresses. However, spam messages still make it to the end-users either through bypassing these filters or mobile messages or social network messaging applications. When the question is about filtering out spam messages based on their internal texts, NLP (Natural Language Processing) [1] is the biggest tool we can use. NLP has been used for a long time to process the information inside a message along with various machine learning algorithms to filter out

M. Singh et al. (Eds.): ICACDS 2022, CCIS 1613, pp. 37–50, 2022.
https://doi.org/10.1007/978-3-031-12638-3_4

spam messages. In this paper, we have used NLP with seven different machine learning algorithms to classify spam and non-spam (ham) messages. We have also compared the results and accuracy of each technique in order to identify which can be the best solution to categorize spam messages.

Spam messages have been there in our life for a long time. Thousands of users have suffered personal and financial losses due to these unwanted messages. To solve these problems, researchers have introduced different techniques from time to time. Machine Learning and more specifically NLP researchers brought out advanced solutions to these problems. N. Govil, K. Agarwal, A. Bansal, and A. Varshney proposed a method in their paper "A Machine Learning based Spam Detection Mechanism" [2] where they created a dictionary of spam words and then used Naive Bayes Classification to detect spam messages. N. Kumar, S. Sonowal, and Nishant [3] proposed a method where they first tokenized the messages, extracted features from them, and then trained them using various machine learning techniques. However, as the number and types of spam messages are growing, it is important to find out the best solution for detecting and classifying spam messages.

In this paper, our objective is to detect spam messages using vectorization and seven machine learning algorithms and compare the results of these classifiers. We will be using TFIDF vectorizer [4] for vectorization as well as seven classifiers - Random Forest, XGBoost, Naive Bayes, Logistic Regression, SVC, Cat boost, and K nearest neighbor. This paper first discusses several techniques and researches for spam detection, then introduces our technique for solving the problem and comparison of results.

2 Literature Review

Processing of Natural language and mining of opinion fields are used to cross the barriers in every single language due to the fact that analysis of opinion is dependent on language. In 2020, Najed et al. gave all their attention to a feature set to develop which can be utilized along with classifiers of numerous types as an input of a reliable source for detecting opinion spam in the Persian language [5]. They used some of the elements discovered during their research that they made while reviewing detection of spam in the English language, modified the definition along with some of the usability to fit the language in Persian, plus added additional novel features as a result. Their investigations show that the classifiers with the best results for detecting opinion spam in Persian are AdaBoost and Decision Tree, getting accuracy percentages of 98.7 and 98, respectively. In relation to the goal of detecting opinion spam, the robustness of their enlarged feature set was also tested by comparing it with other feature sets.

The internet's rapid expansion has made us reliant on it for the majority of our activities. E-commerce is one such rapidly expanding field. Hundreds of e-commerce sites exist today, with millions of products available. [6] These websites allow users to leave feedback on the product or service they received. These reviews can be a valuable source of data for service providers that are analyzing sales or making decisions. It can also help potential customers decide whether or not to purchase a product. These reviews, however, are only beneficial if they are accurate. The rapid The expansion of e-commerce has resulted in fierce competition among businesses and brands. As a result, some parties attempt to post phony reviews on a competitor's website in order to boost the popularity of a product or brand or to disparage it. In their research, Aishwarya et al. in 2019 used the Naive Bayes algorithm to address the issue to put a score on the reviews and give labels to them, thereby distinguishing the real and fake reviews [7].

Peng, Q., and Zhong, M. attempted to integrate sentiment analysis with spam review detection in [8]. The researchers came up with three projects to focus on. To calculate the sentiment score, first, develop a sentiment lexicon and then utilize a shallow dependency parser. The following step is to construct a set of discriminant rules. The final aim is to use a time series technique to develop a strategy for detecting spam reviews. SentiWordNet and MPQA were utilized for sentiment analysis. MPQA achieves a greater classification accuracy than SentiWordNet, which obtains a 61.4% accuracy. The results for Spam Review Detection are all above 0.65, indicating a high level of agreement. As a result, it is clear that evaluators' judgments are consistent and effective. Furthermore, the sentiment score approach has higher accuracy than the rating method. They discovered that certain reviews' ratings conflict with the sentiment stated in natural language text. Because it ignores negation terms in the reviews, the word counting method has low accuracy.

3 Proposed Methodology

It consists of 4 subsections: Section A deals with data collection, Section B describes the dataset we used, Section C deals with the Vectorization and Section D describes the modeling and Evaluation

A. Dataset Collection

The dataset that we have used here in order to train our model has been collected from the UCI [9] datasets, which is a public dataset that contains SMS labeled messages.

B. Dataset Description

Our dataset includes 5574 SMS phone messages in English. Each message is tagged as either SPAM or HAM based on its legitimacy. We have labeled the message texts as 'message' and their tags as 'label' for further usage. In our dataset, there are 4825 messages that are labeled as 'ham' or legit, on the other hand, there are 747 messages that are labeled as spam.

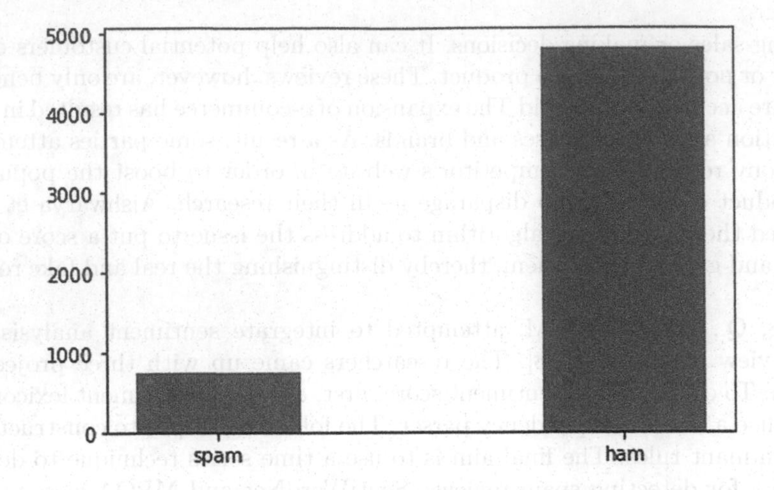

Fig. 1. A bar chart of spam and ham.

We used the technique 'under-sampling majority class (randomly)' in order to combat the class imbalance problems. Initially the majority class 'ham' had 4825 samples (Fig. 1), but 'spam' had only 747 samples. We randomly took 747 samples from 4825 ham samples to balance both classes. In addition to that, we used separate algorithms and metrics. Here, X axis represents spam and ham, on the other hand, Y axis represents counts of spam and ham.

C. Vectorization

It is the base building block of our NLP pipeline. We have used TfidfVectorizer in order to transform our message texts to feature vectors. These transformed feature vectors will be used in various estimators. TF-IDF is an abbreviation for the term frequency-inverse document frequency. In the field of data retrieval, the TF-IDF weight represents a commonly used term. The importance of a word in a text may be gauged using the weight assigned to it in a corpus. The significance of a word is related to how many times it appears in the manuscript. Aiming to reduce the influence of tokens that appear often but are experimentally less informative than those that only appear in a tiny percentage of a corpus, the TF-IDF method replaces raw frequency counts of tokens in a text [10].

The TF-IDF weight is made up of two components. The termed frequency (TF) is computed by dividing the total number of words in the document by the number of occurrences of that word. The Inverse Document Frequency (IDF) is calculated by dividing the logarithm of the number of documents in the corpus by the number of documents containing that particular phrase.

Term frequency calculates how frequently a term occurs in a document. The term frequency is often divided by the document length,

$$TF(t) = M \,/\, N. \tag{1}$$

Here,

M = Number of times term t appears in a document
N = Total number of terms in the document

IDF (Inverse Document Frequency): It measures the importance of a term. All terms are given the same importance while calculating TF. But, some common terms like 'of', 'is', 'that' can occur many times in a document but have little importance. So, we need to weigh down these less important frequent terms and scale-up rare and important ones. IDF is calculated in the following way in

$$IDF(t) = \log_e(P\,/\,Q). \tag{2}$$

Here,

P = Total number of documents
Q= Number of documents with term t in it

D. Modeling and Evaluation

I) Classification Model: In this stage after vectorization, we have our features represented as vectors. Now we can train our classifier for spam detection. First, we shuffled the dataset and then split the data into train and test data. We took 70% training data and 30% testing data. We have used a total of 7 classification algorithms for our research and then observed the results. Classification algorithms that we used here are - Random Forest, XG-Boost, SVC, Naive Bayes, Logistic Regression, Cat Boost, and K-Nearest Neighbor. A brief description of seven algorithms is narrated below. A random forest is one kind of machine learning method. Interestingly, this method can be used to resolve regression as well as classification issues. It is made up of several decision trees. It is a strategy that combines a large number of classifiers to solve complicated issues.

XGBoost is a gradient boosting machine implementation that is designed to tax the computing resources of boosted tree algorithms. It is designed for speed and model performance. By maximizing the distance between the sample points and the hyperplane, the Support Vectors Classifier attempts to discover the optimal hyperplane to partition the different classes. A Naive Bayes classifier is a probabilistic machine learning model for solving classification issues. The classifier's challenge is centered on the Bayes theorem.

In its most basic form, logistic regression is a statistical model that employs a logistic function to represent a binary dependent variable; however, there are various more complex forms available [11].

The K-NN algorithm is a supervised machine learning technique that assumes similarity between new and old cases/data and allocates the new example to the category that is most similar to the existing categories.

CatBoost is an open-source machine learning method that generates cutting-edge results without requiring extensive data training and provides powerful out-of-the-box support for more descriptive data types.

II) Confusion Matrix: The confusion matrix indicates performance measurement in classification issues. It depicts a model's perplexed state during prediction. Depending on the problem statement, the outcome can be two or more.

III) Evaluation Metrics: These metrics are used to assess the quality of a statistical or machine learning model. Precision, F1-Score, and Recall are the metrics we're looking at. We will use these evaluation metrics to seven algorithms to determine which one performs the best.

4 Result Analysis

As previously stated, we vectorized our dataset and trained our model using seven distinct machine learning algorithms. We tested our classifiers using 30% of our dataset. For each classifier outcome, we built a confusion matrix, ROC curve, and precision vs recall curve to examine the findings. Finally, the scores of each classifier were displayed in a single table.

i) Random Forest

We generated a confusion matrix based on predictions made by the Random Forest classifier on our test dataset (Fig. 2). Predictions are given on X-axis whereas actual results are plotted on Y-axis. From the confusion matrix, we can see, out of all spam messages, a total of 195 messages were correctly classified whereas 29 messages were incorrectly classified as 'ham'. All the ham messages were correctly classified [12].

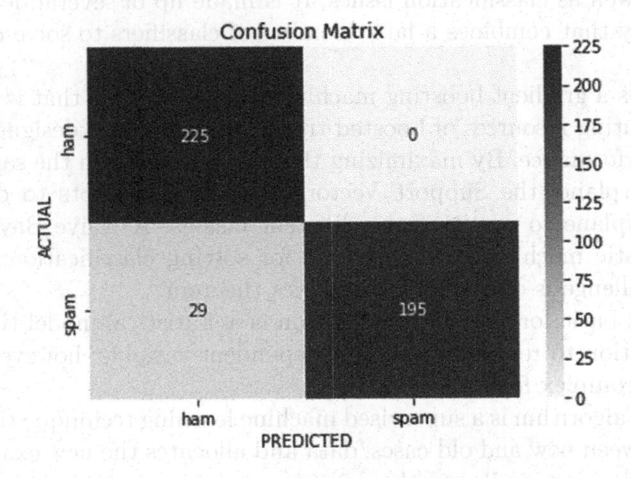

Fig. 2. Confusion matrix for Random Forest

ii) XGBoost

We generated a confusion matrix based on predictions made by the XGBoost library (which uses Gradient boosting) on our test dataset(Fig. 3). For XGBoost, from the confusion matrix, we can see, out of all spam messages, a total of 196 messages were correctly classified whereas 28 messages were wrongfully classified as 'ham'. 224 messages were correctly classified except for one message [13].

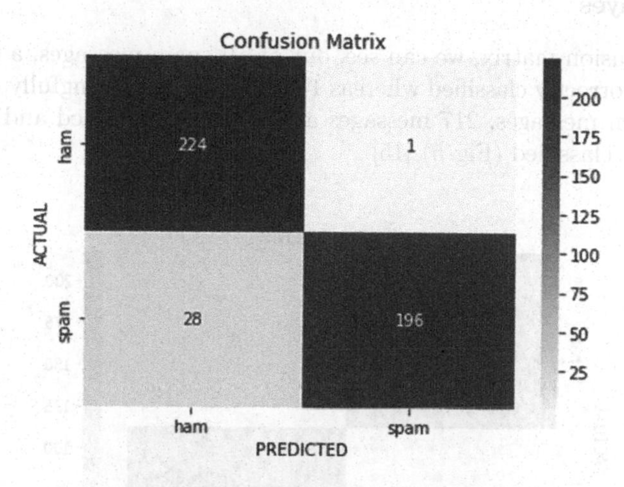

Fig. 3. Confusion matrix for XGBoost

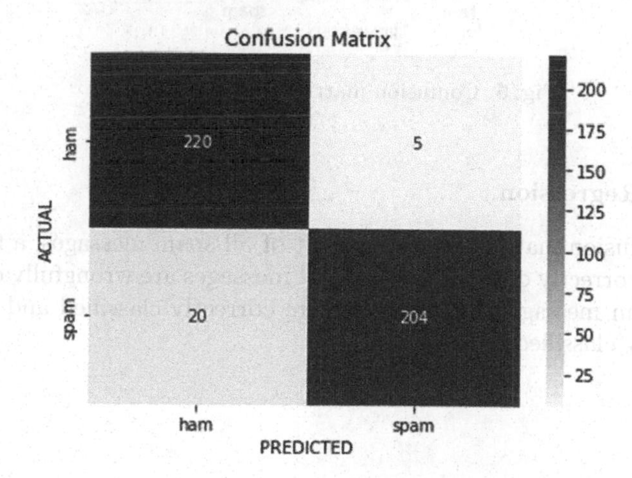

Fig. 4. Confusion matrix for SVC

iii) Support Vector Classification (SVC)

For SVC, from the confusion matrix, we can see, out of all spam messages, a total of 204 messages are correctly classified (Fig. 4). 20 messages are wrongfully classified as 'ham'. For ham messages, 220 messages are correctly classified and 5 messages are wrongfully classified [14].

iv) Naive Bayes

From the confusion matrix, we can see, out of all spam messages, a total of 210 messages are correctly classified whereas 14 messages are wrongfully classified as 'ham'. For ham messages, 217 messages are correctly classified and 8 messages are wrongfully classified (Fig. 5) [15].

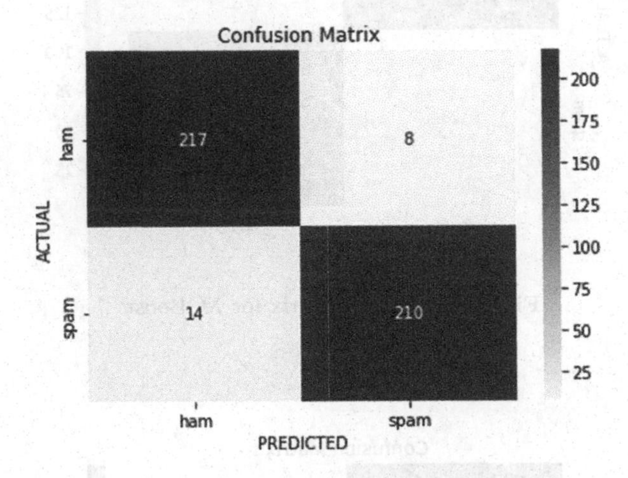

Fig. 5. Confusion matrix for Naive Bayes

v) Logistic Regression

From the confusion matrix, we can see, out of all spam messages, a total of 210 messages are correctly classified whereas 14 messages are wrongfully classified as 'ham'. For ham messages, 217 messages are correctly classified and 8 messages are wrongfully classified (Fig. 6) [16].

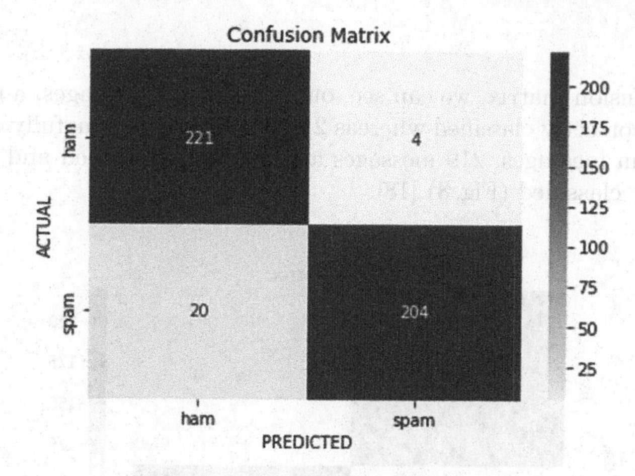

Fig. 6. Confusion matrix for Logistic Regression

vi) CatBoost

From the confusion matrix, we can see, out of all spam messages, a total of 195 messages are correctly classified whereas 29 messages are wrongfully classified as 'ham'. All the ham messages were correctly classified (Fig. 7) [17].

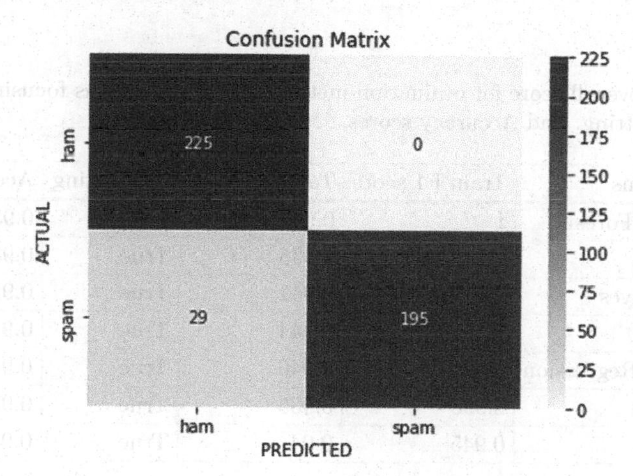

Fig. 7. Confusion matrix for CatBoost

vii) KNN

From the confusion matrix, we can see, out of all spam messages, a total of 203 messages are correctly classified whereas 21 messages are wrongfully classified as 'ham'. For ham messages, 219 messages are correctly classified and 6 messages are wrongfully classified (Fig. 8) [18].

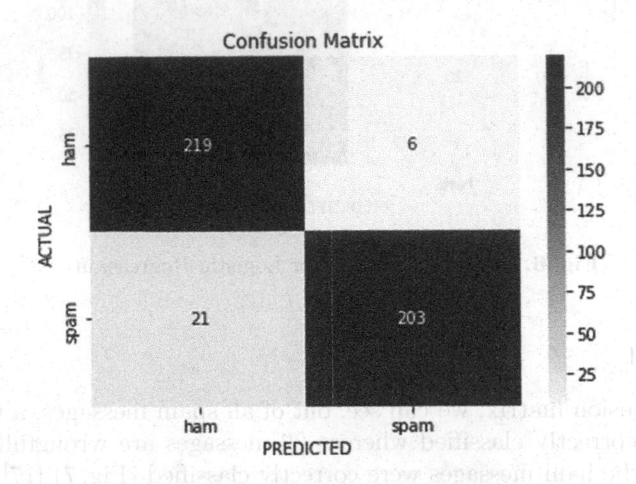

Fig. 8. Confusion matrix for KNN

Table 1. The overall score for evaluation metrics on six algorithms focusing Train F1, Test F1, Overfitting, and Accuracy scores.

Algorithms	Train F1 score	Test F1 score	Overfitting	Accuracy
Random Forest	1	0.935	True	0.935412
XGBoost	0.964	0.935	True	0.935412
Naive Bayes	0.979	0.951	True	0.951002
SVC	0.99	0.944	True	0.944321
Logistic Regression	0.98	0.946	True	0.946548
CatBoost	0.998	0.935	True	0.935412
KNN	0.945	0.94	True	0.939866

From Table 1 above, we find out the train F1 score and test F1 score for the seven algorithms. For train F1 score, random forest shows the best value with 1 and for test F1 score, Naive Bayes shows the best value with 0.951. We also find out the overfitting for all algorithms which is true. There is not much difference in accuracy scores among all the algorithms. Among them, Naive Bayes shows the best accuracy score with 0.951002.

Table 2. The overall score for evaluation metrics on six algorithms focusing on ROC Area, Precision, Recall, and F1 scores.

Algorithms	ROC area	Precision	Recall	F1
Random Forest	0.991389	1	0.870536	0.930788
XGBoost	0.982133	0.994924	0.875	0.931116
Naive Bayes	0.987986	0.963303	0.9375	0.950226
SVC	0.993938	0.976077	0.910714	0.942263
Logistic Regression	0.988958	0.980769	0.910714	0.944444
CatBoost	0.986796	1	0.870536	0.930788
KNN	0.979504	0.971292	0.90625	0.937644

From Table 2 above, we find out ROC area, precision, f1 score, and recall for seven algorithms. For ROC and precision, random forest shows the highest score with 0.991389 and precision shows the highest score with 1 respectively. On the other hand, for Recall and F1 score, Naive Bayes shows the best performance score with 0.9375 and 0.950226 respectively.

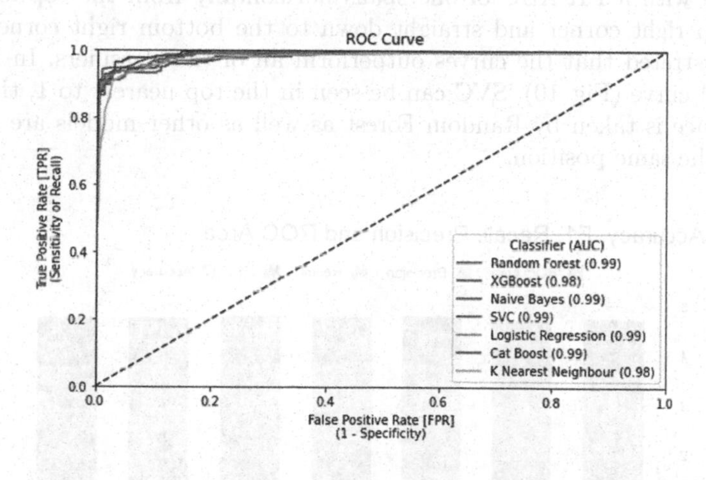

Fig. 9. ROC curve

The ROC curve is a classification task performance statistic with variable threshold values. If the AUC is close to zero, it has the poorest metric of separability. In addition, if the AUC is 0.5, the model has no class separation capacity. Depending on the threshold, the AUC for all algorithms is 0.99 or 0.98 (Fig. 9). As a result, it performs well, and the model can distinguish between positive and negative classes.

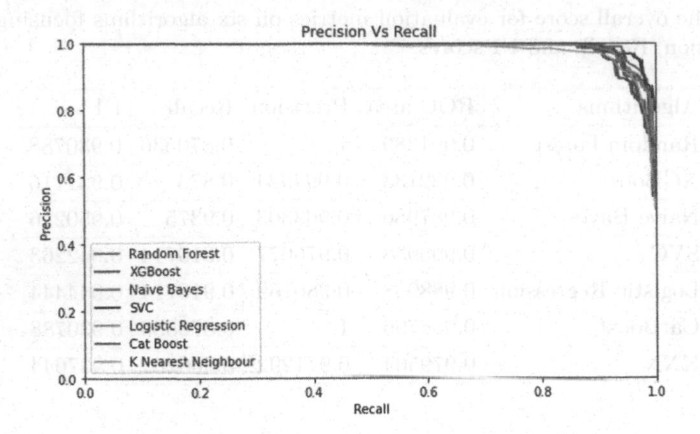

Fig. 10. Precision vs. Recall Curve

The precision-recall curve is used to evaluate the efficacy of binary classification algorithms. It is typically used in situations where the courses are extremely uneven. The picture above depicts the precision vs recall curve for classifiers. A PR-Curve with a PR-AUC of one spans horizontally from the top left corner to the top right corner and straight down to the bottom right corner. It can be demonstrated that the curves outperform all of the classifiers. In Precision Vs. Recall curve (Fig. 10), SVC can be seen in the top nearest to 1, the second topper place is taken by Random Forest as well as other models are in pretty much in the same position.

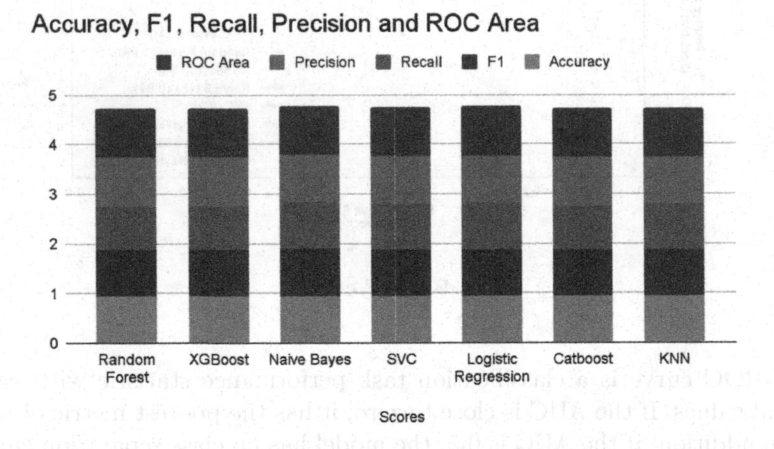

Fig. 11. Comparative analysis of seven classifiers

This stacked column chart (Fig. 11) clearly shows that based only on Accuracy, F1, Precision, ROC Area, Recall ratings, Naive Bayes performs the best;

also, Logistic regression is the second-best performer. SVC performed better than the rest of the approaches except the best two.

5 Conclusion

People communicate through messages, and millions of smartphone users send and receive countless messages every day. Unfortunately, because of the lack of suitable message filtering techniques, that kind of communication is unsafe. Spam is an example of such instability, as it compromises the security of message communication. Spam detection is a significant problem in the modern message exchange. As a result, message transmission is not secure. To combat this issue, an effective and exact mechanism for detecting spam in message transmission is required. In this study, we used TFIDF vectorizer for vectorization and also used seven classifiers, which are Random Forest, XGBoost, Naive Bayes, Logistic Regression, SVC, Cat boost, and K nearest neighbor. After using several classifiers, it is seen that Naive Bayes and Logistic Regression provide the best performance than other machine learning algorithms.

In the future, we want to implement the BERT technique instead of the TF-IDF vectorizer for more precise data preprocessing. We will also try to use other classifier models to show better scores.

References

1. Chowdhary, K.: Natural language processing. In: Fundamentals of Artificial Intelligence, pp. 603–649. Springer, New Delhi (2020). https://doi.org/10.1007/978-81-322-3972-7_19
2. Govil, N., Agarwal, K., Bansal, A., Varshney, A.: A machine learning based spam detection mechanism. In: 2020 Fourth International Conference on Computing Methodologies and Communication (ICCMC), pp. 954–957 (2020). https://doi.org/10.1109/ICCMC48092.2020.ICCMC-000177
3. Kumar, N., Sonowal, S., Nishant: Email spam detection using machine learning algorithms. In: Second International Conference on Inventive Research in Computing Applications (ICIRCA) 2020, pp. 108–113 (2020). https://doi.org/10.1109/ICIRCA48905.2020.9183098
4. Jing, L.P., Huang, H.K., Shi, H.B.: Improved feature selection approach TFIDF in text mining. In: Proceedings. International Conference on Machine Learning and Cybernetics, vol. 2, pp. 944–946. IEEE, November 2002
5. Aagte, A.A., Vrushali, W., Vishwakarma, P., Kamble, S.: Spam detection using sentiment analysis. In: 2020 10th International Conference on Computer and Knowledge Engineering (ICCKE), pp. 209–214. IEEE, May 2019
6. Laudon, K.C., Traver, C.G.: E-commerce, pp. 1–912. Pearson, Boston (2013)
7. Nejad, S.J., Ahmadi-Abkenari, F., Bayat, P.: Opinion spam detection based on supervised sentiment analysis approach. In: 2019 International Journal of Emerging Technologies and Innovative Research, pp. 228–232. JETIR, October 2020
8. Peng, Q., Zhong, M.: Detecting spam review through sentiment analysis. J. Softw. **9**(8), 2065–2072 (2014)

9. Tiago, A.A.: UCI Machine Learning Repository: SMS Spam Collection Data Set, 22nd June 2012. https://archive.ics.uci.edu/ml/datasets/SMS+Spam+Collection
10. Gaydhani, A., Doma, V., Kendre, S., Bhagwat, L.: Detecting hate speech and offensive language on Twitter using machine learning: an N-gram and TF-IDF based approach. arXiv preprint arXiv:1809.08651 (2018)
11. Saif, M.A., Medvedev, A.N., Medvedev, M.A., Atanasova, T.: Classification of online toxic comments using the logistic regression and neural networks models. In: AIP Conference Proceedings, vol. 2048, no. 1, p. 060011. AIP Publishing LLC, December 2018
12. Cutler, A., Cutler, D.R., Stevens, J.R.: Random forests. In: Zhang, C., Ma, Y. (eds.) Ensemble Machine Learning, pp. 157–175. Springer, Boston (2012). https://doi.org/10.1007/978-1-4419-9326-7_5
13. Chen, T., et al.: XGBoost: extreme gradient boosting. R Package Version 0.4-2 **1**(4), 1–4 (2015)
14. Wien, M., Schwarz, H., Oelbaum, T.: Performance analysis of SVC. IEEE Trans. Circuits Syst. Video Technol. **17**(9), 1194–1203 (2007)
15. Leung, K.M.: Naive Bayesian classifier. Polytechnic University Department of Computer Science/Finance and Risk Engineering 2007, pp. 123–156 (2007)
16. Peng, C.Y.J., Lee, K.L., Ingersoll, G.M.: An introduction to logistic regression analysis and reporting. J. Educ. Res. **96**(1), 3–14 (2002)
17. Ibrahim, A.A., Ridwan, R.L., Muhamme, M.M.: Comparison of the CatBoost classifier with other machine learning methods. Int. J. Adv. Comput. Sci. Appl. **11**(11), 738–748 (2020)
18. Yigit, H.: A weighting approach for KNN classifier. In: 2013 International Conference on Electronics, Computer and Computation (ICECCO), pp. 228–231. IEEE, November 2013

A Voice Assisted Chatbot Framework for Real-Time Implementation in Medical Care

Shashi Kant Dargar(✉), Madhavan Vijayadharshini, Gorla Manisha, and Kunchaparthi Deepthi

Department of Electronics and Communication Engineering, Kalasalingam Academy of Higher Education and Research, Krishnankoil 626126, Tamilnadu, India
drshashikant.dragar@ieee.org

Abstract. A chatbot is one of the trending technologies used in many technologies. It is a fast-growing technology that combines hardware and software techniques into a single platform. Many websites take advantage of this technology to communicate directly with their visitors. The chatbot plays a crucial role in a clear-cut conversation with the website. This article provides a comprehensive study of the application-specific implementation of chatbots in real-time applications developed in the past years. The authors propose a framework and implement a medical care specific application using a robotic skull model with IoT, the software programming using embedded C. It leads to opening up numerous possibilities for other technologies to participate. Moreover, the chatbot design is programmed to answer all the queries about first aid prescriptions and offer a reminder for users. As a result, the proposed implementation leads to a framework for integrating intelligent conversational systems with the Android operating system.

Keywords: Chatbot · Conversational user interface · Human-computer interaction · Knowledge base · Software agents · IoT

1 Introduction

A chatbot is a powerful technology in shaping the future by connecting physical devices or things with Android. It also presents various opportunities to intersect other technological trends, allowing it to become even more intelligent and efficient. The introduction of chatbots into a community has brought us to the time of the conversational interface. It is an interface that does not demand a screen or a mouse soon. Instead, the interface is entirely conversational, and all communications are indistinguishable. These chatbots have evolved into a means of direct communication with users and websites. With the help of these chatbots, users can communicate with a particular site and quickly solve their questions. So, we came across an idea of what it would be like if this kind of chatbot were introduced in healthcare. This idea leads to our project, i.e., doctors generally prescribe medicines and tests to patients in the form of papers and files. The doctors also provide doses and a time chart to take medicines and undergo tests. The patients have to see the paper and the file every time they take medicine. It has become a crucial task in

M. Singh et al. (Eds.): ICACDS 2022, CCIS 1613, pp. 51–63, 2022.
https://doi.org/10.1007/978-3-031-12638-3_5

some situations in our current busy lives. So in this work, the authors present a designed chatbot application to answer queries according to the user's need, and further, we have also implemented a reminder system for the user at every point in time.

And in other conditions, in medical care in a hospital for observation and recovery after treatment, some careful medical prescription is usually recommended during post-operative care. However, when a medical attendee, such as a nurse or a caretaker, has to follow a medication schedule for multiple patients, it becomes confusing and tedious to remember the prescribed medicines for each patient. It is then becoming cumbersome to go through the patients' records to recall the medical action, which can sometimes compromise the appropriate care of the patients.

Currently, voice recognition technology has become a subject matter of mass interest. Also, robotics and embedded systems have shown their potential in technology. However, no such system exists to address the issue mentioned earlier by medical attendees. There are very few hospitals with a policy of one attendee per patient. As a result, most patients have become unaffordable to get treatment in such hospitals. Furthermore, some systems create reminders for the medical schedules of patients and transmit the reminders to a user device. However, such systems lack interactive capabilities, and the patients and the attendees often miss the reminders.

2 Background and Related Works

Patel et al. (2019) proposed a chatbot model for student-related information. Chatbot implementation saves both time and energy to solve the problem, and it is significantly better in performance than humans. For the development of chatbot-student-related information, the PHP language was utilized. The chatbot GUI implemented was similar to a messaging application. Pleshkova et al. (2019) have presented cloud technologies, artificial intelligence, and neural networks with universal speech recognition deep learning. The primary reported disadvantage of the project was that the system was able only to understand the words given in the database [1, 2]. By recognizing their mood swings and creating a solid relationship with users, Patel et al. (2019) proposed a social chatbot for better communication with students. As a result, the CNN, RNN, HAN-based chatbot was developed to understand the student's mood swings and change his anxious mood to an optimistic one. Ahmady et al. (2020) designed it to remind the people living in Japan about the occurrence of natural disasters, and it is also capable of capturing pictures of natural disasters for a better understanding of governmental agencies. In a paper, Prabhakar et al. (2020) implemented a real-time application called a chatbot, which is used to communicate with users. It was designed as a Google assistant for immediate response to the user [3–5].

Anbarasi et al. (2018) their paper presents a design application named "Robotic Arm," which could be used in military services. A robotic arm could be used to dispose of bombs from a certain distance. It is designed as a combination of both hardware and software. For successful disposal, the application also includes gesture moments, finger moments, and so on. Mauro et al. (2020) analyzed a real-time application called a "chatbot," which analyzes a person's mental status by understanding their mental health. In the experiment, they surveyed 47 members to understand their work better. Here, data editing, data deleting, and data updating are possible. With the help of the data, we could analyze mental health quickly [6, 7]. Tsao et al. (2020) The paper created a chatbot model that could be used in many supermarkets, entertainment, and digital marketing for entering customer feedback details and purchasing details. It acts as an interactive mode for consumers. Seering et al. (2020) developed a small chatbot machine learning application and used it in online community services. A chatbot being rapidly growing, they used to understand the words the users asked clearly. Bharti et al. (2020) The paper implemented an application named "telemedicine". It was developed to communicate with patients without visiting a hospital during the COVID pandemic. Here they introduced various languages through Natural Language Processing. It serves as a doctor instructing patients about medicines, symptoms, healthcare tips, and preventive measures. The chatbot is very interactive with Indian patients, and they named it the "conversational bot"[8–10].

Mlouk et al. (2020). With increasing semantic web technology, a large amount of information is available on the web in KBs. They developed a chatbot design to answer those questions and challenges in multiple languages and knowledge bases. Ashfaq et al. (2020) created a talk-bot for conversations aimed at drivers' satisfaction and continuance intention toward chatbot-based customer service. Lefteris et al. (2020) Chatbot is a rapidly increasing technology that could be used in education, IT sectors, Medicine fields, and entertainment. They have worked on the chatbot implementation and the challenges faced in implementing a chatbot using machine learning and artificial intelligence [11–13]. Dahia et al. (2017) A chatbot is an application that is designed by programming and implemented with a unique architecture for immediate response to users. The study learned about its working and implementation, and they concluded that the chatbot was user-friendly [14].

Ranoliya et al. (2017) They have designed a chatbot as a virtual assistant to complete various tasks and provide the necessary information asked by the user. In the paper, they used AIML and LSA. For example, AIML provides greeting details, and LSA provides quick responses to satisfy the consumer [15–17]. Vyawahare et al. (2020) In the paper, they designed a chatbot by choosing English as their second language. They have used the IBM Watson model and Google Assistant to develop the application. It provided subtitles and displayed lyrics when the song was playing. Sensuse et al. (2019) developed an advanced ELISA chatbot to respond quickly to online customer service questions. Jangid et al. (2019) article provided a detailed overview of the design and execution of a University Counselling Auto-Reply Bot that can respond to questions about engineering at our university level [18–20].

Luo et al. (2019) discussed that as chatbots and extended reality might be intimidating for instructors, educators must use the innovations to adapt to changes and difficulties in the teaching and learning landscape. So they first reviewed various available methods for instructors to overcome the hurdles. They highlighted the application of artificial intelligence (AI). They shared the achievemnets of constructing multiple chatbots for higher education, utilizing a commercial platform that helps get started with your chatbot in the workshop [21]. Tharewal et al. (2020) stated that advanced tools and technologies such as (AI, NLP) could create a system known as an intelligent-tutoring system or conversation system, such as a chatbot. Designing and producing chatbots involves a variety of strategies and procedures. The purpose of the review article was to study the critical and required strategies employed in the communication and conversational agents [22].

Sands et al. (2020) presented that the use of chatbots to assist, if not completely replace, people in service interactions is growing. Chatbots, like people, can follow service scripts in their interactions, which can influence the customer experience. Service scripts are spoken guidelines that aim to standardize customer service encounters. However, despite the rising usage of chatbots as a service mechanism, little is known about the influence of different service scripts offered during chatbot service interactions on consumers [4, 23, 24]. The above literature study reveals the following summary:

The previous works were implemented using machine learning and artificial intelligence to understand the user emotions used as direct communication between user and chatbot. However, plenty of previous work with wired connections has made the system immobile. All the existing systems are developed to clear user queries, but in our project in advance, we are using an alarming system to remind the users from time to time about their medicines. With the insights obtained from the previous work, attaining our model of medical chatbot application require portability. Therefore, we propose developing a chatbot with Bluetooth connectivity implementation interfacing simple hardware using IoT and embedded programming to get user-touch to clear their queries quickly. It is required to develop using simple language as it can be easily accessible by every user, whether the user is less educated.

3 Methodology

This work is embedded, so we simulated interlinked hardware and software on the same platform. The primary hardware system used in this project is Arduino UNO. The program is developed to interlink with other hardware modules like Bluetooth and (Universal Asynchronous Receiver-Transmitter (UART), and a Liquid Crystal Display (LCD) in the Arduino IDE platform.

An (LCD) is used to display the answer to queries asked by the users. Developing the chatbot in an embedded way is used as an instant responder for users and establishes direct communication. We have developed a reminder system to set a recap for every medicine intake by the users. It makes hospital nursing staff work easier when it is challenging to take care of every patient regarding their medication follow up. With a reminder, the chatbot itself reiterates as it is straightforward to understand as the answer or reply is displayed on both LCD and hearing sound with mouth actions by the skull of chatbot hardware. The block diagram represents the internal as well as external hardware connections of the chatbot is, as indicated in Fig. 1.

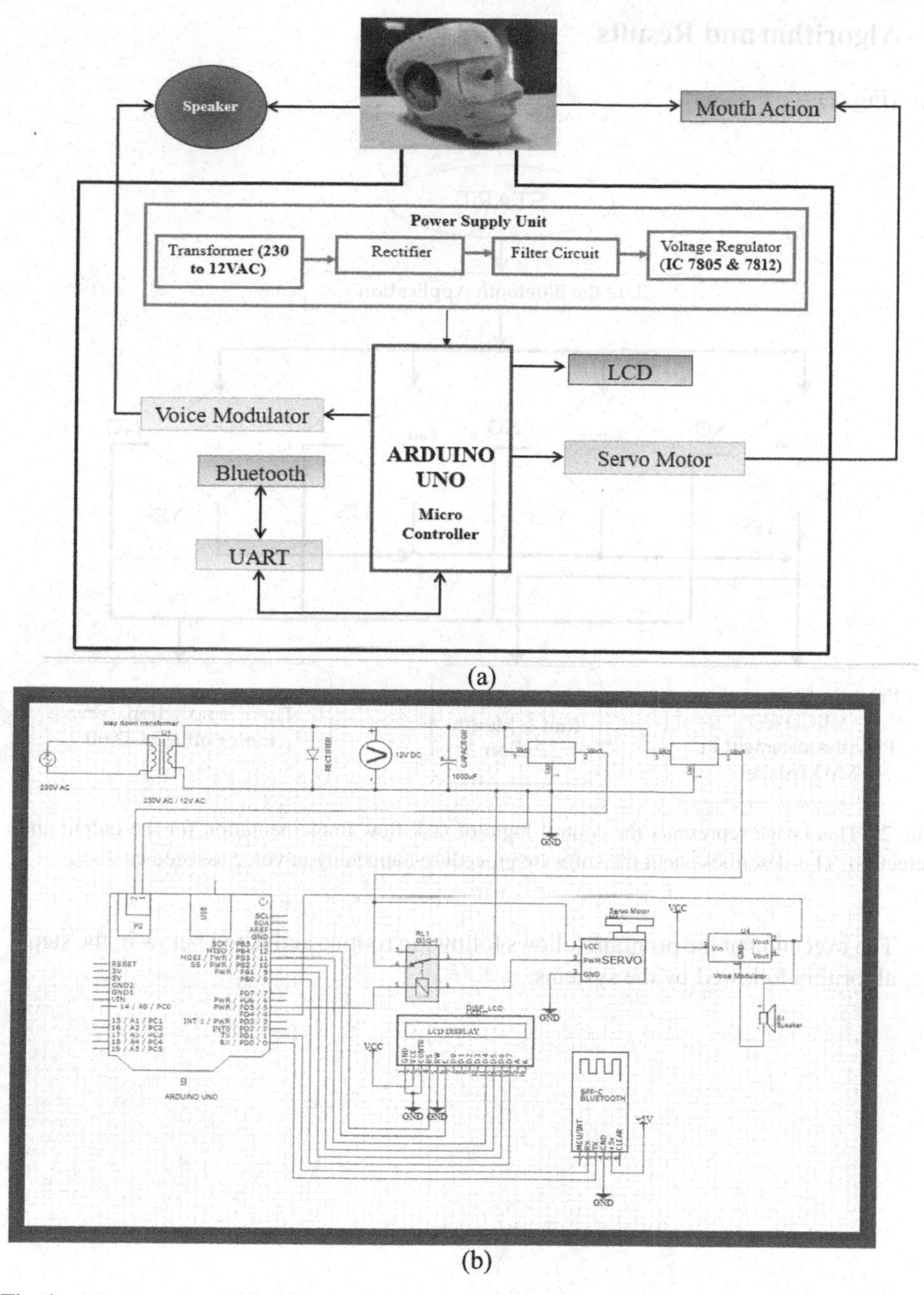

Fig. 1. (a) represents a block diagram which includes a clear process for working of external and internal parts of voice assisted chatbot. (b). This image represents the key to hardware connections. Each and every connections is clearly mentioned in the above pin diagram.

4 Algorithm and Results

See (Fig. 2)

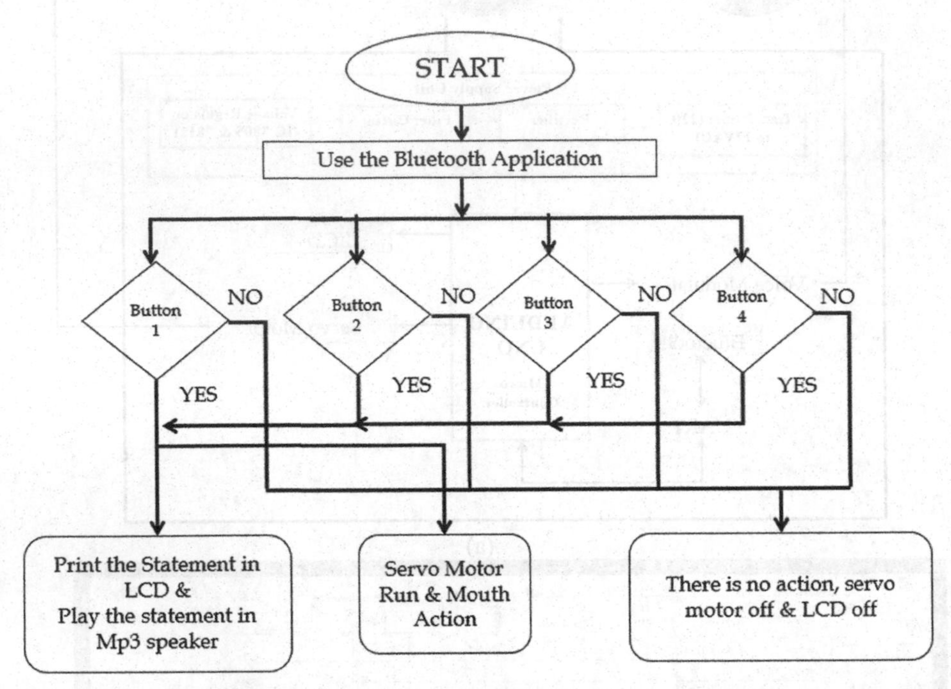

Fig. 2. This image represents the control logic of task flow implementation for the output and detection. This describes about the software procedure happening in voice assisted chatbot.

The execution of the program follows following routine and listed below in the steps i.e. algorithm followed by the systems.

```
Step 1: START program.
Step 2: Use the Bluetooth Application.
Step 3: Button 1:-
        True case: It displays print the statement in LCD
        & play the statement in MP3 speaker and servo motor
        run & mouth action.
   False case: It displays no action, servo motor off & LCD
   off.
Step 4: Button 2:-
        True case: It displays print the statement in LCD
        & play the statement in MP3 speaker and servo motor
        run & mouth action.
   False case: It displays no action, servo motor off & LCD
   off.
Step 5: Button 3:-
        True case: It displays print the statement in LCD
        & play the statement in MP3 speaker and also servo
        motor run & mouth action.
   False case: It displays no action, servo motor off & LCD
   off.
Step 6: Button 4:-
        True case: It displays print the statement in LCD
        & play the statement in MP3 speaker and also servo
        motor run & mouth action.
   False case: It displays no action, servo motor off & LCD
   off.
Step 7: END.
```

As shown in Fig. 3, the fabricated device hardware has been inspected for the input under different health conditions input. Appropriate power input has been checked and found that the action takes place with the glimpse of LED power. It displays print the statement in LCD & play the statement in MP3 speaker and servo motor run & mouth action.

Suppose the no input tt displays no action, servo motor off & LCD off. Then, with the pressing of the operational button, print the statement on the LCD screen and execute the statement in MP3 speaker to turn the servo motor motion for skull mouth action. The obtained result from the designed system is depicted in Fig. 4 (Fig. 5).

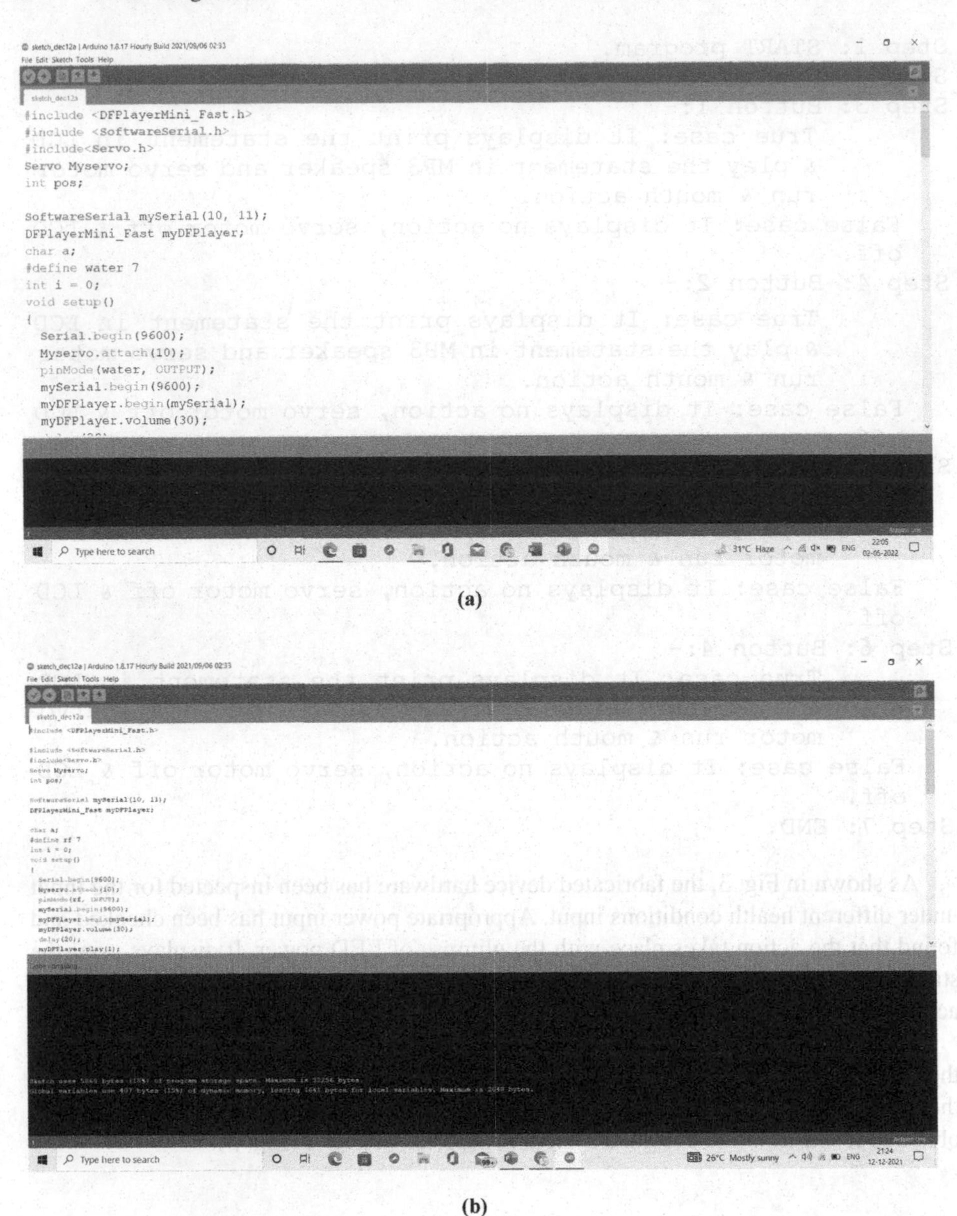

Fig. 3. (a) Insisting the embedded code in arduino IDE compiler. (b) Successful simulation of embedded program in a arduino IDE compiler

Here, we are performing the whole task in an Arduino IDE compiler, where the Arduino IDE is an open-source tool developed by Arduino.cc for writing, compiling, and uploading code to practically all Arduino modules. Thanks to the basic Arduino software, a non-technical person may get their feet wet in the learning process, which makes code compilation so simple. It has methods and instructions for debugging, editing, and compilation. The IDE environment's two most important components are the Editor and the Compiler. The first is for writing the required code, while the second is for compiling and uploading the code to the provided Arduino Module. By compiling the above mentioned embedded code in the Arduino IDE environment, we get

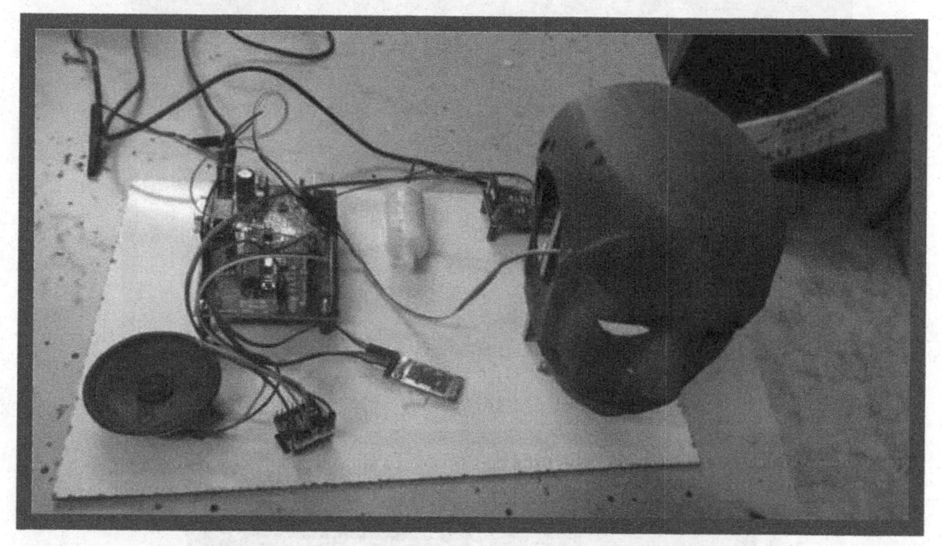

Fig. 4. The designed hardware for and testing for skull speaking mouth action upon input query by the demonstrating user

This system is designed to produce dynamic responses to the queries asked by patients and to remind the patient to take care of his medicine activities. This is done in the form of a voice Mp3 module along with mouth actions as it is connected to a servo motor and text format is displayed on the LCD screen. The control of this system can be implemented in an already existing android application.

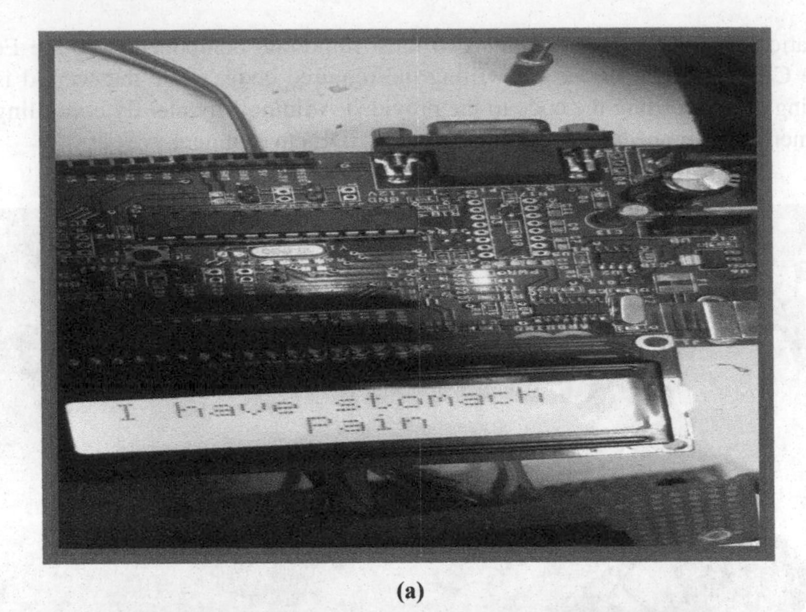

(a)

(b)

Fig. 5. Queries raised by demonstrating user and detected by the designed module for different health condition. (a) This image represents the queries usually asked by user. (b) This image represents the query related cough. (c) This image represents the query related fever

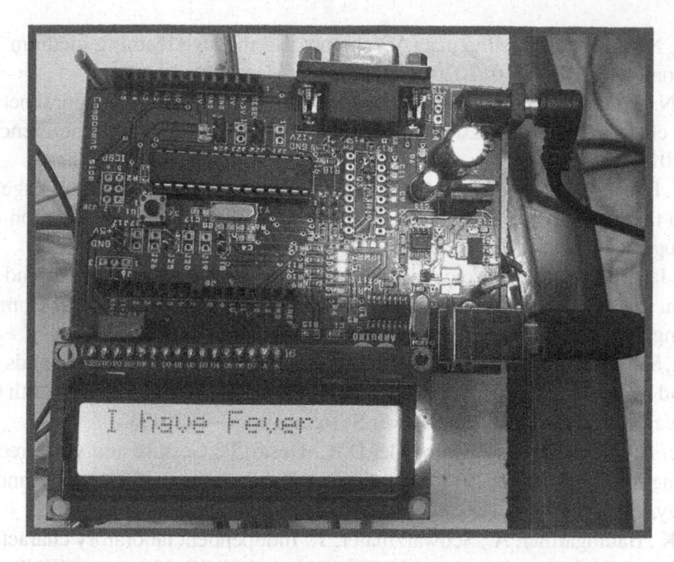

(c)

Fig. 5. continued

5 Conclusion

In this work, we have successfully implemented a real-time chatbot by using embedded and IoT in simulation environment, which is used for medication purposes in hospitals i.e. chatbot eases the nursing care to take care of patients by reminding them to take the prescribed medicine in right time in the form of Mp3 voice module and text in LCD screen, as well as it is used as self-medication purposes as it is very simple gadget to use. The designed hardware and the framework act efficiently on the device input and take less than a 20 ms time for generation of output. A security module can be integrated with implemented device and additional features can be connected for further versatility of chatbot in real time environments.

Acknowledgments. The Authors are thankful to Department of ECE, Kalasalingam Academy of Research and Education for the support and guidance by the faculty. We are thankful to all the authors who contributed in the apstand their research served as literature study used laid foundation of our work.

References

1. Nuruzzaman, M., Hussain, O.K.: A survey on chatbot implementation in customer service industry through deep neural networks. In: International Conference on e-Business Engineering, pp. 54–61 (2018)
2. Kumar, M.N., Chandar, P.L., Prasad, A.V., Sumangali, K.: Android based educational chatbot for visually impaired people. In: 2016 IEEE International Conference on Computational Intelligence and Computing Research, pp. 1–4 (2016)

3. Haristiani, N.: Artificial intelligence (Ai) chatbot as language learning medium: an inquiry. J. Phys. Conf. Ser. **1387**(1), 012020 (1–9) (2019)
4. Bhartiya, N., Jangid, N., Jannu, S., Shukla, P., Chapaneri, R.: Artificial neural network based university chatbot system. In: 2019 IEEE Bombay Section Signature Conference (IBSSC), pp. 1–6 (2019)
5. Purohit, J., Bagwe, A., Mehta, R., Mangaonkar, O., George, E.: Natural language processing based jaro-the interviewing chatbot. In: 2019 3rd International Conference on Computing Methodologies and Communication, pp. 134–136 (2019)
6. Aktar, N., Jaharr, I., Lala, B.: Voice recognition based intelligent wheelchair and GPS tracking system. In: 2019 International Conference on Electrical, Computer and Communication Engineering, pp. 1–6 (2019)
7. Pleshkova, S., Bekyarski, A., Zahariev, Z.: Reduced database for voice commands recognition using cloud technologies, artificial intelligence and deep learning. In: 2019 16th Conference on Electrical Machines, Drives and Power Systems, pp. 1–4 (2019)
8. Kanawade, A., Varvadekar, S., Kalbande, D.R., Desai, P.: Gesture and voice recognition in story telling application. In: 2018 International Conference on Smart City and Emerging Technology, pp. 1–5 (2018)
9. Lenhard, K., Baumgartner, A., Schwarzmaier, T.: Independent laboratory characterization of NEO HySpex imaging spectrometers VNIR-1600 and SWIR-320m-e. IEEE Trans. Geosci. Remote Sens. **53**(4), 1828–1841 (2014)
10. Moshayedi, A.J., Hosseini, M.S., Rezaee, F.: WiFi based massager device with node MCU through arduino interpreter. J. Simul. Anal. Novel Technol. Mech. Eng. **11**(1), 73–79 (2019)
11. Luo, C.J., Gonda, D.E.: Code free bot: an easy way to jumpstart your chatbot!. In: 2019 IEEE International Conference on Engineering, Technology and Education (TALE), pp. 1–3 (2019)
12. Moshayedi, A.J., Agda, A.M., Arabzadeh, M.: Designing and implementation a simple algorithm considering the maximum audio frequency of persian vocabulary in order to robot speech control based on arduino. In: Montaser Kouhsari, S. (ed.) Fundamental Research in Electrical Engineering. LNEE, vol. 480, pp. 331–346. Springer, Singapore (2019). https://doi.org/10.1007/978-981-10-8672-4_25
13. Srivastava, S., Prabhakar, T.V.: Desirable features of a chatbot-building platform. In: 2020 IEEE International Conference on Humanized Computing and Communication with Artificial Intelligence (HCCAI), pp. 61–64 (2020)
14. Sabarivani, A., Anbarasi, A., Vijayaiyyappan, A., Sindhuja, S.: Wireless synchronization of robotic arm with human movements using aurdino for bomb disposal. In: 2018 International Conference on Smart Systems and Inventive Technology (ICSSIT), pp. 229–234 (2018)
15. Singh, A., et al.: Evolving long short-term memory network-based text classification. Comput. Intell. Neurosci. **2022**, 1–15 (2022)
16. Koundinya, H., Palakurthi, A.K., Putnala, V., Kumar, A.: Smart college chatbot using ML and Python. In: 2020 International Conference on System, Computation, Automation and Networking, pp. 1–5 (2020)
17. Vyawahare, S., Chakradeo, K.: Chatbot assistant for English as a second language learners. In: 2020 International Conference on Convergence to Digital World-Quo Vadis (ICCDW), pp. 1–5 (2020)
18. Sensuse, D.I., et al.: Chatbot evaluation as knowledge application: a case study of PTABC. In: 11th International Conference on Information Technology and Electrical Engineering (ICITEE), pp. 1–6 (2019)
19. Moshayedi, A.J., Roy, A.S., Liao, L., Li, S.: Raspberry Pi SCADA zonal based system for agricultural plant monitoring. In: 2019 6th International Conference on Information Science and Control Engineering, pp. 427–433 (2019)

20. Margreat, L., Paul, J.J., Mary, T.B.: Chatbot-attendance and location guidance system (ALGs). In: 2021 3rd International Conference on Signal Processing and Communication, pp. 718–722 (2021)
21. Das, S., Kumar, E.: Determining accuracy of chatbot by applying algorithm design and defined process. In: 2018 4th International Conference on Computing Communication and Automation, pp. 1–6 (2018)
22. Li, L., Lee, K.Y., Emokpae, E., Yang, S.-B.: What makes you continuously use chatbot services? evidence from Chinese online travel agencies. Electron. Mark. **31**(3), 575–599 (2020). https://doi.org/10.1007/s12525-020-00454-z
23. Perez-Soler, S., Guerra, E., de Lara, J.: Creating and migrating chatbots with conga. In: 43rd International Conference on Software Engineering: Companion Proceedings, pp. 37–40 (2021)
24. Seering, J., Luria, M., Ye, C., Kaufman, G., Hammer, J.: It takes a village: integrating an adaptive chatbot into an online gaming community. In: Proceedings of the 2020 CHI Conference on Human Factors in Computing Systems, pp. 1–13 (2020)

Robust Vehicle Detection for Highway Monitoring Using Histogram of Oriented Gradients and Reduced Support Vector Machine

Sarthak Mishra, Dhruv Upadhyay, and P. Saranya(✉)

Department of Computing Technologies, SRM Institute of Science and Technology, Kattankulathur, Chennai, India
saranyap@srmist.edu.in

Abstract. Highway monitoring has been an important yet challenging aspect to address with machine learning methods. The method proposed in this paper addresses essential aspects of highway monitoring vehicle detection systems. A machine learning model is proposed to detect cars from dashboard cameras from the datasets. Images used to train the model are collected and grouped from KITTI vision and GTI datasets. The Region-Based Convolutional Neural Network (RCNN) method fails in providing robust information and generates bad candidate region proposals. On the other hand, Faster-RCNN, able to cover the flaws of RCNN, cannot provide good accuracy. Histogram of Oriented Gradients (HOG) features extractor is used over Color Histogram and Spatial Binning as they lack abundant features resulting in a lack of vital information. The method uses vectors from the vehicle and non-vehicles images to improve the classification. Optimized Support Vector Machine (SVM) 'rbf' kernel classifier, i.e., Reduced Support Vector Machine (RSVM), excludes ambiguity during classification. The classification was performed using multiple tuned Support Vector Classifiers (SVC), random forest and naive Bayes classifiers. Evaluating with other methods based on accuracy score, average precision and F1-score, RSVM presented the highest performance.

Keywords: Vehicle detection · Support vector classifier (SVC) · Feature extraction · Histogram of Oriented Gradients (HOG) · Support Vector Machine (SVM)

1 Introduction

Highway monitoring has been a considerable aspect of drivers' safety and traffic surveillance. Robust Vehicle detection systems have been a great demand to fulfill these requirements. A computer vision-based approach for dashboard cameras is proposed, which is better than other methods. This system can also be applied to automatic-driving systems, vehicle counting and traffic surveillance. The image data gathered is divided into vehicles and non-vehicles, making further computation easy for feature extractors and the classifier.

M. Singh et al. (Eds.): ICACDS 2022, CCIS 1613, pp. 64–73, 2022.
https://doi.org/10.1007/978-3-031-12638-3_6

Traffic video footage from huge directories has been possessed for its study with the prominent setting up of traffic surveillance cameras. Normally, at an upper scrutiny angle, the more isolated and far away road surface can be designed. The accuracy calculated for detection of objects having small sizes and which are long away from the road is minor and the object size of the vehicle changes immensely at this perceiving angle. It is important to conclusively solve the above issues and additionally employ them conveniently for the classifier and feature extractor as in the case of complicated camera footage.

Vehicle Detection is a project based on Machine Learning/Deep Learning and Computer Vision. Many models have been developed to find the best accurate detection method, some of which are RCNN, Faster-RCNN, etc. The vehicle detection system alerts the drivers by detecting overweight vehicles moving towards suspended barriers, such as tunnels, bridges, etc. These systems also have large-scale potential, precision, and utility of different detection types, including accident detection, speed, vehicle classification, number of vehicles, stopped vehicle detection, wrong-way vehicles, and more classical traffic data such as queue detection and space between vehicles etc.

Vehicle detection and type recognition established on stationary images is straightly suitable for various activities in a traffic surveillance system and is greatly pragmatic. Previously, various methods have been seen that introduce the processing of automatic vehicle detection and recognition. Vehicle detection or even object identification itself, is visionary as the objects tend to change their appearance perilously according to the camera's perspective.

A suitable traffic management system is necessary for the easy traffic course on roads and the safety, making the utmost and optimum usage of road facilities to amplify the content of road capacities [9]. In highway monitoring, vehicle detection video scenes are of substantial importance for controlling the highway and intelligent traffic conduct. Our main motive is to use a real-time object detection algorithm to optimize the current traffic management systems. In vehicle detection, many feature extractors are used, such as Color histogram, Spatial binning, Gabor filter etc. Still, HOG is robust among all of these, with less computation as one of the primary reasons [1].

The feature vectors produced by the feature extractor are used to train the SVM classifier [2]. SVM produces optimum results with the HOG feature extractor compared with other classifiers. To produce further optimum results, different kernels in SVM are compared. The proposed method is better than RCNN because it does not produce any bad candidate region proposals and thus improves efficiency. The accuracy and precision are better than Faster-RCNN. And to locate the vehicle in the given image or input frames, the window sliding technique is used.

The remainder of this paper is arranged as follows: In Sect. 2, we examine the relevant work and propose the methodology framework containing feature extraction, detection and classification in Sect. 3. The experimental results and comparisons are presented in Sect. 4, and Sect. 5 concludes the paper.

2 Related Work

There have been many methods for vehicle detection which have lacked either efficiency or execution time. Methods such as RCNN and Faster-RCNN have not been effective due to bad candidate region proposals and lack of efficiency paired with longer execution time.

The dominant feature of histogram-oriented gradient (DPHOG) is one of the feature extractors which use lesser features to give robust information. This feature extraction method excludes non-dominant patterns from each of the images to recognize the images associated with different classes [1]. The selection of non-dominant features is made by selecting the common attributes between the non-vehicles and vehicle images [1]. The vertical histogram of oriented gradients (V-HOG) method would also have faster computational time, and it requires only a handful of features but lacks in accuracy compared to HOG. The kernel extreme learning machine (KELM) classifier used alongside the DPHOG &V-HOG produces desired validation results [1]. Haar-like is another feature extraction method that takes adjacent rectangular regions in a location, sums up the intensities of the pixels in every region, and calculates the difference between the sums used to categorize sections of the image. The drawback of this method is the sensitivity under illumination.

Gabor-filter is also one of the often-used methods whose drawback is an excess number of features to classify an image. An additional method by optimizing the original Gabor-filter method that consolidates genetic algorithms with incremental clustering by merging the filter design with filter selection is also used for vehicle detection [3]. Another well-known method used for vehicle detection is πHOG which is an optimization over HOG [4]. In recent times, deep learning approaches have been implemented to the vehicle detection task. The pooling layer was unable to cope up and filter the different rotations and scales of the image [1]. The histogram of oriented gradients (HOG) was applied to the images before implementing deep learning, which upgrades the image variation [1].

HOG is comparatively well paired with the SVM classifier. However, it can also be suiting up with other classification methods, such as random forest, naive Bayes, AdaBoost, K-nearest neighbor (KNN), etc. [2, 9, 18]. The SVM classifier gives the optimal performance when paired with the HOG [2, 20]. This is because the extracted features from the feature extractor are easily classified and, in less time, using optimal hyperplanes [6]. HOG & SVM pair is comfortable and flexible compared to CNN in preparing training images and hence is easy and widely used for object detection [1].

3 Proposed Methodology

The model used comprises a feature extractor, classifier and detector. Histogram Oriented Gradient is used as a feature extractor. The images passed to the HOG feature extractor are shaped (64, 64, 3). The output vectors are stored as vehicles and non-vehicles and then passed to train the classifier. Using the window sliding method, building blocks are designed to detect and keep track of the vehicles [5]. A pipeline-based implementation is used to execute the model on real-time video. The flow of the proposed model is presented in Fig. 1.

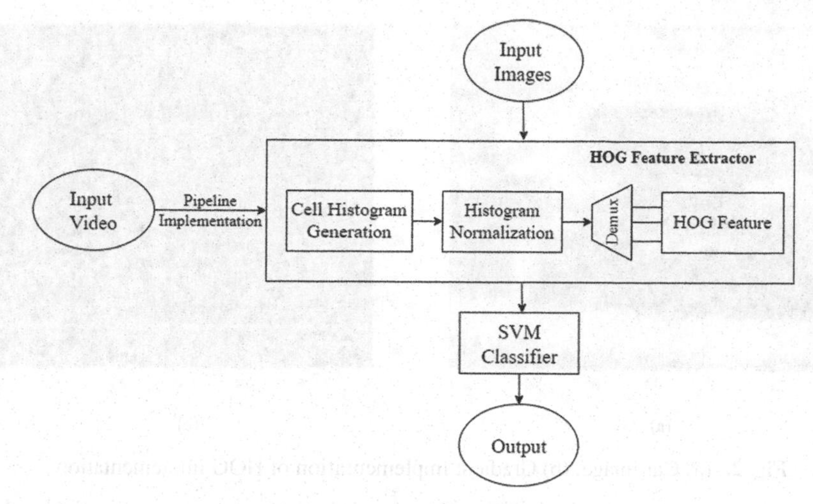

Fig. 1. Architecture/Flow diagram

3.1 Feature Extraction

The first step of vehicle detection is to extract features from images or frames of a video. There have been methods for feature extraction such as Color Histogram, Spatial Binning etc., but none of these are as powerful as HOG [6, 7]. The HOG takes an image and splits it into different blocks containing cells.

With the help of HOG, the objects and shapes in an image can be represented by the intensity gradients distribution. In every cell of an image for every pixel a histogram of gradient direction is determined and the descriptor is formed by the combination of produced histograms.

The pixels are observed, and the feature vectors are extracted in the cells. HOG also executes some computations internally, decreases redundancies in the data, and yields improved feature vectors. These internal computations include image preprocessing done in order to get rid of the image intensities which might further cause problems during feature extraction process, gradient computation done in order to corroborate the normalized color and also the gamma values in particular, cell histogram generation to create histograms based on gradients, histogram normalization done due to sensitivity of image gradients to lightning which will change the values of magnitude resulting in change in values of histogram.

The feature vector length obtained is 324. These feature vectors are stored in two categories, i.e., vehicles & non-vehicles. HOG implementation on an image where the original image of the car (a) and the feature gradients per block of the image (b) are shown (Fig. 2).

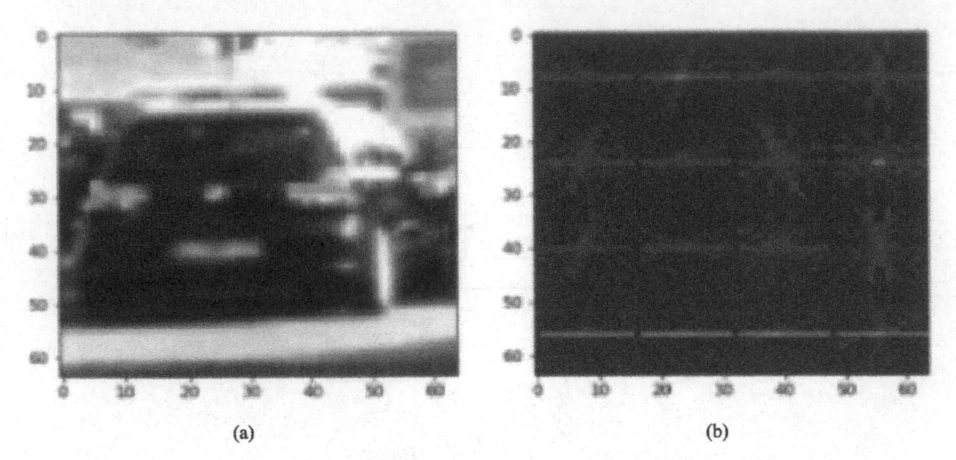

Fig. 2. (a) Car image, (b) Gradient implementation of HOG implementation

3.2 Detection

Car detection in images is an essential part of the model. Thus, a sliding window approach is proposed in which searches for the vehicles in the bottom part of the image as the top part mainly consists of the sky. The base window size is 64 × 64, as the input image length is 64 × 64. Two possibilities of car detection are below the horizon: (1) Distant cars that appear near the horizon and thus far from the dashboard camera. (2) Close cars which appear near the dashboard camera and thus appear big. So, to cover both types of cars, the window size differs from the horizon to the bottom of the image, and the overlap number is increased to avoid ambiguities. As shown in the images, the (c) images are the detection using the window sliding method to which the heat map describes the location on a 2-D plane (d) (Fig. 3).

Fig. 3. (a) Test image 1, (b) Test image 2, (c) Heat map

3.3 Classification

The classifier used in the proposed model is an SVM classifier which is one of the most robust classification methods based on statistical learning [8, 10]. Among various kernel functions in support vector classifiers, the one used in the proposed model is the radial basis function kernel (RBF). The RBF kernel calculates the distance between an input point and a fixed point, either origin or some other fixed point ‘C’, called the center. The distance calculated is Euclidean distance. For all the kernel functions in SVM, the main objective is to find the best hyperplane possible for classification by doing calculations on more than one-dimensional space. These hyperplanes can be a quadratic, polynomial or any other equation resulting from the learning process. The Radial basis function used the exponential function resulting in an infinitely powerful classification hyperplane which is why this function is chosen for the model [11].

The feature vectors extracted by the Histogram Oriented Gradient are used to train the classifier to find the vehicle in the image. Before using the vector list for training the classifier, the data must be pre-processed. The data is split into train and test and normalized and scaled. The scaled data is passed to different classifiers to compare accuracy, precision, and false positive and false negative F1-score.

$$K = exp\left(\frac{-\|p - p'\|^2}{2\sigma^2}\right) \tag{1}$$

$$\gamma = \frac{1}{2\sigma^2} \tag{2}$$

$$K = exp\,(-\gamma||p - p'||^2) \tag{3}$$

In (I) the $-\ ||p - p'||^2$ is the squared euclidean distance between feature vectors and ‘σ’ is a free parameter. The ‘C’ parameter which determines how much you want to eschew misclassification is set to 100 and ‘γ’ parameter which defines how far the sway of training examples reaches is set to 0.3.

4 Results and Discussion

The datasets used to validate the proposed method are KITTI vision & GTI. These are some of the famous, easily accessible and frequently used datasets. Consisting of vehicle images, these datasets are suitable for developing traffic surveillance systems, highway monitoring systems, automated driving systems, etc. The images used for the method proposed are 64 × 64 in size. The validation of the proposed model is based on four matrix accuracy, false-positive rate (FPR), true positive rate (TPR) and F1-score [1].

$$accuracy = \frac{TP + TN}{TP + FP + TN + FN} \tag{4}$$

$$TPR = \frac{TP}{TP + FN} \tag{5}$$

$$FPR = \frac{FP}{FP + TN} \tag{6}$$

$$F1 - score = \frac{TP}{TP + \left(\frac{FP+FN}{2}\right)} \tag{7}$$

Here TP & TN are the numbers of vehicles and non-vehicles correctly classified. In contrast, FP gives the number of vehicles detected as non-vehicles, and FN shows the number of non-vehicles detected as vehicles (Table 1).

Table 1. Performance results of the proposed model using GTI and KITTI dataset.

	GTI	KITTI
Accuracy %	99.70	99.41
TPR	0.9950	0.9941
FPR	0.0000	0.0059
F1-score	0.9963	0.9901

Table 2. Performance comparison of the proposed model with the existing models using GTI dataset

Model	Accuracy%	TPR	FPR	F1-score
DPHOG & KELM [1]	99.01	0.9861	0.0052	0.9890
HOG & KELM [1]	98.75	0.9795	0.0053	0.9867
V-HOG & KELM [1]	98.27	0.9788	0.0146	0.9811
HOG & SVM (linear)	98.83	0.9880	0.0108	0.9878
HOG & Randomforest	98.21	0.9821	0.0188	0.9816
HOG & Naive Bayes	96.40	0.9639	0.0241	0.9628
HOG & RSVM (Proposed method)	**99.70**	**0.9950**	**0.0000**	**0.9963**

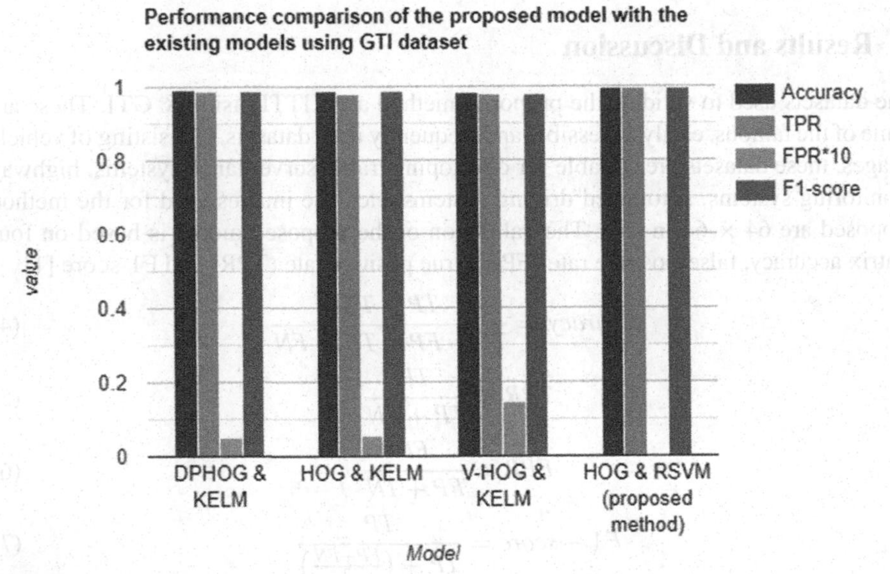

Fig. 4. Visual comparison of the proposed model with the existing models using GTI dataset

Table 3. Performance comparison of the proposed model with the existing models using KITTI Dataset

Model	Accuracy%	TPR	FPR	F1-score
DPHOG & SVM [1]	98.80	0.9798	0.0095	0.9747
HOG & SVM [1]	98.05	0.9539	0.0087	0.9597
V-HOG & KELM [1]	97.48	0.9748	0.0251	0.9456
HOG & SVM (linear)	98.16	0.9812	0.0167	0.9739
HOG & Randomforest	96.98	0.9691	0.0239	0.9622
HOG & Naive Bayes	95.24	0.9520	0.0297	0.9503
HOG & RSVM (Proposed method)	**99.41**	**0.9941**	**0.0059**	**0.9901**

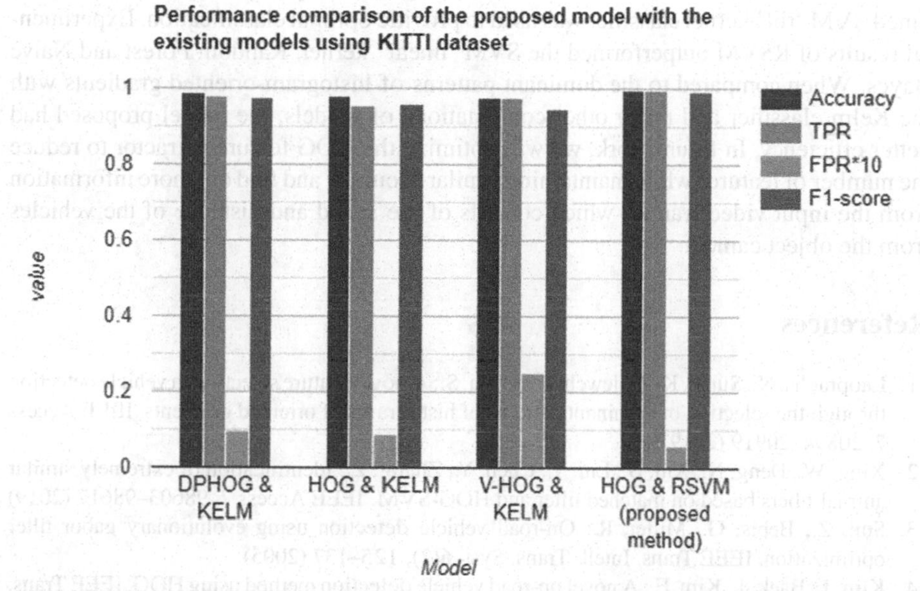

Fig. 5. Visual comparison of the proposed model with the existing models using KITTI dataset

As presented in a brief comparison in Table 2 & Table 3, all of the models seem to perform outstandingly in the given datasets. Yet the proposed models tend to outperform the existing methods in every validation category. Achieving an accuracy of 99.7% and 99.41% in GTI & KITTI vision datasets is a significant improvement accompanied by the best F1-score among all the methods listed of 0.9963 and 0.9901, respectively. With a better understanding of areas in which the proposed model outperforms the other techniques, TPR and FPR reflect the same. The performance comparison graph between the proposed model and a few state-of-the-art models has also been plotted and shown in Fig. 4 and Fig. 5. The proposed HOG & RSVM model is an optimal model that produces the best performance compared to the existing methods on given datasets. The parameter tuning with RSVM allows better classification of the points near the hyperplane.

Having achieved exceptional accuracy, the proposed model can detect vehicles in video inputs from the dashboard camera. This helps manage traffic and significantly avoids accidents through advancements in autonomous driving systems. It addresses the problem of the multi-scale contrast of the object. The proposed model will also aid in developing efficient highway or freeway vehicle traffic monitoring systems.

5 Conclusion

This paper proposes a machine learning-based method for vehicle detection primarily intended for dashboard cameras. Comparisons between various feature extractors have been made. The HOG feature extractor has been used to extract robust information from the images and video frames and convert it into feature vectors. Comparisons between different classifiers were made on combined images from the KITTI & GTI dataset. A tuned SVM 'rbf' kernel classifier was used to provide optimal classification. Experimental results of RSVM outperformed the SVM 'linear' kernel, Random Forest and Naive Bayes. When compared to the dominant patterns of histogram-oriented gradients with the Kelm classifier and many other combinations of models, the model proposed had better efficiency. In future work, we will optimize the HOG feature extractor to reduce the number of features while maintaining similar accuracy and find out more information from the input video frames, which consists of the speed and distance of the vehicles from the object camera.

References

1. Laopracha, N., Sunat, K., Chiewchanwattana, S.: A novel feature selection in vehicle detection through the selection of dominant patterns of histograms of oriented gradients. IEEE Access **7**, 20894–20919 (2019)
2. Xing, W., Deng, N., Xin, B., Liu, Y., Chen, Y., Zhang, Z.: Identification of extremely similar animal fibers based on matched filter and HOG-SVM. IEEE Access 7, 98603–98617 (2019)
3. Sun, Z., Bebis, G., Miller, R.: On-road vehicle detection using evolutionary gabor filter optimization. IEEE Trans. Intell. Trans. Syst. **6**(2), 125–137 (2005)
4. Kim, J., Baek, J., Kim, E.: A novel on-road vehicle detection method using HOG. IEEE Trans. Intell. Trans. Syst. **16**(6), 3414–3429 (2015)
5. Liu, C., Hong, F., Liu, R.R., Lu, C.H., Wei, J.: A traffic surveillance multi-scale vehicle detection object method base on encoder-decoder. IEEE Access **8**, 47664–47674 (2020)
6. Xiang, Z., Tan, H., Ye, W.: The excellent properties of a dense grid-based HOG feature on face recognition compared to gabor and LBP. IEEE Access **6**, 29306–29319 (2018)
7. Wang, X.: Data-driven based tiny-YOLOv3 method for front vehicle detection inducing SPP-Net. IEEE Access **8**, 110227–110236 (2020)
8. Zhang, F., Yang, F., Li, C., Yuan, G.: CMNet: a connect-and-merge convolutional neural network for fast vehicle detection in urban traffic surveillance. IEEE Access **7**, 72660–72671 (2019)
9. Wang, Z., Huang, J., Xiong, N.N., Zhou, X., Lin, X., Ward, T.L.: A robust vehicle detection scheme for intelligent traffic surveillance systems in smart cities. IEEE Access **8**, 139299–139312 (2020)

10. Shao, X., Wei, C., Shen, Y., Wang, Z.: Feature enhancement based on CycleGAN for nighttime vehicle detection. IEEE Access **9**, 849–859 (2020)
11. Awasi, M., et al.: Real-Time surveillance through face recognition using HOG And feedforward neural networks. IEEE Access **7**, 121236–121244 (2019)
12. Ghaffari, S., Soleimani, P., Li, K.F., Capson, D.W.: Analysis and comparison of fpga-based histogram of oriented gradients implementations. IEEE Access **8**, 79920–79934 (2020)
13. Hameed, M.A., Hassaballah, M., Ali, S.A., Ismail, A.: An adaptive image steganography method based on histogram of oriented gradient and PVD-LSB techniques. IEEE Access **7**, 185189–185204 (2019)
14. Luo, A., An, F., Zhang, X., Mattausch, H.J.: A Hardware-efficient recognition accelerator using haar-like feature and SVM classifier. IEEE Access **7**, 14472–14487 (2019)
15. Zhang, Y., et al.: Detecting object open angle and direction using machine learning. IEEE Access **8**, 12300–12306 (2020)
16. Wu, G., Chen, W., Cheng, H., Zuo, W., Zhang, D., You, J.: Multi-object grasping detection with hierarchical feature fusion. IEEE Access **7**, 43884–43894 (2019)
17. Gao, F., Wang, C., Li, C.: A combined object detection method with application to pedestrian detection. IEEE Access **8**, 194457–194465 (2020)
18. Tan, P., Mao, K., Zhou, S.: Image target detection algorithm for smart city management cases. IEEE Access **8**, 163357–163364 (2020)
19. Guo, S., Liu, F., Yuan, X., Zou, C., Chen, L., Shen, T.: HSPOG: an optimized target recognition method based on histogram spatial pyramid of oriented gradients. Tsinghua Sci. Technol. **26**(4), 475–483 (2021)
20. Zhang, Z., Zou, C., Han, P., Lu, X.: A runway detection method based on classification using optimized polarimetric features and reliable re-detection scheme. IEEE Access **8**, 49160–49168 (2020)

A Secure Framework Based on Nature-Inspired Optimization for Vehicle Routing

Righa Tandon[1(✉)], Ajay Verma[2], and P. K. Gupta[2]

[1] Department of Computer Science and Engineering, Chitkara University Institute of Engineering and Technology, Chitkara University, Punjab 140 401, India
righa.tandon@chitkara.edu.in

[2] Department of Computer Science and Engineering, Jaypee University of Information Technology, Solan 173 234, Himachal Pradesh, India
ajayverma322@gmail.com, pkgupta@ieee.org

Abstract. Vehicular networks are gaining popularity day by day due to their numerous applications provided to the users. Vehicles present in the vehicular network sends and receives information related to traffic congestion, vehicle speed, location, etc. So, security of the information that will be shared in the network among vehicles is a matter of concern. In order to tackle this security issue in the vehicular network, a secure framework has been proposed in this paper. This framework helps in secure communication among vehicles and other components of the network. Further, finding the best and optimal routing path for vehicles is another challenging factor. In this paper, a nature-inspired algorithm has been implemented for vehicle routing which uses Ad-hoc on demand distance vector (AODV) routing protocol to enhance its performance. Performance evaluation shows that the proposed algorithm performs better than the existing schemes for vehicle routing.

Keywords: VANET · Encryption · Roadside units · Ant colony optimization · Intelligent water drop

1 Introduction

The term intelligent transportation system (ITS) is combining different technologies which includes information such as communication information, geographic information, computer information, traffic congestion related information, etc. ITS helps in enhancing the safety and quality of traffic management in vehicular networks [15]. Vehicular network is the part of ITS that is responsible for sharing information among vehicles and other sensors in the network. Vehicular ad-hoc network is one of the most interesting and popular topic of research. This is because of its low cost, flexibility, fault tolerance application areas. Vehicular ad-hoc network consists of different registered vehicles, sensors, trusted authority, wireless connectivity, etc. In vehicular ad-hoc network,

M. Singh et al. (Eds.): ICACDS 2022, CCIS 1613, pp. 74–85, 2022.
https://doi.org/10.1007/978-3-031-12638-3_7

vehicles are connected through wireless medium in an ad-hoc manner. Those vehicles can send messages in the network using internet connectivity. Mostly, two types of communication takes place in the vehicular network i.e. vehicle-to-vehicle and vehicle-to-infrastructure. Vehicle can directly communicate with other vehicles so as to send critical or safety messages in the network and the type of communication is known as vehicle-to-vehicle (V2V) communication. If vehicle communicates with the infrastructure such as roadside units or other sensors that is known as vehicle-to-infrastructure (V2I) communication. The types of vehicular communication is shown in Fig. 1. The various messages that will be shared among vehicles in the network may or may not contain emergency or critical information. The critical messages comes under safety application of vehicular network. These messages may contain information related to accidents, passing of ambulance vehicles, etc. Routing and infotainment applications of vehicular network contains information such as location of vehicles, direction, speed of the vehicle, watching movies, listening to music while traveling on the road etc. [10]. The major topic of concern in any vehicular network is security of information to be shared in the network and finding the optimal routing path for vehicles. In this paper, we have covered both the challenging factors and proposed an algorithm that results in better performance and efficiency when compared with existing work by taking different parameters and number of vehicles. The main key points of this paper are:

- A secure framework is proposed for securing the information that is to be shared in the vehicular network.
- An algorithm is proposed to find the best and optimal path for vehicle routing in VANET.

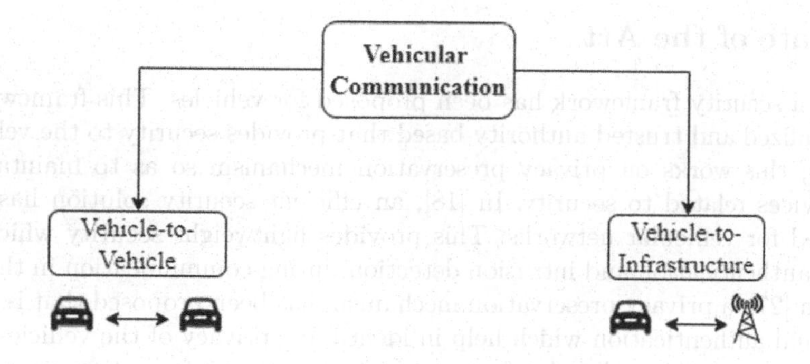

Fig. 1. Types of vehicular communication

In Sect. 2, various different existing techniques and approaches related to proposed work are discussed. Section 3 discusses the different processes of the proposed framework and describes the proposed algorithm with nature inspired technique for vehicle routing. Section 4 evaluates the performance of proposed work by considering different parameters. Also, the proposed work is compared with different existing techniques. Finally, Sect. 5 concludes the overall paper. The various notations that are used in this work are shown in Table 1.

Table 1. Notations

Notation	Illustration
ACO	Ant colony optimization
$AODV$	Ad-hoc on demand distance vector
CA	Central authority
IWD	Intelligent waterdrop
RSU	Roadside unit
$VANET$	Vehicular ad-hoc network
SK_{CA}	Secret key
$Soil(a, b)$	Soil between point a & b added
V_N	Vehicle node
$Vel_{(iwd)}(t)$	Velocity at source node
$Vel_{(iwd)}(t+1)$	Velocity value from point a to b
$\delta Soil(a, b)$	Change in amount of soil density
$\vartheta_{(m,n)}$	Change in pheromone value
$\sum(\Upsilon_{dp})$	Total pheromone value
$Avg_{(\Upsilon_{dp})}$	Average pheromone value

2 State of the Art

In [19], a security framework has been proposed for vehicles. This framework is decentralized and trusted authority based that provides security to the vehicles. Further, this works on privacy preservation mechanism so as to maintain all the services related to security. In [18], an efficient security solution has been proposed for vehicular networks. This provides lightweight security which can handle authentication and intrusion detection during communication in the network. In [27], a privacy preservation mechanism has been proposed that is based on mutual authentication which help in identifying privacy of the vehicles. Further, signature scheme has been proposed for sending information among the vehicles securely. In [11], privacy preservation information transmission scheme has been proposed that is based on cluster-based vehicular network. This helps in reducing the overhead and blockchain is used for secure information transmission in the network. This scheme minimises the delay and throughput and enhances the information packet delivery ratio. In [16], a flexible framework has been proposed that is based on security and privacy. In this, vehicle's information remains secret during communication in the network. A decision tree has been designed for acquiring trust and preserving privacy in the vehicular network. In [8], a privacy preservation authentication scheme has been proposed for vehicular networks. This scheme worked on conditional privacy preservation mechanism which helps in fast and efficient authentication during communication in the network. Authors in [14] have proposed an approach that focuses on providing enhanced security to VANETs and other transport networks by using swarm algorithm. They discuss various problems related to the information security in VANETs. This algorithm is designed to tackle two of the most common attacks in VANETs that are: wormhole attack and black hole attack. In [17] a hybrid dragonfly swarm technique along with chaotic particle algorithm has been proposed. It is claimed to be faster and secure and can detect and prevent DDoS attacks in vehicular networks. Chaotic algorithm further enhances the performance of the dragonfly swarm technique. This hybrid algorithm proves to be efficient and meets the standard requirements of VANET security and privacy. [20] tackles the problem of overburdening of a cluster head in the existing cluster based communication. This paper has proposed a multi cluster head technique in which multiple nodes act as a cluster head hence not overburdening a specific node. In [5], a swarm inspired delay tolerant network has been proposed. It hybrids firefly swarm technique along with glowworm swarm optimisation in order to optimise the vehicle routing. Comparative analysis of the proposed scheme shows that it has better efficiency than existing schemes. In [13], a clustering algorithm based on Moth Flame swarm technique has been proposed for IoV. This helps in finding optimal path for routing and providing robust transmission of messages in the network. It further provides efficiency and effectiveness to the algorithm. In [1], dragonfly-based clustering swarm technique has been proposed for routing of vehicles in IoV. This technique helps in making more stable topology for high traffic density. This has also enhanced the network availability by incorporating the 5G-enabled network infrastructure. In [2], an ACO-based clustering algorithm has been proposed for VANET. This algorithm helps in offering robust communication among vehicles in the network. In [9], bio-inspired particle swarm optimization has been implemented for VANET. This algorithm uses clustering technique for effective vehicular communication in the network. In [25], a routing algorithm that is based on ACO has been proposed for handling traffic congestion by minimizing the average travel time of vehicles on the road. In [24], a secure framework for vehicles has been proposed. This framework is based on lightweight authentication scheme which resists various integrity attacks and also reduces the time delay during vehicular communication in the network. In [21], a privacy preservation scheme based on fully homomorphic encryption has been proposed for securing the information in the vehicular network. This algorithm tackles several integrity attacks and has reduced the computational overhead in the network. In [23], RR-AES with enhanced cuckoo filter has been proposed for security of messages in vehicular network. Further, for providing privacy to the vehicles, pseudonyms have been assigned to the registered vehicles in VANET. In [22], a novel pseudonym and encryption scheme has been proposed for preserving and securing vehicles information in the network. This helps in secure communication among vehicles and preserve their location so as to protect them from any intruder attack. In [6], a genetic-based routing technique has been implemented for vehicular networks. This technique helps in better routing and communication among vehicles on the road. Further, this has reduced the accident rate on the road. In [7], GIS system has been used for managing traffic congestion and finding shortest path for vehicles. A cluster based routing protocol has been implemented for vehicles [4]. This protocol results in reduced network traffic, improved performance of routing and minimized vehicular communication overhead. In [3], a multi-hop broadcasting protocol has been implemented for vehicle routing in VANET. In this, intelligent forwarding concept has been used for fast and reliable communication among vehicles in the network.

3 Proposed Framework

This section describes the proposed framework for vehicle routing and the processes involved in the same. The main components in this framework are central authority(CA), vehicles, roadside units and other wireless sensors as shown in Fig. 2. The CA is responsible for managing the registration of the authenticated vehicles. Roadside units are the sensors that can sense information which is required for the vehicles so as to decide their optimal path. Here, vehicles can communicate to share the information related to routing in the vehicular network using wireless connectivity. This section is further classified into the following subsections:

3.1 Registration Process

Vehicles are not allowed to communicate in the network if they are not registered and authenticated. The CA registers the vehicle on to the network before they send any information in the network. Each vehicle uses its original identity for authentication and registration with the CA. Once the vehicle is registered, only then it can send/receive information in the vehicular network. To protect the original information of the vehicles, central authority uses a secret key Sk_{CA} to encrypt the vehicles original information. This is done so as to maintain the privacy of the vehicles in the network. During the registration process, each requesting vehicle that wants to be the part of the network is assigned with a copy of the Sk_{CA}. This helps the vehicles to retrieve their original information back when they leave the network.

3.2 Communication Process

During communication process Vehicles communicate with other vehicles by sending information in the network and that communication is known as V2V

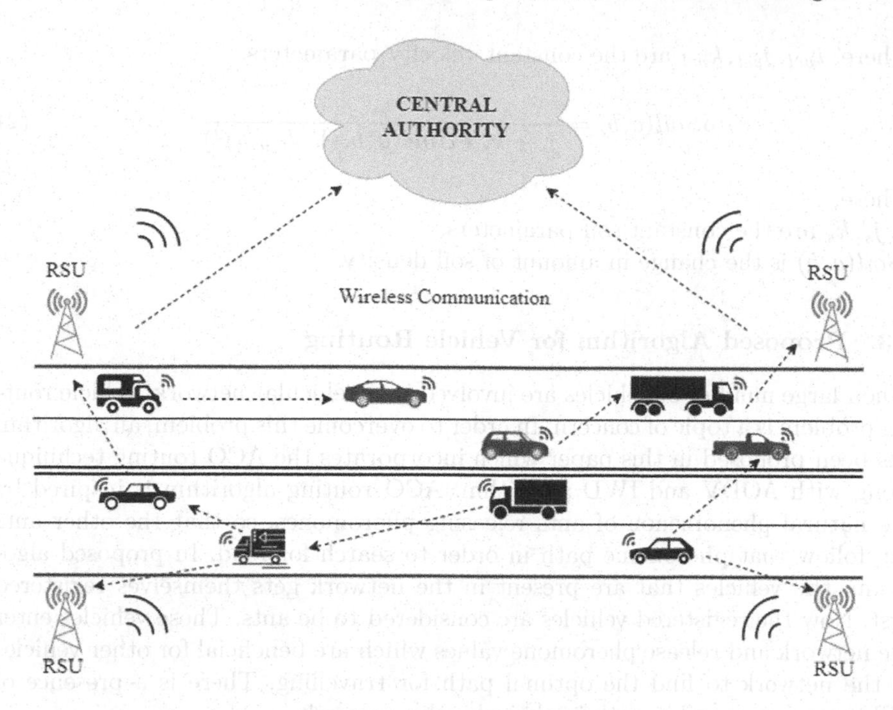

Fig. 2. Proposed framework

communication. Vehicles also communicate with other sensors like RSUs and that communication is known as V2I communication. In the proposed work, Vehicles transfer their pheromone values to the RSU. RSU uses these values to calculate the average pheromone density. This value is encrypted before it is sent to the vehicle that requested the routing information. The RSU uses IWD algorithm to encrypt the information. The information is then transferred using the AODV routing protocol. This protocol is responsible for avoiding duplicate messages to be sent again and again in the network. IWD algorithm follows the natural phenomenon of soil erosion resulting from the water that flows from point a to b in the river beds. In IWD algorithm, the main parameters are soil and velocity of the water drops. For encryption process, the higher the concentration of the soil eroded, higher is the level of encryption. In decryption, the velocity of the water drops determine the speed of decryption. The soil in IWD algorithm is represented as $Soil(a, b)$, and the velocity of water drops at source vehicle node is represented as $Vel_{(iwd)}(t)$.

$$Vel_{(iwd)}(t+1) = Vel_{(iwd)}(t) + \frac{i_{vel}}{j_{vel} + k_{vel} * soil^2(a,b)} \tag{1}$$

Where, $i_{vel}, j_{vel}, k_{vel}$ are the constant velocity parameters.

$$\delta Soil(a,b) = \frac{i_s}{j_s + k_s * time(a,b,(Vel_{(iwd)})^2)} \tag{2}$$

Where,
i_s, j_s, k_s are the constant soil parameters.
$\delta Soil(a,b)$ is the change in amount of soil density.

3.3 Proposed Algorithm for Vehicle Routing

When large number of vehicles are involved in a vehicular network, vehicle routing problem is a topic of concern. In order to overcome this problem, an algorithm has been proposed in this paper which incorporates the ACO routing technique along with AODV and IWD algorithm. ACO routing algorithm is inspired by the natural phenomenon of ants releasing pheromones, so that the other ants can follow that pheromone path in order to search for food. In proposed algorithm, the vehicles that are present in the network gets themselves registered first. Now the registered vehicles are considered to be ants. These vehicles enter the network and release pheromone values which are beneficial for other vehicles in the network to find the optimal path for travelling. There is a presence of RSU on every possible path in the vehicular network.

The vehicles communicate the pheromone values to the nearest RSU which then further calculates the average pheromone density on that particular path. After the average pheromone density calculation, the RSU encrypts this value using IWD algorithm. This value is then recirculated to the vehicles so that they can decide their best optimal path for travelling. The path with higher pheromone density implies higher traffic density and vice-versa. The RSUs also make use of the AODV protocol while sending the message to the vehicles. This is done so that the same message is not recirculated in the network repeatedly. This cycles keeps on repeating as the pheromone values of the vehicles keeps on changing. The complete process is shown in Algorithm 1.

4 Performance Evaluation

In this section, the performance of the proposed algorithm for vehicle routing is evaluated. For performance evaluation, we have considered different number of vehicle nodes in the network varying from 0 to 200 with the step size 40. The first evaluation is estimated on the basis of packet delivery percentage. It is observed that the proposed algorithm has a higher packet delivery percentage when compared with the existing algorithms GA [26] and FA [12]. Figure 3 shows that the proposed algorithm has 29.9% and 45% higher packet delivery percentage when compared with GA and FA respectively. Further, the proposed algorithm is evaluated on the basis of throughput performance. Results show that the proposed algorithm has successfully enhanced the overall throughput

Algorithm 1. ACO-AODV Vehicle routing

START
1. V_N enters the network.
2. V_N transmits its pheromone value to the RSU.
3. RSU updates the pheromone values by:
   $\vartheta_{(m,n)} = (1-\delta)\vartheta_{(m,n)} + \eta\vartheta_{(m,n)}$
4. RSU calculates the average pheromone value for a path using:
   $Avg_{(\Upsilon_{dp})} = \frac{\sum(\Upsilon_{dp})}{V_N}$
5. RSU encrypts this $Avg_{(\Upsilon_{dp})}$ value using IWD algorithm.
6. This encrypted value is transmitted to other V_N using AODV protocol.
7. The V_N decrypts the message and decides the optimal path.
8. Update the pheromone values as V_N are routed.
9. Repeat from step 2.

END

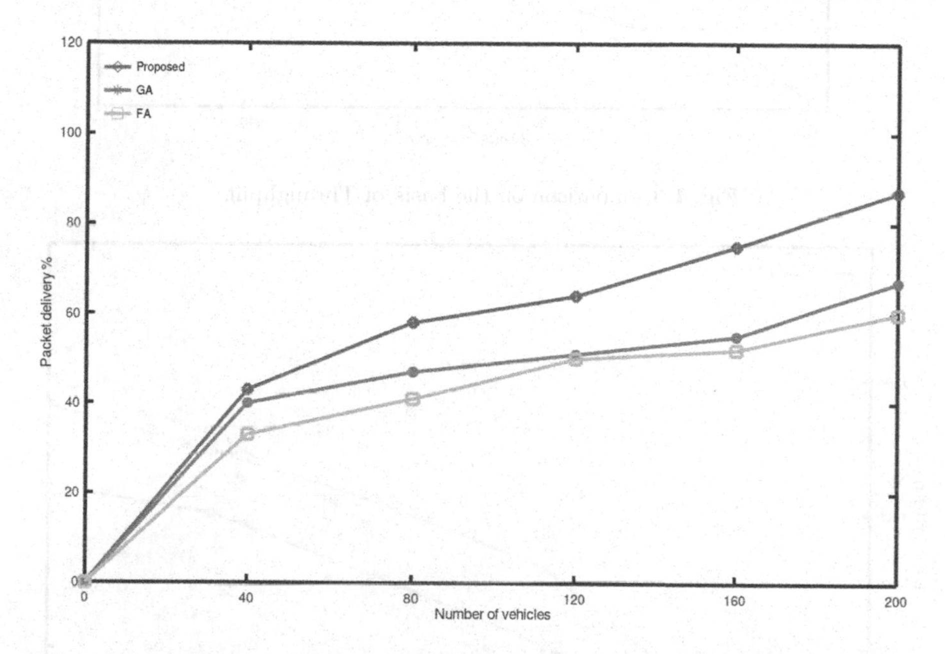

Fig. 3. Comparison on the basis of Packet Delivery percentage

of the vehicular network. It comes out to be 46.86% and 100% higher when compared with the existing work and is shown in Fig. 4.

The performance of proposed algorithm is further evaluated on the basis of end-to-end delay in the vehicular network. It is observed that the proposed algorithm for vehicle routing is capable of reducing the end-to-end delay in the network. On comparison with the existing schemes, the proposed algorithm proves to be 21.42% and 25.63% in reducing the end-to-end delay and is shown in Fig. 5. The routing overhead has also been reduced in the considered vehicular network using the proposed algorithm. The Performance is again evaluated and result

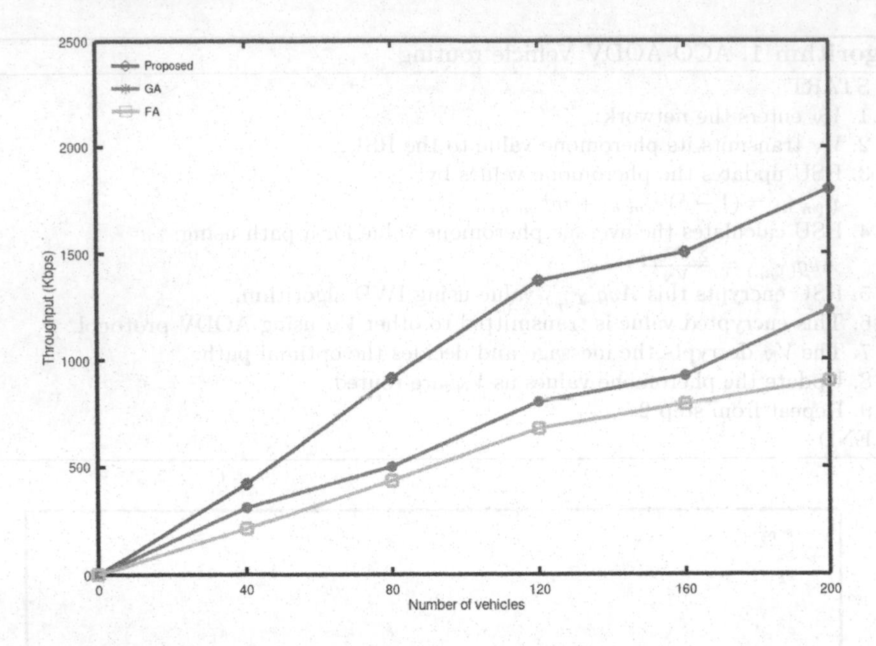

Fig. 4. Comparison on the basis of Throughput

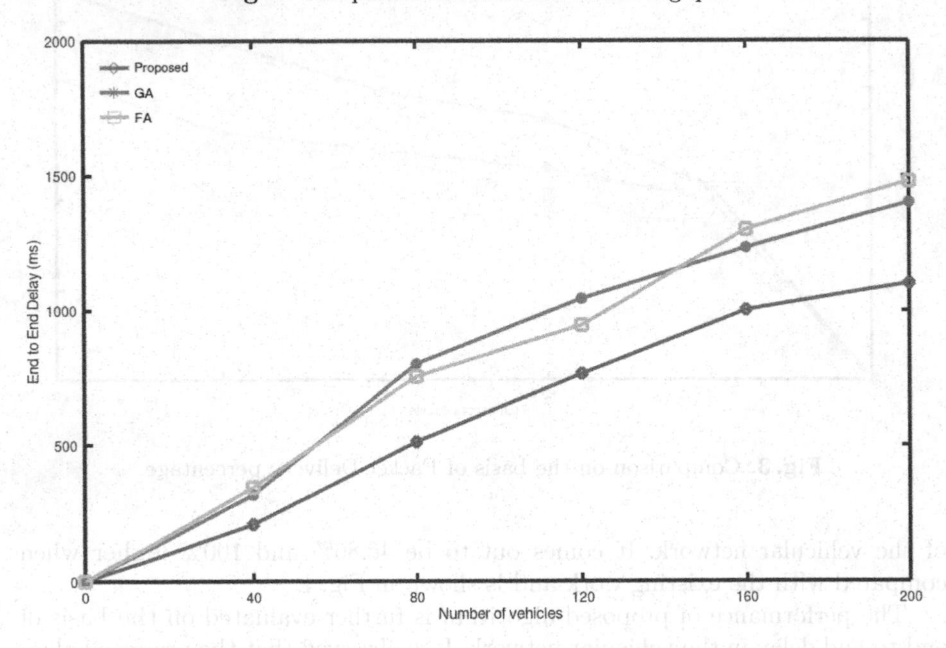

Fig. 5. Comparison on the basis of End-to-end delay

shows that there is a reduction of 60% and 66% in the routing overhead when compared with GA and FA respectively. The resulting graph is shown in Fig. 6.

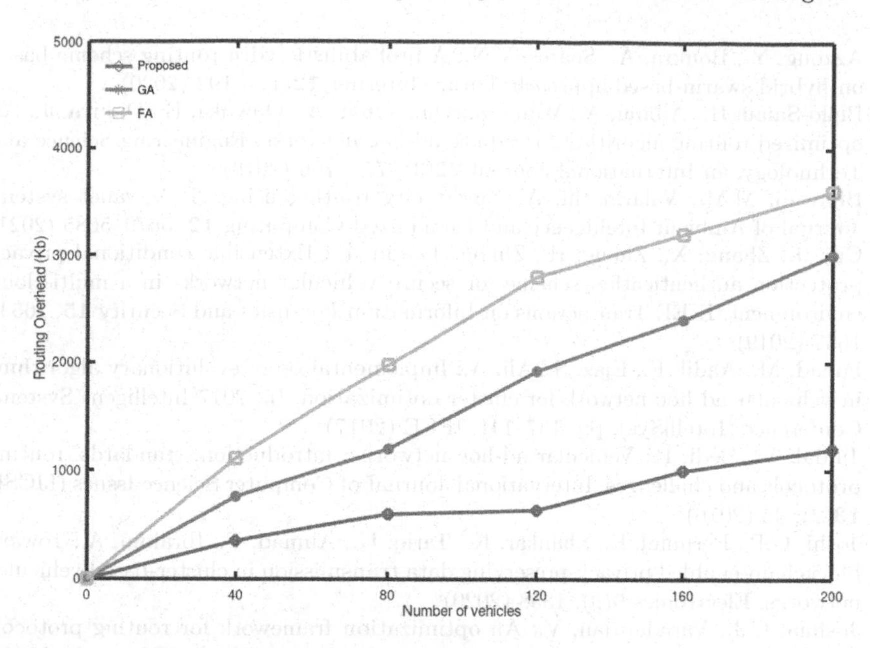

Fig. 6. Comparison on the basis of Routing overhead

5 Conclusion

In this paper, we have proposed a secure framework for vehicle routing that is based on nature-inspired technique. This framework has successfully secured the information that is shared in the vehicular network. Further, an algorithm has been proposed for vehicle routing that helps in finding the optimal and best path for various vehicles in the network. Results show that the proposed algorithm has higher throughout and packet delivery percentage when compared with the existing work. Also, it has reduced end-to-end delay and routing overhead in the network. In future, we will try to incorporate other nature-inspired optimization techniques along with the security aspect for vehicle routing.

References

1. Aadil, F., Ahsan, W., Rehman, Z.U., Shah, P.A., Rho, S., Mehmood, I.: Clustering algorithm for internet of vehicles (iov) based on dragonfly optimizer (cavdo). The Journal of Supercomputing **74**(9), 4542–4567 (2018)
2. Aadil, F., Khan, S., Bajwa, K.B., Khan, M.F., Ali, A.: Intelligent clustering in vehicular ad hoc networks. KSII Transactions on Internet and Information Systems (TIIS) **10**(8), 3512–3528 (2016)
3. Abbasi, H.I., Voicu, R.C., Copeland, J.A., Chang, Y.: Towards fast and reliable multihop routing in vanets. IEEE Transactions on Mobile Computing **19**(10), 2461–2474 (2019)
4. Ardakani, S.P.: Acr: A cluster-based routing protocol for vanet. International Journal of Wireless & Mobile Networks (IJWMN) Vol 10 (2018)

5. Azzoug, Y., Boukra, A., Soares, V.N.: A probabilistic vdtn routing scheme based on hybrid swarm-based approach. Future Internet **12**(11), 192 (2020)
6. Bello-Salau, H., Aibinu, A., Wang, Z., Onumanyi, A., Onwuka, E., Dukiya, J.: An optimized routing algorithm for vehicle ad-hoc networks. Engineering Science and Technology, an International Journal **22**(3), 754–766 (2019)
7. Bhavani, M.M., Valarmathi, A.: Smart city routing using gis & vanet system. Journal of Ambient Intelligence and Humanized Computing **12**, 5679–5685 (2021)
8. Cui, J., Zhang, X., Zhong, H., Zhang, J., Liu, L.: Extensible conditional privacy protection authentication scheme for secure vehicular networks in a multi-cloud environment. IEEE Transactions on Information Forensics and Security **15**, 1654–1667 (2019)
9. Fahad, M., Aadil, F., Ejaz, S., Ali, A.: Implementation of evolutionary algorithms in vehicular ad-hoc network for cluster optimization. In: 2017 Intelligent Systems Conference (IntelliSys). pp. 137–141. IEEE (2017)
10. Jindal, V., Bedi, P.: Vehicular ad-hoc networks: introduction, standards, routing protocols and challenges. International Journal of Computer Science Issues (IJCSI) **13**(2), 44 (2016)
11. Joshi, G.P., Perumal, E., Shankar, K., Tariq, U., Ahmad, T., Ibrahim, A.: Toward blockchain-enabled privacy-preserving data transmission in cluster-based vehicular networks. Electronics **9**(9), 1358 (2020)
12. Joshua, C.J., Varadarajan, V.: An optimization framework for routing protocols in vanets: A multi-objective firefly algorithm approach. Wireless Networks **27**(8), 5567–5576 (2021)
13. Khan, M.F., Aadil, F., Maqsood, M., Bukhari, S.H.R., Hussain, M., Nam, Y.: Moth flame clustering algorithm for internet of vehicle (mfca-iov). IEEE Access **7**, 11613–11629 (2018)
14. Krundyshev, V., Kalinin, M., Zegzhda, P.: Artificial swarm algorithm for vanet protection against routing attacks. In: 2018 IEEE Industrial Cyber-Physical Systems (ICPS). pp. 795–800. IEEE (2018)
15. Li, D., Deng, L., Cai, Z., Yao, X.: Notice of retraction: intelligent transportation system in macao based on deep self-coding learning. IEEE Transactions on Industrial Informatics **14**(7), 3253–3260 (2018)
16. Pham, T.N.D., Yeo, C.K.: Adaptive trust and privacy management framework for vehicular networks. Vehicular Communications **13**, 1–12 (2018)
17. Prabakeran, S., Sethukarasi, T.: Optimal solution for malicious node detection and prevention using hybrid chaotic particle dragonfly swarm algorithm in vanets. Wireless Networks **26**(8), 5897–5917 (2020)
18. Raja, G., Anbalagan, S., Vijayaraghavan, G., Dhanasekaran, P., Al-Otaibi, Y.D., Bashir, A.K.: Energy-efficient end-to-end security for software-defined vehicular networks. IEEE Transactions on Industrial Informatics **17**(8), 5730–5737 (2020)
19. Salem, F.M., Ali, A.S.: Sos: Self-organized secure framework for vanet. International Journal of Communication Systems **33**(7), e4317 (2020)
20. Sharma, S., Kaul, A.: Hybrid fuzzy multi-criteria decision making based multi cluster head dolphin swarm optimized ids for vanet. Vehicular Communications **12**, 23–38 (2018)
21. Tandon, R., Gupta, P.: A novel encryption scheme based on fully homomorphic encryption and rr-aes along with privacy preservation for vehicular networks. In: International Conference on Advances in Computing and Data Sciences. pp. 351–360. Springer (2021)
22. Tandon, R., Gupta, P.: A novel pseudonym assignment and encryption scheme for preserving the privacy of military vehicles. Defence Science Journal 71(2) (2021)

23. Tandon, R., Gupta, P.: Sp-encu: A novel security and privacy-preserving scheme with enhanced cuckoo filter for vehicular networks. In: Machine Learning and Information Processing, pp. 533–543. Springer (2021)
24. Tandon, R., Gupta, P.: Sv2vcs: a secure vehicle-to-vehicle communication scheme based on lightweight authentication and concurrent data collection trees. Journal of Ambient Intelligence and Humanized Computing pp. 1–17 (2021)
25. Verma, A., Tandon, R., Gupta, P.K.: Trafc-antabu: Antabu routing algorithm for congestion control and traffic lights management using fuzzy model. Internet Technology Letters p. e309
26. Zhang, G., Wu, M., Duan, W., Huang, X.: Genetic algorithm based qos perception routing protocol for vanets. Wireless Communications and Mobile Computing 2018 (2018)
27. Zhang, J., Zhong, H., Cui, J., Tian, M., Xu, Y., Liu, L.: Edge computing-based privacy-preserving authentication framework and protocol for 5g-enabled vehicular networks. IEEE Transactions on Vehicular Technology **69**(7), 7940–7954 (2020)

Detection of Bangla Hate Comments and Cyberbullying in Social Media Using NLP and Transformer Models

Md. Imdadul Haque Emon, Khondoker Nazia Iqbal, Md. Humaion Kabir Mehedi(✉), Mohammed Julfikar Ali Mahbub, and Annajiat Alim Rasel

Department of Computer Science and Engineering, Brac University, 66 Mohakhali, Dhaka 1212, Bangladesh

{md.imdadul.haque.emon,khondoker.nazia.iqbal,humaion.kabir.mehedi, mohammed.julfikar.ali.mahbub}@g.bracu.ac.bd, annajiat@bracu.ac.bd

Abstract. Hate speech and cyberbullying detection on social media is one of the most trending natural language processing tasks in the current scenario. However, due to the lack of resources, a few work has been done in detection of cyberbullying in social media for the Bangla language. In this paper, we have applied different transformer models to detect cyberbullying in social media. We have used a Bangla text dataset with 44,001 Bangla comments which are collected from Facebook posts. This dataset is labeled into five categories: sexual, threat, religious, troll, and not-bully. We have applied three transformer models: Bangla BERT, Bengali DistilBERT, and XLM-RoBERTa. Using transformer models, we achieved a very satisfactory score and the best performance we got was using the XML-RoBERTa model which achieved the highest accuracy of 85% and F1-score of 86%.

Keywords: NLP · Bangla BERT · Bengali DistilBERT · XLM-RoBERTa · Cyberbullying · Bangla hate comments

1 Introduction

Today people are more attached to the virtual world compared to their real world. Sudden bloom of the smartphone use and internet connectivity has made people more connected than ever. But every blessing comes with its own consequences. And the most common consequence people commonly face is hate comments, harassments and cyberbullying in their personal social media account.

Online harassment and cyberbullying is now in a very alarming level. It has become a trend among people in social media, especially teenagers. They find it very cool to throw hate comment on people. According to a report [1], around 37% of people in the age range of 12–17 have faced bullying online, and approximately 30% are bullied more than once. The report also adds that 95% of the teenagers of the U.S use mobile as a common medium to bully others and it is

M. Singh et al. (Eds.): ICACDS 2022, CCIS 1613, pp. 86–96, 2022.
https://doi.org/10.1007/978-3-031-12638-3_8

seen that girls are more victims of cyberbullying compared to boys [1]. According to another report published in Dhaka tribune [2], women are mostly harassed in social media and about 70% of the women are from the 15 to 25 years age group. So this is a huge problem ground to detect harassment and cyberbullying in Bengali language.

For being a low resource language, number of works in this domain are very insignificant for Bangla. Several machine learning approaches [3–5] and neural network approach [4] were previously applied in this problem domain. In this paper, we have proposed to use transformer models to detect hate comments and abuse in social media comments, specially Facebook and classify them into 5 categories: threat, sexual, troll, religious and not bully for Bangla language. We have used three transformer models in our research.

In this paper, we have briefly summarized previously done works on this domain in Sect. 2. In Sect. 3 we have provided the information about the dataset used in our research. After that, in Sect. 4 we have described the methodology and preprocessing steps. Next, Sect. 5 includes the results part. And final section has the conclusion and future work part.

2 Related Works

Different hate speech detection works have already been done for high-level languages like English. Only a small amount of work has been done for low-level languages like Bangla. A recent research work [3] presented a binary and multi class classification model to detect cyberbullying from social media comments in the Bangla language. They have built their own dataset, which contains around 44,001 user comments from different Facebook groups and pages. They have classified the dataset into five categories such as sexual, threat, religious, troll, and not-bully in order to multi class classification. At first, they applied their binary classification method to detect if there's any bully text and achieved around 87.91% accuracy. They have also used a multi class approach for classifying the harassment comments. For multi-class classification, they achieved 79.29% accuracy. Finally, in order to improve the classification prediction result, they introduced an ensemble method and tried some different machine learning models like SVM, RF, KNN, and Naive Bayes. After applying the ensemble method, they achieved 85% accuracy using SVM. The author struggled with long training time for their model. Moreover, for long text prediction sometimes their model showed false positive result.

Karim et al. used an method using ensemble and transformer, where the author proposed an explainable approach named DeepHateExplainer to detect hate speech in the Bangla language [4]. They have added 5,000 texts to the existing Bengali Hate Speech dataset and used it in their paper. At first, they classified the Bengali text into four different classes: personal, political, geopolitical, and religious hatred. After that, they applied machine learning methods (SVM, KNN, LR), Neural networks (CNN, Bi-LSTM), and BERT variants (Bangla BERT, XML-RoBERT) in order to detect hate speech. At last, they compared the prediction result of different methods and found that DeepHateExplainer can detect hate texts with an f1-score of 88% when both ML and DNN baselines are performed together. Their proposed paper proved that a combination of several models can detect hate speech better than individual models. Because of having a limited number of labeled data while training their model, the author couldn't rule out the chance of over-fitting.

In another paper [6], the authors prepared three datasets of Bengali text consisting of Bangla hate speeches to do three different experiments: sentiment analysis, hate speech detection, and document classification. They introduced a word embedding model for Bangla language and named it "BengFastText". It contains data based on around 250 million articles. After that, based on the Multichannel Convolutional Long Short-Term Memory Network, they predicted the result of the three experiments as mentioned earlier. After that, they compared their prediction result with other models. Their experiment's outcome showed that BengFastText can detect the texts more correctly than other embedding methods like Word2Vec and GloVe. Using the BengFastText method, they achieved around 92.30% F1-scores in document classification and around 82.25%, and 90.45% F1-scores are achieved in sentiment analysis and hate speech detection.

In 2020 Makhadmeh et al. proposed a model for using NLP and ML approaches combinely to detect abusive comments from social media in the English language [7]. They have collected data from a neo-Nazi website [8], which contains around 10,568 sentences, and each sentence is around 20.39 words in length. They explore the dataset using their proposed method named as "A killer natural language processing optimization ensemble deep learning method (KNLPEDNN)". By using the approach, the dataset is classified into three different classes, which includes hate, offensive, and neutral languages. Their proposed method achieved a maximum accuracy of around 98.71% to predict hate speech from social media texts.

Another research by Akhter et al. [5] proposed an approach to detect social media abuse using different ML algorithms on Bangla language. They have collected data from different social media platforms like Facebook and Twitter. In order to extract the dataset from Facebook and Twitter, they have developed a java program. They have collected 1,000 public Bangla comments from Facebook and 1,400 comments from Twitter. After that, they labeled the dataset into two categories: "bullied" and "not bullied". At last, they applied Machine Learning algorithms like SVM, Naive Bayes, KNN (1-Nearest), KNN (3-Nearest) to predict bullying. After comparing different algorithms, they found that the support vector machine's prediction accuracy is highest than other applied Machine Learning algorithms. They achieved around 97% accuracy while detecting bullying using the SVM approach.

Another proposed method for cyberbullying detection [9] has applied a pre-trained BERT model, which is a very popular transformer model. They have used two publicly available datasets, Formspring [10] and Wikipedia [11] talk pages, to train and evaluate the model. The first dataset contains around 12,773 posts, among which 776 are marked as a bully. On the other hand, the Wikipedia dataset has about 1,15,864 posts, with 13,590 of them labeled as bullies. They have used a hugging face library to work with BERT models. They have compared their models' accuracy with other models like CNN and RNN. From the Formspring and Wikipedia datasets, the BERT model achieved around 98% and 96% accuracy respectively.

3 Dataset

In our research paper, we have collected Bangla online comments dataset from the Mendeley Data website [12]. This dataset contains around 44,001 Bangla comments collected from Facebook posts. The dataset is labeled into five categories: sexual, threat, religious, troll, and not-bully. According to the dataset, 29,950 comments are for females, and 14,051 comments are for males. Besides, the dataset's comments are categorized into five different professional categories: actor, singer, social, politician, and sports. It contains emoticons, special characters, stopwords which are removed from the dataset while preprocessing. In the dataset, the number of reacts to each comments are also mentioned. In our proposed model, we have only used the five different categories of label to detect and classify cyberbullying.

Some examples of the labels of our dataset are shown in below Table 1:

Table 1. Dataset example

Label	Example
Sexual	খানকীর ছবি দেখলেই ঘিন্না লাগে
Threat	তরে কুট্টাইলবাম
Religious	নাস্তিকের বাচ্চা নাস্তিক! জাহান্নামের কিট
Troll	ফইন্নির ঘরের ফইন্নি
Not-Bully	আপনার জন্য ভালবাসা

4 Methodology

We have used three transformer models (Bangla BERT Base [13], Bengali DistilBERT and XLM-RoBERTa Base [14]) from Hugging Face [15] for detection and multi-class classification of hate comments. Before training the models with text data we have done some pre-processing of the data. The workflow diagram is shown the Fig. 1.

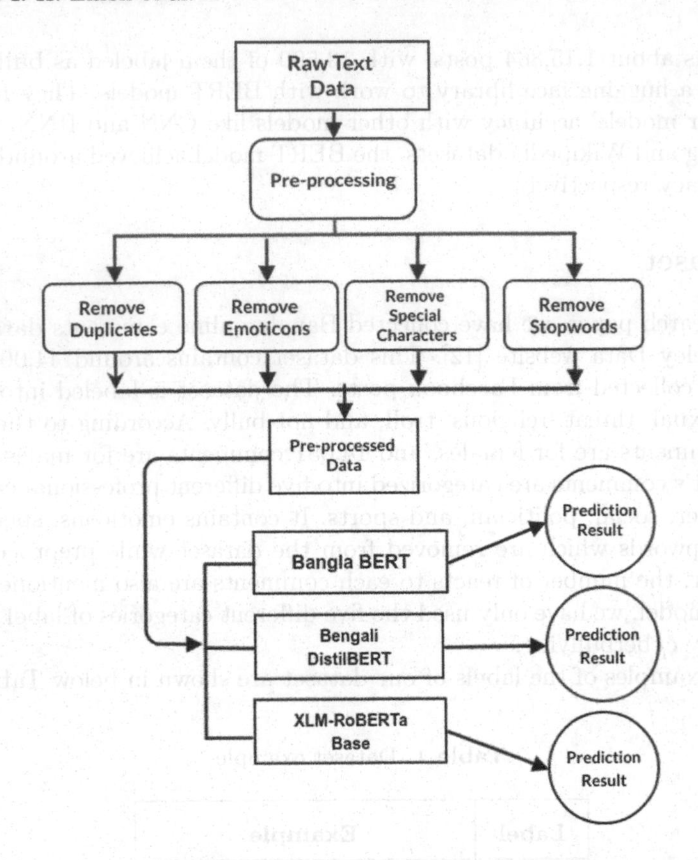

Fig. 1. Workflow diagram

4.1 Preprocessing

Data preprocessing part was mostly conducted with the help of python RegEX. The dataset had a total number of 44,001 comments. The comments consisted of Bengali text with a mixture of numeric, emoticons, special characters(punctuation and other symbols) and some of the comments had some English text. Also, there were some duplicate comments in the dataset. At first we removed the duplicate comments from the dataset. We removed the comments which are exact match to any of the comments available in the dataset. There were a total 434 duplicate comments. After removing the duplicates we were left with 43,567 unique comments. Then we removed all the special characters i.e.: asterisk, dollar sign, brackets etc. and replaced them with empty string. After that all the English text from the comments were removed because we are only concerned about the Bengali texts. All the emoticons, symbols, flags etc. were removed using their Unicode. For the removal tasks, we have used 'sub()' function of the python RegEx. Now we are left with only Bengali text in the comments.

At last we removed the stopwords from the data. Stopwords don't have any significance in a sentence. They are frequently used but they don't consists any useful information. So it is wise to remove them for faster and accurate result. The list of stopwords were taken from a GitHub repository [16]. The author listed a total number of 398 stopwords for Bengali language and all of them were removed from the dataset. The dataset pre-processing example is shown in the below Table 2.

Table 2. Pre-processed text example

Before Pre-processing	After Pre-processing
অন্যরকম .. ভালো লাগলো ..☺	অন্যরকম ভালো লাগলো
অদেখা জিনিসে ত ওর believe নাই,এইটা কেমনে করলো..	অদেখা জিনিসে ত নাই এইটা কেমনে করলো
সাফা কবির কোন **** বাল... ???	সাফা কবির বাল
নাস্তিকের বাচ্চা নাস্তিক! জাহান্নামের কিট।	নাস্তিকের বাচ্চা নাস্তিক জাহান্নামের কিট

4.2 Transformer Models

Bangla BERT Base: Bangla BERT Base [13] is a pre-trained transformer model for Bengali language. This BERT based model is pre-trained with Bengali CommonCrawl corpus from OSCAR1[1] and Bengali Wikipedia Dump[2] Dataset2. This model was pre-trained using mask language modelling.

Bengali DistilBERT: Bengali DistilBERT is another transformer model for Bengali language, and this model was pre-trained on almost six Gigabyte of monolingual training corpus. This is a very lightweight model and provides a very good accuracy for downstream works like POS-tagging and text-classification. Among the three models we applied, the training time of this model was the least.

---

[1] https://oscar-corpus.com/.
[2] https://dumps.wikimedia.org/bnwiki/latest/.

XLM-RoBERTa-Base: XLM-RoBERTa [14] is a very famous transformer model which supports multilingual texts. This model is pre-trained on 2.5TB of filtered data which has 100 different languages. This model was pre-trained on raw text data only using Masked Language modelling. This model has been used in a lot of text-classification and downstream tasks before and it showed excellent accuracy in almost every task.

Model Training Parameters: We wanted to train all the models using the same parameters so that the comparison of the test-score becomes more logical. So we trained all of these transformer models using a learning rate of $2e^{-5}$ in 10 epochs. All of the models were trained using a batch size of 12. Also all of the data was divided using a ratio of 70:20:10 for train, validation and test. So 70% data was used for training while 20% of the data was used for validation and last 10% of the data was used for further testing the model's performance.

5 Result and Analysis

All of the transformer models scored a satisfactory result. The accuracy is calculated by the 10% testing data from the dataset. And the precision, recall and F1 score was obtained from the 20% validation data set while training. The Bangla BERT model scored 83% F1-score, 83% precision and 82% recall score. For the test data, it scored 82.64% accuracy. On the other hand, the Bengali DistilBERT model scored a similar score to Bangla BERT, 83% F1-score, 83% precision and 82% recall score while scoring 82.73% accuracy for the test data. At last, the XLM-RoBERT model gave a most promising score: 86% F1-score, 84% precision and 84% recall. Also this model scored 85% accuracy for the test data.

The Accuracy, Precision, Recall and the F1-score of our applied transformer models are shown in Table 3.

Table 3. Accuracy score comparison

Model	Accuracy	Precision	Recall	F1-score
Bangla BERT	0.82	0.82	0.81	0.83
Bengali DistilBERT	0.84	0.84	0.82	0.83
XLM RoBERTa	**0.85**	**0.84**	**0.84**	**0.86**

From the table, it can be observed that the XLM-RoBERT model scored the highest F1-score among all of the approaches. Here the training vs validation accuracy and loss per epoch for the XLM-RoBERTa model is shown in Fig. 2 and Fig. 3.

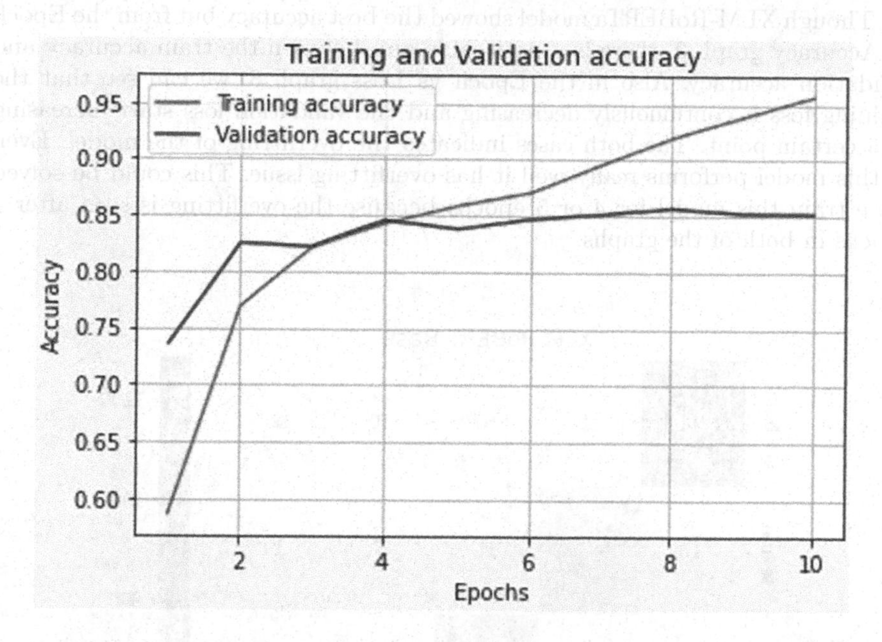

Fig. 2. Epoch vs Accuracy for training and validation (XLM-RoBERTa)

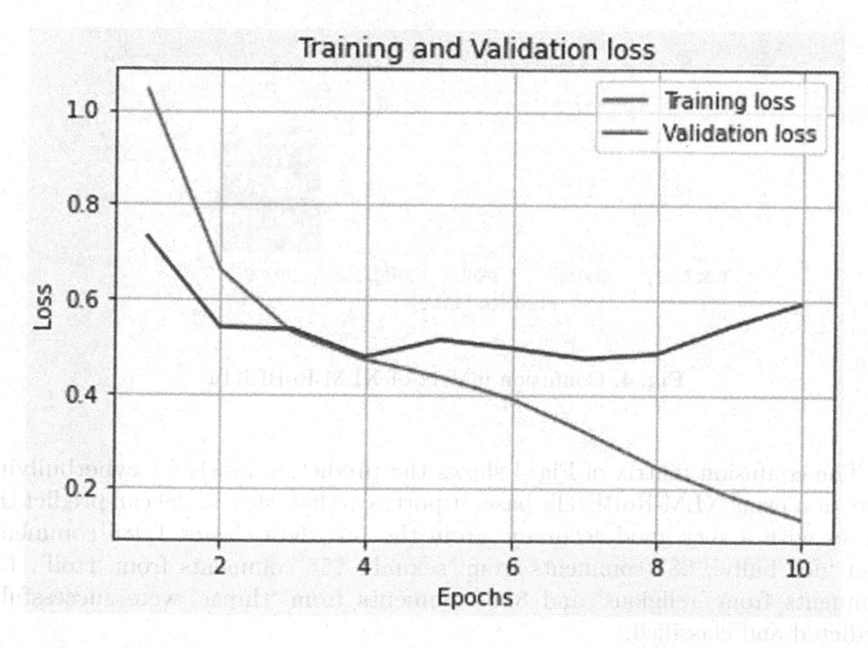

Fig. 3. Epoch vs Loss for training and validation (XLM-RoBERTa)

Though XLM-RoBERTa model showed the best accuracy but from the Epoch vs Accuracy graph 2, there is a noticeable gap between the train accuracy and validation accuracy. Also in the Epoch vs Loss graph 3, we can see that the training loss is continuously decreasing and the validation loss start increasing at a certain point. The both cases indicates the overfitting of the model. Even so this model performs really well it has overfitting issue. This could be solved if we train this model for 4 or 5 epochs because the overfitting is seen after 4 epochs in both of the graphs.

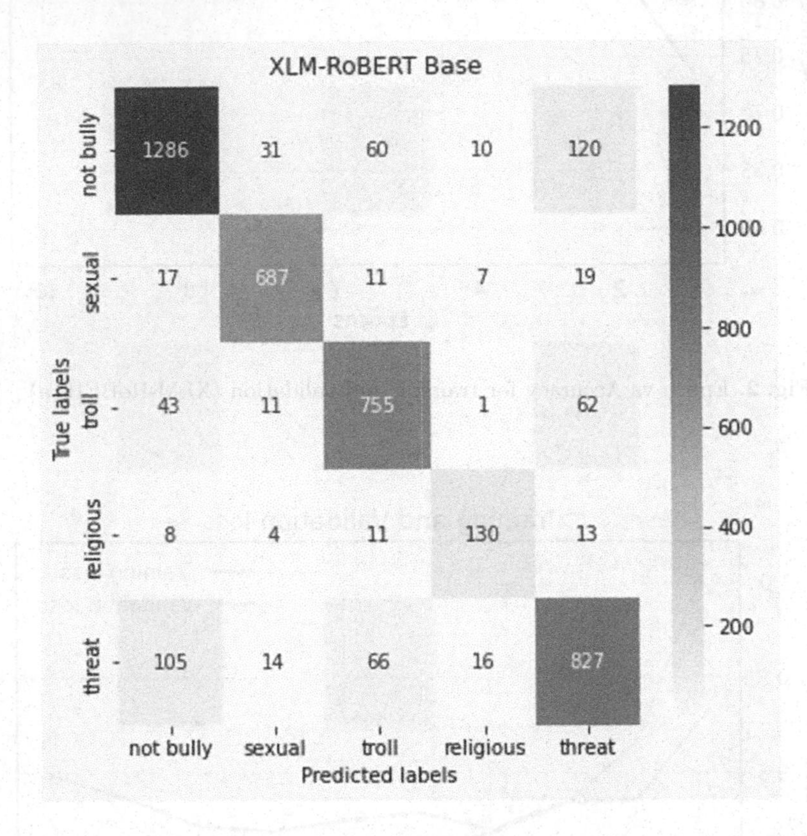

Fig. 4. Confusion matrix of XLM-RoBERTa

The confusion matrix of Fig. 4 shows the prediction labels for cyberbullying detection using XLM-RoBERTa base. It portrays that, this model can predict the classes with a very good accuracy. From the test data classes 1,286 comments from 'not bully', 687 comments from 'sexual', 755 comments from 'troll', 130 comments from 'religious' and 827 comments from 'threat' were successfully predicted and classified.

6 Conclusion

Our research portraits that using transformers approach shows a very good result in detecting and classifying Bangla hate comments and bullies in social media platforms, specially Facebook. Our approach can detect and classify hateful Bangla comments in five classes keeping an accuracy of 85%. With time, the toxicity is increasing and more people are joining everyday to make the task more complex. Transformers can be a very good solution for hate speech detection and classification. The task is still quite challenging because of the low resource availability of Bengali language. This paper can be a dedication for building a more accurate and robust transformer model to filter out the underlying bully and hatred of Bengali social media comments. Also, there is a scope for using transformers models for real life application in social media platforms like Facebook, Twitter etc. to filter out the hateful statements to reduce toxicity in the comment sections. In future, we want to conduct more research in downstream tasks for Bangla language using transformer models.

References

1. 11 facts about cyberbullying. https://www.dosomething.org/us/facts/11-facts-about-cyber-bullying. Accessed 24 Dec 2021
2. 70 percent of women facing cyber harassment are 15–25 years in age 2019. https://archive.dhakatribune.com/bangladesh/dhaka/2019/09/24/70-of-women-facing-cyber-harassment-are-15-25-years-in-age. Accessed 20 Dec 2021
3. Ahmed, M.F., Mahmud, Z., Biash, Z.T., Ryen, A.A.N., Hossain, A., Ashraf, F.B.: Cyberbullying detection using deep neural network from social media comments in Bangla language. arXiv preprint arXiv:2106.04506 (2021)
4. Karim, M.R., Dey, S.K., Islam, T., et al.: DeepHateExplainer: explainable hate speech detection in under-resourced Bengali language. In: 2021 IEEE 8th International Conference on Data Science and Advanced Analytics (DSAA), pp. 1–10. IEEE (2021)
5. Akhter, S., et al.: Social media bullying detection using machine learning on Bangla text. In: 2018 10th International Conference on Electrical and Computer Engineering (ICECE), pp. 385–388. IEEE (2018)
6. Karim, M.R., Chakravarthi, B.R., McCrae, J.P., Cochez, M.: Classification benchmarks for under-resourced Bengali language based on multichannel convolutional-LSTM network. In: 2020 IEEE 7th International Conference on Data Science and Advanced Analytics (DSAA), pp. 390–399. IEEE (2020)
7. Al-Makhadmeh, Z., Tolba, A.: Automatic hate speech detection using killer natural language processing optimizing ensemble deep learning approach. Computing **102**(2), 501–522 (2020)
8. Caren, N., Jowers, K., Gaby, S.: A social movement online community: Stormfront and the white nationalist movement (2012)
9. Yadav, J., Kumar, D., Chauhan, D.: Cyberbullying detection using pre-trained BERT model. In: 2020 International Conference on Electronics and Sustainable Communication Systems (ICESC), pp. 1096–1100. IEEE (2020)

10. Reynolds, K., Kontostathis, A., Edwards, L.: Using machine learning to detect cyberbullying. In: 2011 10th International Conference on Machine Learning and Applications and Workshops, vol. 2, pp. 241–244 (2011). https://doi.org/10.1109/ICMLA.2011.152
11. Wulczyn, E., Thain, N., Dixon, L.: Ex Machina: personal attacks seen at scale. In: Proceedings of the 26th International Conference on World Wide Web, ser. WWW 2017, Perth, Australia, pp. 1391–1399. International World Wide Web Conferences Steering Committee (2017). ISBN: 9781450349130. https://doi.org/10.1145/3038912.3052591
12. Ahmed, M.F., Mahmud, Z., Biash, Z.T., Ryen, A.A.N., Hossain, A., Ashraf, F.B.: Bangla online comments dataset (2021). https://data.mendeley.com/datasets/9xjx8twk8p/1
13. Sarker, S.: BanglaBERT: Bengali mask language model for Bengali language understanding (2020). https://github.com/sagorbrur/bangla-bert
14. Conneau, A., Khandelwal, K., Goyal, N., et al.: Unsupervised cross-lingual representation learning at scale. CoRR, vol. abs/1911.02116 (2019). arXiv: 1911.02116. http://arxiv.org/abs/1911.02116
15. Hugging face, the AI community building the future. https://huggingface.co/. Accessed 15 Dec 2021
16. Diaz, G., Stopwords Bengali (BN) (2016). https://github.com/stopwords-iso/stopwords-bn.git. Accessed 23 Dec 2021

A Modified Pyramid Scale Network for Crowd Counting

Bhawana Tyagi, Swati Nigam, and Rajiv Singh(✉)

Department of Computer Science, Banasthali Vidyapith, Tonk, Rajasthan 304022, India
jkrajivsingh@gmail.com

Abstract. Crowd counting is a complex and strenuous task in the field of computer vision due to illumination, complex background, occlusions, scale variations, non-uniform distribution etc. There are many methods to count the crowd and the multicolumn architecture has been adopted widely to overcome those challenges. But the two major issues with multicolumn methods are feature similarity and scale limitation. To address these issues, pyramid scale network (PSNet) was proposed to deal these issues. In this paper, we propose a modified pyramid scale network (MPSNet) to obtain more detailed features to deal with scale variation. In this work, we have used four modified pyramid scale modules (MPSM) which extracts multiscale features by integrating message passing and attention mechanism in multicolumn architecture. The experiments and performance of the proposed method is compared on the benchmark ShanghaiTech dataset and UCF_CC_50 dataset in terms of mean absolute error (MAE) and root mean square error (RMSE).

Keywords: Crowd counting · Convolutional neural network · Multicolumn neural network · Pyramid scale network

1 Introduction

In today's world the growth of population and urbanization is increasing at a rapid rate due to which the need for crowd counting is increasing at astonishing pace. The major reason for this growth is the security of crowds due to events such as live concerts, political rallies, sports meet, religious functions. Crowd counting has many applications in different fields like disaster management, public space design, safety monitoring [1]. In public space design, it can be used to optimize the architecture of new buildings. In safety monitoring, crowd behavior can be analyzed using crowd counting. In disaster management, it can be used to detect overcrowding.

Earlier the crowd counting has been done manually, which is a very tedious task and the possibility of error while counting is high. So, we need some mechanism which can automatically count the number of people. But crowd counting methods face many challenges, like severe occlusion, high cluttering, varying illumination, scale variation, non-uniform distribution etc. Some of them are depicted in Fig. 1.

M. Singh et al. (Eds.): ICACDS 2022, CCIS 1613, pp. 97–106, 2022.
https://doi.org/10.1007/978-3-031-12638-3_9

a. Occlusion b. Complex background

c. Non-uniform distribution d. Scale variation

Fig. 1. Challenges of crowd counting

Many times, more than one challenge can be there in the single image. Thus, we require a robust crowd counting method which can deal with these challenges. Among all these challenges, scale variation is one of the major challenges [2]. To count the crowd, most of the methods generate the crowd density map. Among them, the most common architecture is multicolumn architecture such as multicolumn convolutional neural network (MCNN) [3], relational attention network (RANet) [4], scale aggregation network (SANet) [5], deep recurrent spatial aware network (DRSAN) [6]. To get multiscale features, the entire network is divided into multiple columns and output layer combines all features. But the major drawback of this network is feature similarity and scale limitation. Considering local perspective, columns are scale specific and works fine when it uses corresponding scale. It cannot work similar on the other scale and its performance automatically decreases. In the global perspective, there is a global scale for each column and it can only work well on the crowded scene. Also, there was the problem of information redundancy with these architectures and it was demonstrated by Li et al. [7]. Based on the above experiments, to measure the similarity of different columns, Cheng et al. [8] proved that the reason behind redundancy is the feature similarity, and cosine similarity was used for this purpose.

The pyramid scale network (PSNet) was proposed by to minimize the problem of feature similarity and scale variation by using message passing mechanism and the attention mechanism. We have further modified the PSNet to make it more appropriate for crowd counting by modifying the pyramid scale module to extract more detailed features. We have demonstrated our method on the ShanghaiTech dataset [3] and UCF-CC-50 dataset [9] with a lesser number of parameters.

2 Related Works

We can categorize deep learning based crowd counting methods on the basis of network properties [2, 10] i.e. basic convolutional neural network (CNN) model, context aware models, multitask models and scale aware models.

Basic CNN Models. The basic CNN model was proposed by Walach and Wolf [11] using layered boosting where training was performed in stages and selective sampling was used. Initially CNN was used for crowd density estimation and later two CNN classifiers were cascaded to improve the accuracy by Fu et al. [12].

Context Aware Models. The end-to-end CNN [13] was proposed, in which computational cost was decreased as in input it takes whole image and returns the final count. The final count was predicted based on local count and global count. To extract the semantic and spatial information, the weighted-vector of locally aggregated descriptor (W-VLAD) [14] was proposed to learn locality aware features. In congested scene recognition network (CSRNet) [7], the first 10 layers of VGG-16 [15] was used in the front end for feature extraction and at the backend they used dilated convolution layers. Contextual pyramid CNN (CP-CNN) was proposed in [16], which comprises following four modules. Global context estimator (GCE) uses VGG-16 to encode global context, local context estimator (LCE) is another CNN that is used to encode the local context, and density map estimator (DME) is another multicolumn based CNN to generate high dimensional feature maps, and to fuse the global and local context a fusion-CNN (F-CNN) was used.

Multitask Models. Deep convolutional neural network (DCNN) was proposed [17] to count number of people in video frames by generating three types of supervision maps: crossing-line crowd counting maps, crowd density maps and crowd velocity maps. End to-end cascaded CNN was proposed [1] for density map estimation.

Scale Aware Models: The scale aware models can be classified as multiscale, attention based and multicolumn methods.

- Multiscale methods: A multiscale adaptive CNN was proposed in [18] to generate density map by combining features that were extracted from multiple layers and then final count of person was computed by integrating density maps. Another multiscale pyramid network was proposed by Chen et al. [19] to extract multi scale information based on single column structure.
- Attention based methods: To deal with a problem of scale variation in the image by using global and local scales, the scale aware attention network [20] was proposed by Hossain et al. They generated density map and estimated final count based on it.
- Multicolumn methods: To deal with the problem of scale variation, the multicolumn based method MCNN [3] was proposed which consists of three different columns with different filter sizes. Switch-CNN [21] was also based on MCNN but it also contains a classification network that can intelligently switch the multiple regressors between different scales. Sam and Babu [22] proposed a top down feedback CNN (TDF-CNN) architecture which consists of bottom up CNN that comprises of two CNN to generate

the crowd density map and to generate feedback. It also consists of top down CNN. To get the contextual information, dilated-attention-deformable ConvNet (DADNet) [23] used dilated CNN with different dilation rate and used adaptive deformable convolutions to generate density maps of crowd. To estimate the density of the crowd without geometric information and perspective map, Hydra-CNN [24] was proposed that takes pyramid of image patches as input.

3 The Proposed Method

3.1 Modified PSNet

The architecture of the proposed network is depicted in Fig. 2. It includes the backbone network which is based on VGG-16, 4 stacked pyramid scale modules followed by 4 convolutional layers to find out density of crowd. In this architecture we have used first 13 layers of VGG-16 to extract low level features like edges and texture. It produces eight times smaller feature maps with respect to input images and the resolution of feature maps will not get changed. Here we are dealing with scale variation problem as head size depends on position of camera, head size of a person is large who is near to the camera and small when person is far from the camera. Thus, to deal with the problem of scale variation, proposed architecture adopts 4 stacked pyramid scale module. It extracts high level features and feeds these features to 4 convolutional layers to generate the density map. Here, we have used multicolumn variance loss to solve the problem of feature similarity.

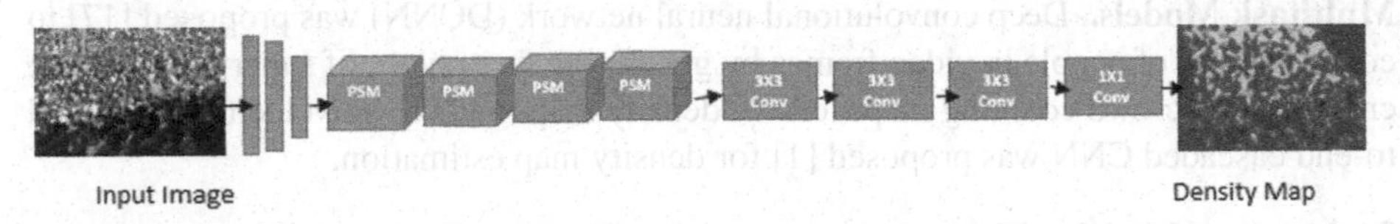

Fig. 2. The architecture of modified PSNet.

3.2 Modified Pyramid Scale Module

As in multicolumn architectures [3, 5], there is problem of scale limitation. In that architecture, features can be extracted from limited number of scales and corresponding column can only perform on the dedicated scale. To deal with the problem of scale limitation, in modified pyramid scale module message passing mechanism and attention mechanism is used. Modified pyramid scale module consists of two components i.e. global attention module and feature pyramid module [8]. The modified pyramid scale module is depicted in Fig. 3.

The global attention module provides global context attention like other attention models [20] to weight the importance of multiscale features generated by feature pyramid module. For given input feature $F_{in} \in R^{C \times H \times W}$, where C represents the channel, H is the height and W is the width of an image. We get the input feature F_{in} from backbone network. Initially, it performs global pooling on F_{in} to produce $F_c \in R^{C \times 1 \times 1}$ channel vector so that each channel get global context softly. In order to reduce the parameter overhead, here 1×1 convolution filter is used to reduce the channel of F_c to obtain $F_c \in R^{C/r \times 1 \times 1}$, where reduction ratio i.e. $r = 16$. To estimate the attention across channels from F_c, another 1×1 convolution filter is used to obtain $F_{att} \in K^{C \times 1 \times 1}$ as global context attention.

Feature pyramid module consists of four branches to extract multiscale features. In the first branch, two filters of 3×3 are used. In the second branch, 5×5 filter is used. In the third branch, 7×7 filter is used and in the last branch there are two filters of 9×9. To deal with the problem of scalability, we have used these four different size filters in the respective branches. To get more feature of small scale size we have used two filters of 3×3 consecutively and for large scale size we have used sequence of two 9×9 filters. With the help of a message passing scheme, the gathered information from different scales is integrated in step by step manner. The 3×3 convolution filter is used to fuse information gathered from lower and adjacent upper branch. In the last, from all the branches extracted features will fuse by 3×3 convolution filter. To get the $F_{out} \in R^{C \times H \times W}$, these features will get multiplied with F_{att} that is obtained from the global attention module. As there is the communication between each branch, so this feature pyramid module deals with the problem of feature similarity also.

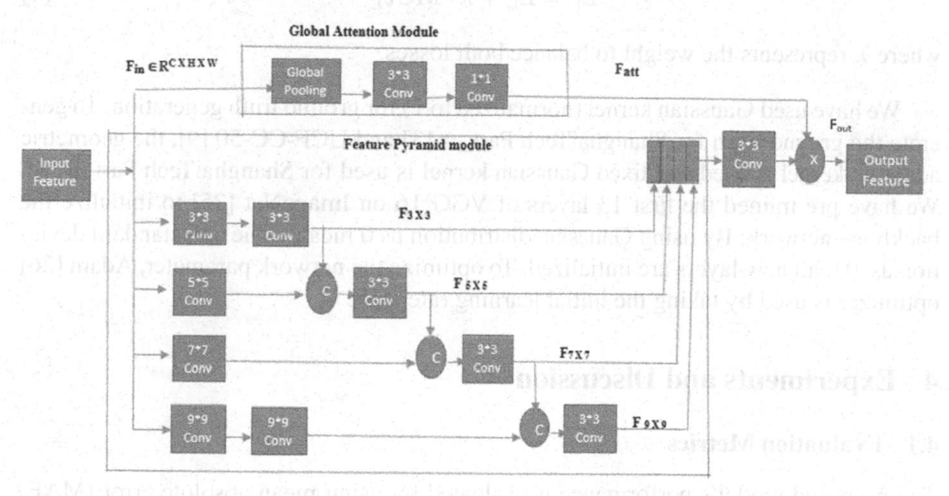

Fig. 3. Architecture of Proposed MPSM

3.3 Implementation Details

Multi-column Variance Loss: To deal with the problem of feature similarity, we have used multi-column variance loss (MC_{VL}) [8].

$$MC_{VL} = 1/(I.P.B) \sum_{(i=1)}^{I} \sum_{(p=1)}^{P} \sum_{(b=1)}^{B} \frac{F_{(att\_pb)} . [F_{att-sum} - F_{(att-pb)}/(B-1)]}{\max(|F_{(att-pb)}| \cdot |(F_{att-sum} - F_{(att-pb)})/(B-1)|, \epsilon)} \quad (1)$$

where *I, P, B* denote number of images in batch, modified PSM in modified PSNet and branches in modified PSM, respectively. F_{att_pb} is the attention vector for relevant branch and $|F_{att\_pb}|$ is the norm of attention vector. The ϵ is taken as $1e^{-6}$ to avoid division by zero error.

$$F_{att\_sum} = \sum_{(b=1)}^{B} F_{att\_pb} \tag{2}$$

The value of MC_{VL} is from 0 to 1. Larger value represents the higher similarity.

Euclidean Loss: To calculate the estimated difference at pixel level between the ground truth and the estimated density map, we have used Euclidean distance.

$$E_L = \frac{1}{N}\sum_{j=1}^{N} |G(Y_i; a) - D_i^G T|_2^2 \tag{3}$$

where, *a* is the model parameters, input image is represented by Y_i, $G(Y_i;a)$ represents generated density map and D_i^{GT} represents ground truth density map.

Final Objective: By weighting the multicolumn variance loss and Euclidian loss, the final objective function for training is formulated as:

$$L_F = E_L + \lambda \cdot MC_{VL} \tag{4}$$

where λ represents the weight to balance both losses.

We have used Gaussian kernel (normalized to 1) for ground truth generation. To generate the ground truth for ShanghaiTech Part_A [3] and UCF-CC-50 [9], the geometric adaptive kernel is used and fixed Gaussian kernel is used for ShanghaiTech Part_B [3]. We have pre trained the first 13 layers of VGG 16 on ImageNet [25] to initialize the backbone network. By using Gaussian distribution as 0 mean value and standard deviation as .01, all new layers are initialized. To optimize the network parameter, Adam [26] optimizer is used by taking the initial learning rate as $1e^{-4}$.

4 Experiments and Discussion

4.1 Evaluation Metrics

The proposed model's performance is evaluated by using mean absolute error (MAE) and root mean squared error (RMSE).

$$MAE = \frac{1}{N}\sum_{j=1}^{N} \left|PC_i - C_i^{GT}\right| \tag{5}$$

$$RMSE = \sqrt{\frac{1}{N}\sum_{j=1}^{N} \left|PC_i - C_i^{GT}\right|^2} \tag{6}$$

where *N* represents total number of images, *PCi* represents predicted count and C_i^{GT} represents ground truth count. To determine the accuracy of estimates *MAE* is used and to determine the robustness of estimates, *RMSE* is used.

4.2 ShanghaiTech dataset

As per the density of crowd ShanghaiTech dataset [3] is divided into two parts: part A and part B. The highly dense crowd images are in part A consists of 482 images, these images are further divided into two sets: train and test set, train set consists of 300 images and rest 182 in test set. The less dense crowd images are in part B consists of 716 images, these images are further divided into train set which consists of 400 images and test set which consists of 316 images. In total it contains 1198 images with 330165 annotations.

The performance comparison of existing models and the proposed model on this dataset is depicted in the Table 1, which shows that proposed model gave the lowest MAE and RMSE on this dataset. The proposed model reports 7.6% better MAE from the second best method [7] and 4.7% better RMSE than second best method [7].

Table 1. Performance comparison on Shanghaitech dataset.

Method	Part A		Part B	
	MAE	RMSE	MAE	RMSE
MCNN [3]	110.2	173.2	26.4	41.3
CSRNet [8]	68.2	115.0	10.6	16.0
Scale Pyramid Network (SPN) [19]	61.7	99.5	9.4	14.4
Crowd CNN [25]	181.8	277.7	32.0	49.8
Switching CNN [21]	90.4	135.0	21.6	33.4
DUBNet [27]	64.6	106.8	7.7	12.5
Contextual Pyramid CNN (CP-CNN) [16]	73.6	106.4	20.1	30.1
Iterative CNN [28]	68.5	116.2	10.7	16.0
PSNet [7]	55.5	90.1	6.8	10.7
Perspective crowd counting (PCCNet) [29]	73.5	124.0	11.0	19.0
Scale aggregation network (SANet) [5]	67.0	104.5	8.4	13.6
CAN [30]	62.3	100.0	7.8	12.2
Feature Pyramid Network for crowd counting (FPNCC) [31]	81.2	139.2	7.6	12.0
Spatial Fully Convolutional Network (SFCN) [32]	64.8	107.5	7.6	13.0
Dual attention aware network (DA2Net) [33]	74.1	128.4	7.9	13.2
Deep Fusion Network (DFN) [34]	77.58	129.7	14.1	21.10
Modified PSNet(Ours)	51.26	84.6	5.9	10.2

4.3 UCF_CC_50

This dataset is the collection of images of extremely crowded scenes. These images are collected from the web. There are 50 images of different resolutions in UCF_CC_50 dataset [9]. There are 94 to 4543 people in the images. Single dot annotation was used to annotate every individual. To perform the fivefold cross validation, we have incorporated the standard settings followed in [9]. The performance comparison of existing models and the proposed model on this dataset is depicted in the Table 2, which shows that [31] had lowest MAE and lowest RMSE on this dataset. The proposed model shows the second lowest RMSE i.e., 241.6 and fifth lowest MAE i.e., 235.2.

Table 2. Performance comparison on UCF_CC_50 dataset.

Method	MAE	RMSE
MCNN [3]	377.6	509.1
CSRNet [6]	266.1	397.5
SPN [19]	259.2	335.9
iterative CNN [28]	260.9	365.5
Switching CNN [21]	318.1	439.2
SANet [5]	258.4	334.9
PSNet [8]	185.3	265.0
PCCNet [29]	240.0	315.5
CAN [30]	212.2	243.7
(FPNCC) [31]	136.4	223.6
SFCN [32]	214.2	318.2
DFN [34]	402.3	434.1
Modified PSNet (Proposed)	235.2	241.6

5 Conclusions and Future Scope

In this paper, we discuss major challenges that are encountered in crowd counting. We focus on scale variation and feature similarity among all the challenges. Here we have proposed the modified PSNet by modifying pyramid scale module. It gives better performance on the UCF-CC-50 and ShanghaiTech dataset. It can be applied to the other existing multicolumn networks. In this work, we have considered the static data which can be applied to the real time data as well. The crowd counting method can be further used for crowd behavior analysis as it is the first step to analyze the behavior of crowd.

References

1. Sindagi, V.A., Patel, V.M.: CNN-based cascaded multi-task learning of high-level prior and density estimation for crowd counting. In: 2017 14th IEEE International Conference on Advanced Video and Signal Based Surveillance (AVSS), pp. 1–6. IEEE

2. Sindagi, V.A., Patel, V.M.: A survey of recent advances in cnn-based single image crowd counting and density estimation. Pattern Recogn. Lett. **107**, 3–16 (2018)
3. Zhang, Y., Zhou, D., Chen, S., Gao, S., Ma, Y.: Single-image crowd counting via multi-column convolutional neural network. In: 2016 Proceedings of the IEEE Conference on Computer Vision and Pattern Recognition, pp. 589–597 (2016)
4. Zhang, A., et al.: Relational attention network for crowd counting. In: 2019 Proceedings of the IEEE/CVF International Conference on Computer Vision, pp. 6788–6797 (2019)
5. Cao, X., Wang, Z., Zhao, Y., Su, F.: Scale aggregation network for accurate and efficient crowd counting. In: Ferrari, V., Hebert, M., Sminchisescu, C., Weiss, Y. (eds.) Computer Vision – ECCV 2018. LNCS, vol. 11209, pp. 757–773. Springer, Cham (2018). https://doi.org/10.1007/978-3-030-01228-1_45
6. Liu, L., Wang, H., Li, G., Ouyang, W., Lin, L.: Crowd counting using deep recurrent spatial-aware network. arXiv preprint arXiv:1807.00601 (2018)
7. Li, Y., Zhang, X., Chen, D.: CSRNet: dilated convolutional neural networks for understanding the highly congested scenes. In: Proceedings of the IEEE Conference 2018 on Computer Vision and Pattern Recognition, pp. 1091–1100 (2018)
8. Cheng, J., Chen, Z., Zhang, X., Li, Y., Jing, X.: Exploit the potential of Multi-column architecture for Crowd Counting. arXiv preprint arXiv:2007.05779 (2020)
9. Idrees, H., Saleemi, I., Seibert, C., Shah, M.: Multi-source multi-scale counting in extremely dense crowd images. In: Proceedings of the IEEE Conference on Computer Vision and Pattern Recognition, pp. 2547–2554 (2013)
10. Ilyas, N., Shahzad, A., Kim, K.: Convolutional-neural network-based image crowd counting: review, categorization, analysis, and performance evaluation. Sensors **20**(1), 43 (2020)
11. Walach, E., Wolf, L.: Learning to count with CNN boosting. In: Leibe, B., Matas, J., Sebe, N., Welling, M. (eds.) Computer Vision – ECCV 2016. LNCS, vol. 9906, pp. 660–676. Springer, Cham (2016). https://doi.org/10.1007/978-3-319-46475-6_41
12. Fu, M., Xu, P., Li, X., Liu, Q., Ye, M., Zhu, C.: Fast crowd density estimation with convolutional neural networks. Eng. Appl. Artif. Intell. **43**, 81–88 (2015)
13. Shang, C., Ai, H., Bai, B.: End-to-end crowd counting via joint learning local and global count. In: IEEE International Conference on Image Processing (ICIP), pp. 1215–1219. IEEE (2016)
14. Sheng, B., Shen, C., Lin, G., Li, J., Yang, W., Sun, C.: Crowd counting via weighted VLAD on a dense attribute feature map. IEEE Trans. Circuits Syst. Video Technol. **28**(8), 1788–1797 (2016)
15. Simonyan, K., Zisserman, A: Very deep convolutional networks for large-scale image recognition. arXiv preprint arXiv:1409.1556 (2014)
16. Sindagi, V.A., Patel, V.M.: Generating high-quality crowd density maps using contextual pyramid CNNs. In: Proceedings of the IEEE International Conference on Computer Vision, pp. 1861–1870 (2017)
17. Zhao, Z., Li, H., Zhao, R., Wang, X.: Crossing-line crowd counting with two-phase deep neural networks. In: Leibe, B., Matas, J., Sebe, N., Welling, M. (eds.) Computer Vision – ECCV 2016. LNCS, vol. 9912, pp. 712–726. Springer, Cham (2016). https://doi.org/10.1007/978-3-319-46484-8_43
18. Zhang, L., Shi, M., Chen, Q.: Crowd counting via scale-adaptive convolutional neural network. In: IEEE Winter Conference on Applications of Computer Vision (WACV), IEEE, pp. 1113–1121 (2018)
19. Chen, X., Bin, Y., Sang, N., Gao, C.: Scale pyramid network for crowd counting. In: IEEE Winter Conference on Applications of Computer Vision (WACV), pp. 1941–1950. IEEE (2019)

20. Hossain, M., Hosseinzadeh, M., Chanda, O., Wang, Y.: Crowd counting using scale-aware attention networks. In: IEEE Winter Conference on Applications of Computer Vision (WACV), pp. 1280–1288. IEEE (2019)
21. Babu Sam, D., Surya, S., Venkatesh Babu, R.: Switching convolutional neural network for crowd counting. In: Proceedings of the IEEE Conference on Computer Vision and Pattern Recognition, pp. 5744–5752 (2017)
22. Sam, D.B., Babu, R.V.: Top-down feedback for crowd counting convolutional neural network. In: Thirty-Second AAAI Conference on Artificial Intelligence, 27 April 2018
23. Guo, D., Li, K., Zha, Z.J., Wang, M.: Dadnet: dilated-attention-deformable convnet for crowd counting. In: Proceedings of the 27th ACM International Conference on Multimedia, pp. 1823–1832, 15 October 2019
24. OñoroRubio, D., LópezSastre, R.J.: Towards perspective-free object counting with deep learning. In: Leibe, B., Matas, J., Sebe, N., Welling, M. (eds.) Computer Vision – ECCV 2016. LNCS, vol. 9911, pp. 615–629. Springer, Cham (2016). https://doi.org/10.1007/978-3-319-46478-7_38
25. Krizhevsky, A., Sutskever, I., Hinton, G.E.: Imagenet classification with deep convolutional neural networks. Adv. Neural. Inf. Process. Syst. **25**, 1097–1105 (2012)
26. Kingma, D.P., Ba, J.: Adam: a method for stochastic optimization. arXiv preprint arXiv:1412.6980, 22 Dec 2014
27. Oh, M.H., Olsen, P., Ramamurthy, K.N.: Crowd counting with decomposed uncertainty. In: AAAI (2020)
28. Ranjan, V., Le, H., Hoai, M.: Iterative crowd counting. In: Ferrari, V., Hebert, M., Sminchisescu, C., Weiss, Y. (eds.) Computer Vision – ECCV 2018. LNCS, vol. 11211, pp. 278–293. Springer, Cham (2018). https://doi.org/10.1007/978-3-030-01234-2_17
29. Gao, J., Wang, Q., Li, X.: PCC net: perspective crowd counting via spatial convolutional network. IEEE Trans. Circuits Syst. Video Technol. **30**(10), 3486–3498 (2019)
30. Liu, W., Salzmann, M., Fua, P.: Context-aware crowd counting. In: Proceedings of the IEEE/CVF Conference on Computer Vision and Pattern Recognition, pp. 5099–5108 (2019)
31. Cenggoro, T.W., Aslamiah, A.H., Yunanto, A.: Feature pyramid networks for crowd counting. Proc. Comput. Sci. **1**(157), 175–182 (2019)
32. Wang, Q., Gao, J., Lin, W., Yuan, Y.: Learning from synthetic data for crowd counting in the wild. In: Proceedings of IEEE Conference on Computer Vision and Pattern Recognition (CVPR) (2019)
33. Zhai, W., et al.: Da 2 net: a dual attention aware network for robust crowd counting. Multimed. Syst. 1–14 (2022)
34. Khan, S.D., Salih, Y., Zafar, B., Noorwali, A.: A deep-fusion network for crowd counting in high-density crowded scenes. Int. J. Comput. Intell. Syst. **14**(1), 1–12 (2021)

Driving Impact in Claims Denial Management Using Artificial Intelligence

Suman Pal(✉), Monica Gaur, Rupanjali Chaudhuri, R. Kalaivanan, K. V. Chetan, B. H. Praneeth, and Uttam Ramamurthy

Cerner Corporation, Bangalore 560045, India
{suman.pal,monica.gaur,rupanjali.chaudhuri,kalaivanan.r, chetan.kv,praneeth.boddupalli,uttam.ramamurthy}@cerner.com
https://www.cerner.com/

Abstract. A healthcare provider's ability to quickly and efficiently process claims and quantify denial rates is critical to ensure smooth revenue cycle management and medical reimbursement. But the hospitals and medical practitioners are receiving more claim denials from payers, with the average rate of denial steadily increasing year over year. The recent COVID-19 pandemic has further accelerated the denial rate. An accurate denial detection algorithm can help to reduce the burden on healthcare providers. In this study, we propose a boosting-based machine learning framework to predict the likelihood of claims being denied along with the reason code at a line level. Prediction at a line level provides a finer-grained explanation to the administrative staff by pointing out the specific line for corrections. The list of important features provides an interpretable solution to the healthcare providers which enables them to create the right edits and correct the claim before going out to the payer. This in turn helps the healthcare provider dramatically improve both net patient revenue and cash flow. They can also put a check on their costs, as fewer denials mean less rework, resources, and time devoted to appealing and recovering denied claims. The denial model showed good performance with Area Under the Curve (AUC) of 0.80 and 0.82 for professional and institutional claims respectively. According to our estimates, the model has the potential to save 15%–50% of the denial cost for a healthcare provider. This in turn would have a tremendous impact on the healthcare costs as well as help make the healthcare process smoother.

Keywords: Machine learning · Claims denial · Boosting · Explainable AI

1 Introduction

Claim denial is a harsh reality in healthcare and poses a serious issue for healthcare providers amid an already complicated reimbursement landscape. Denied claims are defined as claims that were adjudicated by the payer and a negative

M. Singh et al. (Eds.): ICACDS 2022, CCIS 1613, pp. 107–120, 2022.
https://doi.org/10.1007/978-3-031-12638-3_10

determination was made. Each denial to the hospital can cause a delayed cash inflow. Denied claims must be researched to determine why the claim was denied before making an appropriate appeal or reconsideration request. Such claims, if resubmitted without an appeal or reconsideration request, will most likely be denied again as a duplicate, costing even more time and money. Moreover, the claim remains unpaid. The average claims denial rate across the healthcare industry in United States (US) is between 5% and 10%, according to an American Academy of Family Physicians (AAFP) report [1]. With Coronavirus Disease of 2019 (COVID-19) pandemic, the denial rates have increased to 6–13% [2]. However, keeping the claim denial rate closer to the lower end of the industry average is not always easy as the providers engage with multiple payers at once.

According to Change Healthcare, 63% of denied claims are recoverable on the first appeal. Additional labor and manpower associated with the appeal process equal an average of $118 per claim, or $8.6 billion annually for the U.S hospitals, leading to a billion dollars in administrative costs [3]. These statistics make the problem of denied claims imperative and valuable to the healthcare industry and this motivated the work described in this paper.

In the US, the typical revenue cycle workflow, as shown in Fig. 1, begins when a patient walks into the hospital or clinic. Following that the eligibility and benefits verification is done with the insurance companies through Electronic Data Interchange (EDI) transaction sets. After the verification, the patient receives care from the provider in terms of diagnosis, relevant procedures, and services. This care is captured through medical transcription which is then mapped with industry coding standards. The charges are then billed based on the services rendered by the patient and a claim file is generated in the form of an EDI transaction. This claim file passes through the scrubbing process before sending it to payers to identify the errors that could lead to rejection. In case of any errors, this claim file becomes an invalid claim file and is then sent to the edit queue for claims fix. If the claim file passes the scrubbing process at the provider end, it is then forwarded to the payer side scrubber to identify it as a clean claim which is error-free. If the claim does not pass the payer side scrubber, the claim is rejected and is communicated to the provider with the help of an EDI transaction. If the claim clears the payer's scrubbing process, then it goes for payer processing where the payer decides on the reimbursement of the claim amount. A claim may get fully paid, partially paid, or get denied with zero payment and is communicated to the provider using Explanation of Benefits (EoB) through an EDI Transaction.

In this paper, we propose a boosting-based ML model that aids the healthcare provider to prevent claims from denial. The proposed model was placed in the Revenue Cycle Management (RCM) workflow, after the claim clears the provider scrubber and before it goes to the payer scrubber. This is the least expensive stage to review the claim and can help the healthcare provider from a staggering financial loss.

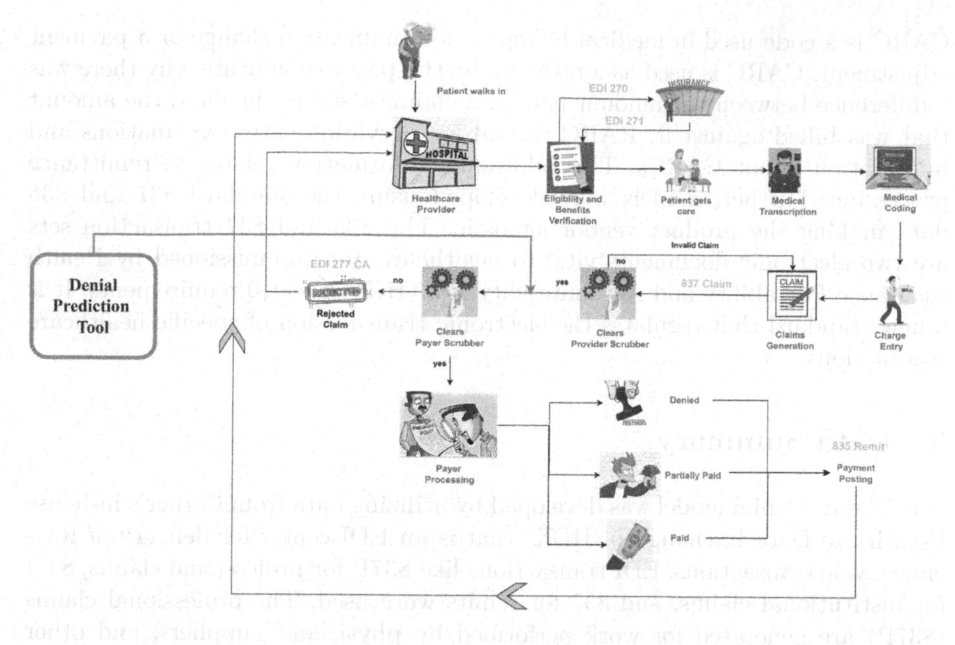

Fig. 1. Detailed RCM workflow

2 Literature Review

In the last few decades, claims were prepared and submitted manually on paper with no electronic transactions involved. In general, claims processing is not an area where machine learning (ML) techniques have been widely used. A lot of the work in claims has focused on fraud detection which is a related but different area. [4] mentioned a system that applies a modified K-Means algorithm to flag out suspicious claims. There do exist few predictive systems to review claims before submission in the revenue cycle but in general, rule-based systems are highly prevalent in the area of claims processing. Accenture has developed a "Rework Prevention Tool" to reduce claim processing errors for two insurance companies in the US [5]. Alpha Health has developed a solution called "Deep Claim" that can predict whether, when, and how much a payer will pay for each claim [6]. NTT Data Services has developed an AI-powered platform that helps healthcare overcome challenges posted by claims denial [7]. GE Healthcare's DenialsIQ uncovers unseen trends from within claims denials to help provider organizations improve financial performance [8]. Optum has developed an application for Denial management (Field-tested techniques that get claims paid) [9]. Change Healthcare have taken an analytics-driven approach to denial prevention to help healthcare providers and their revenue stream [10].

The explainable AI model described in this paper is the first of its kind in identifying claims at a line level that are prone to denial along with CARC (Claim Adjustment Reason Code) and RARC (Remittance Advice Remark Codes). A

CARC is a code used in medical billing to communicate a change or a payment adjustment. CARC is used as a response by the payer to indicate why there was a difference between the amount paid in a claim or service line and the amount that was billed against it. RARC is used for providing extra explanations and information about CARCs. This additional information relates to remittance processing. Further, models were developed using the standard 837 and 835 data making the product vendor agnostic. The 835 and 837 transaction sets are two electronic documents vital to healthcare and commissioned by Health Insurance Portability and Accountability Act (HIPAA) 5010 requirements. It is a new standard that regulates the electronic transmission of specific healthcare transactions.

3 Data Summary

The Claims Denial model was developed by utilizing data from Cerner's in-house Healthcare Data Exchange® (HDX) that is an EDI center for delivery of Revenue Cycle transactions. EDI transactions like 837P for professional claims, 837I for institutional claims, and 835 for remits were used. The professional claims (837P) are generated for work performed by physicians, suppliers, and other non-institutional providers for both outpatient and inpatient services whereas institutional claims (837I) are generated by hospitals and skilled nursing facilities. A claim file contains 'provider, patient, payer & charges' related information whereas a remit file contains 'payer, payment and EoBs like CARCs and RARCs' related information.

The data was collected for the period of January 2019 to October 2021. Table 1 shows the data summary of claims and remit dataset for a healthcare provider. The denial percentage in the dataset was observed between 10%–15% which is aligned with the industry witnessed denial rates.

Table 1. Claims data summary

Characteristic	Professional	Institutional
Denial %	10.78	14.72
[Unique CARCs, Denial CARCs]	[83, 63]	[76, 62]
Unique RARCs	118	114
[Unique Patients, Unique Claims]	[5593, 30386]	[4473, 28898]
Unique Payers	51	50
Payers with high claims	Medicaid outpatient	Medicare outpatient & Part AB
Unique Procedures	731	1262
Claims Submission Range (in days)	[1-1495]	[1-1309]

4 Methodology

The claims denial model utilized the Catboost classifier for predictions. It is a sequential model which comprises of both binary and multi-class classification. The binary classification identifies a denied claim line then followed by the multi-class classification which identifies the CARC & RARC associated with it. The prediction process is shown in Fig. 2.

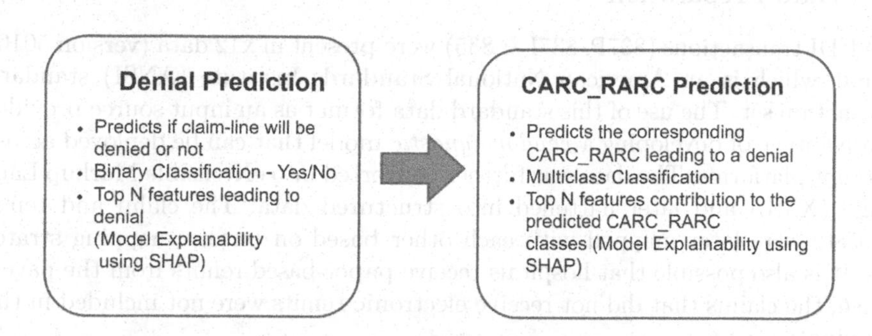

Fig. 2. Prediction process flow

The model was built using the hierarchical modeling concept of having a local classifier per parent node and at the end predictions from each model were combined using joint probability. The CARC-RARC model was developed only on the denial dataset. The predictions are also supported by the top contributing features using SHapley Additive exPlanations also known as SHAP. The model training was done separately for the professional and institutional claims.

4.1 Ground Truth

Denials: The responses about the payments from payers are communicated in the form of CARCs. Some of these CARCs represent denials i.e. zero payment from the payer. 'CARC code 29' is one such example used by the payers when the time limit for filing the claim has expired. Therefore, based on the inputs from the industry experts, the CARCs available in the dataset were labeled as *'Denials'*(1) or *'Non-Denials'*(0) which were then used as target variable for binary classifier.

CARC-RARC: The target variable for the multi-class classifier was CARC-RARC which was created by concatenating the CARC along with the RARC for a given claim line. Many CARC-RARC classes occurred in less frequency which was insignificant for model training. Therefore, the classes occurring less than 1% were grouped into *'other carc rarc'* category. Table 2 shows the CARC-RARC distribution in the dataset.

Table 2. CARC-RARC distribution

CARC-RARC classes	Professional	Institutional
Before grouping	184	170
After grouping	28	25

4.2 Data Preparation

The EDI transactions (837P, 837I & 835) were present in x12 data (version 5010) format which is an American National Standards Institute (ANSI) standard transaction set. The use of this standard data format as an input source provides an advantage of developing a *vendor-agnostic* model that can be deployed across different platforms. The data was further converted into eXtensible Markup Language (XML) and later flattened into structured data. The claim and remit datasets were later mapped with each other based on certain mapping strategies. It is also possible that hospitals receive paper-based remits from the payer. Hence, the claims that did not receive electronic remits were not included in the mapping.

A claim may contain multiple claim lines wherein each line represents the services rendered to the patient by the hospital and the associated charges with those services. The sum of all the line-level charges is equal to the claim level charges. Similarly, a remit may contain multiple remit lines which are responses of payments and associated CARCs and RARCs for each of the services mentioned in the claim. It was observed in the dataset that the majority of the responses from the payers are at line-level. Hence, the model was trained to do predictions at line level rather than at claim level.

The claim files were extracted after the scrubbing process and hence, the dataset was cleaner. The data cleaning step for model building mostly involved the following steps:

- Redundant features and features with no variance were dropped.
- Dummy variables were created for sparse columns to indicate the presence of field value. This was an essential step as in claims few fields are *situational* in nature i.e. such fields provide additional information and their lack of presence may sometimes contribute to a denial.
- Derived features were created based on data analysis and domain understanding.
- Missing values were imputed with 'unknown' value for categorical features and '0' value for numeric features.

Standard Input Features: The standard input features are the fixed variety of features defined and engineered from the information enumerated from 837 EDI files after data cleaning. A wide range of features were considered for model training as shown in Fig. 3. Engineered features like 'days to claim submission' which indicates the time taken for a claim to be submitted to the payer and 'repeated procedures in a claim' which indicates the presence of more than one procedure per claim per day were created based on domain understanding.

4.3 Feature Selection

It is important to have a robust feature selection process when dealing with a large set of variables in order to avoid the inclusion of spurious or random features that do not generalize well. Our goal was to extract a parsimonious set of features that captures the most meaningful information that contributes to the claims denial and CARC-RARC predictions. To achieve this, eXtreme Gradient Boosting (XgBoost) Classifier [11] was used for feature selection with stratified grouped 10-fold outer cross-validation that provided 90:10 split for training and test data. The stratification was grouped using the patient's Medical Record Number (MRN). This ensured no leakage of patient information from the training to the test set.

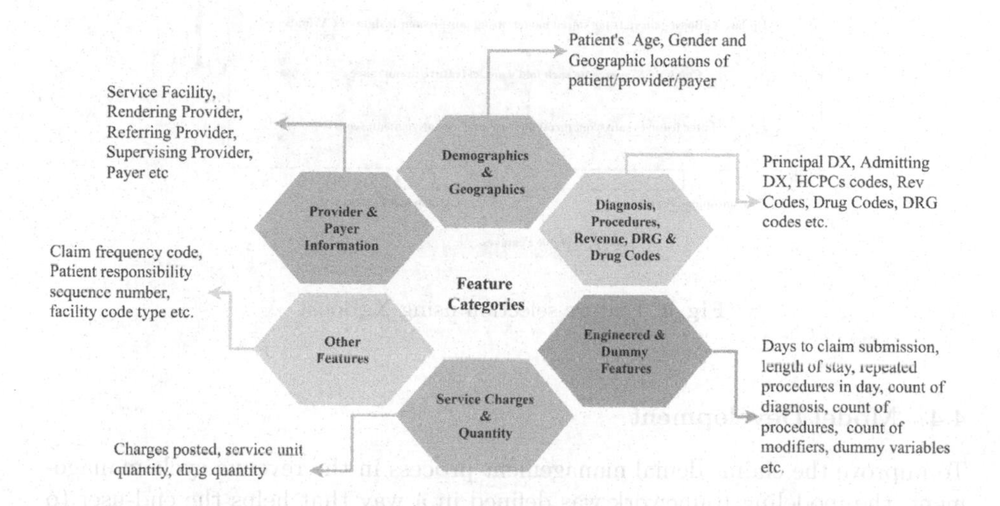

Fig. 3. Standard input features

The cross-validation approach helped in making a fair assessment of the model performance on test sets, ensuring minimal longitudinal correlation between train and test splits. XgBoost, being a tree-based learner, is biased towards features with higher cardinality and hence XgBoost-based feature selection was performed on the categorical and continuous variables separately.

The feature scoring was done using importance type as *total gain*. The evaluation metric used for feature selection for out-of-sample folds data in Denial & CARC-RARC models were 'Area under the ROC Curve (AUC)' & 'mlogloss' respectively wherein the objective was to maximize AUC and minimize mlogloss. Figure 4 shows XgBoost feature selection framework. The feature importances for each model in 10-folds were aggregated. A feature was ranked based on the number of times it was used across all the folds. The negative mean score of a feature represented the number of times it contributed negatively across all the folds. The features that satisfied the threshold rank and threshold negative mean score were selected.

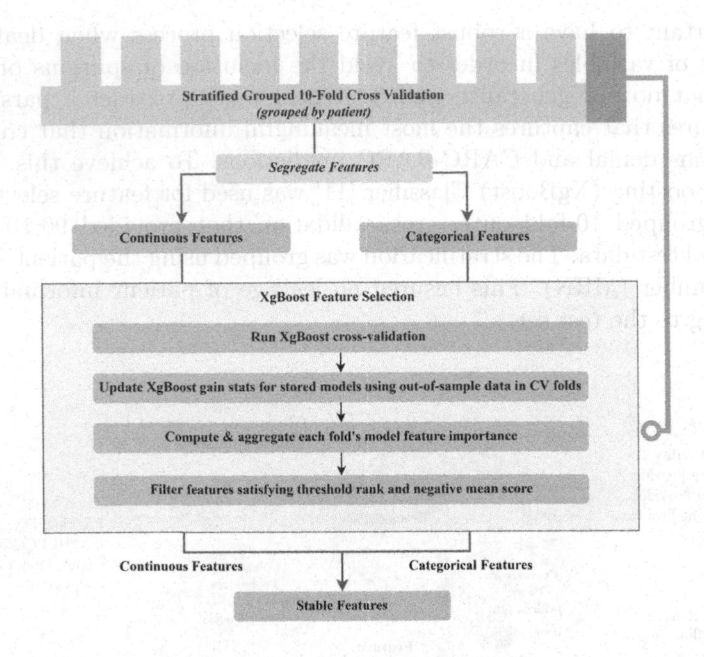

Fig. 4. Feature selection using XgBoost

4.4 Model Development

To improve the claims denial management process in the revenue cycle management, the modeling framework was defined in a way that helps the end-user to prioritize high-risk claims and look into potential areas leading to claim denial thereby preventing it and also saving the cost of claim re-submission.

CatBoost: CatBoost is an implementation of gradient boosting, which makes use of binary decision trees as base predictors [12]. It uses powerful theories like ordered boosting, random permutations that lower the chances of overfitting which leads to a more generalized model. It also implements symmetric trees which eliminate parameters like (min_child_leafs). Further parameters like *iterations, learning_rate, depth, L2_regulariser* are tuned to get the best performance. The past research work done Abdullahi A., et al. [13] and Tanha, J., et al. [14] shows that the catboost algorithm works well with highly imbalanced data having large number of categorical fields and requires less prediction time.

Bayesian Optimization: Hyper-parameters are important for any ML algorithms since they directly control the behaviors of training algorithms and have a significant effect on the performance of ML models. Bayesian optimization [15] was used to optimize the hyper-parameters for training CatBoost classifier. Other techniques like grid search, random search [16] can be extremely time-consuming with less luck in finding optimal parameters for computationally intensive tasks. These methods hardly depend on any information that the model learned during the earlier optimizations. Bayesian optimization, on the other hand, constantly learns from previous optimizations to find a best-optimized parameter list and requires fewer samples to learn or attain the best values. Bayesian optimization works by establishing a posterior distribution of functions that best describes the function to be optimized. As the number of observations increases, the posterior distribution improves, and the algorithm becomes more certain of which regions in hyper-parameter space are worth exploring and which are not. The loss function used for hyperparameter tuning for denial and CARC-RARC models were *LogLoss* & *MultiClass*.

Training and Analysis: The denial & CARC-RARC model training was done using selected features and tuned hyper-parameters with stratified grouped 10-fold cross-validation with CatBoost classifier. The final features for the model training were examined closely by the industry experts to ensure that the features contributions are significant to indicate denials. These models were evaluated on out-of-sample test sets using performance metrics such as AUC for the Denial model and *MultiClass-*AUC for the CARC-RARC model. The Multiclass-AUC was implemented using the approach followed in [17]. Other metrics like precision and recall were also considered as the classes were highly imbalanced.

4.5 Explainability Using SHAP

The model explainability was obtained using the SHAP package. As per S. Lundberg, it is a game-theoretic approach that explains the output of any machine learning model [18] by connecting optimal credit allocation with local explanations using the classic Shapley values from game theory and their related extensions. Each feature contribution is represented by the SHAP value that quantifies magnitude and direction (positive and negative) on a prediction. Figure 5 shows SHAP summary plot with positive and negative relationships of the predictors with the target variable for Denial model. It demonstrates the following information:

- Feature importance: Variables are ranked in descending order.
- Impact: The horizontal location shows whether the effect of that value is associated with a higher or lower prediction.
- Original value: Color shows whether that variable is high (in red) or low (in blue) for that observation.
- Correlation: A high level of the 'days_to_claim_submission' content has a high and positive impact on the claims denial. The 'high' comes from the red color, and the 'positive' impact is shown on the X-axis.

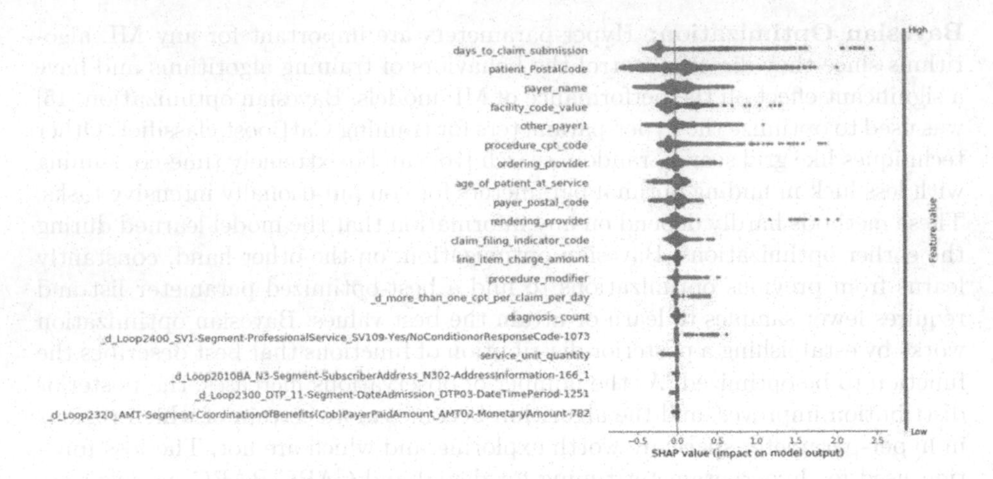

Fig. 5. The SHAP variable importance plot

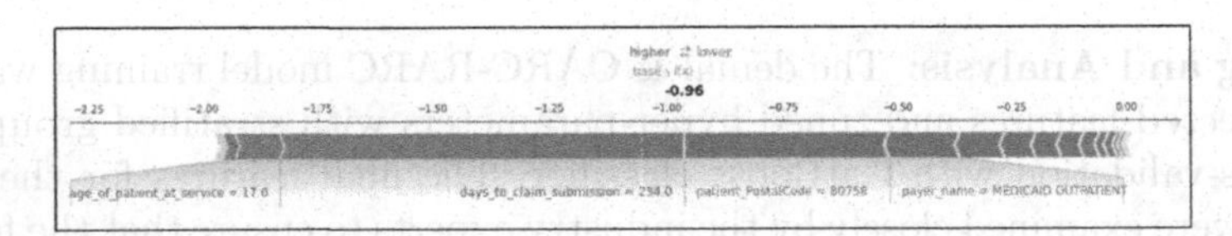

Fig. 6. Explainability using force plot

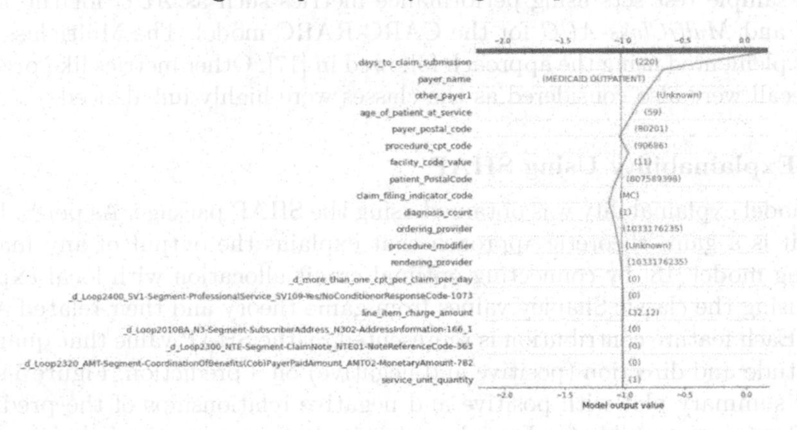

Fig. 7. Explainability using decision plot

The force and decision plot, as shown in Fig. 6 and 7 respectively, are both effective at showing how the model arrived at its decision [19,20]. The *red* indicator represents the feature effects that drive the prediction value higher while *blue* indicator represents the effects that drive the prediction value lower. The model prediction is the output of the sum of *base value* and *shap values*. The base value is the model's average prediction over the training set. Overall, the force plot provides an effective summary for any prediction. On the other hand, the decision plot is more helpful as they are linear, vertical, and interpretable when a large number of significant features are involved. However, both the plots provide similar explanations.

4.6 Joint Probability

The probabilities obtained for each CARC-RARC classes for a claim is the conditional probability as the CARC-RARC model was trained only on denial dataset. The probability of a CARC-RARC class given the claim was denied is represented by Eq. 1. However, the conditional probability is independent of the involved denial risk and may result in false interpretation.

$$P(CARC\_RARC \mid Denial) = P(CARC\_RARC \cap Denial) \div P(Denial) \quad (1)$$

$$P(CARC\_RARC \cap Denial) = P(Denial) * P(CARC\_RARC \mid Denial) \quad (2)$$

Table 3 shows two example claims with different denial risks having the same conditional probability. The joint probability, as represented by Eq. 2, takes into account the denial risk and adjusts the CARC-RARC classes probability.

Table 3. Joint probability

Claim	Denial probability	CARC-RARC conditional probability	CARC-RARC joint probability
Claim line 1	0.99	Class '29_N351': 0.91	Class '29_N351': 0.9009
Claim line 2	0.02	Class '29_N351': 0.91	Class '29_N351': 0.0182

*Only one carc-rarc class is shown for example purpose

5 Results and Discussion

In any ML problem, performance measurement is a crucial task. Over here the model performance was evaluated using stratified grouped 10-fold cross-validation grouped by the patient's MRN. The AUC which represents the degree or measure of separability was used. It tells how much the model is capable of distinguishing between classes. Higher the AUC, the better the model is at predicting 0s as 0s and 1s as 1s. By analogy, the higher the AUC, the better the model is at distinguishing between denied claims and non-denied claims. The AUC value for both the models are shown in Table 4.

Along with AUC, performance metrics such as accuracy, precision, or recall were used, but it was not enough to curtail the number of denied claims, so precision at different Percent Predicted Positive (PPP) values were used to stratify claims. Unfortunately, both precision and recall cannot be made high. It is possible to increase one at the cost of reducing the other. So if we attempt to increase precision, it will reduce recall, and vice versa. This is called the precision-recall tradeoff.

Table 4. Model performance

Model type	Metric	Professional	Institutional
Denial model	AUC (average across 10-folds)	0.80	0.82
CARC-RARC model	MultiClass-AUC (average across 10-folds)	0.95	0.96

5.1 Stratification

As hospitals process hundreds of claims daily, stratification of claims becomes indispensable in order to help the claims department to focus on claims that are at high risk of denial. Based on inputs from industry experts, stratification was provided for the top 20% scored claim lines that maximize the benefits in real scenarios. However, this threshold can be changed based on operational needs. Five stratifications, as shown in Table 5 were generated. Based on workers' capacity, a hospital can choose to focus on 'High' denial risk claims that cover 2% of scored claims or may stretch to 'High-medium' risk claims that cover 5% of scored claims. Also, thresholds for each stratification were defined above the optimal cut-point [21] in the ROC Curve. It is the point having minimum distance to point (0,1) on ROC Curve. It is calculated by minimizing function ER(c), as shown in Eq. 3, where 'Se' denotes *sensitivity* and 'Sp' denotes *Specificity.*

$$ER(c) = (\sqrt{(1 - Se(c))^2 + (1 - Sp(c))^2}) \tag{3}$$

Table 5. Risk score stratification

Stratification	PPP	Professional			Institutional		
		Threshold	Precision	Recall	Threshold	Precision	Recall
High	2	0.74	0.75	0.14	0.93	0.99	0.13
High-medium	5	0.41	0.56	0.26	0.53	0.90	0.30
Medium	10	0.22	0.42	0.39	0.20	0.65	0.44
Medium-low	15	0.15	0.36	0.51	0.15	0.52	0.52
Low	20	0.11	0.31	0.57	0.12	0.44	0.59

*PPP Percent Predicted Positive

5.2 Strengths

The utilization of standard EDI files for developing the model makes this predictive solution *vendor agnostic* with easy integration with hospitals system as the predictor field names are standard across all the US healthcare providers.

The explainable AI adds to the interpretability of the model, thereby, not making the model a 'black-box' anymore. It can help the administrative staffs make appropriate edits to prevent denial. It can also help in streamlining the review process of the claims that require correction at the least expensive stage of the claim life cycle.

6 Conclusion

We have developed and validated the claims denial prediction model that can resolve the major challenges faced by US healthcare providers due to claims denial. The solution is platform-friendly as it utilizes standard EDI files. We ensured that the robust set of features was selected for consistent predictions. The stratification of claims helps the end-users to focus on high-risk claims. We added model explainability to highlight the potential areas that can cause denial to end-users, thereby saving their time to review the entire claim. As a next step, we intend to move further upstream to predict denials even before the claim is generated.

Acknowledgement. We thank John Turek and Dianne Hummel for their extensive inputs on the RCM domain and Claims Denials. We also thank Bob Hansen and Angie Midkiff for their valuable research in claims denial management in the US Healthcare.

References

1. Finances and Your Practice. https://www.aafp.org/family-physician/practice-and-career/managing-your-practice/practice-finances.html. Accessed Dec 2021
2. McKeon, J.: Over Third of Hospital Execs Report Claim Denial Rates Nearing 10% (2021). https://revcycleintelligence.com/news/over-third-of-hospital-execs-report-claim-denial-rates-nearing-10. Accessed Dec 2021
3. Hallock, J.: Change Healthcare Analysis: An Estimated $262 Billion in Healthcare Claims Initially Denied in 2016 (2017). https://www.businesswire.com/news/home/20170626005391/en/Change-Healthcare-Analysis-An-Estimated-262-Billion-in-Healthcare-Initially-Denied-in-2016. Accessed June 2021
4. Wakoli, L.W., Orto, A., Mageto, S.: Application of the K-means clustering algorithm in medical claims fraud/abuse detection. IJAIEM **3**(7), 142–151 (2014)
5. Kumar, M., Ghani, R., Mei, Z.-S.: Data mining to predict and prevent errors in health insurance claims processing. In: Proceedings of the 16th ACM SIGKDD International Conference on Knowledge Discovery and Data Mining, US, KDD, pp. 65–74 (2010). https://doi.org/10.1145/1835804.1835816
6. Kim, B.-H., Sridharan, S., Atwal, A., Ganapathi, V.: Deep claim: payer response prediction from claims data with deep learning. arXiv:2007.06229 (2020). https://doi.org/10.48550/arXiv.2007.06229
7. Harsh, V.: Dealing With Claims Denial the Smart Way 2021 (2021). https://us.nttdata.com/en/-/media/assets/white-paper/hcls-claims-denial-management-whitepaper.pdf. Accessed Dec 2021

8. GE Healthcare (2016). DenialsIQ. https://www.gehealthcare.com/-/media/ec7af20c0a8c4c2283baccd470e985a5.pdf. Accessed June 2021
9. Optum 2016. Denial management: Field-tested techniques that get claims paid. https://www.optum360.com/content/dam/optum3/optum/en/resources/white-papers/2806_Denial_Management_White_paper_r3.pdf. Accessed June 2021
10. Williams, J.: The Denials Challenge: An Analytics-Driven Approach to Denial Prevention and Management. https://www.changehealthcare.com/insights/analytics-driven-approach-to-denial-prevention-management. Accessed Jan 2022
11. Chen, T., Guestrin, C.: XGBoost: a scalable tree boosting system. In: Proceedings of the 22nd ACM SIGKDD International Conference on Knowledge Discovery and Data Mining (ACM SIGKDD), US, KDD, pp. 785–794 (2016). http://dx.doi.org/10.1145/2939672.2939785
12. Liudmila, P., et al.: CatBoost: unbiased boosting with categorical features. In: 32nd Conference on Neural Information Processing Systems, Canada, pp. 6639–6649. NIPS (2018). https://doi.org/10.48550/arXiv.1706.09516
13. Abdullahi, A., et al.: Comparison of the CatBoost classifier with other machine learning methods. IJACSA **11**(11) (2020). http:/dx.doi.org/10.14569/IJACSA.2020.0111190
14. Tanha, J., Abdi, Y., Samadi, N., Razzaghi, N., Asadpour, M.: Boosting methods for multi-class imbalanced data classification: an experimental review. J. Big Data **7**(1), 1–47 (2020). https://doi.org/10.1186/s40537-020-00349-y
15. Snoek, J., Larochelle, H., Ryan, P.: Adams. Practical Bayesian optimization of machine learning algorithms. In: Proceedings of the 25th International Conference on Neural Information Processing Systems (NIPS), vol. 2, pp. 2951–2959 (2012). https://doi.org/10.48550/arXiv.1206.2944
16. Bergstra, J., Bengio, Y.: Random search for hyper-parameter optimization. J. Mach. Learn. Res. **13**(10), 281–305 (2012). http://jmlr.org/papers/v13/bergstra12a.html
17. Kleiman, R.S., Page, D.: AUC μ: a performance metric for multi-class machine learning models. In: Proceedings of the 36th International Conference on Machine Learning, California, vol. 97. PMLR(2019). https://proceedings.mlr.press/v97/kleiman19a.html
18. Lundberg, S., Lee, S.-I.: A unified approach to interpreting model predictions. In: Proceedings of the 31st International Conference on Neural Information Processing Systems (NIPS), pp. 4768–4777. NIPS (2017). https://doi.org/10.48550/arXiv.1705.07874
19. Force Plot (2018). https://shap.readthedocs.io/en/latest/generated/shap.plots.force.html. Accessed Nov 2020
20. Decision Plot (2018). https://shap-lrjball.readthedocs.io/en/latest/generated/shap.decision_plot.html. Accessed Nov 2020
21. Unal, I.: Defining an optimal cut-point value in ROC analysis: an alternative approach. Comput. Math Methods Med. **2017**, 3762651 (2017). https://doi.org/10.1155/2017/3762651

Identification of Landslide Vulnerability Zones and Triggering Factors Using Deep Neural Networks – An Experimental Analysis

G. Bhargavi(✉) and J. Arunnehru

Department of Computer Science and Engineering, SRM Institute of Science and Technology, Chennai 600026, Tamil Nadu, India
bg1064@srmist.edu.in

Abstract. Landslide is the most dangerous type of natural hazard, with devastating consequences for human life and the economy as a whole. Landslides have become an essential responsibility to decrease their harmful consequences, which involves analyzing landslide-related information and anticipating prospective landslides. In this proposed work, the landslide susceptibility zones and the triggering factors were analyzed and classified using a deep neural network (DNNs). A geographical database is created in this study based on 2018 landslide potential points in Kerala, India. 13 districts and 10 parameters are used to create the geographic database: polygon length, polygon width, polygon area, buildings, roads, agricultural data, land use, land cover, longitude, and latitude. A DNN model is generated using fine-tuned parameter with 4728 historic landslide points. This proposed work predominantly concentrates on the experimental aspect, particularly the DNN architecture model has employed for training the dataset. This design employs the adamax optimizer, tanh as an activation function, and four hidden layers with a learning rate of 0.01 to get the highest accuracy with the minimum loss. We found 98.16% as the maximum accuracy after numerous testing operations, implying that the landslide in Kerala was primarily caused by debris flow.

Keywords: Landslide susceptibility mapping · Kerala · Deep neural network · Satellite images · Receiver operating characteristics · Rainfall

1 Introduction

Avalanches are prevalent geographical dangers that have resulted in widespread death and property destruction around the world. Climate change and the accompanying extreme weather patterns have increased natural disasters around the world. Kerala is India's third-most densely inhabited state. Peninsular India's most noticeable environmental feature, the Western Ghats mountain range, takes up 48% of the land as an unstable region [1]. The highlands of Kerala are subject to various landslides, the most common of which are debris flows. Landslides often affect buildings, correspondence constructions, agricultural lands, vegetation, and the environment, a significant reason for

M. Singh et al. (Eds.): ICACDS 2022, CCIS 1613, pp. 121–132, 2022.
https://doi.org/10.1007/978-3-031-12638-3_11

destructiveness [2]. Excluding the beachfront area of the Alappuzha district, 13 districts in Kerala Fig. 1 are inclined to landslides. The Western Ghats of Kerala is recognized as a life-threatening zone for downslope movements in around 9% of the area. Several landslides occur in the region, particularly during the monsoon season in areas where precipitation is the major triggering factor for landslides. Most parts of the Western Ghats district in Kerala are presently casualties of environmental change with concentrated downpours of little length, weakening the grip of mountain soil [3]. Debris flows, rock falls, rock slips, collapses, sneaks, and rotational slides takes part in this region. The prophecy of the event of landslides is often associated with a rainfall threshold condition beyond which landslides are likely to occur. The "debris flows" are the most common, recurring, and disastrous mass movement in Kerala. Landslides are common in Kerala's Wayanad, Kozhikode, Malappuram, Ernakulum, Kannur, Kasaragod, Kollam, Palakkad, Trissur, Thiruvananthapuram, Idukki, Kottayam, and Pathanamthitta districts, which run along the abrupt slopes of the Western Ghats. We ought to recognize the landslide zones as per land use, land cover examination, and undertaking plan developments. Landslide stock planning is the essential and fundamental errand to forestall this risk. In Kerala, we made a landslide weakness planning utilizing memorable landslide focuses and different information. In the wake of finishing this guide, one can identify the landslide-inclined area, order the scope of the landslide seriousness, and foresee the danger assessment and relief.

The proposed work locates slope failures, assesses the causes of failures, and characterizes the types of failures in the research region. The article is organized as follows: Sect. 1 briefly explains the research area as well as the different types of landslides in Kerala, followed by the introduction. Section 2 describes the related study done by various researchers on the occurrence of landslides in and around Kerala. The proposed approach, the database connected to the research activity, and the basic specifics regarding the deep neural network are all illustrated in Sect. 3. Section 4 discusses the experimental setup and results for the specific paper. Finally, Sect. 5 summarizes the landslide susceptibility model's results.

1.1 Study Area

Over 12% of the space in India is touchy to landslides. The significant landslide-inclined regions in India cover the Western Ghats and Konkan slopes Eastern Ghats, North-East Himalayas and North-West Himalayas. Kerala is the study area for the proposed work, and it is one of India's most prone to landslides [4]. According to the Geological Survey of India (GSI) report, Kerala is the second most prone area in India because of the steep slope and thick soil cover that makes this area susceptible to landslides. Both the weather and geographical characteristics contribute to slope collapse. According to the authorities, in 2021, the most exceedingly terrible impacted regions are Kottayam and Idukki, which got 305.5 mm of precipitation. In the study area, 57 lives were lost, and a large number of individuals have been cleared. Somewhere around 100 alleviation camps have been set up. In 2019, Kerala revealed more than 80 landslides in eight districts, and the loss of life crossed 120. In 2018, around 341 significant landslides were seen from ten regions [5]. Idukki area was harmed by 143 landslides, bringing about 104 passings.

The Fig. 1 below depicts the state of Kerala, its 14 districts, landslide spots indicated by black dots, and the rainfall area.

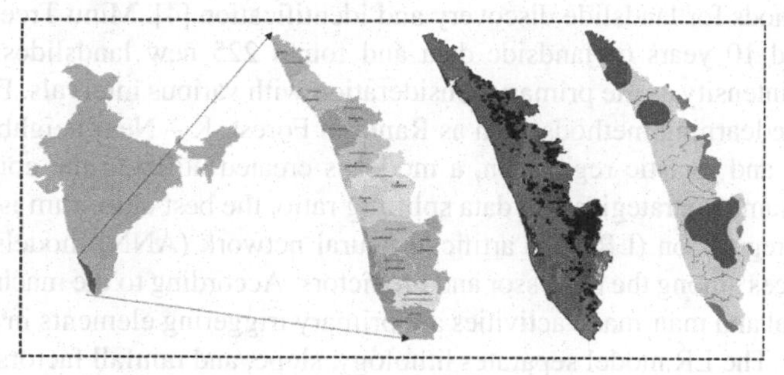

Fig. 1. Satellite image of Kerala's 14 districts with landslide spots marked and excessive precipitation regions emphasized.

1.2 Types of Landslides in Kerala

Landslide is one of the kinds of mass development seen in the scene. Rock fall, landslips, soil channeling, mudslide, debris stream, surficial slide, land subsidence, and so on are various landslides in the state [6]. The most usually seen landslide types in Kerala are debris flow and surficial slide. Out of 4728 historic landslide points, debris flow caused 81 percent of the landslides due to monsoon and climatic change. The Fig. 2 shows the various types of landslide occurs in the study region.

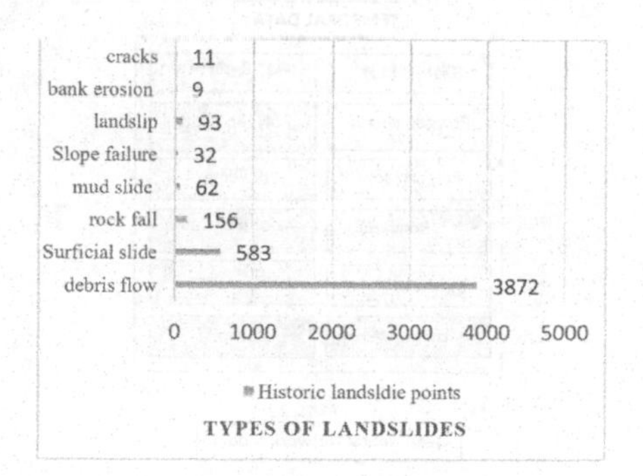

Fig. 2. Types of landslides in the study area.

2 Related Work

Many researchers have shown an intense concern in the classification and identification of geological catastrophes using remote sensing imagery in recent years. East Asian countries and a few European countries have made advancements in landslide inventory, hazard monitoring and analysis, early warning, and other areas [7]. Artificial visual analysis, object-oriented approaches, and statistical model-based methods are the most common methods for landslide discovery and identification [4]. Minu Treesa Abraham [8] considered 10 years of landside data and found 225 new landslides. They took precipitation intensity as the primary consideration with various intervals. Furthermore, using machine learning methods such as Random Forest, K – Near neighbour (KNN), Naive Bayes, and logistic regression, a model is created utilizing the components as input. Using sample strategies and data splitting ratio, the best algorithm is discovered. Using linear regression (LR) and artificial neural network (ANN) models, assess the spatial variances among the regressor and predictors. According to the machine learning model, Natural and man-made activities are primary triggering elements in determining landslide risk. The LR model separates lithology, slope, and rainfall factors as the most influential characteristics for natural landslide occurrence when compared to the ANN model. The ANN model, on the other hand, fails to identify the link between landslide occurrence and conditioning factor.

3 Proposed Work

This work proposes to create a landslide susceptibility model as well as identify the causative component for slide occurrence. The Deep neural network (DNN) approach is

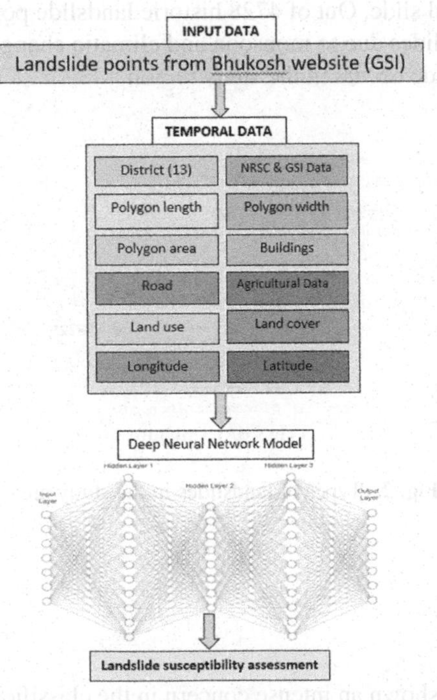

Fig. 3. Proposed methodology for landslide susceptibility assessment.

used to predict landslide risk zones. Other significant important, such as remotely sensed data, soil data, Digital elevation model (DEM) data, and geomorphological information, are used to fabricate the model. To evaluate and predict possible risks, an effective natural catastrophe monitoring system should use a broad range of information. The Fig. 3 illustrates the overview of the proposed methodology.

3.1 Spatial Database

The primary move for generating a landslide database is to collect the potential landslide points using historic landslide data from the Bhuvan website [9] for any type of landslide risk analysis. Landslide inventory is based on 16 different parameters gathered from various sources. This spatial dataset included 13 districts, excluding the Alappuzha district. Initially, the database's structure was based on the landslide inventory information given by the National Remote Sensing Center (NRSC). The Indian Space Research Organization (ISRO) devised a plan to map the Western Ghats' slope failure zones. Landsat satellite images were utilized to extract road, land use, land cover, length, breadth, area, agricultural effect, latitude, longitude, and raster information. ArcGIS tool is essential to extract all of the parameters. The cause of building destructions, roads, and other infrastructure damages (maybe because of debris flow, surficial slide) are also considered in this spatial database preparation [10]. Anthropoid possession and conversion of property are identified by monitoring the land use and land cover values.

3.2 Deep Neural Networks

In this proposed work, we use the deep neural networks (DNNs) to classify the landslide-affected area with the predominant triggering factor which leads to the severe hazard. It's a simple neural network that contains multi-layers between the input and output layers [11] DNN is similar to other Machine Learning methods, excluding the use of many layers. One must first create the architecture before integrating it into a model in Neural Network. The number of layers, neuron counts, and optimizers varies depending on the dataset [12]. However, the enormous number of hyper parameters makes it difficult for researchers to pick which one could utilize to achieve high accuracy. It is not simple to find the appropriate hyper parameter settings. Every dataset requires hyper-parameter tuning for creating the best model [13]. The hyper parameters are the number of neurons, hidden layers, activation functions, learning rate, optimizers, batch size, and epochs [14]. In this DNN architecture, the input layer consists of 11 nodes, where every single node is multiples by the weight with bias according to Eq. 1.

$$Netj = x1 * w1 + x2 * w2 + b \quad (1)$$

where x1, x2 are input nodes, w1, w2 are fixed weights, b is the bias value. Hidden layers are to be precise if too many hidden layers may end up with over fitting if less will bounce back to under fitting. The activation function is a transfer function that acts according to each task [15]. The activation function calculates a weighted sum and then adds bias to it to determine if a neuron should be stimulated or not. There are 6 types of activation functions, namely sigmoid function, tanh, softmax, softsign, Rectified Linear unit (ReLU), and Exponential Linear unit (ELUs) [16].

$$Oj = f(x1 * w1 + x2 * w2 + b) \tag{2}$$

The f is the activation function applies to the convert the neurons into input to the next layer. The input node uses the sigmoid activation function as a weighted sum and then transfers to the next layer [17]. The equation explained in terms of function is mentioned below.

$$F'(x1 * w1 + x2 * w2 + b) = f(x1 * w1 + x2 * w2 + b)$$
$$(1 - (f(x1 * w1 + x2 * w2 + b))) \tag{3}$$

There are seven different optimizers to choose. An optimizer is necessary to achieve the best level of accuracy or the least amount of loss. Each one is based on a unique concept [18]. The optimizer modifies the learning rate and weights of neurons in the neural network to obtain the least loss function. The Fig. 4 illustrates the proposed DNN architecture for Landslide Susceptibility.

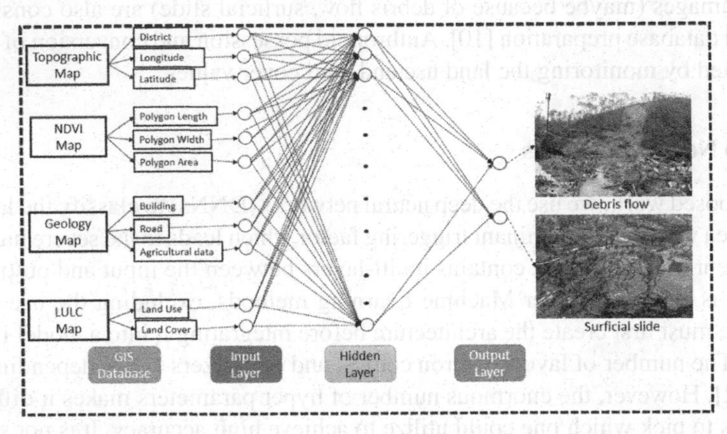

Fig. 4. DNN architecture for the Landslide susceptibility assessment.

4 Experimental Setup

The proposed work uses the following parameters to conduct the experiment. The setup uses Windows 10 with Intel i7 processor with 8 GB RAM. The package Python (3.7.6) with Keras (2.4.3) is constructed on top of Tensorflow (2.3.1). The input layer contains 11 nodes with various hidden layers followed by the output layer to provide a binary classification on landslide zones with the triggering factor [19]. The dataset is undergone a training process with different set of parameter ranges which have been listed in Table 1. Choosing the best hyper parameter with a persistent strategy to enhance the model's

Table 1. Hyper parameter for training the deep neural network

Hyper-parameters	Range/Values
Number of hidden layers	2, 3, 4, 5
Learning rate	0.001 to 0.1
Optimizers	Adam, SGD, RMSprop, Adamax, Adadelta, Adagrad
Activation Function	tanh, sigmoid, softmax, softsign, ReLU, ELUs
Epochs	50-400
Hidden layer nodes	64, 128, 256, 512
Batch size	15, 25, 35
Dropout	10%, 30%, 50% in hidden layers
Output function	Softmax

effectiveness and accurateness is among the considerable challenging parts of creating deep neural networks models [20].

The learning rate is one of the optimizer's hyper parameters. We'll further fine-tune the rate of learning. In our model we tried various learning rate ranges from 0.001 to 0.1 with various optimizers and epochs values. The learning rate decides how large of a stage a model should take to arrive at the least misfortune work. The model advances speedier with a more noteworthy learning rate, however it might miss the least misfortune work and simply arrive at the area around it. A less learning rate improves the probability of observing a base misfortune work. More expanded ages, or more noteworthy time and memory limit assets, are expected as a tradeoff for a lower learning rate. With the assorted scope of learning rates, we attempt to track down the best-fit model by using all of the accessible streamlining agents like adam, SGD, RMSprop, adamax, adadelta, and adaGrad. The model executes in various ages counts from 50 to 400. An epoch is the number of instances a sample is run along through the neural network model [21]. The training dataset is only transmitted forward as well as reverse through the DNN architecture perhaps every epoch [22]. The training data must undergo many times or several epochs are needed. Because the neural network has not had adequate time to train, fewer epochs end in under fitting. An excessive number of epochs, yet then again, will bring about over fitting, in which the model can precisely anticipate the information however not new information. The epoch count must be adjusted to attain the optimal outcome. There are various layers in hidden layers with node ranges from 64 to 512.

5 Experimental Results

After fine-tuning the hyper parameters and layers of the Deep Neural Network, an optimal parameter for locating the landslide susceptibility zone and the triggering event is discovered (Debris flow, surficial slide). The input layer consists of District, longitude, latitude, polygon length, polygon width, polygon area, building, road, agricultural data, land use, and land cover are all nodes in the proposed DNN architecture. The hidden layer is made up of four levels, each with a different node value: hidden layer1 – 64 nodes, hidden layer 2 – 128 nodes, hidden layer 3 – 256 nodes, and hidden layer 4 – 512 nodes. Train the dataset with every conceivable optimizer throughout each learning rate. With the tanh activation function, it delivers the best accuracy in the Adamax optimizer. The best accuracy is attained with a learning rate of 0.01 and 400 epochs for this dataset. The dropout function is also used to regularise the subsequent layers in the hidden layer [23]. This architecture gives 98.16% as the maximum best accuracy with a minimal loss value of 0.17%.

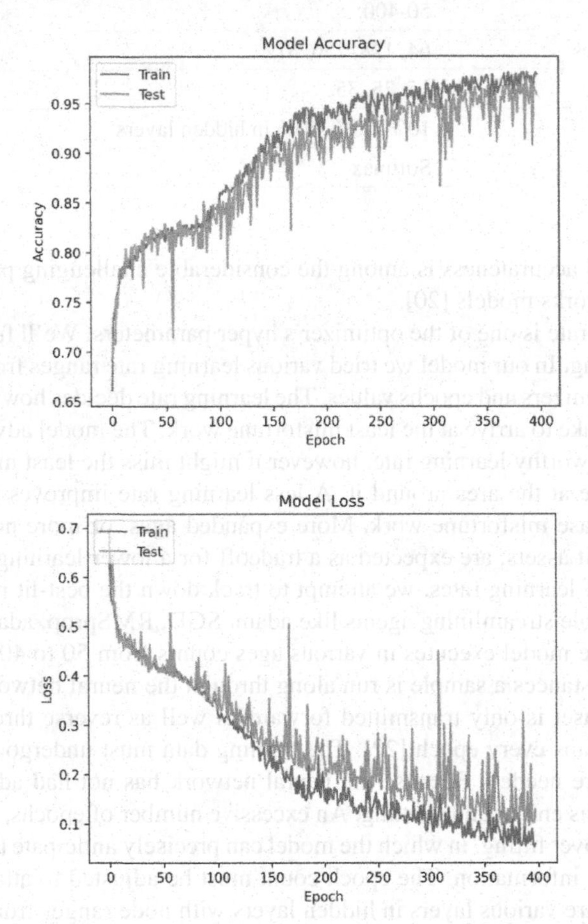

Fig. 5. Model accuracy and loss with 400 epochs for landslide susceptibility.

The Fig. 5 illustrates the model accuracy graph and loss graph for train and test the data. The X axis represents the Epochs value from 0 to 400, the Y axis represents the Accuracy and loss values respectively. True positive (TP), True negative (TN), false positive (FP), and false-negative (FN) values were used to calculate accuracy, specificity, and sensitivity [24]. The true positive (TP) and false-positive (FP) determine the numbers of landslide zones and the triggering factors that are accurately classified as debris flow and surficial slide respectively. The model validates the Accuracy, specificity, sensitivity are some of the statistical criteria used to analyze the DNN landslide susceptibility model.

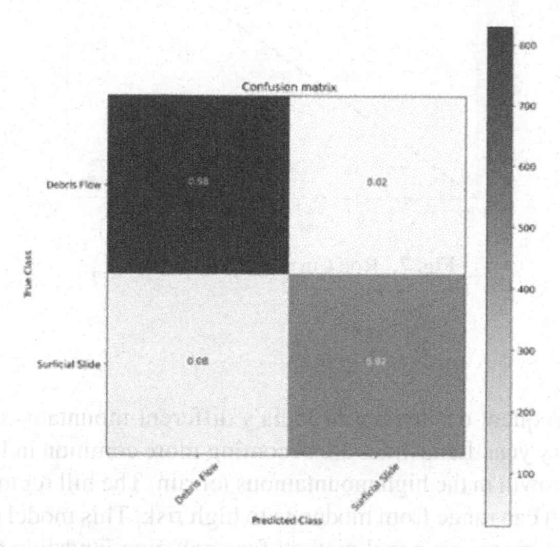

Fig. 6. Confusion matix to classify the derbis flow and surficial slide for the study area.

Figure 6 illustrates the confusion matrix to determine that whether a classification model performs on a set of test data for which the actual values are learned. The Table 2 shows the performance metrics values for the test data.

Table 2. Performance metrics for the landslide classes.

Class	Precision	Recall	Accuracy	F1 score
Debris flow	0.95	0.97	0.98	0.96
Surificial slide	0.95	0.91	0.92	0.93

From the performance table and the confusion matrix, Debris flow and surficial slide are predicted satisfactorily, where 8% of the surficial slide is mispredicted as debris flow which requires an additional interest in our future work.

A traditional method for evaluating model performance is the Receiver Operating Characteristic (ROC). A plot of the false-positive rate on the x-axis and the true positive rate on the y-axis shows the ROC [24]. The appropriate classification of debris flow is the false-positive rate (Sensitivity). The correct classification of the surficial slide is possessed by the true positive rate (Specificity) was predicted well. The training dataset's ROC shows the model's acquisition rate and appropriateness. The Fig. 7 illustrates the ROC curve for the testing dataset.

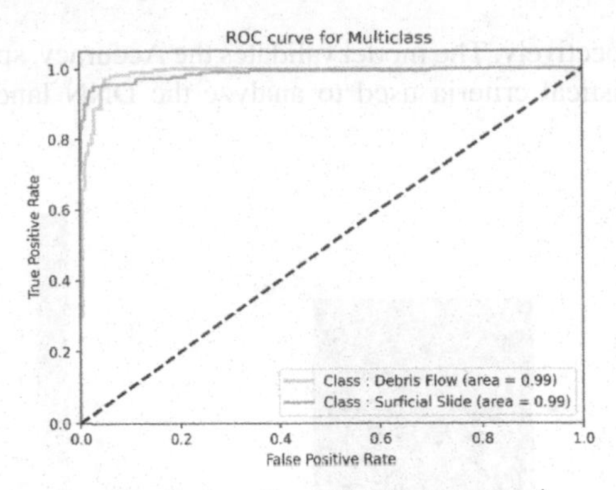

Fig. 7. Roc Curve for the dataset.

6 Conclusion

Landslides are a frequent occurrence in India's different mountains and hilly regions, and they occur every year. Landslides are becoming more common in India as a result of rapid population growth in the high mountainous terrain. The hill regions are vulnerable to landslides, which can range from moderate to high risk. This model employed a Deep Neutral Network to propose a novel method for analyzing landslide susceptibility and detecting the hazard cause's triggering component (DNN). The study region in Kerala, India, was divided into landslide susceptibility zones based on historic landslide potential sites. The spatial database in the study area consists of 13 districts, and additional major temporal data was used to train and evaluate a DNN model. An input layer, a few hidden layers, and an output layer make up the DNN architecture for Kerala landslide susceptibility mapping. Utilizing multiple optimizers, altering hidden layers, adjusting dropout values in the hidden layers, and using the most appropriate activation functions, the dataset was fine-tuned. At varying learning rates, the DNN architecture provides for a wide range of accuracy and loss values. The 0.01 learning rate produces greater accuracy scores when compared to other learning rates. Various parameter combinations were attempted to better understand the dataset and improve the accuracy value. With 98.16% accuracy, the Adamax optimizer with tanh activation function are outperformed well.

References

1. Andrewwinner, R., Chandrasekaran, S.S.: Investigation on the Failure Mechanism of Rainfall-Induced Long-Runout Landslide at Upputhode, Kerala State of India. (2021). https://doi.org/10.3390/land10111212
2. Kuriakose, S.L., Sankar, G., Muraleedharan, C.: History of landslide susceptibility and a chorology of landslide-prone areas in the Western Ghats of Kerala, India. Environ. Geol. **57**(7), 1553–1568 (2008). https://doi.org/10.1007/s00254-008-1431-9
3. Achu, A.L., Aju, C.D., Reghunath, R.: Spatial modelling of shallow landslide susceptibility: a study from the southern Western Ghats region of Kerala, India. Ann. GIS **26**(2), 113–131 (2020). https://doi.org/10.1080/19475683.2020.1758207

4. Jones, S., Kasthurba, A.K., Bhagyanathan, A., Binoy, B.V.: Landslide susceptibility investigation for Idukki district of Kerala using regression analysis and machine learning. Arab. J. Geosci. **14**(10), 1–17 (2021). https://doi.org/10.1007/s12517-021-07156-6
5. Parthasarathy, K.S.S., Deka, P.C., Saravanan, S., Abijith, D., Jacinth Jennifer, J.: Assessing the impact of 2018 tropical rainfall and the consecutive flood-related damages for the state of Kerala, India. Disaster Resil. Sustain., 379–395 (2021). https://doi.org/10.1016/B978-0-323-85195-4.00013-5
6. Bhargavi, G.: J.A.-J. of C. Reviews, and undefined, Land risk susceptibility, hazard and risk factors in western ghats, india–a review. jcreview.com (2020)
7. Mantovani, F., Soeters, R., Van Westen, C.J.: Remote sensing techniques for landslide studies and hazard zonation in Europe. Geomorphology **15**(3–4), 213–225 (1996). https://doi.org/10.1016/0169-555X(95)00071-C
8. Abraham, M.T., Pothuraju, D., Satyam, N.: Rainfall Thresholds for Prediction of Landslides in Idukki, India: An Empirical Approach. (2019). https://doi.org/10.3390/w11102113
9. Abraham, M.T., Pothuraju, D., Satyam, N.: Rainfall thresholds for prediction of landslides in Idukki, India: an empirical approach. Water **11**(10), 2113 (2019). https://doi.org/10.3390/w11102113
10. Chávez-García, F.J., Natarajan, T., Cárdenas-Soto, M., Rajendran, K.: Landslide characterization using active and passive seismic imaging techniques: a case study from Kerala, India. Nat. Hazards **105**(2), 1623–1642 (2020). https://doi.org/10.1007/s11069-020-04369-y
11. Manoharan, K.G., Nehru, J.A., Balasubramanian, S.: Artificial Intelligence and IoT.
12. Khosravi, K., et al.: A comparative assessment of decision trees algorithms for flash flood susceptibility modeling at Haraz Watershed, Northern Iran. Sci. Total Environ. **627**, 744–755 (2018). https://doi.org/10.1016/j.scitotenv.2018.01.266
13. Ragedhaksha, Darshini, Shahil, Arunnehru, J.: Deep learning-based real-world object detection and improved anomaly detection for surveillance videos. Mater. Today Proc. (2021). https://doi.org/10.1016/J.MATPR.2021.07.064
14. Acharjya, D.P., Geetha, M.K. (eds.): Internet of things: novel advances and envisioned applications. SBD, vol. 25. Springer, Cham (2017). https://doi.org/10.1007/978-3-319-53472-5
15. Feng, J., Lu, S.: Performance analysis of various activation functions in artificial neural networks. J. Phys. Conf. Ser. **1237**(2), 022030 (2019). https://doi.org/10.1088/1742-6596/1237/2/022030
16. Buscombe, D., Ritchie, A.C.: Landscape Classification with Deep Neural Networks. https://doi.org/10.3390/geosciences8070244
17. Zintgraf, L.M., Cohen, T.S., Adel, T., Welling, M.: Visualizing Deep Neural Network Decisions: Prediction Difference Analysis. (2017)
18. Hu, X., Zhang, H., Mei, H., Xiao, D., Li, Y., Li, M.: Landslide Susceptibility Mapping Using the Stacking Ensemble Machine Learning Method in Lushui, Southwest China. (2020). https://doi.org/10.3390/app10114016
19. Drusch, M., et al.: Sentinel-2: ESA's optical high-resolution mission for GMES operational services. Remote Sens. Environ. (2012). https://doi.org/10.1016/j.rse.2011.11.026
20. Luo, X., et al.: Mine landslide susceptibility assessment using IVM, ANN and SVM models considering the contribution of affecting factors. (2019). https://doi.org/10.1371/journal.pone.0215134
21. Yao, J., Qin, S., Qiao, S., Che, W., Chen, Y., Miao, Q.: Assessment of Landslide Susceptibility Combining Deep Learning with Semi-Supervised Learning in Jiaohe County, Jilin Province, China. https://doi.org/10.3390/app10165640

22. Quang, V.N.B., Viet, L.D., Chi, C.N., Duc, P.V.N., Quang, B.N.: Predicting Landslide Spatial Probability in Quang Ngai, Vietnam using Deep Learning Technique. In: 4th Asia Pacific Meeting on Near Surface Geoscience & Engineering, vol. 2021, no. 1, pp. 1–5, (2021). https://doi.org/10.3997/2214-4609.202177052
23. Man, A., Pradhan, S., Kim, Y.-T.: Geo-Information Rainfall-Induced Shallow Landslide Susceptibility Mapping at Two Adjacent Catchments Using Advanced Machine Learning Algorithms. https://doi.org/10.3390/ijgi9100569
24. Bradley, A.P.: The use of the area under the ROC curve in the evaluation of machine learning algorithms. Pattern Recognit. **30**(7), 1145–1159 (1997). https://doi.org/10.1016/S0031-3203(96)00142-2

Classifying Offensive Speech of Bangla Text and Analysis Using Explainable AI

Amena Akter Aporna, Istinub Azad, Nibraj Safwan Amlan, Md Humaion Kabir Mehedi(✉), Mohammed Julfikar Ali Mahbub, and Annajiat Alim Rasel

Department of Computer Science and Engineering, Brac University, 66 Mohakhali, Dhaka 1212, Bangladesh
{amena.akter.aporna,istinub.azad,nibraj.safwan.amlan, humaion.kabir.mehedi,mohammed.julfikar.ali.mahbub}@g.bracu.ac.bd, annajiat@bracu.ac.bd

Abstract. The rapid rise of social networking websites and blogging sites not only provides freedom of expression or speech, but also allows people to express society-prohibited behaviors such as online harassment and cyberbullying, which are known as offensive speech or hate speech. Despite the fact that various research work has been done on detecting hate or abusive speech on social networking websites in the English language, the opportunities for research for detecting offensive or abusive speech in the Bengali language remain open due to the computational resource constraints or the lack of standard-labeled datasets for accurate or effective Natural Language Processing (NLP) of Bangla language. In this paper, an Explainable AI approach is used for analysis as well as for detecting offensive comments or speech in the Bengali language is proposed. Moreover, Convolutional Neural Network (CNN) model is used to extract and classify features. Since the Neural Network is time-consuming for extracting features from the dataset, our proposed approach allows people to save time and effort. In the dataset, we classified all user's comments from social media comment sections into four categories: religious, personal, geopolitical, and political. Our proposed model successfully detects Bangla offensive speeches from the dataset (Bengali Hate Speech Dataset) by evaluating Machine Learning algorithms like linear and tree-based models and Neural Networks like CNN, Bi-LSTM, Conv-LSTM, and SVM models. Moreover, we calculate scores for completeness and sufficiency to assess the quality of explanations in terms of fidelity, achieving the results with the accuracy of 78% score, significantly outperforming ML and DNN baselines.

Keywords: Bangla offensive speech classification · Explainable AI · NLP · CNN · DNN

1 Introduction

In the present world, to know about a person is becoming easier day by day as every person is connected through social media. By using social media people

M. Singh et al. (Eds.): ICACDS 2022, CCIS 1613, pp. 133–144, 2022.
https://doi.org/10.1007/978-3-031-12638-3_12

share their lifestyle daily activities. It is helping people to keep in touch with other people. Social media platforms like - Facebook, Google, YouTube, etc. are connecting people and adding new dimensions towards human life. These platforms or websites of social media are giving service by creating their own profiles, enabling the opportunity for interaction and to read what the other people post.

As social media is playing a great role where people are sharing their feelings, it is becoming a place of harassment for some people, groups, ethnicity, culture or nation. Social media platform is becoming the place where people are becoming the victim of cyberbullying, sexual predation, and self-harm practices incitement. People are spreading hatred among the social media. The victim or the target of this type of hatred are an individual or can be in the common people group. Nowadays, people enjoy attacking others through social media.

Bangla offensive speech detection in the social media or any other online platform is important because it helps to visualize and to find out various forms of abusive languages, like which texts or comments or speech are offensive and which texts or comments are non-offensive. Bangla offensive speech detection is also helpful for the Bengali speaking community to detect harassment, abusive texts, hate speeches and offensive comments in social media which are written in Bengali language. Detecting Bangla offensive speech in social media plays a significant role in detecting racism or racial discrimination [1]. Moreover, by detecting offensive speeches in social networking websites, one can get acquainted with cyberbullying. So, it's essential to detect cyberbullying because it is a severe concern in today's age.

The offensive speech spreading is becoming a great concern for social media companies, because they are unable to provide complete security from violence and harassment towards their users. They are devoting a lot of effort in order to resolving security issues. Still their work regarding this field is not sufficient. Distinguishing this type of hate speech is becoming a necessary step for social media. At the present time hate speech in the English language is slowly becoming a familiar scenario. According to the research by Islam (2009) [2], 230 million people are speaking Bangla language in the countries in South Asia. However, while Bangla is a prominent world language, there are fewer resources for detecting offensive Bangla language. Moreover, the detection of Bengali offensive speech on social media websites faces a variety of challenges for the Bengali speaking community, due to the lack of computational resources or standard-labeled datasets for an effective or accurate Natural Language Processing of the Bangla language. Also, the models and datasets are not sufficient enough to detect offensive Bangla comments on social media. We are focusing on distinguishing offensive speech for detecting harassment and abusive texts in Bangla language and classifying the offensive speech. The amount of offensive speech in the Bengali language is increasing daily. The manual detection of Bangla offensive speech is quite a challenging task as it takes a long time and is labor-intensive. So, our proposed approach of detection using the baselines of DNN will allow people to save time and effort. Moreover, we are classifying the offensive speech of Bangla texts and analyzing the result by using Explainable AI (XAI).

In this paper, we have summarized the previous activities which were carried out in the field of offensive speech distinguishing briefly in Sect. 2. In Sect. 3, datasets and its training have been described, Sect. 4, the used methodologies in this paper have been described. Result and analysis is shown in Sect. 5 and lastly, the conclusion and our future plan is provided in Sect. 6.

2 Related Work

Social media is essential for connecting people, for which the growth of social media is undeniable. As in social media people get the freedom to share his/her life but it also enables people to spread hatred about the person. Spreading hatred is now becoming a serious concern around the world. General people are being attacked while using social platforms. So, the detection of abusive speech has become a necessary task globally. There are lots of research works on this topic all around the world. Implementation of NLP in the English language is the most common among all as it is the international language. Scientists and researchers from different nations are working on languages other than English. In the research work by Fabio Dell'Orletta et al. (2009) [3] have worked on detecting hate speech in the Italian language where they selected Facebook as a platform for collecting data. In their proposed paper, they used SVM and LSTM models for hate speech detection. There is another work where the author Malmasi and Zampieri (2017) [4] detected hate speech in social media using sentiment analysis where they collected data from Twitter posts and tweets. In their proposed research paper, they used different classifiers and majority class baselines. Similar works can be found where they carried out their research on abusive words and sentiment analysis, selecting social media platforms like Facebook and Twitter to collect necessary data. Most of the detection of hate or abusive speech works are carried out using ML, neural networks, transformers and so on. In Machine Learning (ML), the commonly used classifiers are Naive Bayes (NB), Support Vector machine (SVM), Linear Regression (LR), etc. Deep Neural Network architecture (DNN) is also used in the field of NLP for detecting offensive speech in social networking sites. In the research paper, the researcher Yin et al. (2017) [5] discussed about that, Natural Language Processing has been transformed by Deep Neural Networks (DNNs). They also mentioned in their paper that, the two primary forms of DNN architecture, Convolutional Neural Network LeCun et al. (1998) [6] and Recurrent Neural Network (RNN) Elman (1990) [7], are widely used to perform a variety of NLP applications. RNN is good at modeling units in sequence proposed by some researchers like Tang, Adel, Gupta et al. (2016) [8], while CNN is good at extracting position-invariant characteristics which is mentioned by the researcher Dauphin et al. (2017) [9]. Due to the struggle of CNNs and RNNs, the state-of-the-art on many Natural Language Processing tasks frequently shifts. The goal of this study is to provide fundamental guidelines for DNN selection by comparing CNN and RNN on a wide range of relevant Natural Language Processing tasks.

In the research paper of Karim et al. (2020) [10] the author has talked about the hate speech detection in Bengali language. From different research works and papers on Bangla language, it is found that the resources for thorough and proper research are very limited. Thus, they created the dataset from different newspapers, image posters and the contents from the social media. For detection, they have used multiple approaches to train their data with ML, DNN and transformers and before that they have preprocessed the Bengali texts and classified them into different genres. A pre-trained model of Fast Bangla text is also used here. For data preprocessing they have used the hashtag normalization, stemming, emojis, duplicates, and tokenization. After the data preprocessing they have trained the baseline models. For training machine learning baseline models, they have used character n-grams and word uni-grams with the two term weighting which is the TF-IDF weighting. They have also trained transformed based models in their paper for getting better results.

With 265 million native and non-native speakers globally, Bangla is the ninth most spoken language. On the other hand, English is the most widely used language for internet resources, technical information, publications, and documentation [11]. For this reason, Bangla language-speaking people, who have a weak grasp of English, may face challenges while trying to use English resources. The research gap occurs because of the limited resources, support, and increasing demand for Bangla Natural Language Processing (BNLP). Researchers also face difficulties in processing Bangla language materials. Furthermore, automated computerized facilities for Bangla text preprocessing and associated research works are scarce, making research on Bangla language text data difficult [12], [13]. Moreover, preprocessing erroneously interpreted Bangla offensive speeches and overlapping Bangla abusive speeches are the typical issues in Bangla speech data preprocessing [14,15]. For the reason of a scarcity of significant resources as well as lack of knowledge for research in this area, and a lack of sufficient tools for developing such systems for the Bangla language, the researchers faced several challenges while developing Bangla word sense disambiguation systems, including differences in the structure and formatting of Bangla words used in different sentences [16]. The lack of a suitable dataset is the source of most of the difficulties in sentiment analysis. Because the sentiment labels only contain a few classes, dealing with a wide range of sentiments is difficult owing to a lack of datasets. The researchers faced numerous challenges while developing Bangla spam and hate speech detection systems. Classifying spam, non-spam and offensive speech detection in Bangla documents is a difficult task because there is a lack of public datasets and there is a lack of significant research on Bangla spam and offensive speech detection [17]. Moreover, when it comes to information extraction, the researchers are limited in their ability to construct a corpus for their model due to a lack of appropriate tools. Bangla also lacks the necessary language skills to manage the various components of information retrieval activities.

3 Dataset

The dataset we have utilized here is in Bangla language, which includes text values. The resources of Bangla Literature in case of NLP based research is very limited. Although, we are using an existing dataset originally generated by Karim et al. [10] in the motive of Hate Speech detection in Bangla Literature we are planning to collect further and precise data to perform the task with better accuracy. The dataset has been created from the collected data from different posts, comments and tweets from the social media. In this work, the datasets are categorized into four sectors depending on the classification type and they are -

- Personal, (Labelled as 0)
- Political, (Labelled as 1)
- Religious, (Labelled as 2)
- Geopolitical, (Labelled as 3)

As we know, abusive and hate speech can refer to any specific sectors or nations or personal and taking this basis the dataset is labeled and classified accordingly. However, there are some speech or comments that can be put in any category based on the linguistic analysis from the specialists. Here the datasets have been collected using the bootstrap approach. Many sentences are present here with some common words. These might be directed towards an individual or generalized targeting a community or group which are considered accordingly. A single word might refer to multiple categories. To handle that, a group of words are selected to differentiate it and used for the analysis in which it is used at what percentage. The details are present in the section of result and analysis.

The Data is preprocessed in such a way that extra connecting words, that is the stop words, extra and unnecessary spaces and other symbols are removed by tokenization with the help of lexicons. As the data are collected from social media, in the contents there are lots of unnecessary symbols and emojis that are not related to our work. These are removed and cleaned through the required processes (Fig. 1).

The existing dataset, which is originally generated by Karim et al. [10] where the collecting data from social media has been re-organized according to our necessity. From Fig. 2 we can see that, the long sentences are broken so that some unnecessary words and phrases are trimmed and separated using the process of tokenization. The removed parts include stop words, extra spaces, hashtags, emojis, html tags, etc. Thus, after that the pre-processed data is prepared for concatenation to train using three DNN model baselines where we used CNN, Bi-LSTM and Conv-LSTM as classifiers. Next, we implement the test data and predict results accordingly. The results are then analyzed thoroughly using the advanced technique of Explainable AI by which it is possible to figure the use of multiple words, specific words for the specific sector of hate speech in Bangla language. The represented value then can be displayed in different ways.

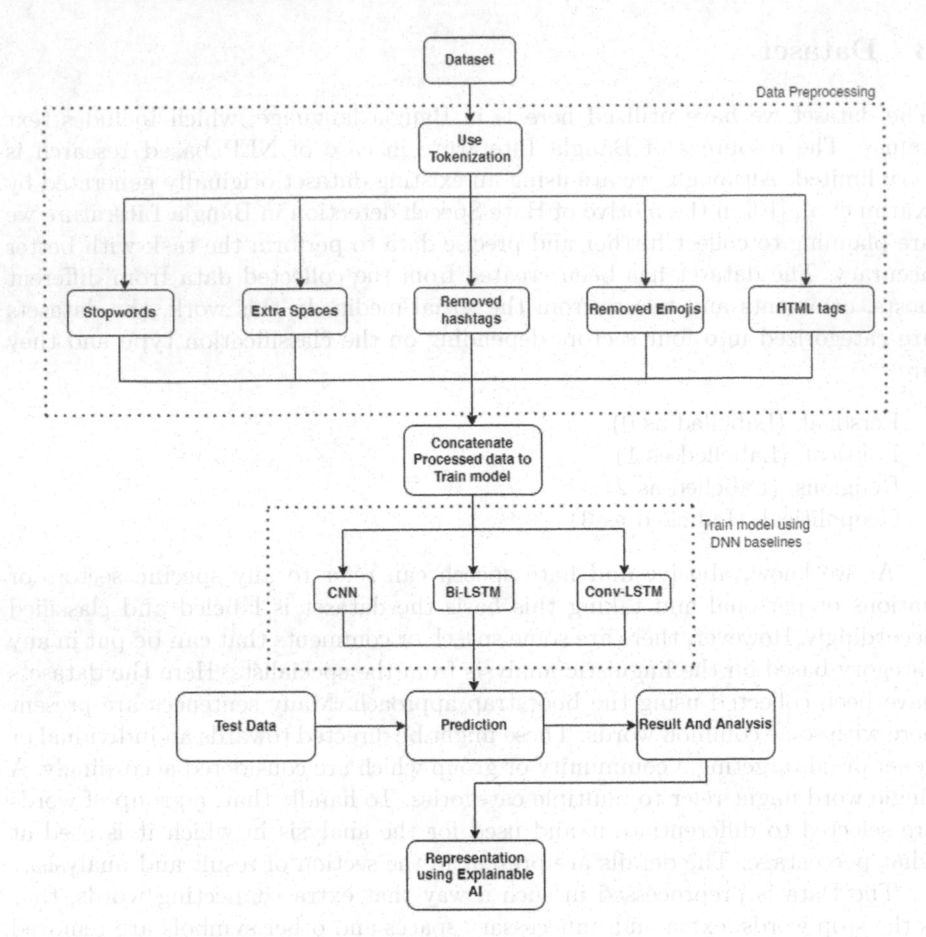

Fig. 1. Workflow diagram

4 Methodology

Multiple kinds of approaches are implemented for segmentation and classification of texts for Natural Language Processing for multiple cases. The most common text classification and segmentation methods are Convolutional Neural Network (CNN), Bi-LSTM, Conv-LSTM, and in the case of Machine Learning approach, Random Forest (RF), Naive Bayes (NB), Support Vector Machine (SVM) and so on models can be used. In our research paper, we thoroughly implement the method of Deep Neural Network (DNN) architecture. We have trained the dataset using three baselines of DNN, which are CNN, Bi-LSTM and Conv-LSTM. In addition, we have also tested and trained using ML approaches where we used NB, SVM, RF.

4.1 Convolutional Neural Network (CNN)

The deep learning method significantly constructed to process the data collected in groups or layers for deep analysis and also for the thorough classification of texts in the form of neural nodes is known as Convolutional Neural Network. It can also be used for one and three dimensions. In deep neural network architecture, it is known as the most popular artificial neural network. It is used in data analysis, language processing, and other classification problems. In the case of Natural Language Processing, its implementation is much popular because of its high accuracy feature. It analyzes text data and is very good at extracting features from the dataset. These hidden layers of the model are called the convolutional layers. There is one input layer in CNN along with multiple convolutional and non-convolutional layers and finally with the output layer. The neurons of these layers are connected with the nearby same weighted neurons. The two primary operations of CNN are convolution and pooling. Convolution is the process of filtering inputs and results in a feature map. These feature maps for different inputs are stacked together, and it provides the output. The pooling operation mainly reduces the number of parameters. This operation is performed in each feature map.

4.2 Long Short Term Memory (LSTM)

Long Short Term Memory is a neural network system which can keep track of the sequence of prediction problems. It is basically a complex context of deep learning which is used for machine translation, speech recognition, anomaly detection, etc. Moreover, it helps to detect the words which are dependent on previous words. It basically keeps track of the words which are used. In LSTM there are basically two states, one state is called the short term memory and another state is called the long term memory.

Bi-directional Long Short Term Memory (Bi-LSTM): Bi-directional Long Short Term Memory (Bi-LSTM) is basically a two layered neural network system [18]. It is a sequence based model which is the advanced form of LSTM (Long Short Term Memory). In this model the input can flow in both the directions which makes it an advance version of the LSTM. In LSTM, the input flows in one direction, it can either flow in the forward direction or in the backward direction. However, in Bi-LSTM it can flow in both the direction means that the input can flow in backward also it can flow in forward. Bi-LSTM is usually applied where we need to use something related to sequence. For the tasks of speech recognition and text classification, Bi-LSTM network model can be used.

4.3 Support Vector Machine (SVM)

The Support Vector Machine, which is also called SVM, is an effective classification approach. Support Vector Machines (SVMs) are incredible yet adaptable directed AI calculations utilized for classification and regression. Several tasks related to Natural Language Processing have benefited from adopting SVM. The SVM method can be used for various Information Extraction (IE) tasks to achieve better performance. This method has high classifying accuracy and excellent performance in case of any fault generalization. For the Information Extraction task, SVM follows some steps. For example, at first, SVM converts the problem into multiple steps for classification tasks. Then transforms the problem of those steps into a series of collective binary classification issues. After that, for each binary classification, an SVM classifier is being trained; and finally, the output of the classifiers are merged to get a suitable result of the original problem. SVM can be helpful for a classification hyperplane in feature space, and the classifier's generalization for which SVM can perform excellently in a good range of NLP tasks. In SVM, the feature vector is formed deliberately from the text using many linguistic properties. The feature vector is mapped into higher dimensional space using the so-called kernel function in many situations. The SVM approach is mainly used for finding an optimal decision boundary for classifying n-dimensional space [19]. For this reason, future categories might easily include more data points. Moreover, SVM determines the extreme points and vectors of hyperplanes to reflect the optimal decision-making boundary. So, for the IE task, SVM significantly provides a good result in NLP applications.

5 Result and Analysis

Here, we discuss the result analysis of models and a comparative explanation of DNN baselines with ML. Although we prefer the approach of Deep Neural Network, we also implement ML and present the results. The following table shows the precision based on the model that we have used.

5.1 Classifiers and Accuracy

Table 1. Performance analysis

Method	Classifiers	Accuracy
DNN baseliners	CNN	0.73
	Bi-LSTM	0.75
	Conv-LSTM	0.78
ML baseliners	SVM	0.67

In the method of Machine Learning approach we have used the SVM to classify the data and train the model to achieve precision (Table 1). After that when we run the test data the accuracy result shows 67%. Now, we implement DNN using it's baselines for classifiers. The prediction was made using the same test data and better accuracy are found than the ML baselines. One by one we have generated results which are for CNN we got 73%, for Bi-LSTM we got 75% and finally for Conv-LSTM we got 78%. So, from this analysis we have deduced that out of these processes we get the best result from Conv-LSTM.

5.2 Explainable AI

After setting up the training model and testing it we represented our data through a special set of framework known as the Explainable AI. Here, the analysed results are interpreted through Artificial Intelligence in the form of graphs and values. It also helps to debug and predict the models so that we can improve it later for further experiments and work in the future. For this procedure we have demonstrated the results and analysis generated by implementing the models of DNN and ML. Here, we presented global and local explanations. For the former, linguist analysis is used to identify a list of the highest and lowest essential words for all the classes.

Individual example explanations are provided by emphasizing the most significant phrases. We have used the leave-one-out experiment to assess word-level relevance for the proper explanation of locales. It is used to implement the backward approach. First, we have chosen a sample hate statement from the test set at random. Then, for the two most likely classes, we provide prediction probabilities for all of them, followed by an explanation of word-level significance.

The weight of the embedding layer is initialized using fastText embeddings for each DNN baseline model. As can be seen, each model outperforms or performs similarly to the ML baseline models. Conv-LSTM, in particular, outperforms the other DNN baselines which is higher than Bi-LSTM.

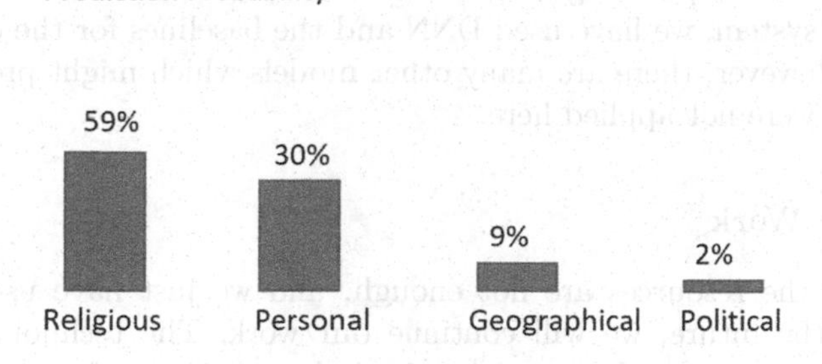

Fig. 2. Relevance test based on prediction

As we mentioned in the data-set Sect. 3 that, a word or some common terms are repeated in multiple sentences as well as it falls into multiple categories. A thorough analysis is performed in our paperwork and representation is displayed in Fig. 2. The data-set named as Bangla Hate speech that we used for model implementation and analysis, contains hate speech of Religious and Personal category most according to the predictions. The model predicted and detected Religious hate speech which is 59%. Then, there is hate speech in the category of Personal is 30%. The least probability we got is the category of Geography which is 9% and Political is 2%. There are some words which are present in multiple categories. These words' predictions are analyzed accordingly. The words are compared in such a way that these are related to a category and how much different it is from a classification. The values are calculated as probability values. The words are compared here on the basis of Religious and Personal classes.

6 Conclusion and Future Work

This chapter concludes with possible areas of future work that could be analyzed to contribute more meaningfully to the problem of the BNLP topic.

6.1 Conclusion

In this paper, we have implemented Deep Neural Network baseline models which are used for detecting abusive speech in social media for Bangla Language. It is basically an under-resourced Language. For further analysis and debugging we have implemented the explainable AI. Applying this framework of AI will help to analyze the results for better understanding and develop the works in this field of research. Thus, it will contribute in future works in the most efficient ways. In this paper, we have implemented the AI explainable model in a way that it shows the results, detailed information of the datasets and label them according to the analysis. Though by applying the methods we got the detailed information by which further research works can be continued. Our paper has some drawbacks as well. For example, we have implemented an existing dataset, which is limited but planning to generate more suitable data for the task. In our proposed system, we have used DNN and the baselines for the classification thoroughly. However, there are many other models which might produce better results which were not applied here.

6.2 Future Work

As we know, the resources are not enough, and we just have used the baseliners, so in the future, we will continue our work. The technology that can be utilized to overcome a language barrier is language processing in the native tongue. Moreover, the success of BNLP might have a significant influence on ordinary people's ability to study and use ICT in Bangla, considerably improving their socioeconomic situation. Many studies on Bangla sentiment analysis have been published in recent years. The accuracy of this section is improving as new datasets become available, and we are going to work to get better precision. Bangla information extraction systems frequently employ modern machine learning and deep learning techniques. Furthermore, future research trends in the Bangla spam and hate speech detection system include developing offensive speech detection in public Facebook pages and other social sites, and the development of Bangla datasets for spam, hate speech and fake data [20]. Implementing other approaches will help us compare the existing approaches and the approaches we will implement in our future work. Thus the result will be more accurate. We want to merge machine learning and deep learning models in the future to construct a significant ensemble model that can assist us in improving our present results.

References

1. Poletto, F., Basile, V., Sanguinetti, M., Bosco, C., Patti, V.: Resources and benchmark corpora for hate speech detection: a systematic review. Lang. Resour. Eval. **55**(2), 477–523 (2021)
2. Islam, M.: Research on Bangla language processing in Bangladesh: progress and challenges. In: 8th International Language Development Conference, pp. 23–25 (2009)
3. Del Vigna, F., Cimino, A., Dell'Orletta, F., Petrocchi, M., Tesconi, M.: Hate me, hate me not: hate speech detection on Facebook. In: Proceedings of the First Italian Conference on Cybersecurity (ITASEC17), pp. 86–95 (2017)
4. Malmasi, S., Zampieri, M.: Detecting hate speech in social media. arXiv preprint arXiv:1712.06427 (2017)
5. Yin, W., Kann, K., Yu, M., Schütze, H.: Comparative study of CNN and RNN for natural language processing. arXiv preprint arXiv:1702.01923 (2017)
6. LeCun, Y., Bottou, L., Bengio, Y., Haffiner, P.: Gradient-based learning applied to document recognition. Proc. IEEE **86**(11), 2278–2324 (1998)
7. Elman, J.L.: Finding structure in time. Cognit. Sci. **14**(2), 179–211 (1990)
8. Vu, N.T., Adel, H., Gupta, P., Schütze, H.: Combining recurrent and convolutional neural networks for relation classification. arXiv preprint arXiv:1605.07333 (2016)
9. Dauphin, Y.N., Fan, A., Auli, M., Grangier, D.: Language modeling with gated convolutional networks. In: International Conference on Machine Learning, PMLR, pp. 933–941 (2017)
10. Karim, M., Dey, S.K., Chakravarthi, B.R., et al.: Deephateexplainer: explainable hate speech detection in under-resourced Bengali language. arXiv preprint arXiv:2012.14353 (2020)
11. Sen, O., et al.: Bangla natural language processing: a comprehensive review of classical, machine learning, and deep learning based methods. arXiv preprint arXiv:2105.14875 (2021)
12. Indurkhya, N., Damerau, F.J.: Handbook of Natural Language Processing. Chapman and Hall/CRC, Boca Raton (2010)
13. Haque, M.M., Pervin, S., Hossain, A., Begum, Z.: Approaches and trends of automatic Bangla text summarization: challenges and opportunities. Int. J. Technol. Div. (IJTD) **11**(4), 67–83 (2020)
14. Kuligowska, K., Kisielewicz, P., Włodarz, A.: Speech synthesis systems: disadvantages and limitations. Int. J. Res. Eng. Technol. (UAE) **7**, 234–239 (2018)
15. Sun, L., et al.: A novel LSTM-based speech preprocessor for speaker diarization in realistic mismatch conditions. In: 2018 IEEE International Conference on Acoustics, Speech and Signal Processing (ICASSP), pp. 5234–5238. IEEE (2018)

16. Moon, S., McInnes, B., Melton, G.B.: Challenges and practical approaches with word sense disambiguation of acronyms and abbreviations in the clinical domain. Healthcare Inform. Res. **21**(1), 35–42 (2015)
17. Zhou, X., Zafarani, R., Shu, K., Liu, H.: Fake news: fundamental theories, detection strategies and challenges. In: Proceedings of the Twelfth ACM International Conference on Web Search and Data Mining, pp. 836–837 (2019)
18. Isnain, A.R., Sihabuddin, A., Suyanto, Y.: Bidirectional long short term memory method and word2vec extraction approach for hate speech detection. IJCCS (Indonesian J. Comput. Cybernet. Syst.) **14**(2), 169–178 (2020)
19. Li, Y., Bontcheva, K., Cunningham, H.: Adapting SVM for natural language learning: a case study involving information extraction. Nat. Lang. Eng. **15**(2), 241–271 (2009)
20. Ishmam, A.M., Sharmin, S.: Hateful speech detection in public Facebook pages for the Bengali language. In: 18th IEEE International Conference on Machine Learning And Applications (ICMLA). IEEE 2019, pp. 555–560 (2019)

Android Malware Detection Using Hybrid Meta-heuristic Feature Selection and Ensemble Learning Techniques

Sakshi Bhagwat(✉) and Govind P. Gupta

Department of Information Technology, National Institute of Technology Raipur, Raipur, India
sakbhagwat14@gmail.com, gpgupta.it@nitrr.ac.in

Abstract. There is wide use of smart mobile phone in modern digital world which is generally operated using open-source software. Being open-source software, it becomes easier to intrude in the system using malicious code. Android malware gets installed into the smart mobile phone without the permission of user and causes harm to user's personal and sensitive information. To detect this malware, various techniques are proposed by researchers. Existing malware detection techniques uses digital signature method which is unable to recognize unknown malware. Thus, this paper has a novel malware framework in which dynamic feature is exploited to detect android malware. In the proposed framework, we aim to select right subset of feature which can increase our performance. In the proposed framework, meta-heuristic feature selection (FS) method using Genetic Algorithm (GA), Gravitational Search Algorithm (GSA) and correlation is used which is named as Correlated Genetic Gravitational Search Algorithm (CGGSA). The optimized features are used by the Adaptive boosting and Extreme Gradient Boosting Classifiers to detect the malware. Performance analysis of the proposed framework is evaluated using real-time CICMalDroid-2020 dataset in terms of accuracy, precision, recall and f1-score. The proposed framework has achieved 95.3% of accuracy.

Keywords: Ensemble technique · Adaptive boosting · Extreme gradient boosting · Feature selection · Correlation · Genetic algorithm · Gravitational search algorithm

1 Introduction

Nowadays mobiles have become basic need of life. Android is the popular operating system used. It is open source software in which applications can be downloaded freely from third party and play store. Due to this android attacks have increased. According to 2020 report of McAfee labs total 121 million malware exists and out of which 49 million malwares are new [15]. Malware is any product which harms mobile system. It is a program written by attacker in order to attack clients PC and harm it in different manners. It damages users confidentiality, can track users location and access personal information of user. Antivirus is enable to protect the system still many systems get

M. Singh et al. (Eds.): ICACDS 2022, CCIS 1613, pp. 145–156, 2022.
https://doi.org/10.1007/978-3-031-12638-3_13

attacked. New tools are available on internet for attacker to develop malware which requires fewer skills. To protect the system of an organization or of individual from malware is responsibility of cybersecurity.

In the past years, researchers have used two different approaches to detect malware in android; one is machine learning based malware detection and other is non-machine learning based malware detection. In non-machine learning based detection signature, permission based approaches are used to detect malware. Drawback of such signature based approach is that it is unable to detect new malware. To overcome this drawback static and dynamic approach for machine learning based detection is used. In static analysis malware is detected without executing code [10, 11, 12] and in dynamic analysis malware detection takes place at the time of execution of code [13]. In our proposed approach we have used dynamic features for malware detection.

Feature selection (FS) plays important role in increasing performance rate of model [14]. It gives right subset of feature which takes less time to execute and gives higher accuracy. In our proposed work we have used hybrid of three feature selection techniques correlation, Genetic Algorithm and Gravitational Search Algorithm. Paper aims to detect 5 categories of malware using ensemble techniques. The main contribution of the paper is as follows:

- A novel android malware detection model is proposed using hybrid metaheuristic-based feature selection and ensemble learning-based classifiers to detect the different categories of malware.
- For optimization of features, a hybridized method by using the correlation, Genetic Algorithm (GA) and Gravitational Search Algorithm (GSA) is proposed which reduces the size of original feature set significantly.

The rest of the paper contain is as follows. Next section is of related work. Section III presents proposed framework for detecting android malware. Section IV contains experimental results observed. Section V presents conclusions.

2 Related Work

Over past years, various research papers have been published for detection of android malware using various feature selection techniques and machine learning schemes. Some of the related papers are discussed in this paper. S. Morales-Ortega et al [4] has developed a model to perform feature extraction, feature selection and ensemble methods for classification on android malware.In this paper there different feature selection techniques are used as chi-square, relief and information gain for selecting important features. Ensemble methods are used to train the dataset. As result proposed method shows accuracy of 94.48%.M Singh et al. (2018) [5] has given both pre and post classification using ensemble method for malware detection. In this paper, filter feature selection methods as chi-square, OneR and Relief are used. Top N features of these feature selection methods are chosen. For pre classification all the features are combined and extreme learning machine classifier is used to train data. While in post classification, the features are not combined but on them individually ELM classifier is applied then the results are given by majority voting. Comparing this classifiers, it is observed that post classification has better accuracy of 94.04%.

M. Dhalaria et al (2020) [6] has proposed android malware detection approach using static and dynamic features. Appropriate feature are chosen using chi-square feature selection algorithm. Then further various classifiers combined using stacking ensemble method for detecting android malware. AndroMD dataset is used in this paper from kaggle. It contains of 675 features. Various base classifiers as SVM, Decision tree, Navie bayes, random forest and K-NN are used. From the base classifiers top 3 model are chosen based on accuracy, this are then combined using stacking. Result shows that KNN_RF has better accuracy of 98.02%. Deepak Gupta et al (2020) [7] has proposed two approaches for detecting malware first one is weighted voting ensemble methods in which weights for base classifiers is computed and second is optimal set of base classifiers used for stacking. Apache spark framework is used to process big data in this paper. From results it is observed that random forest as base classifier is better than other classifiers and have accuracy of 98.1%.

A Mahindru et al (2021) [8] has used five different machine learning algorithm and three ensemble methods to detect android malware. To improve performance feature selection methods are used. Filter approach and wrapper approach is used to select important features. For each feature T-test analysis is carried out to get P value for each feature to tell how efficient feature is to detect malware. Multivariate linear regression stepwise forward selection is used to eliminate irrelevant features. Base learners as Logistic regression analysis (LOGR), Radial Basis Function Neural Network (RBFN), Artifcial Neural Network (ANN) are used to train the models. To combine the result of all the algorithms ensemble methods are used as Best Training Ensemble (BTE), Majority Voting Ensemble (MVE), and Nonlinear Ensemble Decision Tree Forest (NDTF). Best accuracy of 98.8% is obtained by using NDTF.

3 The proposed Android Malware Detection using hybrid meta-heuristic feature selection and Ensemble Learning Techniques

In this section, we present the description of the proposed framework in details. Firstly, the raw dataset is pre-processed. To increase the accuracy feature selection process in include. In feature selection we have used three methods as correlation, genetic algorithm, and gravitational search algorithm. The experiment is divided into two parts where the union of all the subset of feature given by feature selection methods are given another one is where the intersection of all subsets of features is made. The model is trained on two ensemble learning classifiers as Adaboost and Xgboost boost. Then results are compared. The block diagram of the proposed framework is illustrated in Fig. 1.

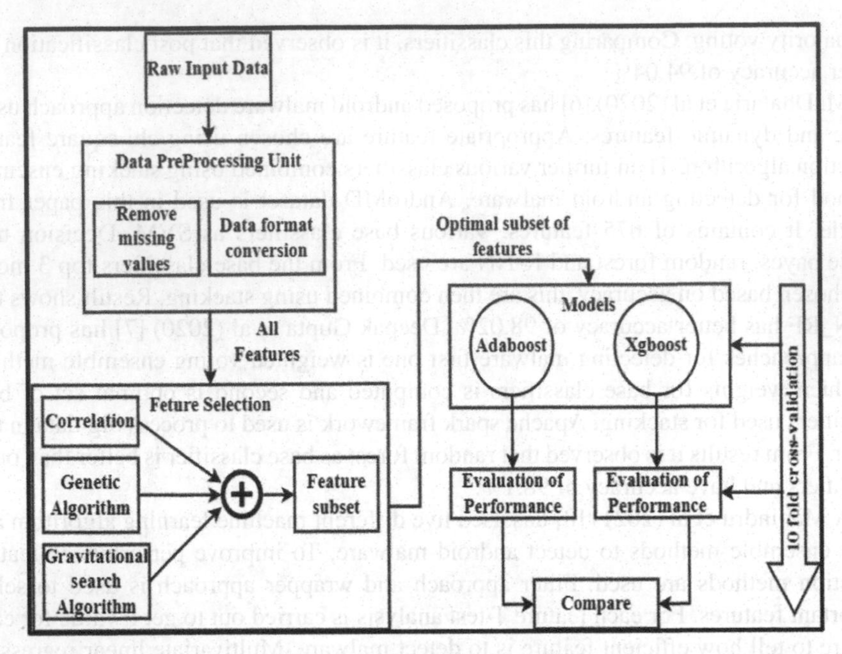

Fig. 1. Proposed Framework

3.1 Feature Selection Process

Feature selection is the most important step in the process of detecting malware. It aims to select the subset data from the complete dataset which are necessary features needed to train the model on and give prediction. All the unnecessary, irrelevant and the feature which may misguide are excluded and the best suited features are considered. Because of this time to train the model decreases and accuracy of the model increases if the right set of data is chosen. This process reduces the complexity of model.

Features which are highly correlated with other feature cause redundancies because of which they get eliminated. When the classifier is very expensive to train his process leads to take minimum time to train the model and increase the performance. Process includes generation stage in which from the original dataset few set of features are selected for training model on them then evaluating the process [1]. In two ways the feature can be selected:

1) Forward selection: At the beginning the generation subset is empty and one by one the features are added to it
2) Backward elimination: From the entire feature already present one by one the feature ae eliminated.

a) **Correlation Based Feature Selection:** Correlation measures the similarity between the two features. Correlation between two features X and Y is expressed as

$$\rho_{P,Q} = \frac{\mathrm{cov}(P, Q)}{\sigma_P, \sigma_Q} \tag{1}$$

cov(P,Q) is a covariance, σ in the standard deviation of P and Q. correlation is covariance divided by product of standard deviations.

When we have n sample we will put covariance and standard deviation in following formula. X and Y are two random features.

$$r_{X,Y} = \frac{\sum_{j=1}^{n}(x_j - \overline{x})(y_j - \overline{y})}{\sqrt{\sum_{j=1}^{n}(x_j - \overline{x})^2}\sqrt{\sum_{j=1}^{n}(y_j - \overline{y})^2}} \quad (2)$$

When features are linearly dependent correlation coefficient value become either +1 or -1. While the features are uncorrelated correlation coefficient becomes 0.

b) **Genetic Algorithm Based Feature Selection:** All the features are given as input to genetic algorithm which gives best subset of features as output. If feature is selected it is represented by 1 while other features are represented as 0 in chromosome [16]. There are following steps in GA feature selection:

- Initial population in GA is formed using fixed number of chromosomes which are having fixed length.
- From the chromosomes two parent chromosomes are selected which are then passed to mutation and crossover to give child chromosomes. To select parent chromosomes roulette-wheel selection is used.
- The child chromosomes are evaluated against fitness function. Fitness function is defined as the chromosome which gives higher accuracy for the classifier are assigned higher value while other features are assigned smaller values.
- These gives next generation which again goes through selection, mutation and crossover process.

c) **Gravitational Search Algorithm Based Feature Selection:** It is the population-based algorithm. Random population is initialized from which the optimal subset of feature is obtained. Initial population should be properly generated otherwise it leads to poor search. In GSA after ever updating of position of agents we will reach to best position. Mass of every agent is the fitness of particle. Particles are the string contains binary values '0' and '1'. '0' and '1' represents the inclusion and exclusion of features. The length of string is the number of features [16].

Fitness function is the ability of particle to give correct prediction or accuracy of the feature subset. Mass of each particle at time t is calculated using following equation:

$$mass_i(\mathrm{t}) = \frac{fitness_i(t) - minimum(t)}{\mathrm{maximum(t)} - minimum(t)} \quad (3)$$

maximum(t) is the maximum fitness value and minimum(t) is the minimum fitness value at time t. The mass are re-calculated for after one unit of time using $mass_i(\mathrm{t})$whichis$Mass_i(\mathrm{t})$. following is the equation for it:

$$Mass_i(\mathrm{t}) = \frac{mass_i(\mathrm{t})}{\sum_{j=0}^{n} mass_i(\mathrm{t})} \quad (4)$$

The mass for candidate keeps on improving for time t.

One mass applies force on another mass. The force applied by p[th] particle on q[th] particle is represented as- $F_{qp}(t)$. Theses force has component for all features. k[th] force is represented as- $F^k{}_{qp}(t)$. The force exerted by k[th] feature on q[th] particle by p[th] particle is calculated by following formula:

$$F_{qp}^{K}(t) = G(t) * \frac{Mass_i(t) * Mass_j(t)}{dist(x_i, x_j)} * (x_j^k(t) - x_i^k(t)) \tag{5}$$

$dist(x_i, x_j)$ is the hamming distance between them, G(t) is a gravitational constant which is calculated as:

$$G(t) = \frac{\Omega * t}{e^{totaltime}} \tag{6}$$

where Ω is taken as -20 and t is current time.

The force on q[th] particle k[th] feature is denoted as $F_q^K(t)$ and calculated by following formula:

$$F_q^K(t) = \sum_{p=1, p \neq 1}^{m} random_j * F_{qp}^K(t) \tag{7}$$

Here $random_j$ is in range of 0 to 1.

The velocity for q[th] particle k[th] feature is calculated as:

$$v_p^k(t+1) = random_j * v_p^k(t) + \frac{F_q^K(t)}{Mass_i(t)} \tag{8}$$

Probability for changing feature state is calculated as:

$$propability = \tanh(v_p^k(t+1)) \tag{9}$$

3.2 Ensemble Learning-based Malware Detection

Multiple learners are trained and combined to solve a problem. In ensemble learning the entire individual learner is trained and combined with some strategies like maximum voting, averaging or weighted averaging. Ensemble learning methods are divided into two categories as boosting and bagging. Boosting is sequential ensemble and bagging is parallel ensemble. Boosting popular algorithms as Adaptive boost, gradient boost and extreme gradient boost. Extreme gradient boosting is advanced version of gradient boost.

Adaptive Boosting: Adaptive boosting is also known as Adaboost, it is a popular sequential ensemble learning method. In Adaboost, weak learners are trained sequentially. After every iteration the sample are reweighted so that in next iteration wrongly classified samples by previous model are more focused. Wrongly predicted instances are given more weight in next iteration in attempt of predicting them correctly. Further is the algorithm of Adaptive Boosting.

Algorithm:

1. For every sample of dataset an initial weight $W_i = \frac{1}{n}$ is assigned, where n is total number of samples.
2. Train a weak model
3. For each observation:
   a) If predicted value is correct then weight of that sample is decreased.
   b) If the value is incorrectly predicted then weight of sample is increased.
4. The samples with higher weight are predicted.
5. Repeat the steps 3 and 4 until the number of trees are trained.

b) **Extreme Gradient Boosting:** Drawback of sequential ensemble is that it is unscalable. This problem is solved by extreme gradient boosting which is scalable tree boosting system. It is also known as Xgboost. It is developed in framework of Gradient Boost but to increase computational speed. Extreme Gradient Boost performs better than Gradient Boosting as it uses more regularized model which predict correct instances. It supports distributed training due to which it can support feature engineering and allocating base predictor.

4 Performance Evaluation and Discussion

We implemented ensemble learning using the jupyter notebook. The experiments are conducted on Windows 10 as an operating system. Intel(R) Core(TM) i5-6200U CPU @ 2.30 GHz 2.40 GHz, 8 GB. Dataset is taken from Canadian institute of cybersecurity, known as CICMalDroid-2020. It is a latest android malware dataset consisting of 17,341 android samples. It has five distinct categories in samples as: Adware, Banking malware, SMS malware, Riskware and benign.

To compare classification performance accuracy, precision, recall, f1-score is used. Results are divided into two parts as before applying feature selection methods and after apply feature selection method. Feature selection method is applied in two ways as union of all the features selected by all the methods and second one is intersection of subset of features selected by methods.

4.1 Without Feature Selection

Without using any feature selection techniques models are trained directly on 470 numbers of features. Adaptive boosting and Extreme Gradient Boost classifier are used to train the dataset. Among both of them Extreme Gradient Boost classifier has highest accuracy of 95.21%. Accuracy is represented in Fig. 2. Table 1 contains the precision, recall and f1-score for both the classifiers.

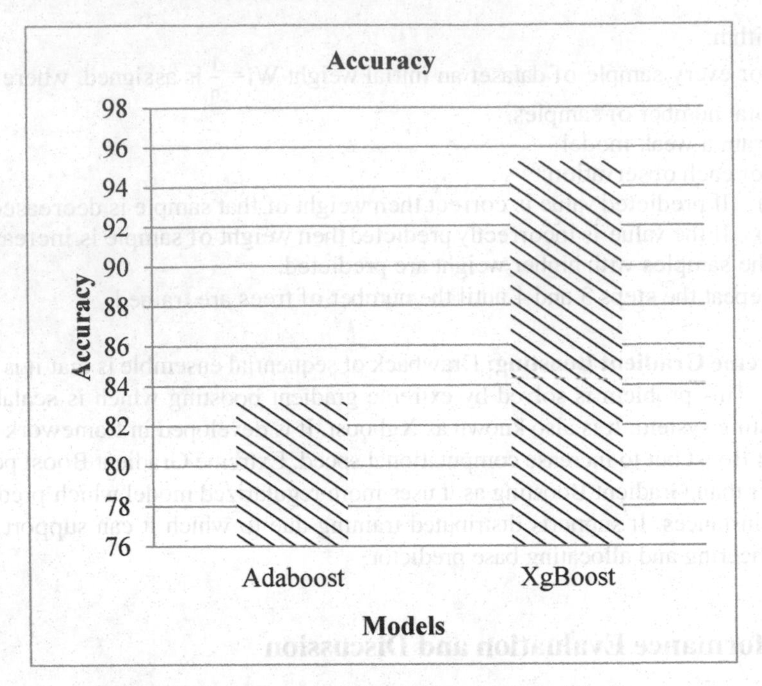

Fig. 2. Accuracy of Models without Feature Selection

Table 1. Results without feature selection

Classification algorithm	Accuracy (%)	Precision	Recall	F1-score
Xgboost	95.21	0.8165	0.8022	0.8086
Adaboost	83.15	0.9418	0.944	0.9428

4.2 With Feature Selection

Individually correlation feature selection technique has selected 302 features, genetic algorithm has selected 145 features and gravitational search algorithm has selected 214 features out of total features. Further union and intersection of all features is made to obtain optimal set of features.

Union of features: Union of all the subset of features given by different feature selection methods is done then the models are trained on the data. In this method 407 features are used to train model. Extreme Gradient boost performs better than Adaptive boost and has accuracy of 95.3%. Accuracy has increased after feature selection. Adaboost has accuracy of 83.2%. Accuracy is represented in Fig. 3. Table 1 contains the precision, recall and f1-score for both the classifiers (Table 2).

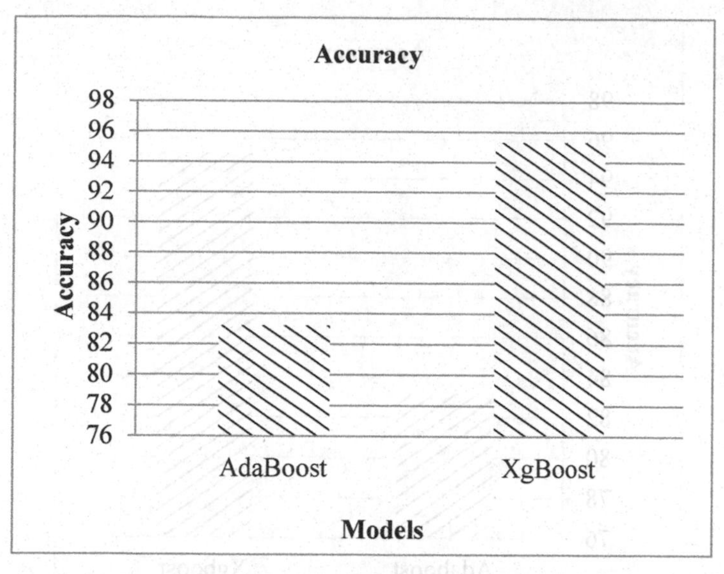

Fig. 3. Accuracy of models with union of features

Table 2. Results with union of features

Classification algorithm	Accuracy (%)	Precision	Recall	F1-score
Xgboost	95.3	0.9437	0.9453	0.9444
Adaboost	83.2	0.8105	0.7962	0.8025

b) **Intersection of features:** Intersection of all the subset of features given by different feature selection methods is done then the models are trained on the data. In this method 51features are used to train model. Extreme Gradient boost performs better than Adaptive boost and has accuracy of 95.3%. Accuracy has increased after feature selection. Adaptive boosting has accuracy of 83.2%. Accuracy of union and intersection feature selection has same accuracy. Accuracy is represented in Fig. 4. Table 3 contains the precision, recall and f1-score for both the classifiers (Table 4).

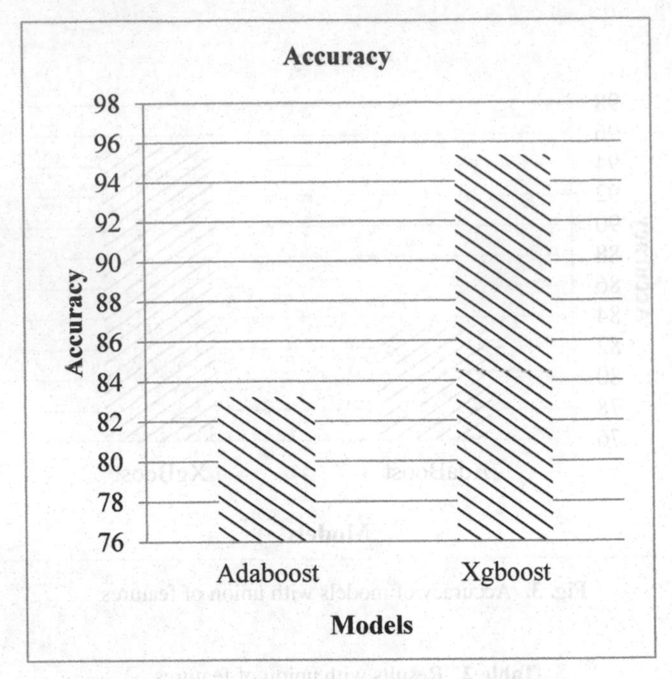

Fig. 4. Accuracy of models with intersection of feature

Table 3. Results with intersection of feature

Classification algorithm	Accuracy (%)	Precision	Recall	F1-score
Xgboost	95.3	0.9438	0.9452	0.9444
Adboost	83.2	0.8104	0.7931	0.8008

Table 4. Feature selection percentage

FS Algorithm	Total Features selected	Percentages of selected features
Correlation	302	64.25%
GSA-based FS	214	45.53%
GA-based FS	145	30.85%
Union of Features	407	86.59%
Intersection of Features	51	10.85%

5 Conclusion

In this paper, we have used popular feature selection algorithm such as correlation, Gravitational Search Algorithm, Genetic Algorithm to give optimal subset of features. We had done union and intersection of all the subsets of features differently. Then these union and intersection of features is taken to train the dataset. Union of features gives 407 features and intersection gives 51 features. Both the union and intersection gives same results. We had used Adaptive boosting and extreme gradient boosting to train the models. Out of these two classifiers extreme gradient boost performs better with any accuracy of 95.3%. We have compared our proposed framework with existing approach which has used same dataset.

References

1. Almomani, A., Alweshah, M., Al Khalayleh, S., Al-Refai, M., Qashi, R.: Metaheuristic algorithms-based feature selection approach for intrusion detection. In: Machine Learning for Computer and Cyber Security, pp. 184–208. CRC Press (2019)
2. Gopika, N., A. Meena Kowshalaya, M.E.: Correlation based feature selection algorithm for machine learning. In: 2018 3rd International Conference on Communication and Electronics Systems (ICCES), pp. 692–695. IEEE (2018). https://doi.org/10.1109/CESYS.2018.8723980
3. Nemati, S., Basiri, M.E.: Particle swarm optimization for feature selection in speaker verification. In: European Conference on the Applications of Evolutionary Computation, pp. 371–380. Springer, Berlin, Heidelberg (2010)
4. Morales-Ortega, S., Escamilla-Ambrosio, P.J., Rodriguez-Mota, A., Coronado-De-Alba, L.D.: Native malware detection in smartphones with android os using static analysis, feature selection and ensemble classifiers. In: 2016 11th International Conference on Malicious and Unwanted Software (MALWARE), pp. 1–8. IEEE (2016). https://doi.org/10.1109/MALWARE.2016.7888731
5. Singh, M., Gupta, P.K., Tyagi, V., Flusser, J., Ören, T., eds.: Advances in Computing and Data Sciences: Second International Conference, ICACDS 2018, Dehradun, India, April 20–21, 2018, Revised Selected Papers, Part II. Vol. 906. Springer (2018). https://doi.org/10.1007/978-981-13-1810-8
6. Dhalaria, M., Gandotra, E.: Android malware detection using chi-square feature selection and ensemble learning method. In: 2020 Sixth International Conference on Parallel, Distributed and Grid Computing (PDGC), pp. 36–41. IEEE (2020)
7. Gupta, D., Rani, R.: Improving malware detection using big data and ensemble learning. Comput. Electr. Eng. **86**, 106729 (2020)
8. Mahindru, A., Sangal, A.L.: HybriDroid: an empirical analysis on effective malware detection model developed using ensemble methods. The Journal of Supercomputing 1–43 (2021).
9. Zhou, Z.-H.: Ensemble learning. In: Machine learning, pp. 181–210. Springer, Singapore (2021). https://doi.org/10.1007/978-981-15-1967-3
10. Aslan, Ö.A., Samet, R.: A comprehensive review on malware detection approaches. IEEE Access **8**, 6249–6271 (2020).
11. Singla, S., Gandotra, E., Bansal, D., Sofat, S.: Detecting and classifying morphed malwares: A survey. International Journal of Computer Applications **122**(10) (2015)
12. Gandotra, E., Bansal, D., Sofat, S.: Malware analysis and classification: a survey. Journal of Information Security 2014 (2014)

13. Fatima, A., Maurya, R., Dutta, M.K., Burget, R., Masek, J.: Android malware detection using genetic algorithm based optimized feature selection and machine learning. In: 2019 42nd International Conference on Telecommunications and Signal Processing (TSP), pp. 220–223. IEEE (2019)
14. Hu, P., Pan, J.-S., Chu, S.-C.: Improved binary grey wolf optimizer and its application for feature selection. Knowl.-Based Syst. **195**, 105746 (2020)
15. McAfee Labs. Threat Predictions Report, McAfee Labs, Santa Clara, CA, USA (2020)
16. Jing, T.W., Murugesan, R.K.: A theoretical framework to build trust and prevent fake news in social media using blockchain. In: International Conference of Reliable Information and Communication Technology, pp. 955–962. Springer, Cham (2018)
17. Mahindru, A., Sangal, A.L.: Deepdroid: feature selection approach to detect android malware using deep learning. In: 2019 IEEE 10th International Conference on Software Engineering and Service Science (ICSESS), pp. 16–19. IEEE (2019)
18. Guha, R., Ghosh, M., Chakrabarti, A., Sarkar, R., Mirjalili, S.: Introducing clustering based population in binary gravitational search algorithm for feature selection. Appl. Soft Comput. **93**, 106341 (2020)
19. Gupta, G.P., Kulariya, M.: A framework for fast and efficient cyber security network intrusion detection using apache spark. Procedia Computer Science **93**, 824–831 (2016)

Scoring Scheme to Determine the Sensitive Information Level in Surface Web and Dark Web

Rahul Singh(✉), P. P. Amritha, and M. Sethumadhavan

TIFAC-CORE in Cyber Security, Amrita School of Engineering, Amrita Vishwa Vidyapeetham, Coimbatore, India

cb.en.p2cys20027@cb.students.amrita.edu, {pp_amritha, m_sethu}@cb.amrita.edu

Abstract. Paste sites are largely used for innocent text sharing but they have grown in popularity as venues for criminal operations such as data leaks and publication. This research examines numerous types of sensitive information and the extent to which each can cause damage if compromised. Our proposal intends to develop an efficient scoring scheme for determining the sensitivity of information included within a paste's body. We designed a scraper to monitor two surface web and two dark web paste sites and extract and score various aspects from the obtained data. The findings indicated that surface web paste sites featured a greater amount of sensitive material than dark web paste sites.

Keywords: Dark web · Surface web · Sensitive information · Paste site

1 Introduction

Global malware campaigns and large-scale data breaches show how everyday life can be impacted when defensive measures fail to protect computer systems from cyberthreats [1, 6]. Although it is anticipated that the dark web has more leaked sensitive information, no comprehensive investigation of the varying amounts of sensitive information shared on the surface web and dark web paste sites has been conducted. Dark web hidden services are often used to carry out activities that are otherwise illegal and unethical on the surface Web [9]. But the Dark Web is not just about illicit and criminal activity, as most people believe. The darknet's primary legitimate actors include ordinary citizens, journalists, activists, business executives, IT specialists, and law enforcement officials [2]. The majority of dark web sites are pastes, hacker forums, carding shops, Internet-Relay-Chat rooms, or marketplaces [15]. Developing a grading system based on common personal sensitive information and resolving the abovementioned issue would assist researchers by providing further insight into the appropriate amount of monitoring for different sites. The purpose of this research is to develop an effective grading system based on the amount of harm that a single paste may do to an individual and to use it to examine how sensitive information changes across the surface and dark webs.

M. Singh et al. (Eds.): ICACDS 2022, CCIS 1613, pp. 157–167, 2022.
https://doi.org/10.1007/978-3-031-12638-3_14

1.1 Personal Information and Sensitive Information

There are many subtle distinctions when understanding personal and personal sensitive information. Even though the inadvertent disclosure of both sort of data causes worries or annoyance, the consequences of sensitive data disclosure are more severe. Additionally, people are extremely context-sensitive when it comes to evaluating their privacy. [10].

Personal data is any data that can be used to identify an individual. Personal information includes first and last names, telephone numbers, home addresses, birthdays, email id addresses, and bank account details. This information is collected more frequently since applications and websites frequently require this information for several reasons.

Sensitive data is a classification of personal data. If released, this material has the potential to cause substantial harm to a person's bodily and mental health. While laws safeguard personal information in general, they place an emphasis on sensitive information due to the potential impact on a person's life. Sensitive information includes information about a person's ethnic background or, political views and criminal background. Biometric, genetic, and medical history information is also considered to be sensitive information.

1.2 Pastes

Pastebin is a content sharing website that enables members to publish raw text via publicly accessible posts dubbed "pastes." The site has a monthly average user base of 17 million unique visitors. Since the original Pastebin was created in 2002, other similar web apps dubbed "paste sites" have developed. Pastebin was created in response to the increased use of Internet Relay Chat (IRC). IRC is a 1988 instant messaging application built for real-time communication between many users and is popular for sharing plain text. Direct code sharing in IRC threads/channels, used to break the message flow and had the potential to inject and modify the code of the IRC service itself. Soon, users needed a third-party site that allows them to distribute plain raw blocks of texts as links that other users may readily view and alter. Paste services are primarily used for harmless text sharing but they have also evolved into popular venues for criminal operations such as data leakage.

Oftentimes, when internet services are compromised, the first indications of the breach are shown on "paste" sites such as Pastebin. On these services, attackers regularly publish samples or whole dumps of compromised data. Tracking and reporting on the availability of information on sites such as Pastebin can help affected people prepare for the aftermath from a breach [3].

Additionally, certain plain-text Dark Web contents found on hacker forums, DNMs, carding shops, and IRC channels may occasionally be available in Paste Sites (e.g., PasteBin) [11].

One of the benefits of paste services is that the structure of the raw content that is allowed to be shared there is unrestricted. As a result, a paste might look self-explanatory or entirely unclear. However, there are some recurring features [4].

Typically, leaks will be in the form of raw database scripts that can be executed in order to reconstruct the database structure. They often contain comma separated fields for distinct columns in the database, and they are frequently protected with a cryptographic hash [7] to ensure that the passwords are secure. However, if the credentials are not sufficiently strong, they can be obtained easily [5]. Example:

```
(`id`, `username`, `name`, `password`,`birthdate`
`email_id`)
(14, 'chester.padberg', 'Chester',
'e5e5878939d5fc893d111517db6e9cab', '318124800',
'chester.padberg@hotmail.com'),
(4, 'anita', 'Anita', '77ffcc64bdf9c16b9e5cc364acc12ab2',
'7862400', 'anita79@hotmail.com'),
```

Compromised credentials are often dumped as lists of usernames and passwords. Example:

```
annabelle.waters@yahoo.com:Kzzg***
addie0@hotmail.com:sFFt***
henderson.schuppe@gmail.com:ieQj***
hiram.adams18@yahoo.com:NgGY***
```

Logs and code blocks: These can take a variety of forms, ranging from tampered system logs to internal system code. Example:

```
list("/temp/node1/","axel10@yahoo.com"),
list("/temp/node5/","keanu_hahn@yahoo.com"),
list("/temp/node2/","osvaldo_abernathy@yahoo.com"),
```

For a paste site, the appearance of an email address does not always indicate that it has been leaked via a breach.

Frequently, some pastes will appear many times on a service such as Pastebin. It may be identical or have minor differences, but it is, for all practical purposes, the same content. This might be because the same person released it several times or because a breach was socialized and then re-published by a large number of individuals.

2 Background

2.1 Personal Information

In the simplest terms, personal data is any piece of information that may be used to accurately identify a live individual. For instance, the email address is deemed personal data since it implies that there can be only one. Similarly, a person's physical address or phone number is deemed personal data since it enables them to be reached. Personal data is also defined as everything that may attest to a person's actual existence in a certain location. Therefore, a person's fingerprints and even a recorded CCTV video are considered personal data, but things become more difficult when it is thought that these individual pieces of information do not have to be taken alone. Typically, organizations gather and keep many bits of information on users, and the resulting information might be called "personal data" if it can be combined to form an individual's unique identity.

Any of the following may be deemed personal data under certain circumstances:

1. A first and last name
2. A residential address
3. An e-mail address
4. A number on an identity card
5. Information about the location
6. An Internet Protocol (IP) address
7. Phone's advertising identifier

2.2 Sensitive Information

Sensitive data can refer to a variety of different things. One simple method to assess is to consider what personal information a person would not prefer to be shared publicly with just anybody. There are governing rules and regulations that specify the categories of sensitive data that must be secured.

1. Credit card numbers, bank account details, and social security numbers are all examples of financial information.
2. Government data: Any material classified as secret or top-secret, restricted, or posing a risk of violation of confidentiality may be considered.
3. Accounting data, trade secrets, financial statements or accounts, and other sensitive information contained in company plans constitute business information.
4. Addresses, medical histories, driver's license numbers, and phone numbers are all examples of personal information.

2.3 Sensitive Personal Information

Sensitive personal data is a subset of an incredibly unique category of information that must be handled with extreme care. This category covers information related to a person's political ideas, ethnic origin, religious beliefs, biometric data, genetic information, sexual orientation, and even trade union memberships.

An individuals' intentions to limit their personal information online appear to be motivated by a desire to strike a balance between their current online exposure and their desire to maintain control over their personal online information [12].

2.4 Types of Sensitive Information

In the United States, *Personally Identifiable Information*, or PII, is defined as "information that can be used to distinguish or trace an individual's identity, such as their name, social security number, or biometric records, either alone or in combination with other personal or identifying information that is linked or linkable to a specific individual, such as date and place of birth, mother's maiden name, or etc." [13].

Personal Information, or PI, is a larger category that includes personally identifiable information (PII). While all PII is PI, not all PI is PII. The following is a more comprehensive definition of PI: "Information that identifies, refers to, characterizes, is capable of being connected with, or may reasonably be associated with, a specific consumer or household". PI might comprise information that is clearly related to an individual's identity—such as a name or date of birth, which is sometimes also included in PII—or it can be read legally in an exceedingly broad sense. IP addresses can and frequently do contain the following types of information: IP addresses, employee records, personal photographs, geolocation data, ethnic origin, political leanings or views, beliefs in religion or philosophy, criminal record, sexual orientation, health-related or genetic data, and even trade union memberships.

Sensitive Personal Information (SPI) is a newly defined word under the impending California Privacy Rights Act (CPRA) that refers to data that is connected to but does not directly identify a person — and that, if made public, may cause harm. Personal information, including a person's passport number, social security number, driver's license, account credentials, credit card numbers, along with the required security codes or passwords allowing access to the financial account, accurate geolocation, beliefs in religion or philosophy, and even trade union memberships, as well as the contents of a person's email, text messages, etc., is included in SPI [14].

3 Implementation

Our proposed solution's architecture is depicted in Fig. 1. Our solution utilizes a custom Node.js scraper to grab pastes from a variety of paste sites. Scraping has gained popularity due to the ease it offers in extracting information from target webpages. It presents the

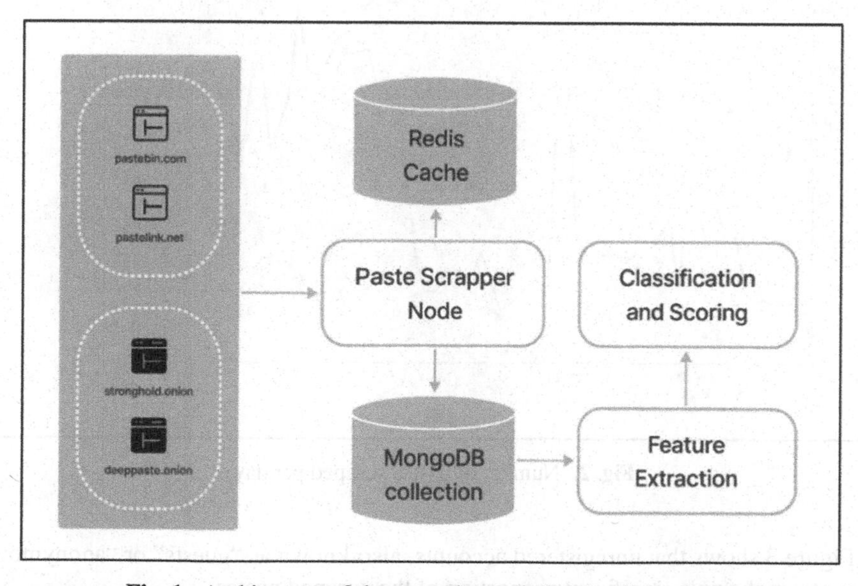

Fig. 1. Architecture of the scrapping and classification system.

information in a structured format with no manual intervention [8]. These pastes are then kept in a MongoDB cloud instance, with the paste IDs also stored in a Redis cache to check and avoid scraping duplicate pastes. After scraping the pastes, they are read by a Python3 NLP based feature extractor, which extracts numerous sensitive information features. These are then used to provide a quantitative score to the paste and to grade it.

Table 1 illustrates the amount of paste scraped. Due to the nature of the dark web paste sites, they are rather unstable, but even though the number of scraped pastes is nearly equal across all four paste sites.

Table 1. Pastes used for extracting features (September 28th, 2021 - December 15th, 2021)

Paste site	Number of pastes
Pastebin	12,316
Pastelink	15,590
Stronghold	10,672
Deeppaste	8,146

The number of pastes scraped by the custom scrapper per day for all sites is shown in Fig. 2. There are occasional decreases in the number of pastes scraped, as well as significant random spikes, but this is to be expected as traffic on the sites changes.

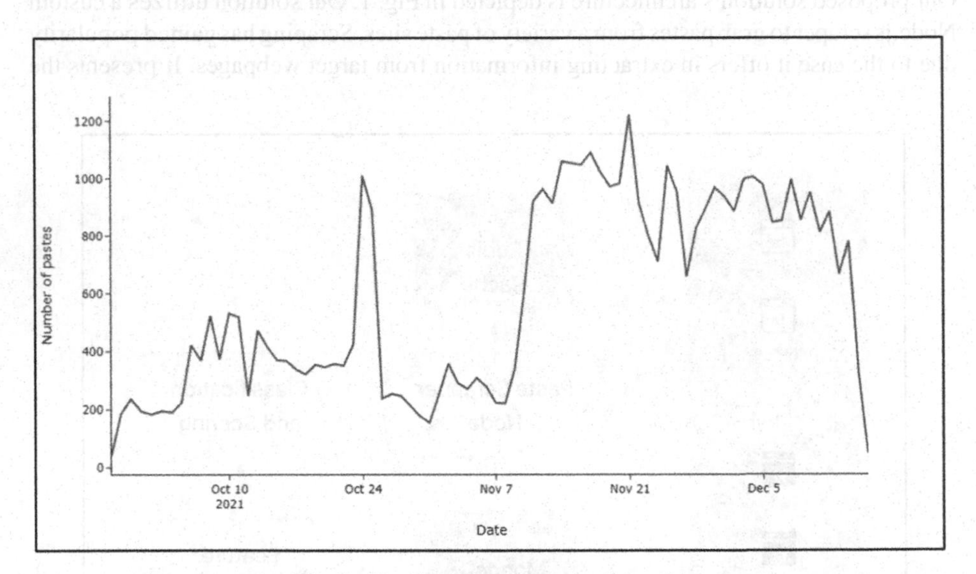

Fig. 2. Number of pastes scraped per day

Figure 3 shows that unregistered accounts, also known as “guests” or “anonymous” accounts, make up a significant proportion of the scraped pastes.

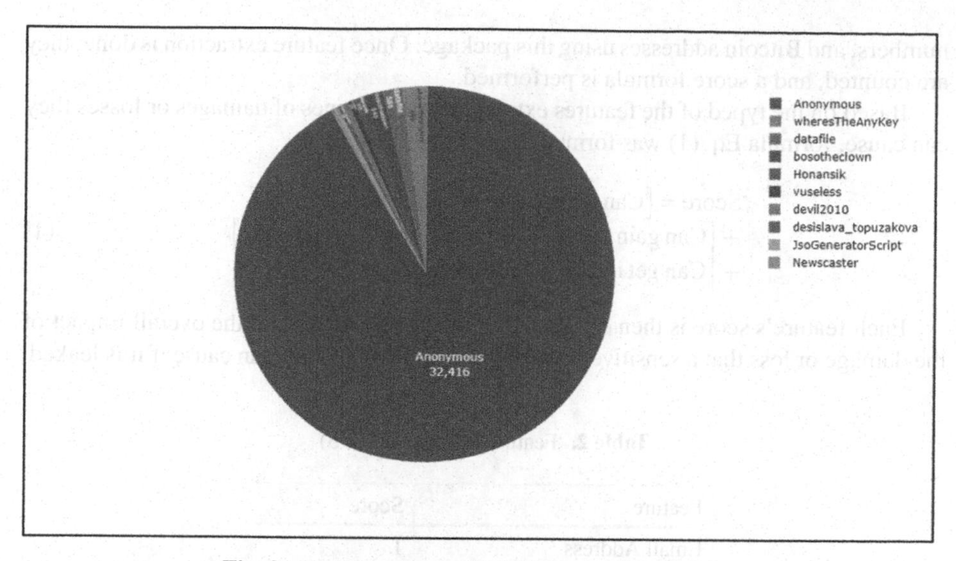

Fig. 3. Top 10 authors with highest number of pastes

Figure 4 shows the number of total features extracted for each paste site. We can see from it that phone numbers are the most common feature found, followed by email addresses, which are mostly popular only on dark web paste sites.

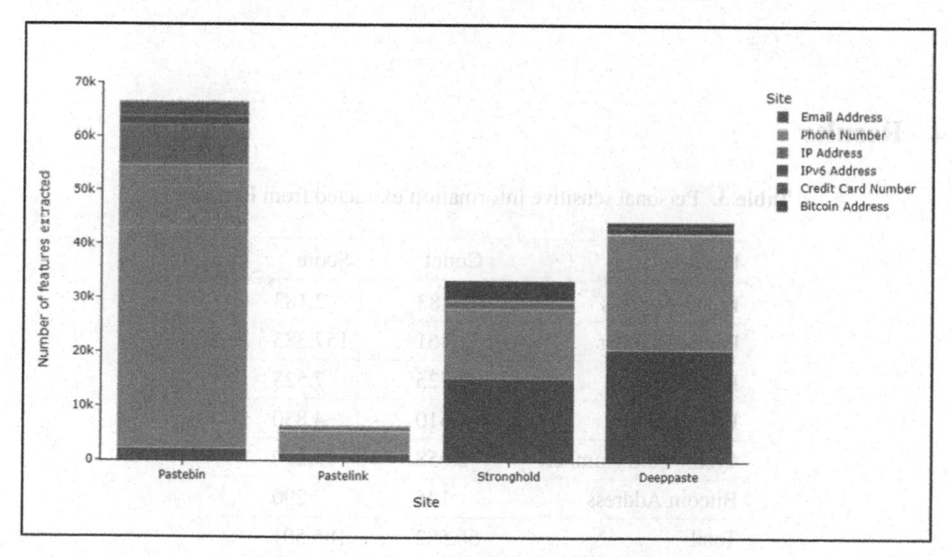

Fig. 4. Features extracted for each paste site

The CommonRegex module for Python was used to extract sensitive information. CommonRegex is a library of frequently used regular expressions packaged in an easy-to-use interface. We retrieved email addresses, phone numbers, IP addresses, credit card numbers, and Bitcoin addresses using this package. Once feature extraction is done, they are counted, and a score formula is performed.

Based on the typed of the features extracted and the types of damages or losses they can cause, formula Eq. (1) was formed.

$$\begin{aligned} \text{Score} &= [\text{Can lead to the physical address? (2)}] \\ &+ [\text{Can gain access to financial information? (2)}] \\ &+ [\text{Can get identity information? (1)}] \end{aligned} \quad (1)$$

Each feature's score is then calculated in the Table 2 based on the overall impact of the damage or loss that a sensitive personal information feature can cause if it is leaked.

Table 2. Feature scores (out of 5)

Feature	Score
Email Address	1
Phone Number	3
IP Address	1
IPv6 Address	3
Credit Card Number	5
Bitcoin Address	2

4 Results

Table 3. Personal sensitive information extracted from Pastebin

Feature	Count	Score
Email Address	2,183	2,183
Phone Number	52,461	157,383
IP Address	7,525	7,525
IPv6 Address	1,610	4,830
Credit Card Number	2,658	13,290
Bitcoin Address	145	290
Total	66,582	185,501

Tables 3, 4, 5, 6 shows the number of each type of SPIs extracted and their relative scores for the four monitored paste sites. This is summarized in the Table 7.

Table 4. Personal sensitive information extracted from Pastelink

Feature	Count	Score
Email Address	1,408	1,408
Phone Number	4,057	12,171
IP Address	61	61
IPv6 Address	13	39
Credit Card Number	700	3,500
Bitcoin Address	214	428
Total	6,453	17,607

Table 5. Personal sensitive information extracted from Stronghold

Feature	Count	Score
Email Address	15,249	15,249
Phone Number	12,793	38,379
IP Address	1,440	1,440
IPv6 Address	30	90
Credit Card Number	164	820
Bitcoin Address	3,616	7,232
Total	33,292	63,210

Table 6. Personal sensitive information extracted from Deeppaste

Feature	Count	Score
Email Address	20,626	20,626
Phone Number	21,377	64,131
IP Address	9	9
IPv6 Address	12	36
Credit Card Number	143	715
Bitcoin Address	2,145	4,290
Total	44,312	89,807

After the analysis, the accompanying data in Table 7 demonstrate the total score for surface web paste sites are greater than the score for dark web paste sites. Additionally, it was also noticed that the "Phone Number" was the most significant feature of these scores.

Table 7. Overall scores

Paste Site	Overall Score
Pastebin	2.78605
Pastelink	2.72850
Stronghold	1.89865
Deeppaste	2.02670

It is noticed that most of our traditional perceptions of the dark web are inaccurate; the dark web is becoming less popular as a means of leaking or compromising sensitive information. This can only indicate one thing: the surface web's effectiveness for dumping compromised sensitive data has increased significantly.

5 Conclusion

The dark web has long been assumed to be the primary platform for sharing data breaches and sensitive data leaks. Due to the rapid evolution of the internet over the last decade, it is critical to investigate how the level of sensitive data found on the surface and dark web differs. By testing the level of sensitivity found on various paste sites, this study determined that the surface web is becoming more prominent in sharing data breaches and sensitive information leaks. Phone numbers were also identified to be the most leaked information. Future studies of sensitive data on paste sites should always consider both the surface and dark web. Furthermore, while this study assigned a calculated fixed score to each paste site, more research is needed to understand how the trend of sensitive information leak dumps evolves over time and across different types of sites, such as hacking forums and shops that sell sensitive personal data.

References

1. Tundis, A., Ruppert, S., Mühlhäuser, M.: A feature-driven method for automating the assessment of OSINT cyber threat sources. Computers Security **113**, 102576 (2022)
2. Kavallieros, D., Myttas, D., Kermitsis, E., Lissaris, E., Giataganas, G., Darra, E.: Using the dark web. In: Dark Web Investigation, Springer, Cham, pp. 27–48 (2021)
3. Vahedi, T., Ampel, B., Samtani, S., Chen, H.: Identifying and categorizing malicious content on paste sites: a neural topic modeling approach. In: 2021 IEEE International Conference on Intelligence and Security Informatics (ISI), IEEE, pp. 1–6 (2021)
4. Guo, Y., Liu, J., Tang, W., Huang, C.: Exsense: Extract sensitive information from unstructured data. Computers & Security **102**, 102156 (2021)
5. Güven, G., Yusuf, E., Boyaci, A., Aydin, M.A.: A novel password policy focusing on altering user password selection habits: a statistical analysis on breached data. Computers & Security **113**, 102560 (2022)
6. Vinayakumar, R., Soman, K.P., Poornachandran, P., Mohan, V.S., Kumar, A.D.: ScaleNet: scalable and hybrid framework for cyber threat situational awareness based on DNS, URL, and email data analysis. J. Cyber Security Mobility **8**(2), 189–240 (2019)

7. Mukundan, P.M., Manayankath, S., Srinivasan, C., Sethumadhavan, M.: Hash-One: a lightweight cryptographic hash function. IET Information Security **10**(5), 225–231 (2016)
8. Bhardwaj, B., Ahmed, S.I., Jaiharie, J., Dadhich, R.S., Ganesan, M.: Web scraping using summarization and Named Entity Recognition (NER). In: 2021 7th International Conference on Advanced Computing and Communication Systems, Vol. 1, pp. 261–265. IEEE (2021)
9. Faizan, M., Khan, R.A.: Exploring and analyzing the dark Web: A new alchemy. First Monday (2019)
10. Wagner, A., Wessels, N., Buxmann, P., Krasnova, H.: Putting a price tag on personal information-a literature review. In: Proceedings of the 51st Hawaii International Conference on System Sciences (2018)
11. Samtani, S., Kantarcioglu, M., Chen, H.: Trailblazing the artificial intelligence for cyber-security discipline: a multi-disciplinary research roadmap. ACM Trans. Manage. Inf. Syst. (TMIS) **11**(4), 1–19 (2020)
12. Punj, G.N.: Understanding individuals' intentions to limit online personal information disclosures to protect their privacy: implications for organizations and public policy. Inf. Technol. Manage. **20**(3), 139–151 (2018)
13. Schwartz, P.M., Solove, D.J.: The PII problem: privacy and a new concept of personally identifiable information. NYUL rev. **86**, 1814 (2011)
14. Rothstein, M.A., Tovino, S.A.: California takes the lead on data privacy law. Hastings Cent. Rep. **49**(5), 4–5 (2019)
15. Du, P.Y., Zhang, N., Ebrahimi, M., Samtani, S., Lazarine, B., Arnold, N., Chen, H.: Identifying, collecting, and presenting hacker community data: Forums, IRC, carding shops, and DNMs. In: 2018 IEEE international conference on intelligence and security informatics (ISI), pp. 70–75. IEEE (2018)

Video Descriptor Using Attention Mechanism

Stuti Ahuja, Aftaabahmed Sheikh(✉), Shubhadarshini Nadar, and Vanitha Shunmugaperumal

SIES Graduate School of Technology, Nerul, Navi Mumbai, India
stutia@sies.edu.in, {aftaabsit118,shubhadarshininit118, vanithasit118}@gst.sies.edu.in

Abstract. This paper is concerned with creation of natural language descriptions using video as an input. Auto-captioning of videos with natural language is a critical challenge of video comprehension. Compared to image, videos have specific time-based structure and various spatial information. In today's world, data is available in different forms and is the most valuable asset. But storing the data in the form of video is not feasible as it will consume huge amount of storage space and extracting any information from such a dataset will cost more processing power. To overcome this, extracting information from the video and storing it in the form of text will allow more efficient usage of storage space. This information can be later processed according to the need of the user. With the objective of generating text from videos we proposed a system which will use a combination of Convolution Neural Network and Recurrent Neural Network followed by Attention mechanism which will provide the text outputs from image and finally the text will be summarized in such a way that it provides a clean description of complete video, the duration of videos is now limited to 1–2 min. The model is tested on Microsoft COCO dataset.

Keywords: Video · Video descriptor · Captioning · CNN · RNN · Attention mechanism · Force teaching · Natural language processing

1 Introduction

In today's world Computer vision technology is in a great demand and it is expected that it could be a future technology and will be helpful in almost every type of field. Making computer to think like human, see like human and make it act like human is the most awaited technology and people across all domains are finding different types of technique and solution for this problem. Computer vision can be divided into 3 different problem statement one is to provide text input and make computer act according to it, second is to provide image input, describe image and make computer act according to it and finally providing a video input and extracting information and storing that in a natural language text so that even computer should understand a video just like how human understand a video. We have focused on the third part of computer vision and tried to find an optimized and improved approach from pre-existing systems so that

M. Singh et al. (Eds.): ICACDS 2022, CCIS 1613, pp. 168–178, 2022.
https://doi.org/10.1007/978-3-031-12638-3_15

the computer can learn and analyse video without human guidance in a very efficient manner.

Auto-captioning of videos with natural language is a critical challenge of video comprehension. Compared to image, videos have specific time-based structure and various spatial information. In today's world, data is available in different forms and is the most valuable asset. But storing the data in the form of video is not feasible as it will consume huge amount of storage space and extracting any information from such a dataset will cost more processing power. To overcome this, extracting information from the video and storing it in the form of text will allow more efficient usage of storage space. This information can be later processed according to the need of the user. This technology can be used in wide domain range from video analysis for security to robotic eyes. It can be used for computer vision where it can be used for smart blind stick so that the video descriptor can give detailed description of scene in front in real time. This system can be very helpful in domains such as security, scanning investigation footage, to describe huge length videos and store it in text so that the video can be understood quickly without spending much time watching the video.

2 Literature Survey

To get a brief idea regarding the existing methods for video caption we reviewed a paper named Video Captioning using Deep Learning: An Overview of Methods, Datasets and Metrics which had all preexisting models and techniques that can be used for captioning of video. This paper is written by M. Amaresh and S. Chitrakala [1]. End-to-end framework or Compositional framework used for both Image captioning as well as in Video Captioning a paper proposed by Vinayals [2] developed an end-to-end framework to generate image descriptions by replacing the RNN encoder with a CNN encoder, which produces a superior image representation for textual descriptions. The suggested Neural Image caption model is a single joint model which trains the CNN for image categorization tasks. The RNN decoder receives the last hidden layer from the CNN as an input, which is utilised to construct the image's textual description. The probability of the target text is increased by utilising stochastic gradient descent to train the images. Since we wanted to do our video processing and labelling in very lest cost so as to reduce the computational time and workload, we reviewed some papers for feature extracting methods.

A paper by Fudong Sun et al. [3]. Where initially the video is downloaded from the internet. And later is processed to get desired features of that particular video. The main motive of the paper is to label each feature of the video with the help of image processing. The technique used for this process was in the following order – (a) Color feature extraction- here the several parameters of color value composition of image need to be captured which is analyzed in the form of histogram. (b) Shape feature extraction – in this process the image is converted to gray-scale image and then the edges of the objects are defined. (c) Texture feature extraction- using frequency spectrum method the type of surface (fine, coarse, smooth, rippled, irregular, or lineated) is identified. The above-mentioned steps are done for continuous stream of images and the result obtained from all the images is then analyzed after which the dominant features of the video are labelled.

Muhammad Usman Ghani Khan et al. [4], proposed a system which constructs natural language tags for video information management which is better than the keyword-based tagging. In their approach, they develop a dataset manually consisting of natural language descriptions of video segments from a small subset of TREC video data. Their work comprises results from experiments pertaining to various categories instead of one. With the help of greedy string tiling algorithms for refining the description, they have combined language models, paraphrasing, and removing redundant sentences. They have focused more on the framework to create descriptions based on HLF's that are extracted from a video stream. Experiments were conducted which can be broadly classified into three categories viz. Automatic evaluation using rogue, task-based evaluation, and application of work to other state of the art datasets. As a result of the experiments, they have concluded that NLP makes it feasible to generate rich, fluent descriptions that arc detail oriented. Their future work focuses on group detection and complex human interaction with objects and behavioral models.

In paper [5], Sadiq H. Abdulhussain et al. proposed a system for extracting features in the most efficient way. Their paper stated as a video consists of a sequence of events made up of several scenes A scene is a collection of semantically connected and temporally contiguous pictures captured from multiple camera angles, allowing us to do shot level analysis, which is an important step in the organization of semantic content. They employed the bock processing method, which involves partitioning an image I of size N1 × N2 pixels into blocks of varying sizes where block size is B1 × B2. It just requires matrix multiplication and is transform independent. This process reduces the cost of extracting features by 30%-40% compared to traditional methods.

If we see a video the video consists of foreground and background and in most cases the background in videos are still and just foreground has moving objects so differentiating foreground and background from a video is still a huge process so to deal with this Thoung-Khanh Tran and his team proposed a method in paper Foreground Extraction in video based on Edge-based Robust Principal Component Analysis [6], The proposed system was based on Robust principal component analysis (RPCA) which is the strong method to extract motion objects from the video frames, in that model it includes two constraints one is for foreground and one is for background. Combined this method with Edge-based RPCA we can extract a clear part of foreground. They deduct the current frame from the low rank component created by the RPCA procedure, which provides them a motion object in frame, but it still includes many noises so to deal noise they added a furthermore step they first do a binary subtraction of image between the required frame and the background model which also have noises in edge the do a XOR operation to cancel out all edge pixels.

Now since we were clear on how to extract features and label it, we reviewed paper which were more targeted on captioning algorithms. Niveda Krishnamoorthy et al. [7] wrote a paper Identifying the optimum subject-verb-object (SVO) triplet for summarizing realistic YouTube videos in order to provide natural-language summaries of short videos. It uses a basic template-based method to candidate phrases given the SVO triplet, which are then rated using a machine translation model trained on web-scale data to achieve the best overall description. We integrate these often-noisy detections with a real-world betway derived from SVO triplets extracted from large-scale online corpora.

They have used Stanford's dependency parser to obtain verbs describing each video and to train their descriptions. The verbs are then arranged in hierarchical order. Followed by object detection, activity recognition, text mining, verb expansion and finally content planning to determine the best possible SVO by combining vision detection and NLP scores. Overall, the results demonstrate that using text-mined knowledge to enhance the selection of an SVO that best describes a video has a significant advantage.

A paper named Generating Natural Language Description of Human Behavior from Video Images by Atsuhiro Kojima et al. [8], they proposed a fresh approach to generate description of human etiquette appearing in real life images using a model-based approach by taking out features of human movements and generating verbs. Starting with extraction of the human region from every video frame, they extract only the head region using chromaticity averages of previously calculated hair and facial skin colors. Movements of a person can be divided into some parts as person does a single act. Here to indicate conceptual features they use a preset value of conceptual features. Followed by which the most suited verb form explains human behavior based on the value calculated with the help of the values predetermined. To produce natural language text, they use an automatic translation methodology based on Fillmore's case grammar, which involves applying case structure patterns of verbs and syntactic rules to behavioral expression. For future development in the field of more complex behaviors they are developing framework to match with motion of hands and objects.

Development of a user-friendly human-robot interface for any service robotics application mainly in particular for the ability to generate natural language descriptions for the scene it observes has been carried out in this paper by Silvia Cascianelli et al. [9]. They have proposed an improved architecture for NVLD based on GRU units to save the time taken for training and present a dataset which features a wide range of contexts for service robot applications. They accomplished it by using a deep recurrent neural network design based entirely on the gated recurrent unit paradigm. The robot can produce full phrases that describe the scenario in visual sequences. The main challenge they encountered was in NLVD to overcome this, they focus on the NLVD tasks and present a full GRU encoder-decoder architecture to address it. This proposed approach is faster to train and less memory consuming than other State-of-the-Art algorithms. The subject of their future work is to Their future work pertains to ability to better cope with videos of greater length. Although the same technique can be used for this purpose by breaking the video into continuous sequences, dealing with much longer videos are in interest as well as to develop effective solutions to this problem.

Video captioning can be used for many field one of which is the surveillance and security. A paper which proposes a method to report hours of surveillance video in a pattern-based text log which was written by Samarjit Kar et al. [10]. Proposers main aim was to generate the textual log for any surveillance scene having some prior region information, they used three types of information to generate the descriptions, namely frame-level information like short-term behavior, long-term scene activities like moving objects e.g., car, pedestrian. The behavior of targets is shown with the help of graph-based patter discovery and then finally a sentence is generated by direct mapping to template provided by human experts. Algorithms used by them was CNN for guided Object tagging, to tag the still regions the first represent a trajectory denoted as $\tau = \{Rentry, Rexit\}$. For creation of natural text they employed a template-based strategy to represent targets and behavior for natural text production. First, they track the target, and then object-specific information such as target type, speed, and color are extracted hierarchically. The text generated was much accurate compared to all older methods and papers which are mentioned. Every paper has a different approach to deal with Natural Language description generations in a paper named Translating Video Content to Natural Language Description by Marcus Rohrbach et al. [11]. The paper suggests that the best way to convert visual information into linguistic expression is to use a 2-step approach. In first step they start learning an intermediate SR (Semantic representation) using a probabilistic approach and the SR is given to generate a sentence. They try to do the verbalization from a parallel training corpus so that only most relevant information should be verbalized to do this they use statistical machine translation. To extract SR from visual content they used Loopy Belied Propagation (LBP) method, generating SR from visual content means, suppose if a video has some objects, then finding the relation between those objects is known as SR extraction and then to translate SR to a description, they first translate SR to a simple English word which is used more often e.g., if SR has a word as HOB we convert it to Stove which is much easier to form a sentence for every types of SR their system tries to relate and try to predict a verb which will be best suited according to the video. To compare their final accuracy, they compared their output with BLUE scores which are available for all approaches, by replacing the raw features with a higher-level representation of attribute classifier and CRF prediction they got improvement of about 16% and the overall prediction of their SR with baseline values on noisy data was around 22.1%. And if the video is more of a human description, then they got improvement of about 50%.

In the paper by Ziwei Yang et al. [12] named Multirate Multimodal Video Captioning, the authors have proposed a Multirate Multimodal Approach for video captioning. The videos contain different modality cues hence they have proposed a fusion method. Using Multirate GRU for encoding video frames and combine visual motion and other information for a well-constructed representation which is then fed into RNN-based language model for generating descriptions. The video representation incorporates frame-level feature, clip level feature and video topic feature where the information from tagged videos are collected and utilized. The fusion method stated before involves fusion of all these three features to form the final video representation. Their dataset included 13000 video clips extracted from YouTube of which some were used for training and the rest for testing purposes. Their experiment resulted in the conclusion that fusing multiple modalities improves the performance of the captioning model.

3 Proposed System

In this paper we are proposing a system which will take video as an input and will give an output as text summary of that video, In today's world, data is available in different forms and is the most valuable asset. But storing the data in the form of video is not feasible as it will consume huge amount of storage space and extracting any information from such a dataset will cost more processing power. To overcome this, extracting information from the video and storing it in the form of text will allow more efficient usage of storage space.

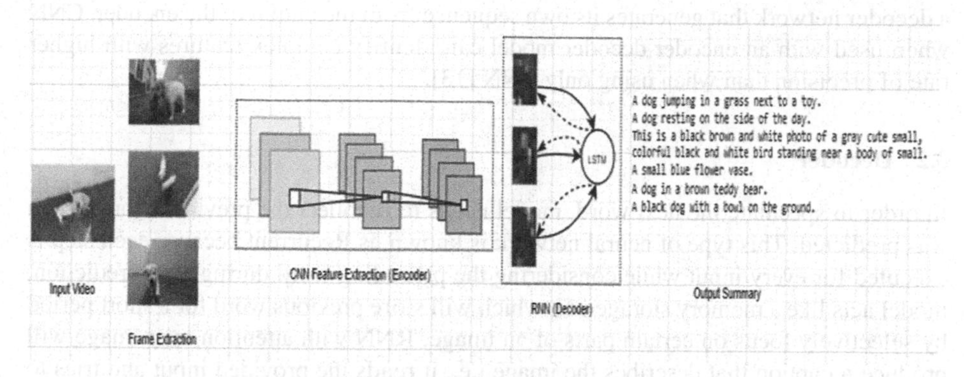

Fig. 1. Workflow of Video Descriptor Model

The input video will be first preprocessed and then we will be extracting frames of video in every 1 s intervals. Since 1 s consists of 30 or 60 frames, and most of the frames in 1 s is almost similar to each other. The frames will be extracted using OpenCV library where we will be first calculating the number of frames present in a 1 s and we will be saving one of these frames in JPEG format. The These extracted frames will be then passed into our Attention based Image captioning model which will provide captions for each of the extracted frames after frame extraction Natural language processing will be done to summarize the caption generated for each frame and final video summary will be displayed as an output. The flow is showcased in Fig. 1.

Our proposed system represents two permutations of our attention-based model that allows us to see what parts of the image the model focuses on as it generates a caption. It is a mechanism by which a network can weigh features by the level of importance to a task and using this weighting to help achieve the task. There are 2 types of attention, hard and soft, in Hard attention the model is forced to focus only on the required area and the rest all area is masked with black filter. So, in implementation we will be focusing on the non-blur areas. Whereas in soft attention the model focuses on specific parts of an image and then generates captions describing what it is focusing on. Our model uses combination of hard and soft attention which slightly black out the rest of the image part, which helps the model to focus on the certain area of the image which provides higher accurate detection of images. Attention plot can be referred from Fig. 4

3.1 Encoder

Our method focuses mainly on attention-based models which can provide better accuracy compared to all other types of captioning. We use Convolutional Neural Networks to encode the image. Each of the CNN layers has filters which detect patterns, and these features are extracted from a convolutional layer, which allows the attention in the decoder to focus on spatially relevant portions of the input image. The features that are extracted is passed through the CNN encoder that comprises of single fully connected layer which understands the given sequence and creates a smaller representation (features) holds information regarding the input and this representation is then forwarded to a decoder network that generates its own sequence from the output of the encoder. CNN when used with an encoder-decoder model can identify complex features with higher rate of precision than when using only CNN [13].

3.2 Decoder

In order to speculate the next word, model needs to recollect the previous word which was predicted. This type of neural network is known as Recurrent because each step is executed for every input while considering the preceding word during next prediction, model acts like a memory storage unit which will store previous word for a short period by selectively focus on certain parts of an image. RNN with attention over image will produce a caption that describes the image i.e., it reads the provided input and tries to predict the variable target sequence. Just like how the encoder retrieves features from an image, the decoder learns to generate its sequence with a process of training called teacher forcing. This process helps to decide what word would be passed as the next input to the decoder.

3.3 NLP

This model will be further enhanced by combining all the caption outputs which will be then fed into the Natural language processing algorithm which will remove any grammatical mistakes and then it will provide a better form of description which will be simple and readable format.

4 Training Procedure

Dataset used for training the model was MS COCO dataset which has total image of 82000 along with 5 different captions of each of the image. The software requirements for our model are TensorFlow 2.5 with NumPy version of 1.19.0 due to heavy dataset it is recommended to have a GPU onboard so as to train the model quickly for testing purpose the model was trained with random 30000 images.

To keep images of same features we did some preprocessing on image where we resized all the images to 299 x 299 pixels and normalized the image pixels in range of −1 to 1. To train the model Keras InceptionV3 model was used which is pretrained on Imagenet to classify each image, and extract features from Last convolution layer. The output of last convolution layer is 8 x 8 x 2048. Above all the part was done for image processing after this step caption tokenization was done where all the text caption was converted into integer sequence using Text Vectorization Layer. Each word was converted into tokens and top 15,000 words was selected. After tokenization word to index and index to word mapping was done. The model was trained with 20 epochs refer Fig. 2 for loss plot. The model works in following way when a input image is given the model converts the image into 299 x 299 size and then it is passed to convolution neural network of InceptionV3 where the final layer out given by CNN is (8, 8, 2048) which is then compressed to shape of (64, 2048) and this vector is sent as an input for CNN Encoder which is made up of only one fully interconnected layer, and finally the output of CNN encoder is passed to RNN decoder which examines the image in order to predict the next word.

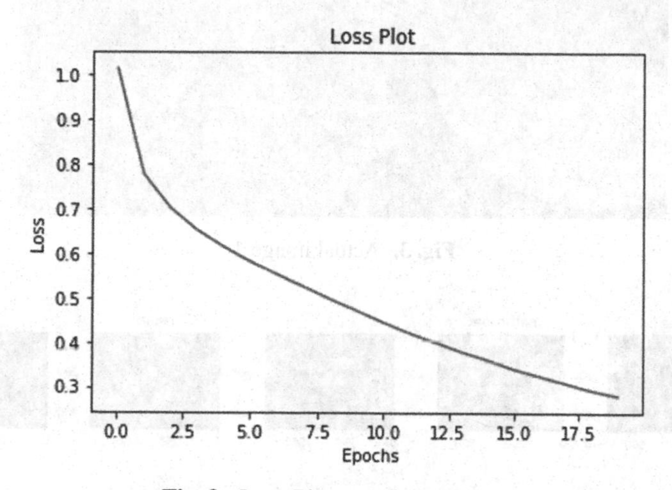

Fig. 2. Loss Plot on training dataset

Time taken by model to train on 30,000 images with 78,000 image captions was around 19.5 h. Time taken by each epoch was around 500–700 s. The model was trained on a computer with following specification Intel i5 processor (2.4 Ghz) 8 GB ram, Nvidia 1650 4 GB GPU. The accuracy of this model can be increased when trained on complete dataset with much better configuration systems.

5 Results

For real world evaluation of our model, we collected some random videos from internet and passed through our model which provided an accuracy of approx. 60% compared to real caption. Figure 3 shows one frame of a random video chosen from internet Fig. 4 shows the attention plot of that frame and Fig. 5 shows the comparison of actual and predicted caption. Similarly, another random video was referred for evaluation refer Fig. 6, 7, 8 for results of that frame.

Fig. 3. Actual image 1

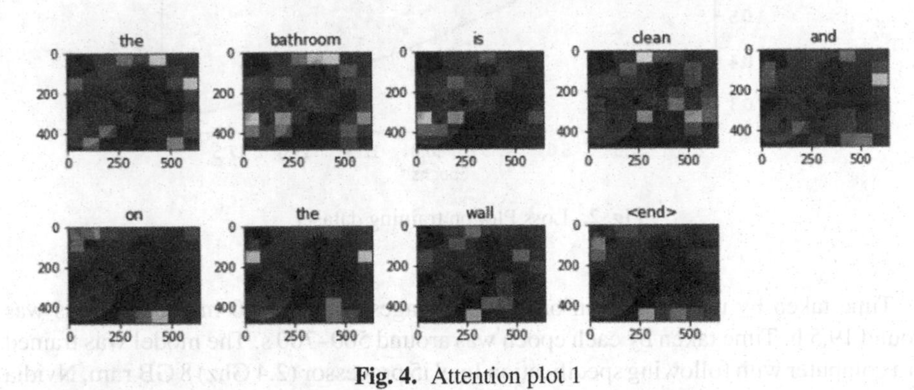

Fig. 4. Attention plot 1

Real Caption: <start> a bathroom sink and shower stall with a basket of toiletries <end>

Prediction Caption: the bathroom is clean and on the wall <end>

Fig. 5. Real caption vs predicted caption

Fig. 6. Actual image 2

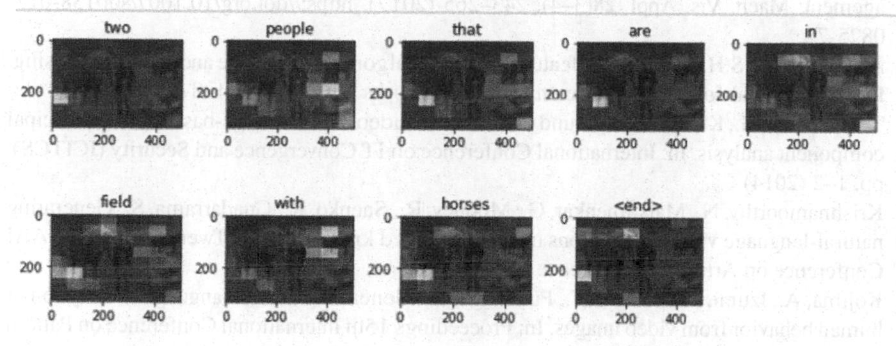

Fig. 7. Attention plot 2

```
Prediction Caption: two people that are in field with horses <end>
```

Fig. 8. Predicted caption 2

6 Conclusion

Video Descriptor is a project where if we provide videos of any kind, it will generate a natural language description of that video which will be in a simple English format so that any naïve user can also understand the meaning of a video without watching complete video. In this paper we discussed about different types of technologies which can be used for implementing this system. We also found some pre-existing system for every pre-existing system all have their own advantage and disadvantage we tried to use their favorable points and we tried to overcome the drawbacks of the pre-existing models. This model now can provide the output which has a **bleu4 score of 0.66**. In conclusion this system can be used for every type of domain which will help user to save time and learn more and can also save the description of any videos so that he/she can refer the text and get a brief idea of any video which they saw long time ago. This work can be extended in future for surveillance and robotic automation, where summarized text of videos can be converted to speech.

References

1. Amaresh, M., Chitrakala, S.: Video captioning using deep learning: an overview of methods, datasets and metrics. In: International Conference on Communication and Signal Processing (ICCSP). 0656–0661 (2019)
2. Vinyals, O., Toshev, A., Bengio, S., Erhan, D.: Show and tell: a neural image caption generator. In: Proceedings of the IEEE Conference on Computer Vision and Pattern Recognition, pp. 3156–3164 (2015)
3. Sun, F., Shi, M., Lin, W.: Feature Label Extraction of Online Video International Conference on Computer Science and Electronics Engineering, pp. 211–214 (2012)
4. Khan, M.U.G., Gotoh, Y.: Generating natural language tags for video information management. Mach. Vis. Appl. **28**(3–4), 243–265 (2017). https://doi.org/10.1007/s00138-017-0825-7
5. Abdulhussain, S.H., et al.: A fast feature extraction algorithm for image and video processing. In: International Joint Conference on Neural Networks (IJCNN), pp. 1–8 (2019)
6. Tran, T., Bui, N., Kim, J.: Foreground extraction in video based on edge-based robust principal component analysis. In: International Conference on IT Convergence and Security (ICITCS), pp. 1–2 (2014)
7. Krishnamoorthy, N., Malkarnenkar, G., Mooney, R., Saenko, K., Guadarrama, S.: Generating natural-language video descriptions using text-mined knowledge. In: Twenty-Seventh AAAI Conference on Artificial Intelligence (2013)
8. Kojima, A., Izumi, M., Tamura, T., Fukunaga, K.: Generating natural language description of human behavior from video images. In: Proceedings 15th International Conference on Pattern Recognition, pp. 728–731 (2000)
9. Cascianelli, S., Costante, G., Ciarfuglia, T.A., Valigi, P., Fravolini, M.L.: Full-GRU natural language video description for service robotics applications. In: IEEE Robotics and Automation Letters, vol. 3., pp. 841–848 (2018)
10. Ahmed, S.A., Dogra, D.P., Kar, S., Roy, P.P.: Natural language description of surveillance events. In: Information Technology and Applied Mathematics, pp. 141–151. Springer, Singapore https://doi.org/10.1007/978-981-10-7590-2_10
11. Rohrbach, M., Qiu, W., Titov, I., Thater, S., Pinkal, M., Schiele, B.: Translating video content to natural language descriptions. In: IEEE International Conference on Computer Vision, pp. 433–440 (2013)
12. Yang, Z., Xu, Y., Wang, H., Wang, B., Han, Y.: Multirate multimodal video captioning. In: Proceedings of the 25th ACM International Conference on Multimedia, pp. 1877–1882 (2017)
13. Ji, Y., Zhang, H., Zhang, Z., Liu, M.: CNN-based encoder-decoder networks for salient object detection: a comprehensive review and recent advances. Inf. Sci. **546**, 835–857 (2021)

Informative Software Defect Data Generation and Prediction: INF-SMOTE

G. Rekha[1(✉)], K. Shailaja[2], and Chandrashekar Jatoth[3]

[1] Department of Computer Science and Engineering, Koneru Lakshmaiah Education Foundation, Hyderabad, India
gillala.rekha@klh.edu.in

[2] Department of Computer Science and Engineering, Vasavi College of Engineering, Hyderabad, India

[3] Department of Computer Science and Engineering, National Institute of Technology, Raipur, India

Abstract. Highly imbalanced data typically make accurate predictions difficult. Unfortunately, software defect datasets tend to have fewer defective modules than non-defective modules. Synthetic oversampling approaches, namely SMOTE, address this concern by creating new minority defective modules to balance the class distribution before a model is trained. Despite its success, these approaches come with the following shortcomings such as 1) over-generalization problem and generate near-duplicated data instances (less diverse data) due to oversampling of noisy samples, and 2) increasing the overlaps between different classes around the class boundaries. This paper introduces INF-SMOTE (Informative- Synthetic Minority Oversampling Technique), a novel and efficient synthetic oversampling approach for software defect datasets, simultaneously targeting all the shortcomings. INF-SMOTE identifies the informative minority samples that are appropriate for over-sampling. The process is in two way 1.) it identify and remove the noisy and overlapping samples from borderline minority instances based on the sampling seeds, and 2) synthetic samples are generated from the informative minority samples. Experiments were conducted on 12 releases of SDP (Software Defect Prediction) Datasets from the NASA repository. By comparing with the state-of-the-art techniques, we observe that the INF-SMOTE improves the defect prediction performance.

Keywords: Class imbalance problem · Imbalanced classification · Imbalanced datasets · Over-sampling · SMOTE

1 Introduction

The classifier can predict the unknown class with high accuracy if the training dataset is balanced. However, this is not always the case. Software Defect Prediction is an example of an area where the number of non-defective occurrences is much higher than the number of defective occurrences [14,23]. As a

M. Singh et al. (Eds.): ICACDS 2022, CCIS 1613, pp. 179–191, 2022.
https://doi.org/10.1007/978-3-031-12638-3_16

result, the classes are skewed. Based on the values of various attributes existing in the given data, the machine learning algorithms may be used to predict if a piece of software is flawed or not. However, trained on skewed datasets may be biased towards non-defective cases and fail to detect defects [1,9]. Hence, the primary issue with applying machine learning algorithms to such datasets, and as a result, many researchers have proposed various approaches to improve algorithm performance [10,18]. The three primary kinds of strategies developed by the researchers to handle this challenge include sampling-based methods, cost-sensitive approaches, and approaches incorporating ensemble learning [7,19]. The two common and popular sampling techniques are undersampling and oversampling techniques. Cost-sensitive algorithms have varying misclassification costs, resulting in non-uniform training penalties for the classifiers. Ensemble learning is the process of combining the outcomes of many learning algorithms.

Software Defect Prediction (SDP) models frequently include extremely unbalanced data, making it challenging for classifiers to detect faulty instances [6]. Many approaches have recently been presented to resolve this problem, with the over-sampling method being one of the most well-known strategies to fix the problem of class imbalance [13,16]. This strategy generates new faulty instances to balance the amount of defective and non-defective samples. However, the synthetic minority examples should not be too random, as this might cause the classifier to fail to identify the majority class examples, nor should they be too confined around existing minority cases, as this might cause the classifier to overfit the dataset [2]. In most cases, the approaches created to fall short of achieving both requirements. Furthermore, oversampling methods must guarantee that the created instances do not result in an uncertain decision boundary, such as forming a minority sample with the majority of the nearest neighbours belonging to the majority class. It would also have an impact on the performance of the classifier.

Further, after applying the most standard and popular approaches called SMOTE [5], areas with a lot of minority instances acquire a lot more new minority instances, whereas sparse regions remain sparse [8]. Furthermore, the distinction between defective and non-defective classes are ambiguous, making it challenging to categorize classifiers. Therefore, it follows i) new synthetic instances may be non-diverse and spread in a narrow region, making classifiers vulnerable to overtake. ii) The K-nearest neighbour examples evaluated would be in the majority class, causing boundary ambiguity between the majority and minority classes and resulting in noise instances.

To address the shortcomings above, we present a strategy based on a two-way approach: 1. Removal of uninformative samples from minority regions (overlapped and noisy instances). 2. Generation of synthetic minority samples from diverse and informative samples.

Mainly, to prevent new defective instances from being non-diverse and dispersed in a narrow region, the proposed method creates new ones after removing uninformative samples from the existing defective ones.

The remainder of this paper is structured as follows. Section 2 present the literature on the prediction of imbalanced software defect datasets. A detailed description of the proposed INF-SMOTE model is introduced in Sect. 3. The experimental settings and results interpretation are summarized in Sect. 4. Finally, in Sect. 5, conclusions are drawn.

2 Literature Review

Since the 1970s, one of the most widely discussed software engineering research topics has been software defect prediction technologies. With the fast growth of machine learning in recent years, several machine learning approaches have been widely used to enhance SDP performance. Furthermore, faulty modules are significantly less common in software projects than non-defective modules, resulting in a class imbalance problem. This issue has a significant impact on classifier performance.

In SDP, there have been several studies on the problem of class imbalance, which may be classified as sampling approaches, ensemble techniques, and cost-sensitive learning techniques. Sampling methods are often used to balance the datasets, such as random over-sampling or under-sampling strategies. Multiple weak supervised classifiers are combined in ensemble learning methods to produce a more robust supervised classifier. Bagging, boosting, and stacking are three common ensemble learning strategies. Finally, based on the class name, cost-sensitive learning algorithms give varying misclassified costs to instances [17,22]. Over-sampling techniques are one of the sampling strategies that involves replicating existing minority instances or creating new minority instances.

Random oversampling (ROS) is the most straightforward technique, wherein it randomly replicates minority cases to balance datasets. However, the ROS method is prone to focusing on insignificant samples or noise, resulting in over-fitting. To deal with these issues, [5] devised the synthetic minority oversampling method (SMOTE), which uses interpolation to generate new minority instances in the k closest neighbourhoods of the selected one. While SMOTE decreases the risk of over-fitting, it is oblivious to recent occurrences while synthesizing them. Because most imbalanced datasets, in reality, including overlaps between various classes, randomly picking minority examples for sampling will increase the overlap between classes. Even if the dataset is severely skewed, most new instances will be assigned to the majority class, compromising the dataset's separability. SMOTE is also susceptible to noise in addition. Many other algorithms have been developed to address SMOTE's shortcomings above. In [11] the author presented the Borderline-SMOTE (BSMOTE) oversampling technique, which first calculates the distribution of their k nearest neighbours to find the minority border instances, then conducts SMOTE specifically at the borders. In [12] the author presented the Adaptive Synthetic (ADASYN) oversampling technique, which determines the probability of sampling minority occurrences by computing the distribution of their k nearest neighbours. That is, minority examples with a higher number of majority examples in their immediate surroundings are more likely to be chosen. While BSMOTE and ADASYN can mitigate SMOTE's blindness and limit the effects of noise to some level, they may not detect all minority boundary occurrences and may even see unsuitable boundary instances. So motivated by this, we propose a strategy based on a two-way approach: 1. Removal of uninformative samples from minority regions (overlapped and noisy instances). 2. Generation of synthetic minority samples from diverse and informative samples.

Mainly, to prevent new defective instances from being non-diverse and dispersed in a narrow region, the proposed method creates new ones after removing uninformative samples from the existing defective ones.

3 Proposed Method

This section introduces the proposed sample generation from the informative sampling seeds (INF-SMOTE) technique for the imbalanced software defect dataset. By considering the problems encountered by oversampling methods as discussed, we propose the INF-SMOTE approach, which generates the new samples in the data space by considering the informative sampling seeds from minority regions.

Given the binary input dataset 'D', the minority or positive class (Min), the majority or negative class (Maj) is represented as Min $= (mi_1, mi_2, ..., mi_m)$, Maj $= (ma_1, ma_2, ...ma_m)$ and mi_m and ma_m represents the number of minority and majority samples, respectively. The main goal of the proposed approach is to identify and not to consider the noise and overlapping minority instances while generating the new synthetic samples. However, not all minority class instances are important for generating the synthetic instances because some of them cannot provide much discriminate information in the learning process. The primary concern is not to consider such samples as overlapping and noisy samples for synthetic data generation. Accurate identification and elimination of these instances maximize the visibility of the informative minority sampling seeds. Second, we can generate efficient minority samples from the informative minority samples by employing the SMOTE technique.

3.1 Elimination Uninformative Samples

As shown in Fig. 1, the instance x is a borderline minority instance, which has a similar distribution to its nearest neighbour minority instance. The instance N is a noisy instance in the majority region. i.e., this can be taken as a redundant point. In worse cases, if x and N are considered sampling seeds, it may lead to uninformative samples generation. The process of selecting the uninformative samples from minority class is as follows. For each instance $x_i (i = 1, 2, \dots, x_n)$ we calculate its K nearest neighbors represented as $NN_{(x_i)}$ and store them in nearest count table. The neighbors in $NN_{(x_i)}$ can be minority or majority class instances. In the nearest count table, for a given sample the number of majority

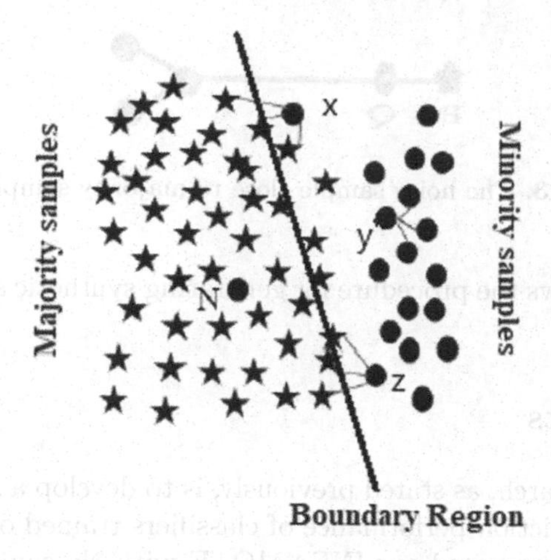

Fig. 1. The possible way of synthetic sample generation process

instances and minority instances that exist among its neighbor is represented as $(x_i)_{maj'}$ and $(x_i)_{min'}$ respectively. Next, to overcome the uninformative samples from the minority region we calculate the difference of each minority sample x_i between the $(x_i)_{maj'}$ and $(x_i)_{min'}$ for each x_i respectively as given in Eq. 1 and store the resultant value in Dist table.

$$Dist_i = (xi)_{maj'} - (xi)_{min'} \tag{1}$$

Now, if $Dist_i$ value is too big, then the sample is closer to the majority region and cannot provide any informative seed when considered for synthetic sample generation. For example, consider Fig. 2. If Q is selected as a sampling seed for artificial sample generation, it leads to redundant and non-informative samples. Similarly, consider Fig. 3, Q may generate noisy samples. To identify the benchmark value for eliminating instances by considering them as uninformative, we calculated the mean mi of distance from every instance x_i to its k nearest neighbours. And let sigma equal to the average m_i plus the standard deviation std of all m_i. Then, the informative samples are identified based samples dist table and the sigma value. For every instance x_i in S, if its dist values are more significant than sigma, consider them uninformative samples for synthetic data generation. Later, only the informative sampling seeds are used in synthetic sample generation using SMOTE technique.

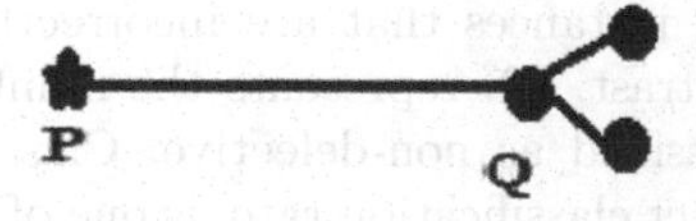

Fig. 2. The redundant sample close to minority samples

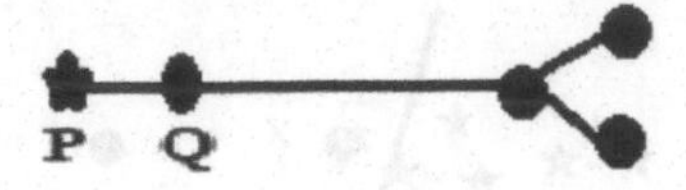

Fig. 3. The noisy sample close to majority samples

Algorithm 1 shows the procedure for generating synthetic samples using INF-SMOTE.

4 Experiments

The goal of our research, as stated previously, is to develop a synthetic approach to improve the prediction performance of classifiers trained on class-imbalanced SDP datasets. We compared our INF-SMOTE with No sampling, SMOTE [5], ADASYN [12], Borderline-SMOTE [11] technique using five classification algorithms: three single classifiers such as Decision Tree (J48) [3], Naive Bayes (NB) [20] and Support Vector Machines(SVM) [15] and two ensemble classifiers namely Adaboost [21] and bagging [4]. It should be emphasized that five classifiers are trained after these sampling strategies provide a balanced training dataset.

4.1 Datasets

We conduct tests on 12 projects from NASA repositories that are often employed in SDP research to create and validate the effect of our INF-SMOTE on imbalanced datasets of SDP. Table 1 lists the detail information of these experimental data description. The first five columns show the module name, language used, number of attributes, total instances, and defect ratio. These projects are written in Java, C, and C++, and the number of cases varies significantly, with most projects having a ratio of instances far below 25%. As a result, these projects may be utilized for evaluating these class-imbalanced techniques.

4.2 Evaluation Metric

In this section, we present true positive (TP), true negative (TN), false positive (FP), and false-negative (FN) measures for software defect prediction. For example, the number of faulty software instances correctly categorized as defective is denoted by TP. In contrast, TN indicates the number of non-defective software instances correctly categorized as non-defective. Likewise, the number of non-defective software instances that are incorrectly identified as defective is denoted by FP. In contrast, FN represents the number of defective software instances incorrectly classified as non-defective. Classification accuracy, commonly known as the correct classification rate, is one of the most basic measures for evaluating the effectiveness of predictive models. It calculates the proportion of correctly categorized samples in the total number of samples. Another metric

Algorithm 1. Identification of informative and Uninformative Samples

1: **INPUT:** Samples of entire dataset 'D'.
2: Identification of noise and redundant samples from the dataset.
3: **Output:** differentiate the samples into informative and uninformative data set.
4: **for** (each instance x_i in the given dataset) **do**calculate K nearest neighbors;
5: The k nearest can be majority or minority or combination of both majority and minority class instance;
6: **end for**
7: **for** (each attribute of minority sample) **do**
8: Calculate distance based on the equation 1
9: **end for**
10: **for** (each sample x_i) **do**
11: compute its mean and SD;
12: calculate the Sum based on mean and SD (α = mean + SD);
13: Now based on the dist value and α label the samples.
14: if dist of x_i is greater than α mark it as uninformative otherwise informative samples
15: **end for**

Table 1. Software defect prediction dataset description

Module	Lang	Number of attributes	Number of instances	Defect%
CM1	C	21	498	9.84
KC1	Java	21	2109	15.46
KC2	Java	21	522	20.5
KC3	Java	39	194	18.56
MC1	C++	38	1988	2.31
MC2	C++	39	125	35.2
MW1	C	37	253	10.67
PC1	C	21	1109	6.94
PC2	C	36	745	2.15
PC3	C	37	1077	12.44
PC4	C	37	1458	12.21
JM1	C	21	10885	19.35

is precision, which is determined by dividing the number of instances classified as faulty (TP + FP) by the number of cases accurately categorized as defective (TP + FP). Furthermore, recall is the proportion of instances accurately categorized as defective (TP) to the number of faulty examples (TP + FN). The F-score is a harmonic mean of accuracy and recall, and it has been utilized in numerous research. By computing trade-offs between TPR and FPR, ROC-AUC determines the area under the receiver operating characteristic (ROC) curve. We evaluated the performance using F-measure and AUC-ROC metrics for the proposed model (Table 2).

Table 2. Performance metrics with its formula

Metric	Formula
Recall	$\frac{TP}{TP+FN}$
Precision	$\frac{TP}{TP+FP}$
Specificity	$\frac{TP}{TP+FN}$
Sensitivity	$\frac{TN}{TN+FP}$
F-Measure(FM)	$\frac{2*Precision,\times Recall}{Precision+Recall}$
ROC(AUC)	$\frac{Specificity+Sensitivity}{2}$
GM (GM)	$\sqrt{Specificity \times Sensitivity}$

4.3 Experimental Results

To prove the experiment's effectiveness, we apply five classifiers that have been used for SDP. These classifiers include DT, NB, SVM, AB and BB. Note that our experimental environment is Python3.6 and WEKA. Here we want to demonstrate how effective is our proposed INF-SMOTE over existing techniques in addressing the problem of class imbalance.

As mentioned, the experimental results are compared with existing oversampling methods such as No sampling, SMOTE [5], ADASYN [12], Borderline-SMOTE [11] technique using F-measure and AUC-ROC metrics based on five classifiers over 12 projects. Table 3 shows the results using five classification algorithms based on F-measure and AUC-ROC metrics.

Figures 4 and 5 show that the proposed approach produces the highest AUC-ROC value using an SVM classifier on MC1, PC2, CM1, MW1, PC5, PC4, PC3 projects. In the same way Fig. 6 and 7 shows F-measure with 99.85% on PC2 dataset and also for CM1, MC1, PC5, PC4, MW1, PC3, MC2 the proposed method achieved 98.39%, 98.27%, 96.74%, 94.49%, 93.45%, 92.80% and 86.37% respectively using J48 classifier. We conclude from the above result that the proposed algorithm achieves the best AUC-ROC and F-Measure when applied to all the classification algorithms. We can conclude that our proposed algorithm can build a more robust classifier for software defect prediction.

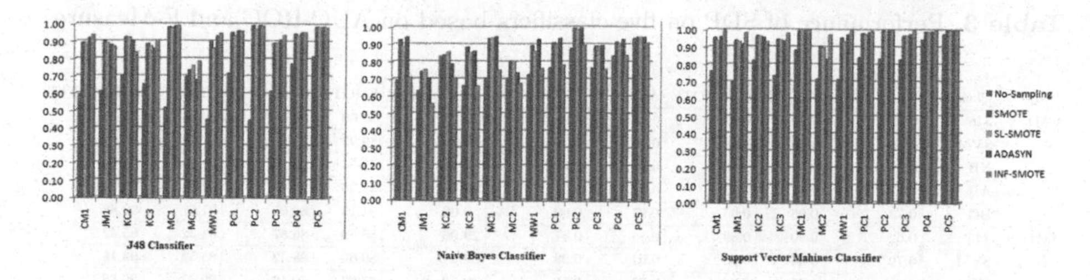

Fig. 4. The comparison of various sampling methods with INF-SMOTE on three single classifiers using AUC-ROC

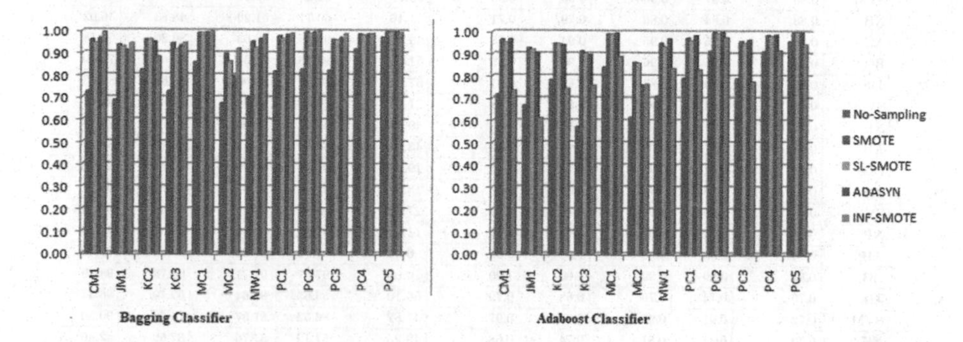

Fig. 5. The comparison of various sampling methods with INF-SMOTE on two ensemble classifiers using AUC-ROC

The existing methods generate new synthetic instances in a non-diverse way and spread in a narrow region, making classifiers vulnerable to falling behind. Furthermore, the K-nearest neighbour examples evaluated would be in the majority class, causing boundary ambiguity between the majority and minority classes and resulting in noise instances. To address the above shortcomings, the proposed INF-SMOTE works by removing uninformative samples from minority regions (overlapped and noisy instances) and generating synthetic minority samples from diverse and informative samples.

Furthermore, Wilcoxon's signed ranks test is applied to the proposed method. The results are compared with state-of-the-art approaches and achieve the following p-value. SMOTE Vs INF-SMOTE with .00001, Borderline SMOTE Vs INF-SMOTE with .0008 and ADASYN Vs INF-SMOTE with .00016. The test detects a significant difference between the methods. The statistical analyses are used to equate the efficiency of the proposed method to that of other techniques. The p-values of the associated methods are compared, and the result shows an improvement of the proposed method over existing approaches.

Table 3. Performance of SDP on five classifiers based on AUC-ROC and F-Measure

		AUC-ROC					F-Measure				
Projects	Classifier	No-Sampling	SMOTE	SL-SMOTE	ADASYN	INF-SMOTE	No-Sampling	SMOTE	SL-SMOTE	ADASYN	INF-SMOTE
CM1	J48	0.59	0.89	0.89	0.92	0.94	30.57	86.60	85.67	89.70	98.39
	SVM	0.76	0.96	0.94	0.96	1.00	0.00	89.93	87.56	91.38	96.85
	NB	0.69	0.93	0.90	0.94	0.71	31.45	85.54	78.15	89.25	43.60
	AB	0.72	0.97	0.95	0.97	0.74	0.00	91.36	88.41	92.13	75.51
	BG	0.73	0.96	0.95	0.97	0.99	0.00	91.13	88.05	91.61	92.88
JM1	J48	0.62	0.90	0.89	0.88	0.87	29.06	88.80	86.87	84.32	85.42
	SVM	0.70	0.94	0.93	0.92	0.98	29.91	89.60	88.42	85.52	93.15
	NB	0.63	0.75	0.76	0.70	0.56	26.72	37.62	40.19	29.37	20.88
	AB	0.67	0.93	0.92	0.90	0.61	0.00	88.10	86.47	83.57	57.88
	BG	0.69	0.94	0.93	0.91	0.94	25.76	89.44	87.90	85.22	86.58
KC2	J48	0.70	0.93	0.92	0.91	0.84	52.19	91.51	90.45	87.16	76.79
	SVM	0.83	0.97	0.96	0.96	0.93	53.99	91.55	91.13	88.75	82.64
	NB	0.83	0.84	0.86	0.79	0.71	51.16	60.71	61.29	48.83	36.02
	AB	0.78	0.95	0.95	0.94	0.75	49.22	90.16	88.63	86.20	64.18
	BG	0.83	0.96	0.96	0.95	0.89	51.98	90.96	90.64	88.61	77.60
KC3	J48	0.65	0.89	0.87	0.86	0.91	37.50	88.96	85.90	85.62	86.13
	SVM	0.74	0.94	0.94	0.93	0.98	21.75	89.65	85.84	85.98	90.61
	NB	0.66	0.89	0.85	0.86	0.66	40.57	86.52	72.93	83.74	39.55
	AB	0.57	0.91	0.90	0.90	0.76	43.34	88.61	86.13	86.44	63.44
	BG	0.73	0.94	0.92	0.94	0.95	19.61	87.67	87.50	85.51	86.92
MC1	J48	0.52	0.98	0.98	0.99	0.99	13.34	97.94	97.62	98.75	98.27
	SVM	0.88	1.00	1.00	1.00	1.00	27.73	98.14	97.83	98.84	99.55
	NB	0.71	0.94	0.94	0.95	0.75	11.93	81.93	82.50	89.34	49.16
	AB	0.84	0.99	0.99	1.00	0.89	0.00	97.92	97.67	98.69	71.40
	BG	0.86	0.99	0.99	1.00	1.00	0.00	97.97	97.71	98.79	98.79
MC2	J48	0.70	0.74	0.76	0.68	0.79	56.10	83.65	82.64	69.70	86.37
	SVM	0.72	0.91	0.90	0.83	0.97	47.87	86.53	87.37	74.32	90.50
	NB	0.70	0.81	0.81	0.74	0.68	49.25	57.12	57.76	37.69	42.46
	AB	0.62	0.86	0.86	0.76	0.76	48.64	82.67	82.07	69.67	78.09
	BG	0.68	0.90	0.87	0.80	0.92	47.19	84.99	82.59	71.85	86.97
MW1	J48	0.45	0.91	0.86	0.93	0.95	19.51	91.39	86.99	93.32	93.45
	SVM	0.72	0.96	0.94	0.97	1.00	24.37	91.23	87.79	94.03	95.94
	NB	0.73	0.90	0.86	0.93	0.76	38.97	74.08	70.62	78.50	63.48
	AB	0.71	0.95	0.93	0.97	0.84	34.02	91.01	88.77	93.67	74.28
	BG	0.71	0.95	0.93	0.97	0.99	18.17	91.07	85.89	94.07	94.60
PC1	J48	0.72	0.95	0.94	0.96	0.96	30.44	87.82	84.07	90.83	84.64
	SVM	0.84	0.98	0.98	0.99	1.00	30.32	93.07	89.86	94.96	94.68
	NB	0.77	0.91	0.90	0.94	0.78	19.50	93.54	90.23	94.95	98.01
	AB	0.79	0.97	0.96	0.98	0.83	33.12	59.24	59.56	68.71	48.32
	BG	0.82	0.98	0.97	0.99	0.99	0.00	93.25	90.28	95.23	77.35
PC2	J48	0.45	0.99	0.98	0.99	0.99	14.29	93.12	90.71	95.39	95.18
	SVM	0.84	1.00	0.99	1.00	1.00	0.00	99.35	97.02	99.40	98.79
	NB	0.88	1.00	0.99	1.00	0.90	0.00	99.45	97.59	99.45	99.85
	AB	0.91	1.00	1.00	1.00	0.98	12.23	94.71	96.74	95.05	52.58
	BG	0.83	1.00	0.99	1.00	1.00	10.06	99.45	97.45	99.50	93.61
PC3	J48	0.62	0.89	0.89	0.91	0.94	0.00	99.45	97.59	99.50	99.40
	SVM	0.83	0.97	0.97	0.98	1.00	30.15	88.79	87.53	91.64	92.80
	NB	0.77	0.89	0.89	0.90	0.76	17.45	90.91	89.10	92.74	96.01
	AB	0.79	0.96	0.95	0.97	0.78	26.23	68.74	67.59	72.15	63.86
	BG	0.82	0.96	0.96	0.97	0.99	0.00	90.77	89.24	92.50	74.30
PC4	J48	0.78	0.95	0.94	0.96	0.96	17.55	90.74	88.65	92.22	91.51
	SVM	0.95	0.99	0.99	0.99	1.00	50.17	91.99	90.28	93.85	94.49
	NB	0.84	0.92	0.91	0.93	0.84	49.77	93.60	91.86	94.99	96.81
	AB	0.91	0.99	0.98	0.99	0.92	41.86	82.00	79.41	86.30	58.34
	BG	0.92	0.99	0.99	0.99	0.99	34.13	92.18	90.18	93.39	83.14
PC5	J48	0.82	0.99	0.99	0.99	0.99	53.63	93.32	91.67	94.19	94.09
	SVM	0.98	1.00	1.00	1.00	1.00	52.31	98.55	98.55	98.60	96.74
	NB	0.94	0.94	0.95	0.95	0.91	52.77	98.70	98.65	98.80	97.39
	AB	0.96	1.00	1.00	1.00	0.95	42.87	76.29	77.14	76.30	60.52
	BG	0.98	1.00	1.00	1.00	1.00	23.36	98.09	97.95	98.23	84.98

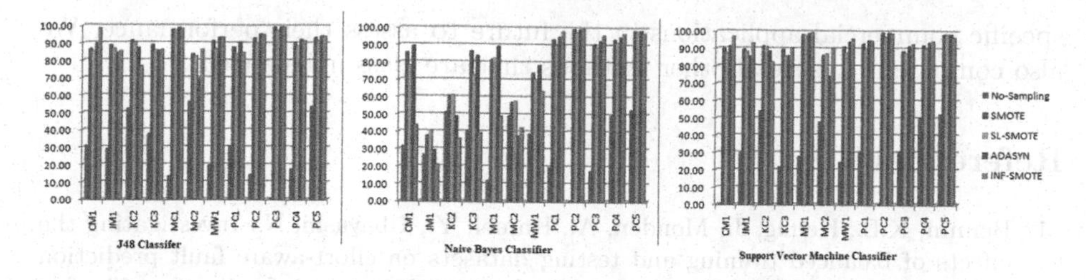

Fig. 6. The comparison of various sampling methods with INF-SMOTE on three single classifiers using F-Measure

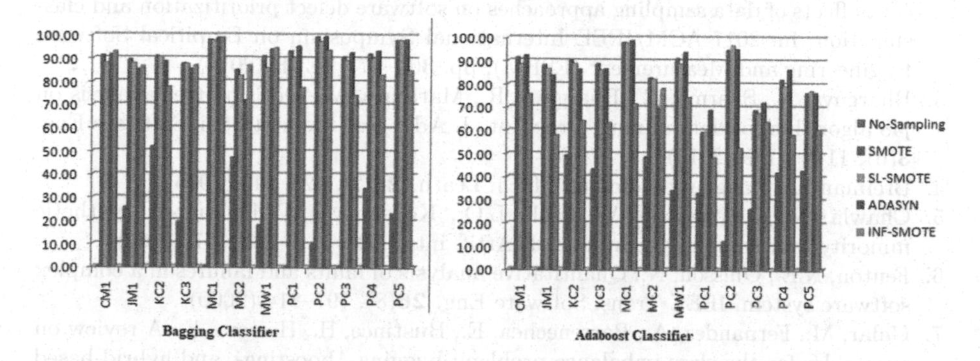

Fig. 7. The comparison of various sampling methods with INF-SMOTE on two ensemble classifiers using F-Measure

5 Conclusion and Future Work

In real-world software development, defect datasets contain more non-faulty instances than defective ones, hampering classifiers' effectiveness. Various sampling strategies have been utilized to mitigate the impact of unbalanced datasets. Some approaches employed resampling or created new instances to balance the unbalanced datasets. As a result of the unpredictability of these sampling approaches, new synthetic samples would be non-diverse and spread across a narrow region. At the same time, they failed to account for noise. They may enhance classifiers' ability to detect more faulty examples, but they may also cause them to misclassify more non-defective instances. Taking advantage of these difficulties, we offer the INF-SMOTE approach, which uses informative samples from minority classes to produce new instances widely distributed in the space of faulty space. INF-SMOTE determines which samples are uninformative based on distance then uses the informative samples to generate new ones. We compare INF-SMOTE to other over-sampling approaches (No-sampling, SMOTE, ADASYN, Borderline-SMOTE) using five classifiers. The experimental findings show that our INF-SMOTE outperforms the different techniques. We will apply the INF-SMOTE technique to additional software defect prediction models and specific commercial applications in the future to assess their performance. We also compare it to several other methods that are class unbalanced.

References

1. Bennin, K.E., Keung, J., Monden, A., Kamei, Y., Ubayashi, N.: Investigating the effects of balanced training and testing datasets on effort-aware fault prediction models. In: 2016 IEEE 40th Annual Computer Software and Applications Conference (COMPSAC), vol. 1, pp. 154–163. IEEE (2016)
2. Bennin, K.E., Keung, J., Monden, A., Phannachitta, P., Mensah, S.: The significant effects of data sampling approaches on software defect prioritization and classification. In: 2017 ACM/IEEE International Symposium on Empirical Software Engineering and Measurement (ESEM), pp. 364–373. IEEE (2017)
3. Bhargava, N., Sharma, G., Bhargava, R., Mathuria, M.: Decision tree analysis on j48 algorithm for data mining. Proc. Int. J. Adv. Res. Comput. Sci. Software Eng. **3**(6), 1114–1119 (2013)
4. Breiman, L.: Bagging predictors. Mach. Learn. **24**(2), 123–140 (1996)
5. Chawla, N.V., Bowyer, K.W., Hall, L.O., Kegelmeyer, W.P.: Smote: synthetic minority over-sampling technique. J. Artif. intell. Res. **16**, 321–357 (2002)
6. Fenton, N.E., Ohlsson, N.: Quantitative analysis of faults and failures in a complex software system. IEEE Trans. Software Eng. **26**(8), 797–814 (2000)
7. Galar, M., Fernandez, A., Barrenechea, E., Bustince, H., Herrera, F.: A review on ensembles for the class imbalance problem: bagging-, boosting-, and hybrid-based approaches. IEEE Trans. Syst. Man Cybern. Part C (Appl. Rev.) **42**(4), 463–484 (2011)
8. Gosain, A., Sardana, S.: Handling class imbalance problem using oversampling techniques: a review. In: 2017 International Conference on Advances in Computing, Communications and Informatics (ICACCI), pp. 79–85. IEEE (2017)
9. Gray, D., Bowes, D., Davey, N., Sun, Y., Christianson, B.: The misuse of the NASA metrics data program data sets for automated software defect prediction. In: 15th Annual Conference on Evaluation & Assessment in Software Engineering (EASE 2011), pp. 96–103. IET (2011)
10. Hall, T., Beecham, S., Bowes, D., Gray, D., Counsell, S.: A systematic literature review on fault prediction performance in software engineering. IEEE Trans. Software Eng. **38**(6), 1276–1304 (2011)
11. Han, H., Wang, W.-Y., Mao, B.-H.: Borderline-SMOTE: a new over-sampling method in imbalanced data sets learning. In: Huang, D.-S., Zhang, X.-P., Huang, G.-B. (eds.) ICIC 2005. LNCS, vol. 3644, pp. 878–887. Springer, Heidelberg (2005). https://doi.org/10.1007/11538059_91
12. He, H., Bai, Y., Garcia, E.A., Li, S.: ADASYN: adaptive synthetic sampling approach for imbalanced learning. In: 2008 IEEE International Joint Conference on Neural Networks (IEEE World Congress on Computational Intelligence), pp. 1322–1328. IEEE (2008)
13. Kamei, Y., Monden, A., Matsumoto, S., Kakimoto, T., Matsumoto, K.I.: The effects of over and under sampling on fault-prone module detection. In: First International Symposium on Empirical Software Engineering and Measurement (ESEM 2007), pp. 196–204. IEEE (2007)
14. Menzies, T., Greenwald, J., Frank, A.: Data mining static code attributes to learn defect predictors. IEEE Trans. Software Eng. **33**(1), 2–13 (2006)

15. Meyer, D., Wien, F.T.: Support vector machines. Interface LIBSVM Package e1071 **28**, 20 (2015)
16. Pelayo, L., Dick, S.: Applying novel resampling strategies to software defect prediction. In: NAFIPS 2007–2007 Annual Meeting of the North American Fuzzy Information Processing Society, pp. 69–72. IEEE (2007)
17. Potharaju, S.P., Sreedevi, M.: A novel LtR and RtL framework for subset feature selection (reduction) for improving the classification accuracy. In: Pati, B., Panigrahi, C.R., Misra, S., Pujari, A.K., Bakshi, S. (eds.) Progress in Advanced Computing and Intelligent Engineering. AISC, vol. 713, pp. 215–224. Springer, Singapore (2019). https://doi.org/10.1007/978-981-13-1708-8_20
18. Provost, F.: Machine learning from imbalanced data sets 101. In: Proceedings of the AAAI 2000 Workshop on Imbalanced Data Sets, vol. 68, pp. 1–3. AAAI Press (2000)
19. Rekha, G., Tyagi, A.K., Krishna Reddy, V.: A wide scale classification of class imbalance problem and its solutions: a systematic literature review. J. Comput. Sci. **15**, 886–929 (2019)
20. Rish, I., et al.: An empirical study of the Naive Bayes classifier. In: IJCAI 2001 Workshop on Empirical Methods in Artificial Intelligence, vol. 3, pp. 41–46 (2001)
21. Schapire, R.E.: Explaining AdaBoost. In: Schölkopf, B., Luo, Z., Vovk, V. (eds.) Empirical Inference, pp. 37–52. Springer, Heidelberg (2013). https://doi.org/10.1007/978-3-642-41136-6_5
22. Thirugnanasambandam, K., Prakash, S., Subramanian, V., Pothula, S., Thirumal, V.: Reinforced cuckoo search algorithm-based multimodal optimization. Appl. Intell. **49**(6), 2059–2083 (2019). https://doi.org/10.1007/s10489-018-1355-3
23. Wang, S., Yao, X.: Using class imbalance learning for software defect prediction. IEEE Trans. Reliab. **62**(2), 434–443 (2013)

2D-CNN Model for Classification of Neural Activity Using Task-Based fMRI

Sudhanshu Saurabh(✉) and P. K. Gupta

Jaypee University of Information Technology, Waknaghat, Solan 173234, H.P., India
ssmiete@gmail.com, pkgupta@ieee.org

Abstract. In the recent years, several deep learning techniques have been applied to classify the cognitive states of the human brain based on the. We consider the 2D-CNN approach to classify the neural activities with the task-based fMRI. Task-based fMRI displays the neuronal activation by the blood oxygen level dependence (BOLD) response to a specific task. Spontaneous signal fluctuation with low frequency (< 0.1 Hz.) in fMRI occurs in the BOLD signal in a specific region of the human brain during the cognitive task, and voxel changes the regulation of blood flow in the brain causes the hemodynamic signal. In this paper, we have proposed a 2D-CNN model that extract the feature maps and classify the neural activity from fMRI data. The neural activation of seed to voxel connectivity in fMRI voxel are used for the classification of neural responses from the voxel. The classification performance of the proposed 2D-CNN model has been achieved from the task-evoked fMRI data with classification accuracy of 85.3%, sensitivity of 89.5%, and F1-Score of 87.2%. The experimental results shows that the proposed model effectively distinguishes the neuronal response under the task evoked stimuli.

Keywords: Task fMRI · Functional connectivity · Neural activity · CNN · Classification

1 Introduction

The human brain is a complex organ which is made up of cells known as neurons. These neurons have tiny branches which are responsible for the information processing between each other through chemical and electrical signals that propagate along the axon. The mental states developed from the brain and constructs are emotions, body states, and thoughts by creating situated conceptualization that combines the sources of stimulation [4,21]. The human brain actively processes sensory input (e.g. hearing, touch and sight) the sensory input conveying the conditional stimulus(CS) to the cerebellum originates in the various subcortical sensory nuclei, it is a cluster of neurons that relay information to and from the neocortex [7,17]. The human brain and the human body are related

M. Singh et al. (Eds.): ICACDS 2022, CCIS 1613, pp. 192–206, 2022.
https://doi.org/10.1007/978-3-031-12638-3_17

together. Cortical somatosensory processing would be affected from movie watching by a subject and direction of effect (enhancing/suppressing) in the subject's brain [30].

The sensory can depend heavily on the instantaneous level of attention to specific stimulus features and the hemodynamic responses in the cerebral cortex associated with repeated delivery of the stimulus [1,31]. To observe the hemodynamic response of voxel we obtained the fMRI data while the participants from two groups i.e. children and adults watched a 5.6-minute silent animated movie (Partly Cloudy) [28]. The study of cognitive behaviors and functional activities of the brain is based on the task that is derived from the BOLD [8,31].

1.1 Deep Neural Network Framework

The functional network of the brain is constructed from the fMRI data and the deep learning model is widely used in medical image processing [22,32]. The CNN has been proven to be an end-to-end learning and a powerful deep learning framework and it is discriminate the features from a given raw dataset [20,24]. A CNN has multiple convolution layers each of these convolution layers generates many alternate convolutions. The first successful implementation of the CNN model is LeNet to identify the digits. The success of the AlexNet model for the classification of images achieved significant results. U-Net model is one of the most influential deep learning architectures it is outperformed the sliding window CNN [10]. The task-based voxel-wise fMRI data classification using 2D-CNN was performed by [14].

In this work, we have classified the 3D voxel-wise fMRI response from movie watching based brain dataset with 4 mm resolution and a number of subjects are 155 using a two dimensional convolutional neural network (2D-CNN) model as shown in Fig. 1. We aim to discriminate specific stimulation of neuronal activity from fMRI response of the training dataset. Since a large amount of dataset required to train the 2D-CNN model is simply not available in fMRI. Mainly due to small sample sizes and high dimensionality of fMRI datasets [3]. Each fMRI sample (volume) consist thousands dimension i.e. voxel. We use 2D-CNN in predicting the neural activity from the brain response of fMRI images data of subjects, specifically, we trained and tested the model with the fMRI data for the classification of brain activity. The major findings are 1. Visualize voxel-wise signal extracted region of brain activity from proposed 2D-CNN model. 2. In particular, we observe the performance of model classification, training and validation loss, accuracy, precision and F1-score that a CNN model that has been trained on the task-fMRI dataset.

2 Related Work

Brain generates the pattern of spontaneous activity and the fluctuation in these activity affects the neural activity related to stimulation for the task-specific response. Based on task, recent studies used the fMRI to shows the developmental imaging to detected the changes in activation fMRI response.

Some studies discuss and contribute to the neural activity in healthy human brain during the event where the stimuli were presented [2,17,31]. In the developing brain the region is examined for the stimulus and neural mechanism [7].

2.1 Neural Activity

The fMRI is the most common and widely used to study the human brain function that is based on changes in the BOLD signal that causes the neuronal activity occurs, therefore, changes in brain state and produces a stimulus or task [8,11]. Task-based fMRI is commonly used to identify the brain regions and to explore the functionality of the brain, involved in the task performance. The presence of deoxyhemoglobin in the blood changes the proton signal from water molecules surrounding a blood vessel in gradient echo MRI producing a blood-oxygen-level dependent (BOLD) contrast [18]. The sensory can depend heavily on the instantaneous level of attention to specific stimulus features and the hemodynamic responses in the cerebral cortex associated with repeated delivery of the stimulus [1]. The importance of functional connectivity (FC) using the transfer function between the neural activity and BOLD signal shows the impact of HRF on the FC [26]. Authors in [19] have discussed the linear model and variation in neural activity are necessary to convey the information.

2.2 Deep Learning Approaches

Deep Learning models use end-to-end learning models to decode and classify the cognitive states from fMRI data. Authors in [20] have developed a volumetric CNN based approach for classification of neuroimaging data. They have used class saliency visualization method to detect the biomarkers on each classification task. Authors in [15] have designed a new CNN structure called deep convolutional auto encoder (DCAE) to model the brain network based on task fMRI data. Authors in [14] have proposed a multichannel 2D-CNN (M2D CNN) model to classify the 3D fMRI data. Authors in [32] have proposed a Deep Learning model known as C3D-LSTM model that consist of 3D-CNN and LSTM where 3D-CNN is used to extract the spatial feature of 4D fMRI data, and LSTM is used for time varying information of the data for the detection of Alzheimer. Authors in [3] have explored the Deep Learning models to train the fMRI dataset of the Human Connectome Project database (HCP) for decoding the cognitive states and unrelated tasks.

3 Materials and Methods

Here, for classification of neural activity of various fMRI subjects watching a silent animated movie, we have captured the dataset at 168 time points. We have also considered the various statistical parameters are shown in Table 1.

Table 1. Overview of the statistical parameters of age used for the study.

Age group	Age count	Mean	Std. Dev.	Min. Age	25%	50%	75%	Max. Age
3yo	17	3.75	0.17	3.51	3.60	3.78	3.93	3.98
4yo	14	4.43	0.28	4.05	4.17	4.36	4.71	4.85
5yo	34	5.50	0.28	5.01	5.31	5.46	5.73	5.99
7yo	23	7.53	0.36	7	7.14	7.66	7.91	7.96
8–12yo	34	9.76	1.18	8.02	8.63	9.69	10.53	12.3
Adults	33	24.77	5.30	18	21	23	28	39

3.1 Dataset Analysis

In this work, we have used publicly available fMRI and behavioral data through the OpenfMRI database with accession number $ds000228$. The fMRI data consist of total 155 subjects that includes 71 males, and 84 females, out of which there are 122 children and 33 adults. Obtained 4D-fMRI images are high dimensional data of size $(50 \times 59 \times 50 \times 168)$ with 50 slices and the voxel size is 3mm isotropic for the classification of neural activity and behavioural activity between the adult and children.

3.2 Methodology

The aim of fMRI to measure the changes in BOLD fMRI signal provides the neuronal activity in the brain region that is corresponding to voxel and hemodynamic responses in the cortical regions [2,6]. In this work, we have used seed-based functional connectivity to analyse the fMRI data in which all the participant performs a cognitive task (continuous task such as watching a movie) during fMRI scan. BOLD activation from fMRI data have been pre-processed and the voxels are included for the further analysis, The BOLD signal intensities as shown in Fig. 2a have been calculated for the concerned brain region from the obtained fMRI with $TR = 1.0$ and $numberofslice = 50$. We have extracted the fMRI time series for the all activated voxels as an input to the CNN. The BOLD intensities signal of the task related fMRI voxel time series have been normalized further to change the signal which is relative to the average BOLD signal [15].

In the brain region, the specific seed point and cortical voxel stimulation from the fMRI images of 3 to 12 years children and adults during a movie are used to visualize and predict the fMRI responses from the CNN. The identification of neural and behavioural activity using fMRI data based features from movie stim-

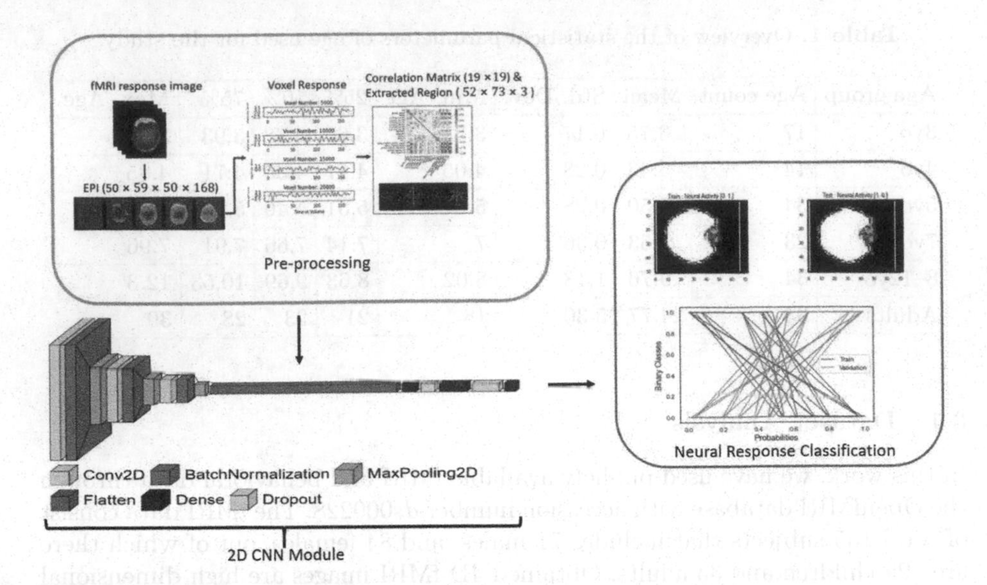

Fig. 1. Building block of the proposed classification model

uli seed based correlation to extract the activities in the brain regions, and it is defining the temporal activities in the brain dynamics. Figure 2b shows the correlation matrix has shape (19,19) for each subject of brain development dataset (fMRI) which shows the 19 regions of brain areas, which includes cerebrospinal fluid (CSF) to be indicator of stimuli to vascular system [16,29]. CSF signal is generally considered as physiological noise, and uses a linear regression to remove the noise that improves the specificity of the FC. The for mean, Standard Deviation (SD), and the variance of CSF signal are considered as 896.30, 8.80 and $\sigma^2 = 77.44$ respectively. White-matter (WM) activity signals have BOLD changes associated with neuronal activity. Temporal derivative of time-course to RMS variance over voxel (DVARS) measures the changes in intensity from previous timepoint (as opposed the global signal: GS) while the Framewise displacement (FD) measure the indexes movement of head from one voxel to the adjacent voxel [23]. The functional network from Fig. 4a shows the correlation between seed and voxel with size of the BOLD time series $= 32504 \times 168$. We have obtained the seed to voxel dimension of input $= 32504 \times 1$. Using a selection of seed based analysis has an impact on the accuracy of the CNN.

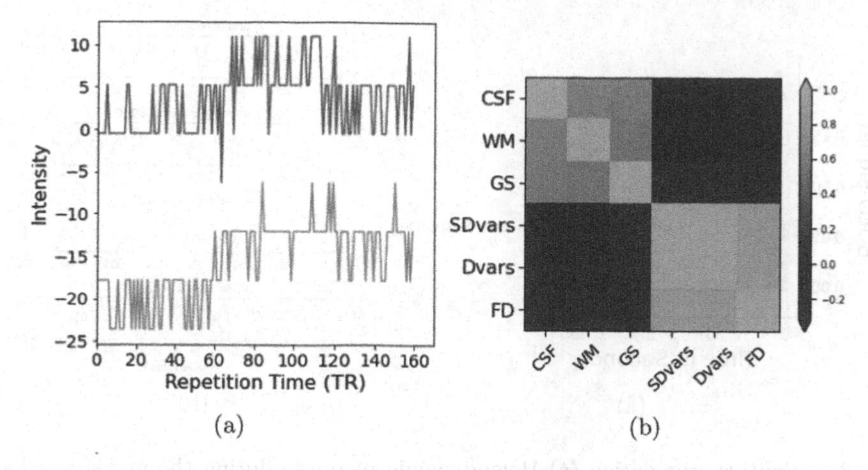

Fig. 2. Functional connectivity (a) BOLD signal intensities (b) Correlation matrix of developmental brain

3.3 Functional Connectivity Analysis

The intensities of fMRI voxels in a cognitive stimulation represent the changes in local concentration that is a hemodynamic response at brain voxels. We have extracted the voxel intensities that gets changes over time as shown in Fig. 2a. Here, fluctuations can be visualized from Fig. 3b as a variation in the neural activity, necessary to process the information [9,19]. For all the 155 participants, we have considered the repetition time ($TR = 1$) second and stimulus presentation at every 2 s. Consider a predicted BOLD response $a_i(t)$ that arises due to neural activity $v_i(t)$ as the BOLD signal in the output of Linear Time Invariant (LTI) [5]. This form of response is independent of time. Here, we can define the $a_i(t)$ as a convolution of neural activity with hemodynamic response function (HRF) $h_i(t)$ that can be expressed as: $h_i(t) = v_i(t) \otimes a_i(t) + \epsilon(t)$ $\epsilon(t)$ assumed to be $N(0, \sigma^2)$ where σ^2 represents a variance of the noise [25]. We have observed the BOLD response to an event (continuous stimulus) and duration (less than 5 s) of neural activity where the BOLD signal is output and response to continuous stimuli as shown in Fig. 3a.

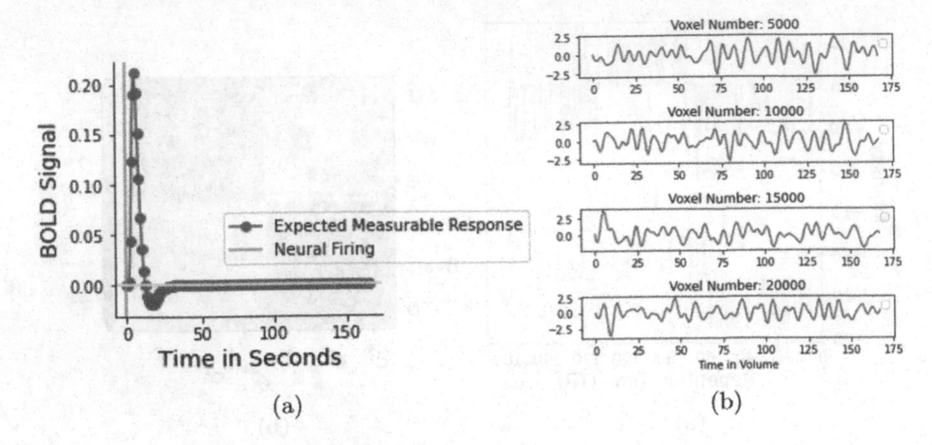

Fig. 3. Cognitive stimulation (a) Hemodynamic response during the movie in which positive response in the same region (b) Voxel signal

Seed Based Functional Connectivity. The seed-based approach generates a spatial map and its time courses that regress the fMRI dataset, in this approach voxel within a region of interest (ROI) is selected and their BOLD signal is correlated with other voxels in the region of the brain [11]. In the task-based condition for all participants, the extraction of time series for all voxels within a seed region [27]. The connectivity is calculated from the correlation of time series for the priori seed and time course of all other voxels in the brain [12]. The relationship between spatially adjacent voxels using a time series of BOLD signal that is measured at voxels during the stimulus [13]. This forms the local network (functional connection) among each voxel or seed voxel as shown in Fig. 4a which extract the seed time-series and brain time-series. The influence of seed voxel dimension is covariate. Therefore, we obtain the min and max connectivity values are −0.624 and 0.958 respectively as shown in Fig. 4b of the cognitive covariate on seed to voxel functional connectivity.

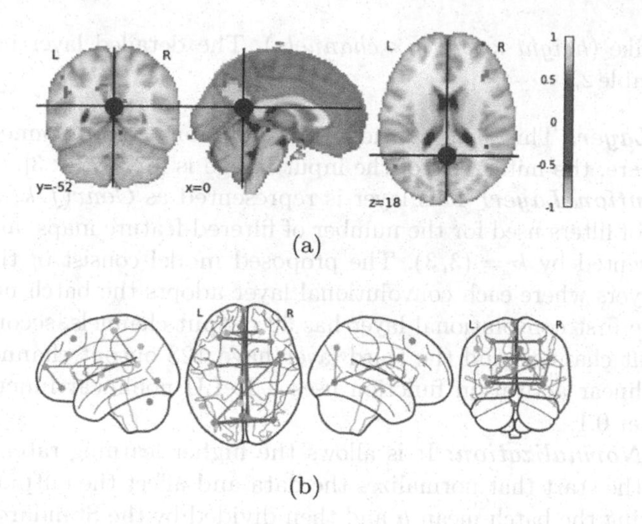

Fig. 4. Seed-voxel connectivity (a) Seed to voxel coorelation (b) Seed to voxel functional connectivity

Seed Based Approach for CNN. In this proposed method the brain region considers as a node and normalized time-course signal of voxel from each brain region as a time-course signal of voxel. On the other hand, the desired objective through the interaction among the voxels. The BOLD signal intensities were calculated of the brain region from fMRI with and, we extracted the fMRI time series for the all activated voxels were considered as input to the CNN. Brain mask that is overlapped across all the volume and the resulting in a total voxel 32504 in the brain. In the brain region the specific seed point and cortical voxel stimulation from the movie is used to visualize and predict the fMRI responses from the CNN. A specific seed region (ROI) is used to calculate the correlation between the other regions of the brain by the mean time series and seed time series shape = 168×1. The seed-voxel connectivity from Fig. 4 shows the correlation between seed and voxel and size of the BOLD time series = 32504×168. We obtain the BOLD changes and neural activity with the dimension of input = $67 \times 77 \times 3$ to the 2D CNN. Using a selection of seed based analysis has an impact on the accuracy of the 2D CNN.

4 Proposed Network Architecture

We have proposed a 2D-CNN architecture for functional connectivity data as it uses 2-D kernels and could provide the features for the time series data. Here, 2D images are used in the CNN model along with three convolution layers disseminated with batch normalization, max-pooling, followed by the densely connected layer. The Brain image of seed to voxel connectivity of brain region as shown in Fig. 4b is submitted as an input to the convolutional layer. Input shape of fMRI image is 2D but we have to pass a 3D array at the time of fitting a data that should be like (*height* × *width* × *channels*). The detailed layer information is shown in Table 2.

- ***Input Layer:*** This layer considers $2D$ functional brain connectivity as an input. Here, the initial size of the input image is $[67 \times 77 \times 3]$.
- ***Convolution Layer:*** This layer is represented as $Conv(f, k, a)$, where f is number of filters used for the number of filtered feature maps, and kernel size is represented by $k = (3, 3)$. The proposed model consist of three convolutional layers where each convolutional layer adopts the batch normalization. Here, the first convolutional layer has 32 output channels, second layer have 64 output channels and the third layer have 128 output channels. The output nonlinear activation function uses a ReLU nonlinear function with the parameter 0.1.
- ***Batch Normalization:*** It is allows the higher learning rate. This layer is used at the start that normalizes the data and affect the output of ReLU by subtracting the batch mean μ and then divided by the Standard Deviation σ which reduces the data loss between processing layers.
- ***Max Pooling Layer:*** This layer pooled the input volume of size $[65 \times 75 \times 32]$ with the filter size= (3×3) applied with $stride = 1$ downsamples into output of size $[32 \times 37 \times 32]$. The proposed $2D - CNN$ model have learnable weights and consist of three $2D$ convolutional layer. Here, each convolutional layer extract the feature of the initial size of image is $[67 \times 77 \times 3]$ followed by a ReLU activation function.
- ***Flatten:*** This layer converts the series of convolution and entire pooled 3D feature map matrix from the third Max pooling layer into 1D vector, which is then fit the input to the fully connected layer for the classification.
- ***Fully Connected Layer:*** In the proposed model, 1D vector from flatten layer connect to the dense layers in which each input is connected with each output by learnable weight followed by a ReLU activation function. In the baseline model, we have used three fully connected layers. The first FC layer has 256 neurons and second FC layer has 512 neurons with ReLU activation function.
- ***Classification Scheme:*** The classification scheme of fMRI response is performed at the end of the CNN block that categorizes the fMRI image into a neural activity, and in no neural activity. For this purpose, we have added dense layer with 512 neurons followed by a ReLU activation function. Output layer has two neurons contains Softmax activation function which output a probability for each class in binary classification for the fMRI image.

Table 2. Proposed architecture of 2D-CNN model.

Operation	Kernel size	Data dimension	Weights (N)	Weights (%)
Input	3 × 3	67 × 77 × 3	–	–
Conv2D	3 × 3	65 × 75 × 32	896	0.10%
ReLU	3 × 3	65 × 75 × 32		
Batch normalization	3 × 3	65 × 75 × 32	128	0.00%
Max pooling 2D	3 × 3	32 × 37 × 32	0	
Conv2D	3 × 3	30 × 35 × 64	18496	1.20%
ReLU	3 × 3	30 × 35 × 64		
Batch normalization	3 × 3	30 × 35 × 64	256	0.00%
Max pooling 2D	3 × 3	15 × 17 × 64	0	
Conv2D	3 × 3	13 × 15 × 128	73856	4.60%
ReLU	3 × 3	13 × 15 × 128		
Batch normalization	3 × 3	13 × 15 × 128	512	0.00%
Max pooling 2D	3 × 3	6 × 7 × 128	0	
Flatten	–	5376		
Dense	–	256	1376512	85.9%
ReLU	–	256		
Dropout	–	256	0	0.00%
Dense	–	512	131584	8.20%
ReLU	–	512		
Dropout	-	512	0	0.00%
Dense	–	2	1026	0.10%
Softmax	–	2		

5 Experimental Results

5.1 Prediction Accuracy of the Proposed $2D - CNN$ Model

The $4D - fMRI$ time series data were randomly shuffled and 134 training datasets including 80% of the fMRI image for the fMRI responses which includes two classes (Neural activity and No Neural activity) and 34 validation dataset including 20% of fMRI images. $2D - CNN$ model adjusted for 200 epochs and initialized Stochastic Gradient Descent for one update in training, learning rate ($lr = 0.000001$). The mean of fMRI images was calculated and subtracted from each image. The $2D-CNN$ uses (3 × 3) filters with three convolution layers (16, 32, 64 filters and ReLU activation function) and one softmax function and these layers were all trained using Batch size of 16 and the parameters are fine tuned. In this work, we have used a categorical cross entropy as a loss function that uses Adaptive Moment Estimation (ADAM) optimizer for the faster convergence of the algorithm. We have calculated the prediction accuracy of the proposed

$2D-CNN$ model based on voxel response of $4D-fMRI$ data. The prediction accuracy of fMRI voxel responses are shown in Table 3 of the proposed model that yields the higher accuracy on the validation data as shown in Fig. 5.

Table 3. Accuracy and loss for training and validation data.

Model	Kernel	Activation	Training accuracy	Training loss	Validation accuracy	Validation loss
2D-CNN	3×3	ReLU	84.32	41.85	85.29	36.19

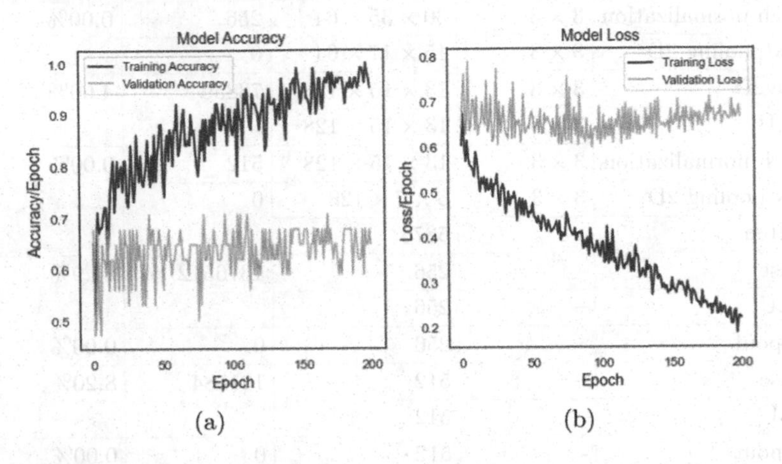

Fig. 5. Model performance (a) Training and validation accuracy (b) Training and validation loss.

5.2 Evaluation of Model Performance

Evaluation of the proposed 2D-CNN model shows the findings in the terms of accuracy, precision, and F1-score for 168 sample data acquired from 155 participants. Evaluation of a learned CNN model is traditionally performed by the maximization of the accuracy metric. The empirical analysis includes the metrics like Precision, F1-Score, Recall, Sensitivity, Specificity, and Accuracy. In the binary classification problem let TP, TN, FP, FN true positive, true negative, false positive and false negative class labels and let Neural Activity, No Neural Activity be the predicted true positive and false negative class labels. These metrics can be computed as per following:

$$Sensitivity = \frac{TP}{TP+FN},\ Precision = \frac{TP}{TP+FP},\ Specificity = \frac{TP}{FP+TN}$$

$$F1-Score = \frac{TP}{TP+\frac{1}{2}(FP+FN)},\ Accuracy = \frac{TP+TN}{(TP+FP)+(TN+FN)}$$

The statistical evaluation of the classification performance of the CNN model is formulated and plotting the relationship between the observed values and predicted values by using the confusion matrix. The performance of the model is obtained by using the four categories of confusion matrix as shown in Fig. 6a and the predicted classification probabilities are shown in Fig. 6b. We have obtained the classification Information from the confusion matrix: Matthews correlation coefficient (MCC) between observed and predicted neural activity are shown in Table 4. MCC ranges in the interval [−1, +1] and Sensitivity, Specificity and MCC is also dependent on the prevalence (P) and Bias (B) which depend upon classifier and dataset. The prevalence (P) is the rate of all predicted positive neural activity in the test dataset and bias measure the how likely classifier predict for the test dataset and can be defined as:

$$MCC = \frac{(TP \times TN) - (FP \times FN)}{\sqrt{(TP + FP)(TP + FN)(TN + FP)(TN + FN)}}, \; P = \frac{TP + FN}{N}, B = \frac{TP + FP}{N}$$

For the original fMRI image predicted class of neural activity is shown in Fig. 7b where we can see predicted binary class corresponding to Fig. 6b of the training and validated data of the fMRI image.

Table 4. Classification report of proposed 2D-CNN model.

Classifier	Accuracy	Sensitivity	Specificity	Precision	F1-Score	Prevalence	Bias	MCC
2D - CNN	85.3	89.5	80	85	87.2	55.9	58.8	70.1

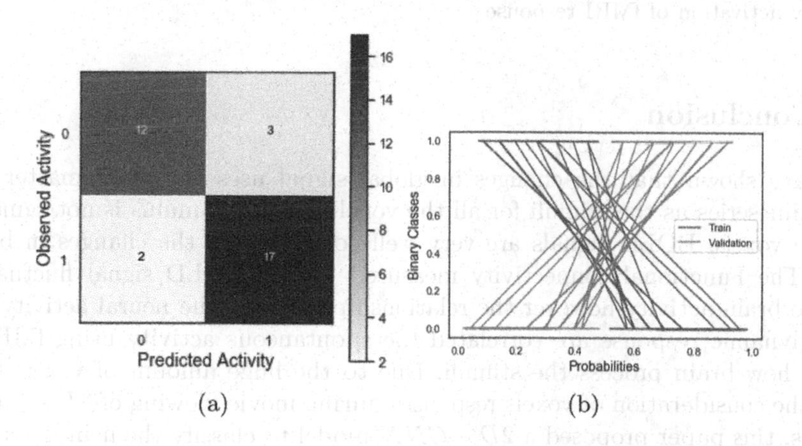

Fig. 6. Classification performance (a) Confusion matrix fMRI response 2D-CNN model (b) Predicted classification value for training and validation

5.3 Classification of Voxel Response fMRI Images

The 2D-CNN model was implemented by using the Tensorflow and keras open source library and the model is trained on the Nvidia Geforce GTX 1080 GPU. We trained the model to classify the voxel response between the neural activity activation and no activation using the Adam optimizer. The input of our CNN model was the whole brain voxels (volume $50 \times 59 \times 50$ and to extract timeseries signal, define the mask from extraction of a brain mask for filling the brain volume to extract the data from the brain region. Here we made the fMRI Image classification based on the feature decoded from fMRI response. Original fMRI images with of the train and test set to distinguish the neural activity from fMRI response data. Our proposed 2D-CNN model achieved 85.3% accuracy on the neural activity activation classification. The visualization of the classification of the neural activity activation from the original fMRI image of brain FC region to posterior cingulate cortex as shown in Fig. 7b.

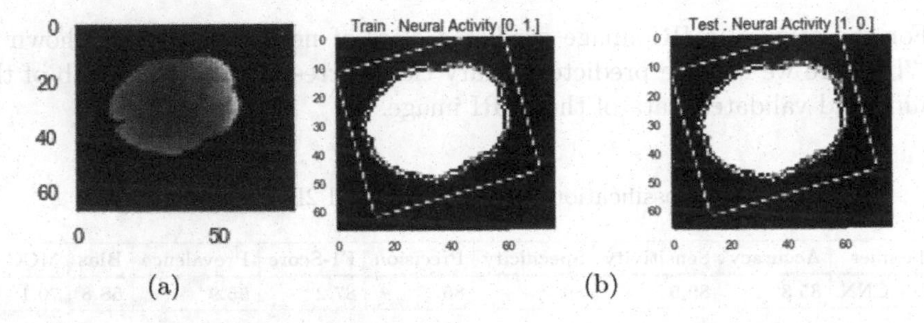

Fig. 7. Active state of brain fMRI (a) Convolution Layer activated fMRI (b) Neural activity activation of fMRI response

6 Conclusion

We have shown that the changes in global signal uses the white-matter and CSF timeseries as the stimuli for all the voxel and the stimulus is not same for all the voxels. BOLD signals are very well correlated to the changes in blood flow. The Functional connectivity measured on the BOLD signal fluctuation for the brain activity however the relationship between the neural activity and hemodynamic response are correlated the spontaneous activity using fMRI to shows how brain process the stimuli. Due to the huge amount of voxel signal with the consideration of voxels responses during movie viewing of $3D - fMRI$ images, this paper proposed a $2D - CNN$ model to classify the neural activity based on the voxel signal extracted from the fMRI data. The proposed $2D - CNN$ model analysed the activation maps of the original image and visualizing the state of art image classification of neural activity with the validation accuracy is 85.3%. The feature maps making a prediction with this proposed 2D-CNN model plot all 128 two dimensional images.

References

1. Andermann, M., Kauramäki, J., Moore, C., Hari, R., Jääskeläinen, I., Brain, S.M.: Brain state-triggered stimulus delivery: an efficient tool for probing ongoing brain activity. Open J. Neurosci. **2**(5) (2012)
2. Annabel, D., Nijhof, L.B., Marcel, B., Jan, R.W.: Brain activity for spontaneous and explicit mentalizing in adults with autism spectrum disorder: an fMRI study. NeuroImage Clin. **18**, 475–484 (2018)
3. Armin, W., Thomas, Klaus-Robert, M., Wojciech, S.: Deep transfer learning for whole-brain fmri analyses. arXiv e-prints, p. arXiv:1907.01953 (2019)
4. Barrett, L.F.: Solving the emotion paradox: categorization and the experience of emotion. Pers. Soc. Psychol. Rev. **10**(1), 20–46 (2006)
5. Boynton, G.M., Engel, S.A., Glover, G.H., Heeger, D.J.: Linear systems analysis of functional magnetic resonance imaging in human v1. J. Neurosci. **16**(13), 4207–4221 (1996)
6. Calhoun, V.D., Adali, T., Pearlson, G.D., Pekar, J.J.: Spatial and temporal independent component analysis of functional MRI data containing a pair of task-related waveforms. Hum. Brain Mapp. **13**(1), 43–53 (2001)
7. Freeman, J.H.: The ontogeny of associative cerebellar learning. Cerebellar Conditioning Learn. **117C**, 53–72 (2014)
8. Friston, K.J., Jezzard, P., Turner, R.: Analysis of functional MRI time-series. Hum. Brain Mapp. **1**(2), 153–171 (1994)
9. Friston, K., Fletcher, P., Josephs, O., Holmes, A., Rugg, M., Turner, R.: Event-related fMRI: characterizing differential responses. Neuroimage **7**(1), 30–40 (1998)
10. Gao, Y., Zhang, Y., Wang, H., Guo, X., Zhang, J.: Decoding behavior tasks from brain activity using deep transfer learning. IEEE Access **7**, 43222–43232 (2019)
11. Gore, J.C.: Principles and practice of functional MRI of the human brain. J. Clin. Investig. **112**(1), 4–9 (2003)
12. Haochang, S., et al.: Shrinkage prediction of seed-voxel brain connectivity using resting state fMRI. NeuroImage **102**(2), 938–944 (2014)
13. Hasson, U., Nir, Y., Levy, I., Fuhrmann, G., Malach, R.: Intersubject synchronization of cortical activity during natural vision. Science **303**(5664), 1634–1640 (2004)
14. Hu, J., et al.: A multichannel 2D convolutional neural network model for task-evoked fMRI data classification. Comput. Intell. Neurosci. **2019**, 1–9 (2019)
15. Huang, H., et al.: Modeling task fMRI data via deep convolutional autoencoder. IEEE Trans. Med. Imaging **37**(7), 1551–1561 (2018)
16. Jahanian, H., et al.: Print study research group.: measuring vascular reactivity with resting-state blood oxygenation level-dependent (BOLD) signal fluctuations: a potential alternative to the breath-holding challenge? S. J. Cereb. Blood Flow Metab. **37**(7), 2526–2538 (2017)
17. Jung, W.M., Ryu, Y., Park, H.J., Lee, H., Chae, Y.: Brain activation during the expectations of sensory experience for cutaneous electrical stimulation. NeuroImage: Clin. **19**, 982–989 (2018)
18. Lindquist, M., Geuter, S., Wager, T., Caffo, B.S.: Modular preprocessing pipelines can reintroduce artifacts into fMRI data. Hum. Brain Mapping **40**(8), 1634–1640 (2019)
19. Miller, P., Cannon, J.: Combined mechanisms of neural firing rate homeostasis. Biol. Cybern. **113**, 47–59 (2019)

20. Oh, K., Chung, Y.C., Kim, K.W., Kim, W.S., Oh, I.S.: Classification and visualization of Alzheimer's disease using volumetric convolutional neural network and transfer learning. Sci. Rep. **9**(1), 18150 (2019)
21. Oosterwijk, S., Lindquist, K.A., Anderson, E., Dautoff, R., Moriguchi, Y., Barrett, L.F.: States of mind: emotions, body feelings, and thoughts share distributed neural networks. NeuroImage **62**(3), 2110–2128 (2012)
22. Poonam, R., Pradeep, K.G., Sharma, V.: A novel deep learning-based whale optimization algorithm for prediction of breast cancer. Braz. Arch. Biol. Technol. **64** (2021)
23. Power, J., Mitra, A., Laumann, T., Snyder, A., Schlaggar, B., Petersen, S.: Methods to detect, characterize, and remove motion artifact in resting state fMRI. Neuroimage **84**, 320–41 (2014)
24. Prakhar, B., Gupta, P.K., Harsh, P., Mohammad, K.S., Ruben, M.M., Bhaik, A.: Application of deep learning on student engagement in e-learning environments. Comput. Electr. Eng. **91**, 1–11 (2021)
25. Rajapakse, J., Kruggel, F., Maisog, J., von Cramon, D.: Modeling hemodynamic response for analysis of functional MRI time-series. Hum. Brain Mapp. **6**(4), 283–300 (1998)
26. Rangaprakash, D., Wu, G., Marinazzo, D., Hu, X., Deshpande, G.: Hemodynamic response function (HRF) variability confounds resting-state fMRI functional connectivity. Magn. Reson. Med. **80**(4), 1697–1713 (2018)
27. Reid, A.T., et al.: A seed-based cross-modal comparison of brain connectivity measures. Brain Struct. Funct. **222**(3), 1131–1151 (2016). https://doi.org/10.1007/s00429-016-1264-3
28. Richardson, H., Lisandrelli, G., Riobueno-Naylor, A., Saxe, R.: Development of the social brain from age three to twelve years. Nat. Commun. **9**(1), 1027 (2018)
29. Saurabh, S., Gupta, P.K.: Functional brain image clustering and edge analysis of acute stroke speech arrest mri. Thirteenth International Conference on Contemporary Computing (IC3-2021), New York, NY, USA, pp. 234–240. ACM (2021)
30. Svenja, E., et al.: The effect of movie-watching on electroencephalographic responses to tactile stimulation. Neuroimage **220**, 117–130 (2020)
31. Trulsson, M., Francis, S.T., Kelly, E.F., Westling, G., Bowtell, R., McGlone, F.: Cortical responses to single mechanoreceptive afferent microstimulation revealed with fMRI. Neuroimage **13**(4), 613–622 (2001)
32. Wei, L., Xuefeng, L., Xi, C.: Detecting Alzheimer's disease based on 4D fMRI: an exploration under deep learning framework. Neurocomputing **388**(C), 280–287 (2020)

Real-Time Multi-task Network for Autonomous Driving

Vu Thanh Dat, Ngo Viet Hoai Bao, and Phan Duy Hung(✉)

Computer Science Department, FPT University, Hanoi, Vietnam
{datvthe140592,baonvhhe141782}@fpt.edu.vn, hungpd2@fe.edu.vn

Abstract. End-to-end Network has become increasingly important in multi-tasking, especially a driving perception system in autonomous driving. This work systematically introduces an end-to-end perception network for multi-tasking and proposes several key optimizations to improve accuracy. First, we propose efficient segmentation head and box/class prediction networks based on weighted bidirectional feature network. Second, we propose automatically customized anchor for each level in the weighted bidirectional feature network. Third, we propose an efficient training loss function. Based on these optimizations, we develope an end-to-end perception network to perform multi-tasking, including traffic object detection, drivable area segmentation and lane detection simultaneously which achieves better accuracy than prior art. In particular, our network design achieves the state-of-the art **77 mAP@.5** on BDD100K Dataset, outperforms lane detection with 0.293 **mIOU** on **12.83** parameters and **15.6** FLOPs. The network can perform visual perception tasks in real-time and thus is a practical and accurate solution to the multi-tasking problem.

Keywords: Deep learning · Multi-task learning · Detection · Segmentation · Autonomous-driving

1 Introduction

Recent advances in embedded systems' computational power and neural networks' performance have made autonomous driving an active field in computer vision. From time to time, it has been shown that such vehicles can make relatively good driving decisions with just the assistance of a single camera attached to the front. There is a general consensus that the three most critical tasks in guiding intelligent vehicles are: traffic object detection, drivable area segmentation, and lane line segmentation.

Each one of these tasks has got its state-of-the-art networks, including but not limited to SSD [1], YOLO [2] for object detection; UNet [3], SegNet [4] for semantic segmentation; LaneNet [5] and SCNN [6] for lane line detection. Many researchers (MultiNet [7], DLT-Net [8], YOLOP [9]) have thought about combining the networks into a simple encoder-decoder architecture. We propose that this architecture can be improved even further with proper selection of the feature extractor and fusing lane line with drivable area into one segmentation head. Our experiment achieves the highest recall of

M. Singh et al. (Eds.): ICACDS 2022, CCIS 1613, pp. 207–218, 2022.
https://doi.org/10.1007/978-3-031-12638-3_18

94.2% and segmentation IoU of 0.698, outperforming existing multi-task networks on the challenging BDD100K dataset [10].

Our overall network architecture is fairly simple. We make improvements upon the excellent multi-scale feature fusion BiFPN in EfficientDet [11], together with an EfficientNet [12] backbone pre-trained on ImageNet with its balanced trade-off between accuracy and computational overhead. We also construct a BiFPN decoder to utilize existing multi-scale features into the newly designed segmentation head. For an input resolution of 640×384, the entire network comes in at 14.6 BFLOPS on 12.2M parameters, comparable to the latest multi-task network YOLOP at 18.6 BFLOPS on 7.9M parameters. To finetune our results even further, we also tinkered with anchor box generation [13].

To sum it up, the main contributions of this research are:

- An end-to-end perception network, achieving state-of-the-art results in real-time on the BDD100K dataset.
- Automatically customized anchor for each level in the weighted bidirectional feature network, on any dataset.
- An efficient training loss function.

2 Related Works

This section will review some of the best networks in each respective task, then conclude with the latest multi-task networks to emphasize the strength of this unified architecture.

Traffic Object Detection
Current developments in improving detectors' performance have nearly split the area into two distinct branches: region-based and one-stage detectors. While region-based methods are more accurate, one-stage detectors gained more attraction due to their efficiency in embedded systems with limited hardware constraints. FPN [14] and BiFPN officially showed the performance boost of bidirectional feature fusion to one-stage detectors.

Drivable Area Segmentation
FCN [15] sparked the flame with the first fully convolutional segmentation network. From then on, researchers have found various ways to improve the performance, such as encoder-decoder architecture with UNet [3], the pyramid pooling module of PSPNet [4], or even semisupervised learning based on generative adversarial networks [16] with many data augmentation techniques [17].

Lane Line Segmentation
LaneNet [5] proposed individual lane lines as instances to be segmented. Spatial CNN [6] preferred slice-by-slice convolutions over deep layer-by-layer convolutions, emphasizing objects with heavy spatial relationships but barely noticeable appearances, such as poles, traffic lights, or lane lines. ENet-SAD [18] created self attention distillation, a technique allowing models to self-learn. It works by using attention maps generated

in earlier training points as a form of supervision for later, surpassing SCNN by a large margin.

Multi-task Networks

Many published papers attempted to combine perception tasks into a unified network. Mask R-CNN [19], BlitzNet [20], LSNet [21], MultiNet achieved outstanding results at the time. Then YOLOP became the first real-time state-of-the-art on the BDD100K dataset on three perception tasks: vehicle detection, drivable area, and lane line segmentation.

3 Methodology

Based on these challenges, this research has proposed an end-to-end network architecture that can multi-task. As shown in Fig. 1, our one-stage network includes one sharing encoder and two separated decoders to solve distinct tasks.

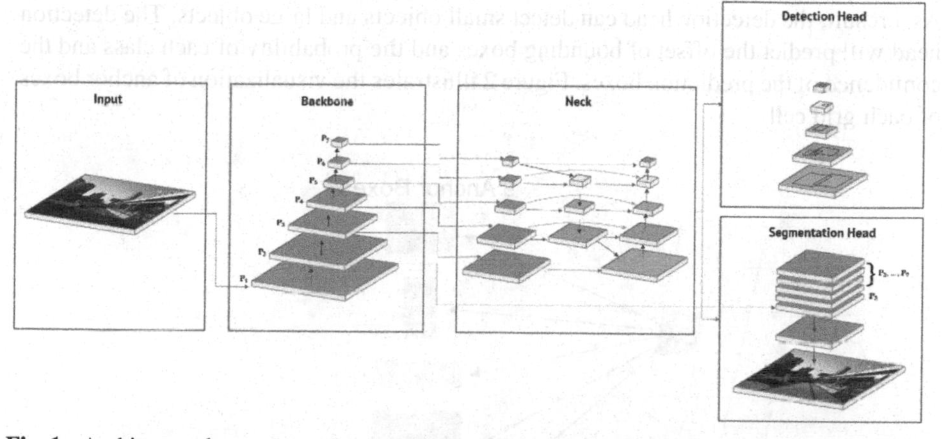

Fig. 1. Architecture has one encoder: backbone network and neck network; two decoders: Detection Head and Segmentation Head. The backbone network generated 5 feature maps from P_1 to P_5. By down-sampling the feature map P_5, we obtain two feature maps P_6 and P_7.

3.1 Backbone Network

The feature extracting, serving as a backbone, is an essential part of the model that can help a variety of networks achieve excellent performance to various tasks. Many modern network architectures currently reuse networks that have good accuracy in the ImageNet dataset to extract features. Recently, EfficientNet showed high accuracy and efficient performance over existing CNNs, reducing FLOPs by orders of magnitude. We choose EfficientNet-B3 as the backbone, which solves the problem of network optimization by finding depth, width, and resolution parameters based on neural architecture search to design a stable network. Therefore, our backbone can reduce the cost computation of the network and obtain several vital features.

3.2 Neck Network

The feature maps from the backbone network are fed to the neck network pipeline. Multi-scale feature representation is the main challenge; FPN recently proposed a feature extractor design to generate multi-scale feature maps to obtain better information. However, the limitation of FPN is that information feature is inherited by a one-way flow. Therefore, our neck network uses a BiFPN module based on EfficientDet. BiFPN fuses feature at a different resolution based on cross-scale connection for each node by each bidirectional (top-down and bottom-up) path and adds weight for each feature to learn the importance of each level. We adopt the method to fuse features in our work.

3.3 Detection Head

Each grid of the multi-scale fusion feature maps from the Neck network will be assigned nine prior anchors with different aspect ratios. Similar to YOLOv4 [22], the aspect ratios are calculated by k-means [23]. In addition, each feature map level will focus on detecting a particular boundary region, so we adopt the specific scaling for each level. As a result, the detection head can detect small objects and large objects. The detection head will predict the offset of bounding boxes and the probability of each class and the confidence of the prediction boxes. Figure 2 illustrates the visualization of anchor boxes of each grid cell.

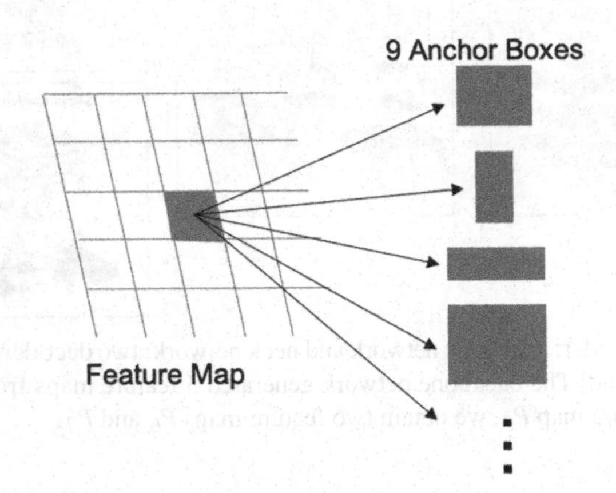
9 Anchor Boxes
Feature Map
Anchor Boxes is calculated by Anchor Scales and Anchor Ratios
based on K-Means algorithm.

Fig. 2. Anchor box ratios is calculated by K-Means algorithm.

Each grid cell has nine anchor boxes, and this research follow the anchor box calculation way from EfficientDet [11], the formula is described as:

$$A(h, w) = 2l.\,base.\,scale.\,ratio$$

where l is the feature map i^{th} level, base scale is constant which is adapted to various datasets, scale and ratio is computed by K-Means algorithm. Figure 3 demonstrates anchor boxes visualization, the green boxes are the boxes matching to ground truth boxes, the cyan blue boxes are the boxes have IoU score value from 25% to 50% compared to ground truth boxes, whereas the red boxes are boxes mismatching to ground truth boxes.

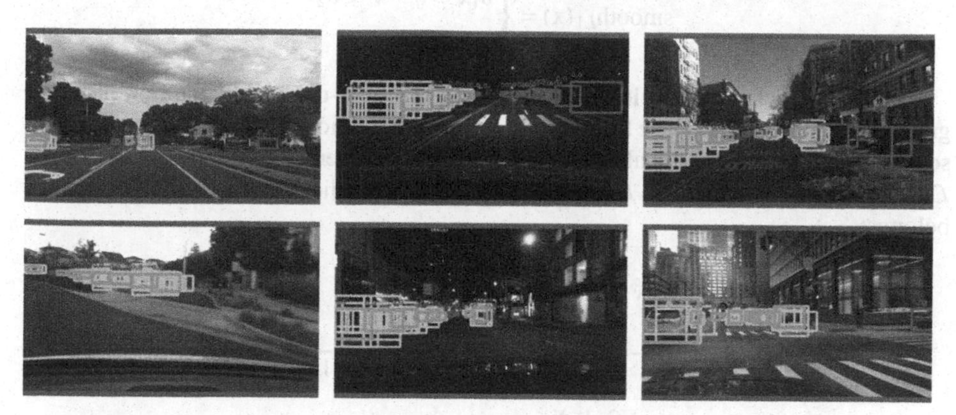

Fig. 3. Anchor boxes visualization.

3.4 Segmentation Head

Segmentation head has 3 classes for output, which are background, drivable area and lane line. This paper keeps 5 feature levels $\{P_3, \ldots, P_7\}$ from Neck network to segmentation branch. First, we up-sample each level to have the same output feature map with size $(\frac{W}{4}, \frac{H}{4}, 64)$. Second, we feed P_2 level to convolution layer to have the same feature map channels with other levels. Then, we combine them to obtain a better feature fusion by summing all levels. Finally, we restore the output feature to the size $(W, H, 3)$, representing the probability of each belonging pixel class. This research scale feature maps to the size of P_2 level, because P_2 level is a strongly semantic feature map. Additionally, we feed P_2 feature map from backbone network which represents low-level feature into the final feature fusion that helps network improve output precision.

3.5 Loss Function

The loss function is described as:

$$\mathcal{L}_{all} = \alpha \mathcal{L}_{det} + \beta \mathcal{L}_{seg}$$

where α, β are tuning parameters to balance the total loss, $\mathcal{L}_{det}$ is the loss for object detection task and $\mathcal{L}_{seg}$ is the loss for segmentation task, the formulation can be written as follow:

$$\mathcal{L}_{det} = \alpha_1 \mathcal{L}_{class} + \alpha_2 \mathcal{L}_{obj} + \alpha_3 \mathcal{L}_{box}$$

$\mathcal{L}_{class}$ and $\mathcal{L}_{obj}$ are focal loss [24], which is implemented for classifying class and the confidence of objects, respectively. The focal loss reduces the slope of loss function and focuses on misclassified examples. $\mathcal{L}_{box}$ is computed by smooth L1 loss, which takes absolutely between the predicted box and ground truth box:

$$\text{smooth}_{L1}(\mathrm{x}) = \begin{cases} \delta_1 x^2 & if \quad x < \delta_2 \\ x - \delta_1 \end{cases}$$

$\mathcal{L}_{seg}$ is multiclass hybrid loss that is utilized for multi-class segmentation of background, drivable area and lane line. Small object segmentation is a challenge in semantic segmentation caused by imbalanced data distribution. Therefore, this paper combines $\mathcal{L}_{Tversky}$ tversky loss [25] and $\mathcal{L}_{Focal}$ focal loss [24] to predict the class to which a pixel belongs:

$$\mathcal{L}_{seg} = \mathcal{L}_{Tversky} + \lambda\mathcal{L}_{Focal}$$

$$\mathcal{L}_{Tversky} = C - \sum_{c=0}^{C-1} \frac{TP_p(c)}{TP_p(c) + \varphi FN_p(c) + (1-\varphi)FP_p(c)}$$

$$\mathcal{L}_{Focal} = -\lambda\frac{1}{N}\sum_{c=0}^{C-1}\sum_{n=1}^{N} g_n(c)(1-p_n(c))^{\gamma}\log(p_n(c))$$

4 Experiments

4.1 Experiment Settings

The Berkeley DeepDrive Dataset (BDD100K) is used in training and validating our model. Since the test labels of 20K images are unavailable, we opt to evaluate on the validation set of 10K images. We prepare the dataset for three tasks according to existing multi-task networks trained on BDD100K to aid in comparison. Of all the ten classes in object detection, we only select {car, truck, bus, train} and merge them into a single class {vehicle} since DLT-Net and MultiNet can only detect vehicles. We also merge two segmentation classes {direct, alternative} into {drivable}. We follow the practice of calculating two lane line annotations into a central one, dilating the annotations in training set to 8 pixels while keeping validation set intact [18]. Images are resized from 1280x720 to 640x384 due to three main reasons, in order of importance: respecting the original aspect ratio, maintaining a good trade-off between performance and accuracy, then making sure the dimensions are divisible by 128 for BiFPN. We use basic augmentation techniques such as rotating, scaling, translating, horizontal flipping, and HSV shifting. While training detection head specifically, we utilize mosaic augmentation that was first introduced in YOLOv4 with great results [22].

Thanks to the proven benefits of transfer learning, we jump-start our model by using EfficientNet-B3 weights pre-trained on ImageNet. Our custom anchor box settings found automatically have scales of $(2^0, 2^{0.7}, 2^{1.32})$ and ratios of $[(0.62, 1.58), (1.0, 1.0), (1.58, 0.62)]$. The chosen optimizer is AdamW [26] with $\gamma = 1e^{-3}$, $\beta_1 = 0.9$, $\beta_2 = 0.999$, $\varepsilon = 1e^{-8}$, $\lambda = 1e^{-2}$. When the model stucks around for 3 epochs, learning rate is decreased tenfold. For object detection, we use smooth L1 loss with $\delta_1 = 4.5$, $\delta_2 = \frac{1}{9}$ for regression and focal loss with $\alpha = 0.25$, $\gamma = 2.0$ for classification. When matching anchor boxes to annotations, we set an IoU threshold of 0.5 for annotations larger than 100 pixels in area but only 0.25 for those smaller. We emphasize regression 4 times more than classification because one-class classification is easy to converge. For drivable area and lane segmentation, we use a combination of tversky loss with $\alpha = 0.7$, $\beta = 0.3$ and focal loss with $\alpha = 0.25$, $\gamma = 2.0$, weighted the same. We train with a batch-size of 16 on a RTX 3090 for 200 epochs.

4.2 Results

In this section, we run experiments with other models on all three tasks and compare on specific tasks.

We present the vehicle detection results and compare them to six models on the BDD100K dataset. Table 1 compares our models to previous networks on the BDD100K dataset. Our model achieves **5.0%** better recall and achieves the best mAP50 at **77%**. We can outperform all previous networks on recall and mAP50 metrics because our network can detect awfully small objects ranging from 3 pixels to 10 pixels with input size 640x384x3 thanks to our automatically customized anchor aspect ratio and scale. Figure 4 illustrates the visualization of traffic object detection.

Table 1. The comparison result on traffic object detection task. The experiment settings include confidence threshold of 0.001 and NMS threshold of 0.6. This paper mainly focuses on obtaining highest Recall IoU

Model	Recall (%)	mAP50 (%)
MultiNet	81.3	60.2
DLT-Net	89.4	68.4
Faster R-CNN	77.2	55.6
YOLOv5s	86.8	77.2
YOLOP	89.2	76.5
Ours	**94.2**	**77.0**

(a) Day-time result

(b) Night-time result

Fig. 4. Visualization of the traffic object detection results. (a) shows results in day-time series with different weather conditions such as clear, heat stroke and heat-wave. (b) shows results in night-time series with different weather such as cool and flurries.

Next, we evaluate the drivable area segmentation task. IoU metric is used to evaluate the segmentation performance of various networks. Table 2 shows the Drivable IoU of five networks. Our network achieves 83.2% IoU, pale in comparison to PSPNet (89.6%) and YOLOP (91.5%). We built a decoder network for multi-classes, whereas YOLOP constructed two decoders for specific tasks. Therefore, our architecture is more flexible and optimistic than theirs. Figure 5 visualizes the semantic segmentation output of drivable areas in various conditions.

Table 2. Performance comparison on drivable area segmentation task

Model	Drivable IoU (%)
MultiNet	71.6
DLT-Net	71.3
PSPNet	89.6
YOLOP	91.5
Ours	**83.2**

(a) Day-time result

(b) Night-time result

Fig. 5. Visualization of the drivable area segmentation results. (a) shows semantic segmentation results in day-time with various views. (b) shows results in night-time series with various brightness views.

Finally, lane detection is one of the main challenges in autonomous driving. The evaluation metrics we use for lane detection are accuracy and IoU. As shown in Table 3, our network outperforms all previous models with accuracy **87.1%** and IoU **29.3%**. Figure 6 visualizes the lane detection results in various conditions.

Table 3. Performance comparison on lane detection task

Model	Accuracy (%)	Line IoU (%)
ENet	34.12	14.64
SCNN	35.79	15.84
Enet-SAD	36.56	16.02
YOLOP	70.50	26.20
Ours	**87.10**	**29.30**

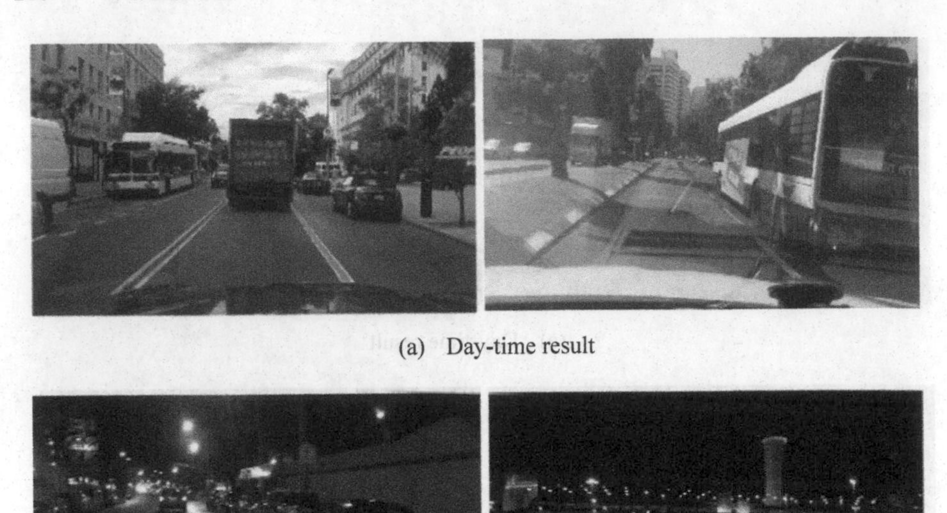

(a) Day-time result

(b) Night-time result

Fig. 6. Visualization of the lane detection results. (a) shows results in day-time series with various weather conditions. (b) shows results in night-time series with various brightness views.

5 Conclusion

In this paper, we systematically study network architecture design choices for multi-tasking, propose an efficient end-to-end perception network, customize automatic aspect ratios for each level in the weighted bidirectional feature network, and build efficient training loss function and training strategy to improve accuracy and performance. Based on these optimizations, we develop a new end-to-end multi-network, which achieves better accuracy and efficiency than prior art across a broad spectrum of resource constraints. Especially, our network achieves state-of-the-art accuracy with fewer FLOPS than previous multi-network models. In future works, we will try to ameliorate lane lines performance as well as context of structures in drivable area segmentation.

This paper also can be a reference for many fields in Machine Learning, for example, Pattern Recognition [28, 29], Deep learning [30–32], etc.

References

1. Liu, W., et al.: SSD: single shot multibox detector. In: Leibe, B., Matas, J., Sebe, N., Welling, M. (eds.) ECCV 2016. LNCS, vol. 9905, pp. 21–37. Springer, Cham (2016). https://doi.org/10.1007/978-3-319-46448-0_2
2. Redmon, J., Divvala, S., Girshick, R., Farhadi, A.: You only look once: unified, real-time object detection. In: 2016 IEEE Conference on Computer Vision and Pattern Recognition (CVPR) (2016)

3. Ronneberger, O., Fischer, P., Brox, T.: U-Net: convolutional networks for biomedical image segmentation. In: Navab, N., Hornegger, J., Wells, W.M., Frangi, A.F. (eds.) MICCAI 2015. LNCS, vol. 9351, pp. 234–241. Springer, Cham (2015). https://doi.org/10.1007/978-3-319-24574-4_28
4. Badrinarayanan, V., Kendall, A., Cipolla, R.: SegNet: a deep convolutional encoder-decoder architecture for image segmentation. IEEE Trans. Pattern Anal. Mach. Intell. **39**, 2481–2495 (2017)
5. Wang, Z., Ren, W., Qiu, Q.: LaneNet: Real-Time Lane Detection Networks for Autonomous Driving (2018)
6. Pan, X., Shi, J., Luo, P., Wang, X., Tang, X.: Spatial As Deep: Spatial CNN for Traffic Scene Understanding (2017)
7. Teichmann, M., Weber, M., Zollner, M., Cipolla, R., Urtasun, R.: MultiNet: real-time joint semantic reasoning for autonomous driving. In: 2018 IEEE Intelligent Vehicles Symposium (IV) (2018)
8. Qian, Y., Dolan, J., Yang, M.: DLT-Net: joint detection of drivable areas, lane lines, and traffic objects. IEEE Trans. Intell. Transp. Syst. **21**, 4670–4679 (2020)
9. Wu, D., Liao, M., Zhang, W., Wang, X.: YOLOP: You Only Look Once for Panoptic Driving Perception (2021)
10. Yu, F., et al.: BDD100K: a diverse driving dataset for heterogeneous multitask learning. In: 2020 IEEE/CVF Conference on Computer Vision and Pattern Recognition (CVPR) (2020)
11. Tan, M., Pang, R., Le, Q.: EfficientDet: scalable and efficient object detection. In: 2020 IEEE/CVF Conference on Computer Vision and Pattern Recognition (CVPR) (2020)
12. Tan, M., Le, Q.: EfficientNet: Rethinking Model Scaling for Convolutional Neural Networks (2019)
13. Redmon, J., Farhadi, A.: YOLO9000: better, faster, stronger. In: 2017 IEEE Conference on Computer Vision and Pattern Recognition (CVPR) (2017)
14. Lin, T., Dollar, P., Girshick, R., He, K., Hariharan, B., Belongie, S.: Feature pyramid networks for object detection. In: 2017 IEEE Conference on Computer Vision and Pattern Recognition (CVPR) (2017)
15. Shelhamer, E., Long, J., Darrell, T.: Fully convolutional networks for semantic segmentation. IEEE Trans. Pattern Anal. Mach. Intell. **39**, 640–651 (2017)
16. Han, X., Lu, J., Zhao, C., You, S., Li, H.: Semisupervised and weakly supervised road detection based on generative adversarial networks. IEEE Signal Process. Lett. **25**, 551–555 (2018)
17. Munoz-Bulnes, J., Fernandez, C., Parra, I., Fernandez-Llorca, D., Sotelo, M.: Deep fully convolutional networks with random data augmentation for enhanced generalization in road detection. In: 2017 IEEE 20th International Conference on Intelligent Transportation Systems (ITSC) (2017)
18. Hou, Y., Ma, Z., Liu, C., Loy, C.: Learning lightweight lane detection CNNs by self attention distillation. In: 2019 IEEE/CVF International Conference on Computer Vision (ICCV) (2019)
19. He, K., Gkioxari, G., Dollar, P., Girshick, R.: Mask R-CNN. 2017 IEEE International Conference on Computer Vision (ICCV) (2017)
20. Dvornik, N., Shmelkov, K., Mairal, J., Schmid, C.: BlitzNet: a real-time deep network for scene understanding. In: 2017 IEEE International Conference on Computer Vision (ICCV) (2017)
21. Liu, B., Chen, H., Wang, Z.: LSNet: Extremely Light-Weight Siamese Network For Change Detection in Remote Sensing Image
22. Bochkovskiy, A., Wang, C., Mark Liao, H.: YOLOv4: Optimal Speed and Accuracy of Object Detection (2020)
23. MacQueen, J.: Some Methods for Classification and Analysis of Multivariate Observations (1966)

24. Lin, T., Goyal, P., Girshick, R., He, K., Dollar, P.: Focal loss for dense object detection. In: 2017 IEEE International Conference on Computer Vision (ICCV) (2017)
25. Salehi, S.S.M., Erdogmus, D., Gholipour, A.: Tversky loss function for image segmentation using 3D fully convolutional deep networks. In: Wang, Q., Shi, Y., Suk, H.-I., Suzuki, K. (eds.) MLMI 2017. LNCS, vol. 10541, pp. 379–387. Springer, Cham (2017). https://doi.org/10.1007/978-3-319-67389-9_44
26. Loshchilov, I., Hutter, F.: Decoupled Weight Decay Regularization (2019)
27. Chollet, F.: Xception: Deep learning with depthwise separable convolutions. In: 2017 IEEE Conference on Computer Vision and Pattern Recognition (CVPR) (2017)
28. Hung, P.D., Loan, B.T.: Automatic vietnamese passport recognition on android phones. In: Dang, T.K., Küng, J., Takizawa, M., Chung, T.M. (eds.) FDSE 2020. CCIS, vol. 1306, pp. 476–485. Springer, Singapore (2020). https://doi.org/10.1007/978-981-33-4370-2_36
29. Hung, P.D., Linh, D.Q.: Implementing an android application for automatic vietnamese business card recognition. Pattern Recognit. Image Anal. **29**, 156–166 (2019). https://doi.org/10.1134/S1054661819010188
30. Hung, P.D., Su, N.T., Diep, V.T.: Surface classification of damaged concrete using deep convolutional neural network. pattern recognit. Image Anal. **29**, 676–687 (2019). https://doi.org/10.1134/S1054661819040047
31. Hung, P.D., Kien, N.N.: SSD-Mobilenet implementation for classifying fish species. In: Vasant, P., Zelinka, I., Weber, G.-W. (eds.) ICO 2019. AISC, vol. 1072, pp. 399–408. Springer, Cham (2020). https://doi.org/10.1007/978-3-030-33585-4_40
32. Su, N.T., Hung, P.D., Vinh, B.T., Diep, V.T.: Rice leaf disease classification using deep learning and target for mobile devices. In: Al-Emran, M., Al-Sharafi, M.A., Al-Kabi, M.N., Shaalan, K. (eds.) ICETIS 2021. LNNS, vol. 299, pp. 136–148. Springer, Cham (2022). https://doi.org/10.1007/978-3-030-82616-1_13

Cardiovascular Disease Classification Based on Machine Learning Algorithms Using GridSearchCV, Cross Validation and Stacked Ensemble Methods

Satyabrata Pattanayak and Tripty Singh(✉)

Department of Computer Science and Engineering, Amrita School of Engineering, Amrita Vishwa Vidyapeetham, Bengaluru, India
{p_satyabrata,tripty_singh}@blr.amrita.edu

Abstract. Cardiac illness is a foremost research area in medicine that has recently gotten a lot of interest around the world. In the medical profession, there is an enormous volume of data that can be eviscerated and used for numerous determinations. In the prediction of cardiovascular illness, machine learning algorithms play a critical role. Many studies have been conducted throughout the years in order to deal with the early detection of diseases. Based on the clinical data, our study article determines if the patient is likely to be analyzed with cardiovascular disease. Various strategies for data preprocessing will be employed throughout this work, and performance analysis will be performed on the distinct classification algorithms in order to predict if the patient has heart disease or not. For the UCI cardiovascular dataset, the suggested study offered many forms of machine learning and deep learning approaches that will tackle the heart disease prediction challenge. In addition, the proposed model takes into account not only various machine learning algorithms, but also hyper tweaking the parameters using GridSearchCV, Cross Validation, and Stacked Ensemble approaches. The suggested technique provides a good interpretation of the model validation through accuracy, AUC, precision, recall, KS statistics, and cumulative gain, lift curve, learn curve, calibration curve, and cross validation curve in terms of relative accurateness.

Keywords: Cardiovascular disease · Random forest · Logistic regression · K Neighbors · Accuracy · ROC · Cumulative gain · Lift · KS Statistics · Calibration curve

1 Introduction

Any disorder in artery with respect to heart or blood, we can consider it as the cardiovascular diseases. The common types of cardiovascular diseases are coronary artery disease, arrhythmia, heart valve disease, and heart failure. Coronary artery disease is a disorder in which the heart muscles do not receive enough blood and oxygen due to a blockage in the coronary arteries. The hardening of the arteries causes coronary artery and vascular disease (atherosclerosis). Coronary artery disease is caused by constricted

M. Singh et al. (Eds.): ICACDS 2022, CCIS 1613, pp. 219–230, 2022.
https://doi.org/10.1007/978-3-031-12638-3_19

or blocked arteries in the heart. It's the most frequent type of heart disease, and it's what causes the majority of heart attacks and angina (chest pain). Problems in other blood channels cause vascular disease, which reduces blood flow and impairs heart function.

There are multiple open challenges for the researchers in heart diseases like Is it possible to forecast heart attacks using high-resolution images of blood cells? How can we make the Health Check more accurate? What kind of exercise is best for your heart? Who is at risk of MI drug side effects? How do some regions of our genes influence our risk of cardiovascular disease? How do drugs impact calcium signal in the heart?

Many machine learning techniques are utilized to detect cardiac disease. Not only can we use a single algorithm, but we can also combine two or more ML algorithms to accurately diagnose heart disease. Machine Learning's fundamental goal is to create predictions based on past experience [16]. To begin, the Machine Learning algorithms create a model using an acceptable training dataset. The model then accepts the user's input and analyses and provides an accurate response based on the training data. Currently, a number of researchers are utilizing machine-learning algorithms to assist the health-care industry and specialists in the detection of cardiac-related disorders. ML is thought to be a useful tool since it creates rules and then makes decisions based on those rules.

2 Literature Survey

Many researchers have worked in the field of cardiovascular diseases. Primarily these works can be categorized into statistical techniques and learning techniques. Pronab et. al. [1] implemented a tree-based model by combining basic classifiers. It took mostly bagging and boosting techniques. By using these techniques, new hybrid model created. It looks like Decision tree with bagging, Random Forest with bagging etc. to solve the machine learning cardiovascular problems [2]. The model is based on the extended version of deep learning. It added with CNN model with existing advanced DL model. The model is also used the Multilayer perceptron and also regularization techniques to classify the heart disease prognostics [3]. During this investigation, the study devised a framework for comprehending the concepts of forecasting the risk profile of patients using clinical data characteristics. Deep Neural Networks and a 2-statistical model are used to build the proposed model. The issue of under fitting and overfitting is no longer an issue. The effectiveness of the model that accurately predicts the presence or absence of heart disease was evaluated using DNN and ANN [4] this research describes a machine learning-based automatic categorization system for diagnosing cardiac diseases based on heart sounds. The system takes each frame of heart sound as the dimension for the model and classifies the cardiovascular disease [5]. AI has been used in agriculture and healthcare to improve agricultural production, prognostication. Similar to agricultural study, healthcare also provides a classification heart disease [6]. The study focuses on the chronic disease. Due to this chronic disease, patients are spending a huge amount of the money. Patients with such conditions require lifelong treatment. Predictive models are now commonly used in the diagnosis and prediction of various diseases. The machine learning model will help to predict the chronic disease [7]. Gen-Min Lin et. al. offers a strategy that effectively detects more than 70% of RVH in young adults using simple physiological markers and ECG data. Clinical Impact: This approach allows for a quick, accurate, and practical diagnosis of RVH [8]. Machine learning (ML) may be employed for CVD diagnostic screening based on the gut microbiome. To test the hypothesis, researchers used 5 supervised machine learning algorithms to analyses the data. These algorithms included random forest, support vector machine, decision tree, and elastic net, and neural networks. Between the CVD and non-CVD groups, 39 distinct bacterial taxa were discovered [9]. The study is focused on the machine learning model on ECG Signal data. The model considers moving average and Fourier transformation techniques. It provides the peak in the signal and will classify the heart disease [10]. The study occurred on ECG data of approx. thirty-nine thousand individuals. A deep learning model with LSTM conducted on the data. It detects the heart measure from the data and classify the heart related disease [11]. The study conducted on the hundreds of the patients with angiography images. Different arteries extracted from the angiography images for the further analysis [17]. A machine learning model created which was extracting the feature automatically. Subsequently boosting based ensemble method created to detect the blockage on the artery [12]. The study was conducted for the heart disease by considering machine learning model more over the information gain. With this approach automatic feature selection extracted and result was focused on the area under curve method [13]. The study conducted on heart illness through optimization method which shows how all different features focus to the disease through SWAM optimization method [14]. Swarm Intelligence utilizing data reduction in healthcare was investigated and the results were presented in this research. Swarm Intelligence algorithms were used to anticipate major diseases like cancer, tumors, cardiology, and heart disease. The Swarm Intelligence approach encompasses applications in the fields of illness diagnosis and therapy.

3 Research Methodology

3.1 Proposed Approach

The proposed method is to define a spontaneous system architecture that analyses many machine learning models and regulates which model is the best. The first step is to process the UCI cardio disease dataset by treating the missing data, outlier detection, checking correlation and variance. Then there's data preparation, which includes, feature engineering, feature selection, and dimensional reduction. Subsequently using GridSearchCV, Cross Validation and Stacked ensemble method, to find the best model to evaluate. The next stage is to define and test a variety of models that could improve bi-directional categorization results. Following that, model validation, which can be done using several statistical tests and many model parameters. Models will be compared and the optimal output model will be chosen based on the results.

3.2 Cross Validation, Stacked Ensemble, GridSearchCV Method

We really cannot fit the models on the training examples in machine learning, thus we can't say that it will perform accurately on real data [18]. It is required to remove the noise from the data and also should create a desired pattern to provide the accurate prediction. Cross Validation performs better result when the data is unseen. It will not consider the whole dataset. It will consider a subset of the whole dataset and conduct the test [15]. Most importantly, it will take subset of whole data as sample data. And will consider all the remaining data as train data. Again, it will do the resampling and same procedure will repeat for the multiple times. Many cross-validation techniques are available like K-fold cross validation, leave p-out cross validation, leave one-cut cross validation etc. There are many more different cross validation techniques are available like Time Series cross-validation, Stratified k-fold cross-validation, Repeated random subsampling validation etc. Leave-p-out is exhaustive. K-fold is non-exhaustive.

Stacked model is the ensemble techniques. It uses classification for numerous nodes. It folds the data and each part considers the training data. It passes the model into another model generated data. So, it created the model which provides the best result and pattern.

GridSearchCV analyses all the available combinations of variables to consider and their values to optimize, this will take quite a long time [19]. GridSearchCV is an algorithm for detecting the optimal parameter values from a given set of parameters in a grid. It's essentially a cross-validation technique. The model as well as the parameters must be entered. After extracting the best parameter values, predictions are made. A clean Stratified Fold or K-Fold cross-validator will be used by GridSearchCV. Shuffle = False is the baseline for such cross-validators. GridSearchCV's cv parameter description also contains some valuable knowledge.

4 Experimental Design

4.1 Dataset Description

The UCI dataset has 1190 rows and 12 features in total. Age, sex, chest pain type, resting blood pressure, cholesterol, fastingbloodsugar, rest-ecg, max heart rate achieved, exercise induced angina, st-depression, and st-slope are all included in the dataset. There are 629 records for cardiac disease patients and 561 records for healthy patients. Typical angina, atypical angina, non-angina pain, and asymptomatic angina are the different types of chest pain. Normal resting ECG, ST-T wave abnormalities, and left ventricular hypertrophy. ST Slope is divided into four categories: normal, upsloping, flat, and down sloping. 483 people with cardiac disease are asymptotic in the collection, with ages ranging from 28 to 77.

4.2 Data Pre-processing

Missing data was examined during data preprocessing, and no missing records were discovered. During the outlier discovery process, several cholesterol level records were discovered during the box plot detection and were spotted by deleting the Z-score criteria. Before moving on to the model, normalize the data to reduce variance. To check the relationship, run a correlation check within the data. There is no correlation exists between the variables before moving to the model (Fig. 1).

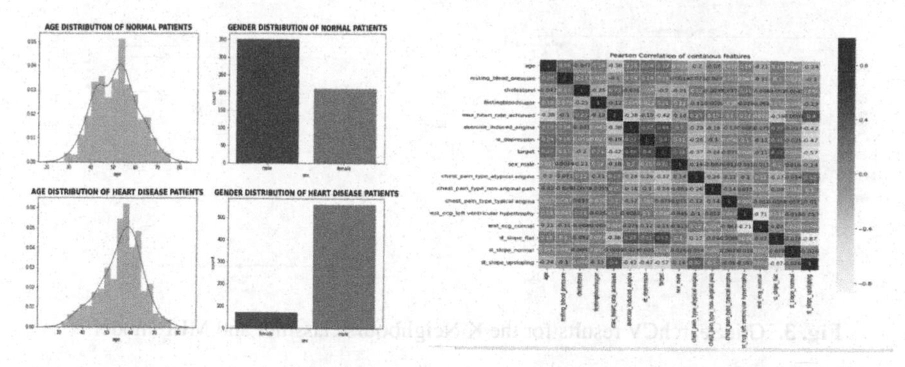

Fig. 1. Age and gender distributions among the heart disease patients and normal people.

4.3 Baseline Models

Different classification machine learning algorithms like Logistic regression, Random Forest Classifier, K Neighbors Classifier is implemented to derive the best accuracy for the conversion.

Different parameters such as l1 penalty, l2 penalty, elastic net, and multiple solvers such as newton-cg, lbfgs, liblinear, sag, saga, and others are taken into account when implementing the Logistic Regression model. Different estimators (10,25,50,100), different criterion (Gini, entropy), maximum depth, minimum sample split, minimum impurity decrease, and Random Forest provides the better result while criterion defined as gini and n_estimators = 200 (Fig. 2).

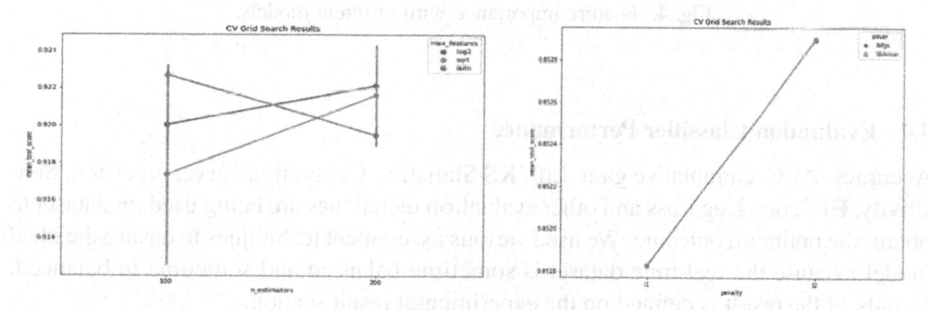

Fig. 2. GridSearchCV results for the random forest and logistic regression classifier model.

Different parameters are taken into account when implementing the K Neighbors Classifier model, such as the number of neighbors (2,3,5), different weights (uniform, distance), different algorithms for computing neighbors (auto, ball tree, kd tree, brute), and other parameters such as leaf size, metric, and so on. K Neighbors Classifier provides a better result with algorithm as Ball Tree, distance metrics as minkowski and no of neighbors as 9. Multilayer perceptron classifier provides a better result with activation function as relu, auto batch size, solver as adam, learning rate as constant with early stopping False and maximum iteration is 100 (Fig. 3).

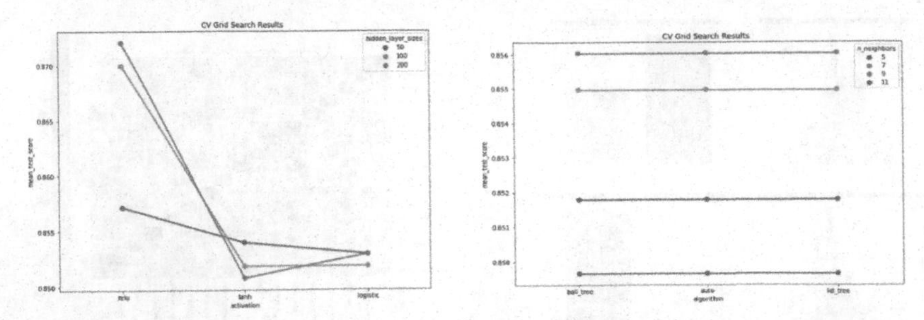

Fig. 3. GridSearchCV results for the K Neighbours classifier and MLP model.

During the model, there are feature selections are implemented with some of the models. Different feature selection algorithms has been used to find out best features like Pearson, LightGBM, Chi-2, Logistic, Random Forest etc (Fig. 4).

	Feature	Pearson	LightGBM	Chi-2	RFE	Logistics	Random Forest	Total
1	st_slope_upsloping	True	True	True	True	True	True	6
2	exercise_induced_angina	True	True	True	True	True	True	6
3	st_slope_flat	True	False	True	True	True	True	5
4	st_depression	True	True	True	True	False	True	5
5	max_heart_rate_achieved	True	True	True	True	False	True	5
6	age	True	True	True	True	False	True	5
7	sex_male	True	False	True	True	True	False	4
8	cholesterol	True	True	False	True	False	True	4
9	chest_pain_type_non-anginal pain	True	False	True	True	True	False	4
10	chest_pain_type_atypical angina	True	False	True	True	True	False	4
11	fastingbloodsugar	True	False	True	False	True	False	3

Fig. 4. Feature importance with different models.

4.4 Evaluation Classifier Performance

Accuracy, AUC, Cumulative gain, Lift, KS Statistics, Calibration curve, precision, Sensitivity, F1 Score, Log Loss and other evaluation techniques are being used on dataset to obtain the optimum outcome. We used various assessment techniques to ensure the ideal model because the real-time dataset is sometime balanced and sometime imbalanced. Details of the result is defined on the experimental result section.

4.5 Implementation Specifics

The model can be run on a basic system with a configuration of 16GB RAM. The model was created using Python and a Jupyter notebook. Numpy, pandas, and other python packages are used for data preprocessing and feature engineering.

5 Experiment Results

The final analysis was done to evaluate the efficiency of different models and establish how the Random Forest model understands the nature of the UCI cardiovascular disease dataset. The results of the analysis, based on the performance measures like Accuracy, ROC, Cumulative gain, Lift, KS Statistics, Calibration curve etc., are shown in the subsections below.

5.1 Accuracy

Table 1 shows that the Random Forest model outperforms compare to the other models by providing the highest accuracy 96.01 percent. We explored the Random forest model with various parameters such as the number of trees, the criterion, the class weight, and so on, but it produced good results with the criterion of Gini impurity and 200 trees. Furthermore, the Logistic regression model was experienced with various solvers, penalties, and class weights, but it performed best with the lbfgs (Limited-memory Broyden–Fletcher–Goldfarb–Shanno) solver and the l2 penalty by providing the accuracy of 92.43%. Even MLP classifier provides 93.56% of accuracy while K Neighbor classifier shows accuracy 91.24%.

Table 1. Model accuracy

Model	Accuracy	AUC
Logistic regression	92.43	.90
Random forest	96.01	.93
K Neighbours	91.24	.89
MPL classifier	93.56	.91

5.2 AUC, ROC, Precision and Recall

AUC is a validation technique. It will plot the graph having x-axis as false positive rate and y-axis is true positive rate. Bu plotting this graph it will separate the signals from the noisy data. As always area under the curve lies between 0 to 1, so the score is above .9 considered to be the good accuracy.

Evaluating the AUC is an excellent way of confirming the outcome at a greater level. So, to test the probability estimate for a classifier ranking a randomly selected positive occurrence greater than a randomly selected negative occurrence, it will be better to check the AUC along with accuracy. As according Table 1, Random Forest does have a higher AUC of .93 over Logistic Regression (.92) as well as other machine learning models such as K-Neighbor (.89). In Random Forest, the micro average of area in the ROC curve was .93, whereas the macro average of area in the ROC curve was.93. Similarly, the Random Forest model has an AUC of.93 for class 0 and.93 for class 1, which demonstrates the maintainability spanning across the classes. Precision is the figure of essential entries obtained by research separated by the total figure of specific pertinent entries found, whereas recall is the figure of pertinent IDs recovered by a exploration separated by the total figure of IDs found.

In the below Precision-Recall curve, area is .933 for class 0 and .922 for class 1 (Fig. 5).

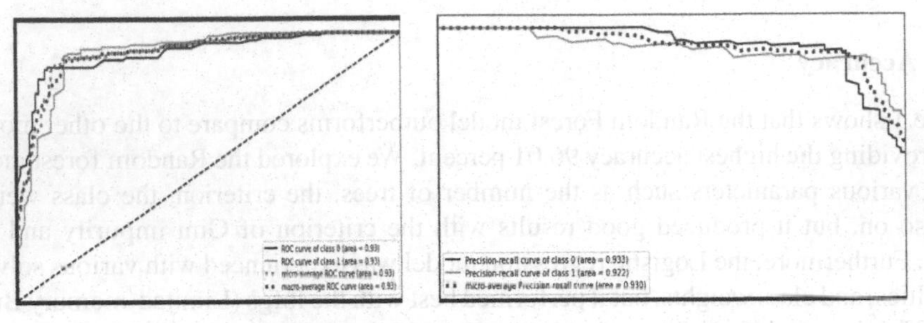

Fig. 5. AUC, ROC, precision and recall curve.

5.3 KS Statistics

As the dataset created with intention of imbalance [20], so to verify the result in the statistical distribution level, the Kolmogorov-Smirnov Goodness of Fit Test (K-S test) is the best test to verify across the classes. The test is commonly used as a normality test to determine if the distribution is normal, regardless of the fact that it is non - parametric and does not imply any specific underlying pattern. Random Forest provides a K-S statistics score of .794 at a threshold level of .570 which clearly states that the conversion prediction is correctly distributed even also the data is imbalanced (Fig. 6).

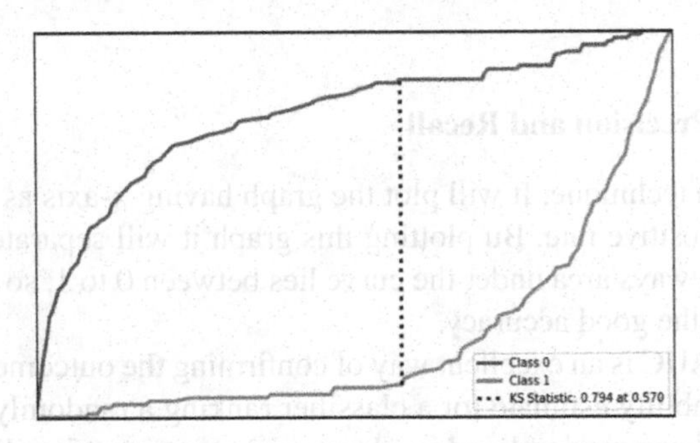

Fig. 6. K-S statistic result of random forest.

5.4 Cumulative Gain and Lift Curve

The sum of the scaled significance values among all outcomes in a search result list is called Cumulative Gain (CG). The rank (position) of a result in the result list is not taken into account when determining the usefulness of a result set in this predecessor of DCG. The cumulative gains graph depicts the proportion of total cases in a specific category that were "gained" by targeting a percentage of total instances. Lift is the predicted rate divided by the Average rate. In direct marketing, the lift curve is a popular approach. Consider a data mining technique that seeks to identify the most likely responders to a mailing by assigning a probability of responding" score to each case. The Lift curve depicts the relationship between the fraction of real positive data instances and the classifier's threshold, or the number of examples we categories as positive (Fig. 7).

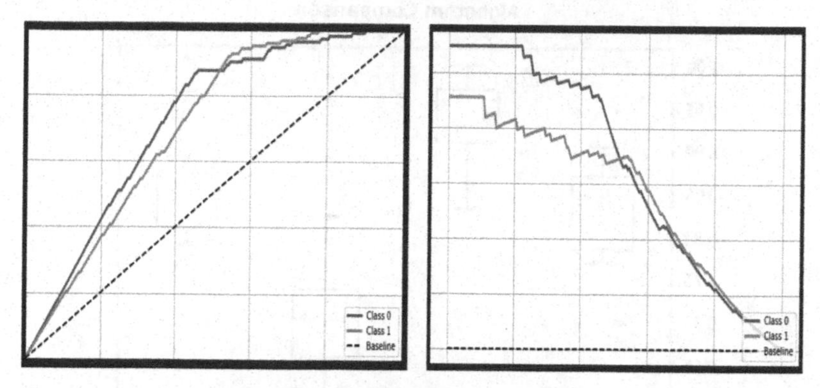

Fig. 7. Cumulative gain and lift curve.

5.5 Learning Curve and Calibration Curve

In machine learning, a learning curve compares the optimal value of a model's loss function. For a training set to the same loss function measured on a validation data set with the same parameters. Calibration curves (sometimes called dependability diagrams) assess how well a binary classifier's probabilistic predictions have been calibrated. It graphs the true incidence of the positive label against its expected likelihood for binned forecasts. The x axis depicts the average predicted chance in each bin. On the y axis is plotted the fraction of positives, or the proportion of samples in the positive class (in each bin). In terms of calibration plot outputs, Random Forest performs better. The smoothness of the curve from 0 to 1 is plainly visible. It states unequivocally that the model is a good fit for the data (Fig. 8).

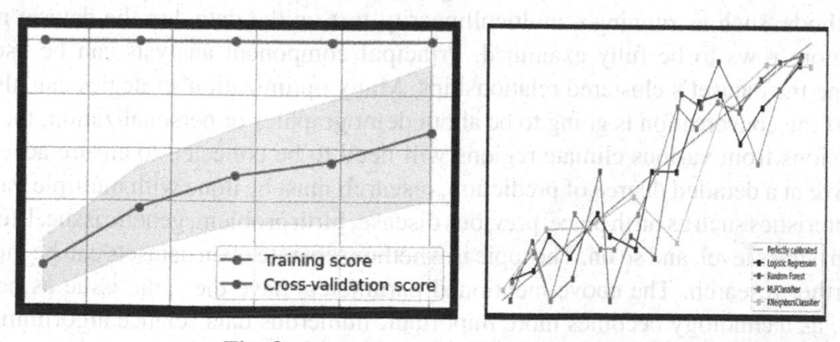

Fig. 8. Learning and calibration curve.

5.6 Cross Validation Score

In Cross validation score, Random Forest provides a better result of .96 where Logistic regression provides .91. In the same time KNN provides below .9 score. As Random Forest is the ensemble method and the problem statement is binary, so it fits properly for the correct result (Fig. 9).

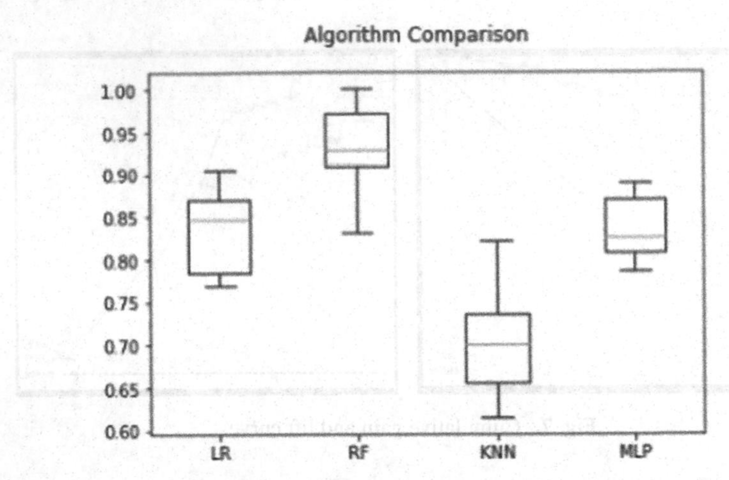

Fig. 9. Cross validation score of each model.

6 Conclusion and Future Work

The proposed paper presented different types of machine learning and deep learning methodologies for the UCI cardiovascular dataset which will solve the heart disease prediction problem. The suggested model, in addition, takes into consideration not only the various machine learning algorithm but it took the consideration of hyper tuning the parameters through GridSearchCV, Cross Validation and Stacked Ensemble methods. In terms of relative inaccuracy, the suggested method performs a good interpretation of the model validation through accuracy, AUC, precision, recall, KS statistics, and cumulative gain, lift curve, learn curve, calibration curve and cross validation curve.

Future work will take a variety of intriguing paths. The task can be done in a variety of methods, such as running a multicollinearity test on the data, but the dataset needs have more rows to be fully examined. Principal component analysis can be used to examine the dataset's clustered relationships. Many optimization strategies can also be used. If the conversation is going to be about demographics or personalization, then the dimensions from various climate regions will need to be collected to ensure accuracy. To arrive at a detailed degree of prediction, research must be done with multiple various characteristics such as birth place, previous disease, birth problem, genetic issue, lifestyle pattern, stress level, and so on. The topic is whether separate open datasets can be merged for further research. The above-mentioned parameters have the same issue as before. Again, as technology becomes more important, numerous data science algorithms can be combined with numerical and image parameters.

References

1. Ghosh, P., et al.: Efficient prediction of cardiovascular disease using machine learning algorithms with relief and LASSO feature selection techniques. IEEE Access **9**, 19304–19326 (2021). https://doi.org/10.1109/ACCESS.2021.3053759
2. Pan, Y., Fu, M., Cheng, B., Tao, X., Guo, J.: Enhanced deep learning assisted convolutional neural network for heart disease prediction on the internet of medical things platform. IEEE Access **8**, 189503–189512 (2020). https://doi.org/10.1109/ACCESS.2020.3026214
3. Ramprakash, P., Sarumathi, R., Mowriya, R., Nithyavishnupriya, S.: Heart disease prediction using deep neural network. Int. Conf. Inventive Comput. Technol. (ICICT) **2020**, 666–670 (2020). https://doi.org/10.1109/ICICT48043.2020.9112443
4. Yadav, A., Singh, A., Dutta, M.K., Travieso, C.M.: Machine learning-based classification of cardiac diseases from PCG recorded heart sounds. Neural Comput. Appl. **32**(24), 17843–17856 (2019). https://doi.org/10.1007/s00521-019-04547-5
5. Pallathadka, H., Mustafa, M., Sanchez, D.T., Sajja, G.S., Gour, S., Naved, M.: Impact of machine learning on management, healthcare and agriculture. Mater. Today: Proc. (2021). https://doi.org/10.1016/j.matpr.2021.07.042
6. Battineni, G., Sagaro, G.G., Chinatalapudi, N., Amenta, F.: Applications of machine learning predictive models in the chronic disease diagnosis. J. Pers. Med. **10**(2), 21 (2020). https://doi.org/10.3390/jpm10020021
7. Memon, M.S., Lakhan, A., Mohammed, M.A., Qabulio, M., Al-Turjman, F., Abdulkareem, K.H.: Machine learning-data mining integrated approach for premature ventricular contraction prediction. Neural Comput. Appl. 1–17 (2021)
8. Hossain, M.E., Uddin, S., Khan, A.: Network analytics and machine learning for predictive risk modelling of cardiovascular disease in patients with type 2 diabetes. Expert Syst. Appl. **164**, 113918 (2021). https://doi.org/10.1016/j.eswa.2020.113918
9. Aziz, S., Ahmed, S., Alouini, M.-S.: ECG-based machine-learning algorithms for heartbeat classification. Sci. Rep. **11**(1), 1–14 (2021)
10. Chang, K.-C., et al.: Usefulness of machine learning-based detection and classification of cardiac arrhythmias with 12-lead electrocardiograms. Can. J. Cardiol. **37**(1), 94–104 (2021). https://doi.org/10.1016/j.cjca.2020.02.096
11. Dey, D., et al.: Integrated prediction of lesion-specific ischaemia from quantitative coronary CT angiography using machine learning: a multicentre study. Eur. Radiol. **28**(6), 2655–2664 (2018). https://doi.org/10.1007/s00330-017-5223-z
12. Motwani, M., et al.: Machine learning for prediction of all-cause mortality in patients with suspected coronary artery disease: a 5-year multicentre prospective registry analysis. Eur. Heart J. **38**(7), 500–507 (2017)
13. Habib, M., Aljarah, I., Faris, H., Mirjalili, S.: Multi-objective particle swarm optimization: theory, literature review, and application in feature selection for medical diagnosis. In: Mirjalili, S., Faris, H., Aljarah, I. (eds.) Evolutionary Machine Learning Techniques. AIS, pp. 175–201. Springer, Singapore (2020). https://doi.org/10.1007/978-981-32-9990-0_9
14. Narmatha, P., Ramesh, M., Theivanayaki, S.: Data mining and swarm intelligence in healthcare applications. J. Comput. Theor. Nanosci. **18**(3), 1100–1106 (2021)
15. Nair, R.R., Singh, T.: An optimal registration on shearlet domain with novel weighted energy fusion for multi-modal medical images. Optik **225**, 165742 (2021)
16. Yaramalla, D., Singh, T.: A Machine learning paradigm for explanatory cases with CKD. In: 2021 12th International Conference on Computing Communication and Networking Technologies (ICCCNT), pp. 1–7. IEEE (July 2021)
17. Sahay, A., Amudha, J.: Integration of prophet model and convolution neural network on wikipedia trend data. J. Comput. Theor. Nanosci. **17**(1), 260–266 (2020)

18. Soman, K.P., Amudha, J., Kiran, Y.: Feature selection in top-down visual attention model using WEKA. Int. J. Comput. Appl. **975**, 8887 (2011)
19. Maheswari, K.U., Shobana, G., Bushra, S.N., Subramanian, N.: Supervised malware learning in cloud through System calls analysis. In: 2021 International Conference on Innovative Computing, Intelligent Communication and Smart Electrical Systems (ICSES), pp. 1–8. IEEE (Sep 2021)
20. Chandini, A.A.: Improved quality detection technique for fruits using GLCM and multiclass SVM. In: 2018 International Conference on Advances in Computing, Communications and Informatics (ICACCI), pp. 150–155. IEEE (Sep 2018)

Human Emotion Recognition from Body Posture with Machine Learning Techniques

S. Vaijayanthi(✉) and J. Arunnehru

Department of Computer Science and Engineering, SRM Institute of Science and Technology, Vadapalani, Chennai, Tamilnadu, India
vaijayanthisekar@gamil.com

Abstract. Recently, increasing attention in the field of gesture recognition, has become a key strategy in analyzing the emotional states of human body movements for social communication. Most real-life scenarios include identifying emotions from facial expressions, vocal synthesis, hand recognition and body gestures. The body posture powerfully conveys the micro emotions of a person in depth. The prediction of human - gait is significantly harder, because the pattern of the human pose estimation has additional degrees of self-determination than the facial emotions, and the overall shape varies robustly during the articulated motion. In this paper, we propose a novel method to recognize 17 different micro emotions from GEMEP dataset based on human upper body gestures dynamics features extracted from the abstract representations of patterns from videos. In the experimental results, KNN exhibit the proposed architecture's effectiveness with an accuracy rate of 97.1% for the GEMEP dataset, 95.2% for SVM, 51.6% for Decision Tree and 49.7% Naive Bayes, respectively.

Keywords: Body movements · Gesture analysis · Human emotion recognition · Machine learning

1 Introduction

In today's world, the interaction of emotional intelligence plays an effective role in computer-mediated communication. Human emotion analysis is an emerging subject of study in many scientific disciplines, including cognitive science, psychology, and neuroscience. Emotions are elicited by various forms through facial expressions [1], the voice of tone, speech, hand gestures and body posture [2]. The various modalities in verbal communication yield a better interpretation of human behaviour, in which nonverbal cues is highly essential that characterize the emotional expressions [3]. According to psychological research, facial expressions [4] transmit 55% of the total information and speech features [5] expresses 38% in conversation, so works based on facial expressions and spectral features have proliferated in the last decade. However, recognizing the emotions solely through the use of facial regions and voice tone are too complex and body

M. Singh et al. (Eds.): ICACDS 2022, CCIS 1613, pp. 231–242, 2022.
https://doi.org/10.1007/978-3-031-12638-3_20

language remains less explored [6]. For example, some humans tend to rely on body gestures to express their intuitive feelings rather than facial expression. Gesture recognition has a high level of resiliency and maybe a better alternative for emotion recognition at a distance. A collaborative system like a robot that can recognise emotions based on body movement can be extremely useful in a variety of applications, including bio metric security, healthcare and virtual reality [7].

Emotions are primarily classified based on facial expressions, such as movements of the upper lip, jawline, and cheek, but hardly only a few of them relates the face features with body movements. Another core aspect of emotion recognition is the use of EEG signals from brain activity. When it comes to human-computer interaction, the knowledge of gesture and body pose estimation is essential for conveying the relevant emotional key information. Gestures are a group of action cues performed by the human body with the joint movement of the head, hand and arm to convey the interactive emotional feelings in depth. Most basic gestures are universal: when we are happy, we enjoy the extreme part of happiness; when we are upset, we ultimately frown. Experimental psychology has also shown that motion qualities are linked to specific individual emotions: for instance-body attempting to turn is a characteristic of joy, distress, and surprise; panic causes the body to contract; joy causes open-mindedness and acceleration of the arms and shoulders; fear and sadness cause the body to look away; The human skeleton is perhaps the most straightforward method for portraying human actions, as it shapes the body movement with a series of joint positions.

The wide-scale applications of gesture recognition help estimate the human behaviour [8] in public spaces such as railway stations, telecommunications, airports, animation, automobile security, parking slots, and security assistance in real-time analysis, and estimate the emotional state of students in e-learning programs. In the medical domain, diagnosing the patient behaviour disorders (Psychiatry) and monitoring the stress level and detecting depression has evolved greater extent in healthcare systems. With the recent progress and rapid increase of motion capture techniques, the quantity and quality of research in automatic recognition of expressive movements has increased. Despite growing interest in this Human-Computer Interaction (HCI), affective body movement patterns in automatic analysis of cognitive science remain underestimated.

Recently, many machine learning models have been emerging based on gesture emotion recognition and gained several advantages over traditional recognition techniques. Convolutional Neural Network (CNN) [9] has a wide area, applied in the field of emotion recognition, and is helpful in the extraction of global features with a tendency to classify emotions with different transformations and into different classes. In this paper, we propose human body emotion recognition through the representation of gesture movements based on the sequence of joint position from the head, hand and upper body motions. The research objective is to recognize emotion from visual body clues (Facial actions, hands, body gestures, etc.). The main contribution of this research is to describe different categories of frontal view emotions regarding human body movements using 19-dimensional body expressive features like distance, angle and velocity are computed with the nose as the midpoint, left hand, right hand and torso. Then the combined feature vector is fed to the four different machine learning algorithms to predict the accuracy rate of the classifier.

The paper is organized as follows: Sect. 2 provides the previous work carried in gesture-based emotion recognition. Section 3 details the pipeline of proposed emotion recognition process through body gestures and technical aspects are discussed. Section 4 explains the classification approach. Section 5 describes the dataset used, performance metrics and experimental results on various machine learning algorithms on the GEMEP dataset and Finally, Sect. 6 recaps the paper with conclusion and future work.

2 Prior Research

This section presents a study of various existing approaches for recognising human upper body gesture emotions. Furthermore, the success factors of existing techniques, as well as their limitations, are presented. J. Arunnehru et al. [10] Proposed activity recognition in videos by extracting the region of interest in the dynamic gestures sequence in the KTH dataset by specifying the motion intensity code feature classified with Support Vector Machine (SVM). Santhosh Kumar et al. [11] focus on recognizing four human emotions like angry, fearful, happy and sad in surveillance video using the point position and orientation of CMI (Cumulative Motion Images) and SURF (Speeded-Up Robust Features) key points with cumulative frame differencing. Presented [12] a dynamic body pattern of recognizing the actions in the frame like walking, sitting and jumping sequence behaviour in videos by modelling the features in SVM, Naive Bayes (NB) and Dynamic Time Wrapping (DTW).

Holden, Daniel et al. [13] developed a framework on high-level parameters of human body motion data ruled by the convolutional autoencoder's hidden units. The Convolutional Neural Network (CNN) [14] plays a significant role in the emotion recognition system, preferably using a supervised learning strategy with reduced overfitting and dropout. Proposed [15] a framework for recognizing torso actions like walking, boxing, jogging, clapping, running and waving in the moving frame using Difference Intensity Distance Group Pattern (DIDGP) by a distance relationship for analyzing human actions based on SVM. A detailed survey [16] of recent works in stating the six emotions like fear, sadness, happiness, disgust, surprise and anger associated with body pose estimation of the 2 D kinematic models with culture and gender differentiation, modelled for automatic emotional body gesture recognition. A neural architecture model using six emotional affective states is recognized by the human pose motion pattern in emotion recognition.

A self-organizing neural model develops emotion recognition from the video sequence in the single CNN and a combination of CNN and RNN models is proposed with various state of art performances [17]. Proposed [18] a novel block-based intensity value (BBIV) feature, a model for classifying five emotional state instances with tree-based classifiers by extracting the regions in the difference image. From the above discussion, the recognition pose estimation from the human body motion pattern has greater enhancement in the field of HCI. Considerably less emotional states are utilized by researchers for identifying the affect from the body posture and gestures. However, recent advances in motion capture technologies and their increasing dependability resulted in a significant increase in the amount of literature on automatic recognition of expressive movements. Our method is primarily concerned with identifying the upper body motion patterns with 17 different emotional states in the GEMEP Dataset.

3 Proposed Approach

This work deals with the extraction of Geometrical features from the body pose representing vital information that is required for emotion recognition. To understand human emotional expressions, the motion pattern and shape information is extracted from the GEMEP video sequences consisting of 17 different micro expressions. The overall workflow of proposed emotion recognition is depicted in Fig. 1.

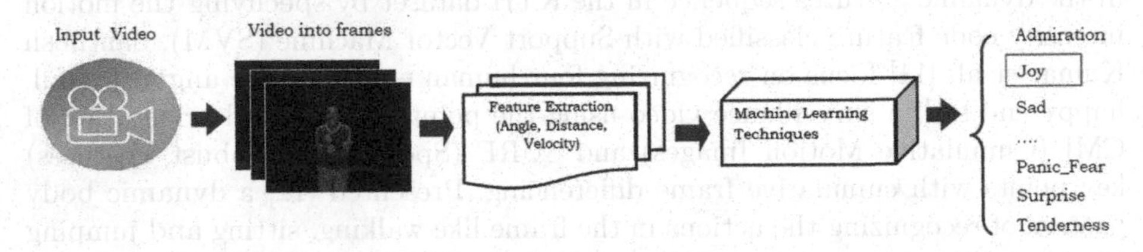

Fig. 1. Proposed approach for gesture emotion recognition

Pose Based Feature Extraction. Initially, the emotional information from the movements of the head, shoulders, hand and waist region from the GEMEP corpus dataset is taken. The variation in the kinetics, motion and shape of the body pose information is extracted from the video sequences to recognise emotion. The versatile information in the input video sequence is preprocessed and converted to frames. From the frames sequence, the media pipe mesh predictor [19] is imposed for human pose estimation which outputs the 33 landmarks points. To ensure the correspondence between frames, the pre-defined pixel locations are mapped with the detected pose landmarks. The skeleton-based model help to extract the micro variations of the body pose along with the motion changes by tracking the salient movements in the head, neck and torso. Therefore, we considered the feature vector based on the distance, angle and velocity of skeleton points to discriminate the 17 different emotions in this work.

The landmarks in the kinematic model help to estimate the distance, angle and velocity measure using Pythagoras Theorem. These (x_i, y_i) coordinates represent the distance parameters between each key point in the body regions. To calculate the indices of the points, we have chosen 19 - dimensional feature points that have the most subtle changes in the upper body parts. This pairwise coordinates of the two landmark points help to identify the feature vector for emotion recognition.

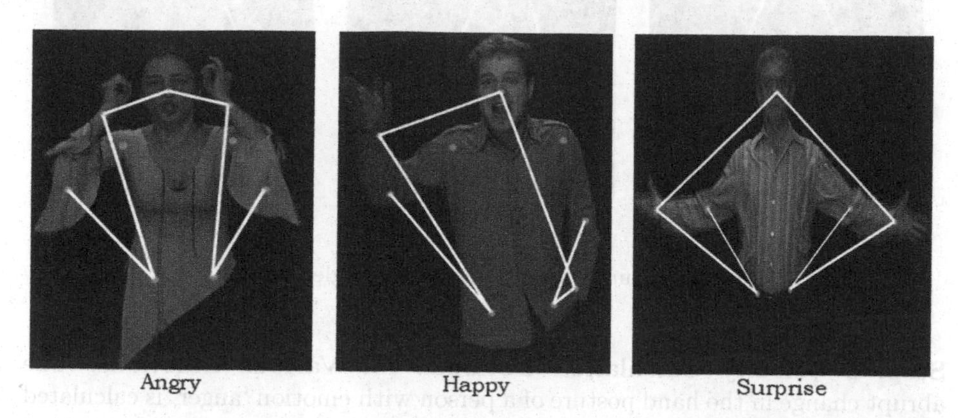

Fig. 2. Feature points extracted for distance measure

Step: 1 Distance Measure Obtained for Various Emotions. The Euclidean distance between the hand, elbow, wrist and hip landmark points is taken for consideration. The six landmark points taken for distance mapping from head to hip are considered as follows {(0, 16), (0, 15), (16, 24), (15, 23), (13, 23), (14, 24)}. Because certain motion characteristics like 'happy' and 'joy' have the dynamic elbow movement of clapping hands, 'sad' and 'panic_fear' bring the hand closer to the 'torso'. Figure 2 shows the Euclidean distance $D(pt_1, pt_2)$ calculated between each pair of cartesian coordinate feature points $pt_1(x_1, y_1)$ and $pt_2(x_2, y_2)$ as shown in Eq. 1.

$$D_{(pt_1,pt_2)} = \sqrt{pt_1(x_2 - x_1)^2 + pt_2(y_2 - y_1)^2} \quad (1)$$

Step: 2 Angle Measure Obtained for Various Emotions. The recognition of affine body gestures can be judged easily by the cosine angle formed by two non-zero vector landmarks is calculated as shown below in Eq. 2. The four angle points considered for the measuring is as follows {(16, 14, 12), (11, 13, 15), (14, 12, 24), (13, 11, 23)}. The joint movements express 'pleasure' and 'surprise' relates are likely to widen the 'torso' in the video sequences. The emotion 'pleasure' gives the pleasant sensation that arises from one's senses whereas 'surprise' relates to the unexpected happening of something. Figure 3 represents the angle calculation between the shoulder and wrist features for differentiating the emotions in depth.

$$A_{(pt_1,pt_2,pt_3)} = \arccos\left(\frac{(pt_3 - pt_2).(pt_1 - pt_2)}{||pt_3 - pt_2||.||pt_1 - pt_2||}\right) \quad (2)$$

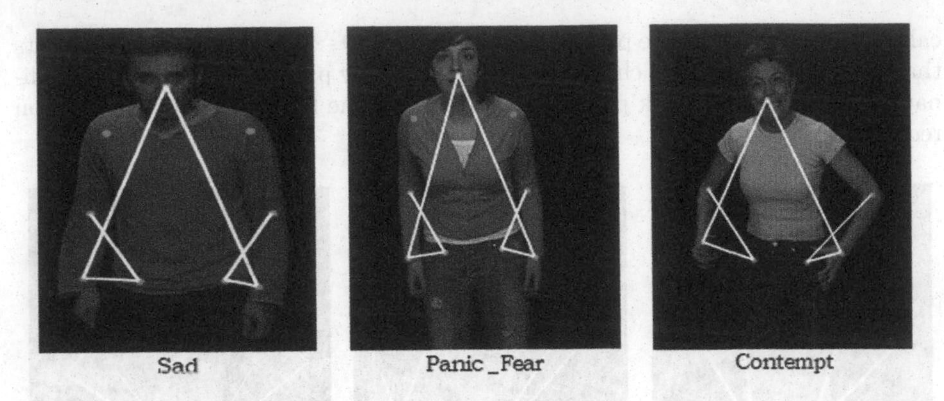

Fig. 3. Feature points extracted for angle measure

Step: 3 Velocity Calculation Obtained for Various Emotions. The abrupt change in the hand posture of a person with emotion 'anger' is calculated by the rough changes in the three landmark points. Because 'anger' sometimes corresponds by 'to and fro' vibrations of palm actions. In addition to the change in position over the change in time is referred to as velocity. As a result, the average changes in the nine landmarks velocity points {0, 11, 12, 13, 14, 15, 16, 23, 24} of three-point co. ordinates are viewed as the feature vectors.

$$V_{avg} = \left(\frac{dx_f - dx_i}{t_f - t_i} \right) \tag{3}$$

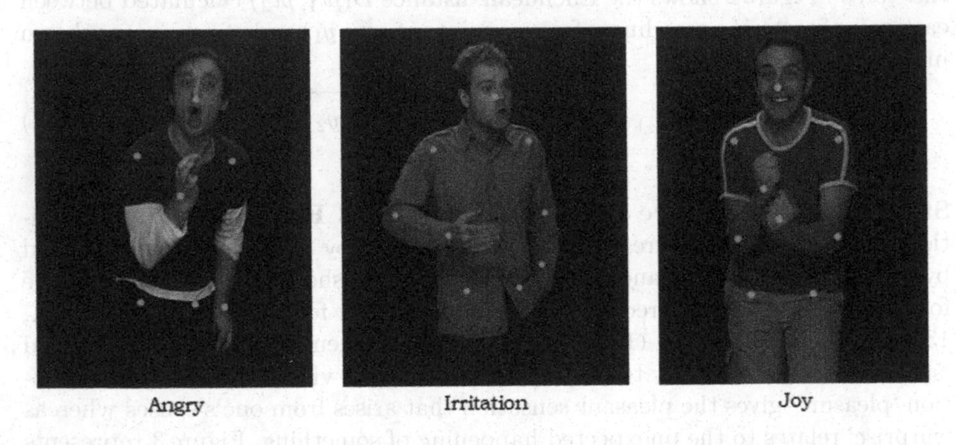

Fig. 4. Sample frames of points chosen for velocity measure

Figure 4 represents the velocity points of sample frames as shown in Eq. 3 the 'f' and 'i' represents the final and initial time of a centre point nose. Similarly, the average velocity of a frame between hand, wrist and hip skeleton joints are calculated for recognizing the emotions. First, the distance between each pair of landmarks coordinates within a frame is determined then the angle between the three pairs of points and velocity of the joints are estimated. Secondly, the feature fusion combines the Cartesian distances, angles points and velocity, helps to identify the different actor's gestures from the neutral state to the apex state yielding 19 feature vectors for training the machine learning algorithms.

4 Emotion Classification Using Machine Learning Models

This section briefs the four best classification techniques used in this paper for recognizing body gesture emotions. We also compare the accuracy prediction performance of the classification techniques. The top suited model for predicting the accuracy of Angle vs. Distance vs. Velocity were selected from the six K Nearest Neighbor (KNN) classifiers, six Support Vector Machine (SVM) Classifiers, three Decision Tree (DT) classifiers and two Naive Bayes (NB) classifiers. KNN [20] is exclusively a statistical analysis and classification algorithm used in computer vision for calculating the nearest number of neighbours. The six KNN classifiers are Fine KNN, Medium KNN, Coarse KNN, Cosine KNN, Cubic KNN and Weighted KNN. This supervised learning algorithm classifies the K features with the help of votes taken from the neighbours of high probability. SVM is an efficient supervised learning algorithm recognizes the bodily gesture emotions by handling outliers in the GEMEP dataset. The training set of SVM is not linearly separable and for classifying the hyperplane into 17 emotional categories, it uses RBF kernel function in high dimensional space. The six SVM classifiers involved here are listed as follows: the Linear, Quadratic, Cubic, Fine Gaussian, Medium Gaussian and Coarse Gaussian.

The primary use of the Decision Tree [21] algorithm is to predict the value of the target emotions by learning the features vector rules inferred from the prior trained data. The significant challenge in the DT is the identification of the attribute for the feature split. The attribute selection is achieved easily through Entropy or Gini Impurity methods. The three DT classifiers used for estimating the accuracy are the Fine tree, Medium and Coarse tree. Naive Bayes classifier, influenced by Bayes theorem, is most suitable for kinematic model-based learning applications well adapted for classification problems. The pre-processed 19 features from the GEMEP dataset is passed through the probabilistic naive for classifying the 17 emotions with k features. Gaussian and Kernel are the two analyses used by Naive Bayes for predicting the accuracy values.

5 Experimental Results and Discussion

The demonstrations are implemented in Windows 10 operating system with Intel Core i7 processor with 3.40 GHz and 16 GB RAM using MATLAB 2019b. As explained in Sect. 3, the 19 feature vectors extracted from the input frames are normalized using min-max normalization techniques and tested on different machine learning algorithms like KNN, SVM, Decision Tree and Naive Bayes to recognize the accuracy of 17 different emotional states from the GEMEP corpus dataset.

5.1 GEMEP Dataset

The Emotional Corpus Dataset GEMEP (Geneva Multimodal Emotion Portrayals) [22] is a publically available multimodal dataset, portraying 17 different micro expressional body gesture states with audio, face and gesture information. For our experimental studies, 1823 instances are chosen with 720 × 576 pixel rate of 25 frames per second. It includes admiration, amusement, anger, anxiety, contempt, despair, disgust, interest, irritation, joy, panic_fear, pleasure, pride, relief, sadness, surprises, and tenderness performed by 10 subjects. The sample frames of the 17 emotions from the GEMEP Dataset is shown in Fig. 5.

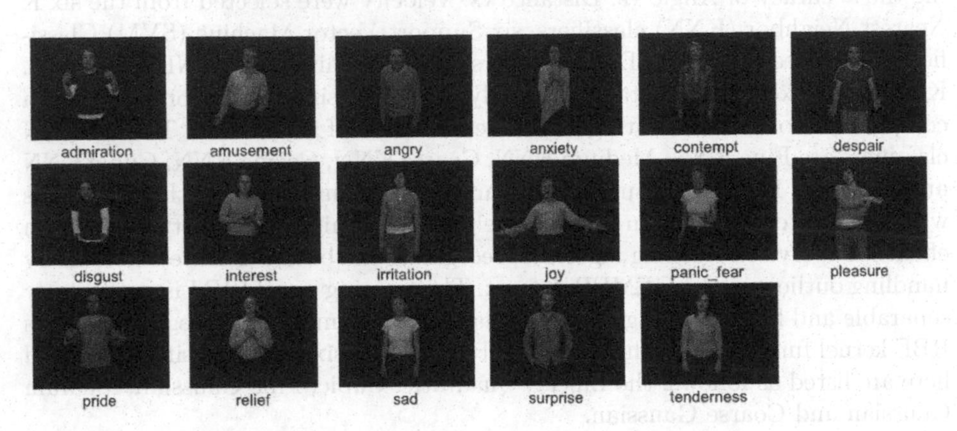

Fig. 5. Sample frames of 17 emotions from the GEMEP dataset

5.2 Performance Metrics

The proposed feature vector with angle, distance and velocity uses a ten fold cross-validation strategy for training and testing the machine learning models. The statistical metrics like Precision, Sensitivity, Specificity and F-score has opted for estimating the performance. Precision estimates the total predicted positive samples. Specificity limits the true positive rates to the overall accurate recognition of emotions. F-score pursues a suitable measure of performance by taking the harmonic mean of precision and recall. From the results, KNN performs well, compared to other classifiers. Where tp (positive prediction) is true positive, tn (Negative prediction) is true negative, fp (mispredicted as positive) is the false positive, fn (mispredicted as negative) is the false negative.

Accuracy (A) $= \left[\frac{tp+tn}{tp+fp+tn+fn}\right]$ predicts the overall correctness of the upper-body emotions classified.

Precision (P) $= \left[\frac{tp}{tp+fp}\right]$ gives the measure of perfection.

Recognizing emotions in the exact way defined by Specificity (S) $= \left[\frac{tp}{tp+fn}\right]$. F-Score $= 2\frac{P\times R}{P+R}$ gives the detailed mean of both Precision and specificity.

5.3 Results and Discussion

To exhibit the performance of the proposed approach for human emotion recognition using gestures, we conducted comprehensive experiments on the GEMEP datasets. The different combinations of the model performance result with angle, distance and velocity are listed in Table 1.

Table 1. Model performance for the test datasets of four classification algorithms

Classifiers		Accuracy (%)						
		A	D	V	A+D	A+V	D+V	A+D+V
KNN	Fine	77.1	95.2	10.6	**97.1**	51.9	67.6	78.9
	Medium	71.3	89.4	12.5	91.7	48.9	64.3	75.9
	Coarse	49.7	63.5	14.5	64.8	39.2	51.6	59
	Cosine	65	86.5	12.9	91.8	46.4	61.3	73.2
	Cubic	70.9	89.1	12.5	91.1	45.9	58.9	69.4
	Weighted	76.2	93.3	12.3	95	53.4	69.4	79.1
SVM	Linear	18.1	31.6	15.1	45.5	23.4	35.4	48.6
	Quadratic	41.9	77.3	15.2	89.1	48.8	70.8	81.9
	Cubic	52.6	91	14.8	94.8	55.8	78.2	87.3
	Fine Guassian	70.8	93	14.5	**95.2**	44.7	59.4	67.5
	Medium Gaussain	46.6	70.4	16.8	79.4	44.4	62.3	72.9
	Coarse Gaussain	18.8	32.1	15	35.9	21.7	30.7	33.8
DT	Fine Tree	41.9	51.2	14.5	51.4	39.5	**51.6**	51.6
	Medium Tree	25.1	30.8	14	30.9	25.6	31.7	31.7
	Coarse Tree	16.2	19.7	13	19.7	16.8	19.8	19.4
NB	Gaussain	14	20.3	10.3	20.2	13.8	16.5	20
	Kernel	35.3	44.4	9.1	**49.7**	18.7	31.5	39.5

Angle (A), Distance (D), Velocity (V), Angle & Distance (A+D), Angle & Velocity (A+V), Distance & Velocity (D+V), Angle & Distance & Velocity (A+D+V)

The recognition accuracy obtained by Fine KNN is 97.1% is higher when compared to other models for the combined angle and distance features. SVM shows an accuracy of 95.2% for predicting the frontal view of emotions from the collected Dataset. Decision Tree predicts the maximum of 51.6% and Naive Bayes gives an moderate of 49.7%. The middle instance in the confusion matrix predicts the sample of correctly classified. The Confusion matrix for the combined angle and distance features with the KNN classifier is shown in Fig. 6.

Confusion Matrix for Validation Data

True Class \ Predicted Class	admiration	amusement	angry	anxiety	contempt	despair	disgust	interest	irritation	joy	panic_fear	pleasure	pride	relief	sad	surprise	tenderness
admiration	**423**	0	1	0	0	0	1	0	1	2	0	0	0	3	0	0	0
amusement	0	**669**	0	1	1	1	2	2	3	1	5	1	1	0	0	1	1
angry	2	9	**580**	2	0	8	2	1	1	6	6	4	7	6	3	0	0
anxiety	1	0	1	**469**	0	2	0	0	1	2	1	0	2	0	1	0	0
contempt	0	0	0	0	**368**	0	0	0	0	0	0	0	0	0	0	2	0
despair	0	0	3	0	0	**605**	0	1	5	2	2	2	0	0	2	0	0
disgust	0	3	1	0	0	0	**321**	0	0	0	0	0	0	0	1	0	0
interest	0	1	1	0	0	0	0	**693**	0	0	1	0	0	0	0	0	0
irritation	0	0	0	0	0	2	0	0	**723**	3	0	0	1	1	0	0	0
joy	6	3	7	0	1	4	2	0	9	**557**	9	2	7	4	3	1	0
panic_fear	1	1	2	0	0	2	0	0	1	5	**433**	2	4	0	0	0	0
pleasure	0	1	0	1	0	0	0	0	2	0	0	**938**	1	0	0	1	1
pride	6	0	6	0	0	1	0	2	3	8	8	2	**433**	1	0	0	0
Relief	1	0	0	0	0	0	0	0	1	2	0	0	1	**605**	2	0	0
sad	0	2	0	1	0	0	0	0	1	0	0	1	1	0	**592**	0	1
surprise	0	0	0	0	0	0	0	0	0	0	0	0	0	0	1	**288**	0
tenderness	0	1	0	0	1	1	0	0	1	0	0	0	0	1	0	0	**305**

Fig. 6. Confusion matrix obtained for the GEMEP dataset for the fine KNN classifier

Table 2. Performance measure of fine KNN classifier

Emotions	Precision	Sensitivity	Specificity	F-score
Admiration	0.961	0.981	0.998	0.970
Amusement	0.969	0.970	0.997	0.970
Angry	0.963	0.910	0.997	0.936
Anxiety	0.989	0.977	0.999	0.983
Contempt	0.991	0.994	0.999	0.993
Despair	0.966	0.972	0.997	0.969
Disgust	0.978	0.984	0.999	0.981
Interest	0.991	0.995	0.990	0.993
Irritation	0.961	0.990	0.996	0.975
Joy	0.947	0.905	0.996	0.926
Panic_fear	0.931	0.960	0.996	0.945
Pleasure	0.985	0.992	0.998	0.988
Pride	0.945	0.921	0.997	0.933
Relief	0.974	0.988	0.998	0.981
Sad	0.978	0.988	0.990	0.983
Surprise	0.982	0.996	0.999	0.989
Tenderness	0.990	0.983	0.999	0.987
Mean (μ)	0.971	0.971	0.998	0.971

Table 2 represents the average emotion performance of Precision, Sensitivity, Specificity and F- Score values for KNN classifier. From the results, KNN performs good in identifying the 17 micro emotions when compare to SVM, DT and NB.

6 Conclusion

In this paper, we present a novel method for recognizing human bodily emotions based on gesture dynamic features. Demonstrations were carried on the GEMEP dataset to identify the 17 different micro emotional states. The results show that KNN gives an overall recognition accuracy of 97.1% for the GEMEP dataset and 95.2% for SVM, and 51.6% for Decision Tree and 49.7% for Naive Bayes respectively. The proposed combined features of gesture dynamics with different combinations are modelled using machine learning algorithms to find the accuracy of Quantitative metrics like precision, sensitivity, specificity and F-Score. From the experiment, it is concluded that the system could not able to differentiate joy, pleasure and admiration with high accuracy. The proposed work requires less processing time comparing to the existing models. Hence, this model is perfect for recognizing visual bodily emotions with increase in system's efficiency. Our future research extends to recognize the various different emotional states by combining the visual body gestures with vocal spectral features using real-time datasets.

References

1. Ko, B.C.: A brief review of facial emotion recognition based on visual information. Sensors **18**(2), 401 (2018)
2. Baltrušaitis, T., et al.: Real-time inference of mental states from facial expressions and upper body gestures. In: Face and Gesture 2011, pp. 909–914. IEEE (2011)
3. Arunnehru, J., Chamundeeswari, G., Prasanna Bharathi, S.: Human action recognition using 3D convolutional neural networks with 3d motion cuboids in surveillance videos. Procedia Comput. Sci. **133**, 471–477 (2018)
4. Michael Revina, I., Sam Emmanuel, W.R.: A survey on human face expression recognition techniques. J. King Saud Univ.-Comput. Inf. Sci. **33**(6), 619–628 (2021)
5. Minaee, S., Bouazizi, I., Kolan, P., Najafzadeh, H.: Ad-Net: audio-visual convolutional neural network for advertisement detection in videos. arXiv preprint arXiv:1806.08612 (2018)
6. Elfaramawy, N., Barros, P., Parisi, G.I., Wermter, S.: Emotion recognition from body expressions with a neural network architecture. In: Proceedings of the 5th International Conference on Human Agent Interaction, pp. 143–149 (2017)
7. Oommen, D.K., Arunnehru, J.: A comprehensive study on early detection of Alzheimer disease using convolutional neural network. In: AIP Conference Proceedings, vol. 2385, pp. 050012. AIP Publishing LLC (2022)
8. Arunnehru, J., Kalaiselvi Geetha, M.: Behavior recognition in surveillance video using temporal features. In: 2013 Fourth International Conference on Computing, Communications and Networking Technologies (ICCCNT), pp. 1–5. IEEE (2013)

9. Bhargavi, G., Vaijayanthi, S., Arunnehru, J., Reddy, P.R.D.: A survey on recent deep learning architectures. In: Manoharan, K.G., Nehru, J.A., Balasubramanian, S. (eds.) Artificial Intelligence and IoT. SBD, vol. 85, pp. 85–103. Springer, Singapore (2021). https://doi.org/10.1007/978-981-33-6400-4_5
10. Arunnehru, J., Geetha, M.K.: Motion intensity code for action recognition in video using PCA and SVM. In: Prasath, R., Kathirvalavakumar, T. (eds.) MIKE 2013. LNCS (LNAI), vol. 8284, pp. 70–81. Springer, Cham (2013). https://doi.org/10.1007/978-3-319-03844-5_8
11. Santhoshkumar, R., Kalaiselvi Geetha, M., Arunnehru, J.: Activity based human emotion recognition in video. Int. J. Pure Appl. Math. **117**(15), 1185–1194 (2017)
12. Arunnehru, J., Kalaiselvi Geetha, M.: Automatic human emotion recognition in surveillance video. In: Dey, N., Santhi, V. (eds.) Intelligent Techniques in Signal Processing for Multimedia Security. SCI, vol. 660, pp. 321–342. Springer, Cham (2017). https://doi.org/10.1007/978-3-319-44790-2_15
13. Holden, D., Saito, J., Komura, T.: A deep learning framework for character motion synthesis and editing. ACM Trans. Graph. (TOG) **35**(4), 1–11 (2016)
14. Krizhevsky, A., Sutskever, I., Hinton, G.E.: ImageNet classification with deep convolutional neural networks. In: Advances in Neural Information Processing Systems, 25 (2012)
15. Arunnehru, J., Kalaiselvi Geetha, M.: Difference intensity distance group pattern for recognizing actions in video using support vector machines. Pattern Recogn. Image Anal. **26**(4), 688–696 (2016)
16. Noroozi, F., Corneanu, C.A., Kamińska, D., Sapiński, T., Escalera, S., Anbarjafari, G.: Survey on emotional body gesture recognition. IEEE Trans. Affect. Comput. **12**(2), 505–523 (2018)
17. Khorrami, P., Paine, T.L., Brady, K., Dagli, C., Huang, T.S.: How deep neural networks can improve emotion recognition on video data. In: 2016 IEEE International Conference on Image Processing (ICIP), pp. 619–623. IEEE (2016)
18. Santhoshkumar, R., Kalaiselvi Geetha, M.: Human emotion recognition in static action sequences based on tree based classifiers
19. Bashirov, R., et al.: Real-time RGBD-based extended body pose estimation. In: Proceedings of the IEEE/CVF Winter Conference on Applications of Computer Vision, pp. 2807–2816 (2021)
20. Arunnehru, J., Nandhana Davi, A.K., Sharan, R.R., Nambiar, P.G.: Human pose estimation and activity classification using machine learning approach. In: Reddy, V.S., Prasad, V.K., Wang, J., Reddy, K.T.V. (eds.) ICSCSP 2019. AISC, vol. 1118, pp. 113–123. Springer, Singapore (2020). https://doi.org/10.1007/978-981-15-2475-2_11
21. Vaijayanthi, S., Arunnehru, J.: Synthesis approach for emotion recognition from cepstral and pitch coefficients using machine learning. In: Bindhu, V., Tavares, J.M.R.S., Boulogeorgos, A.-A.A., Vuppalapati, C. (eds.) International Conference on Communication, Computing and Electronics Systems. LNEE, vol. 733, pp. 515–528. Springer, Singapore (2021). https://doi.org/10.1007/978-981-33-4909-4_39
22. Bänziger, T., Scherer, K.R.: Using actor portrayals to systematically study multimodal emotion expression: the GEMEP corpus. In: Paiva, A.C.R., Prada, R., Picard, R.W. (eds.) ACII 2007. LNCS, vol. 4738, pp. 476–487. Springer, Heidelberg (2007). https://doi.org/10.1007/978-3-540-74889-2_42

A Body Area Network Approach for Stroke-Related Disease Diagnosis Using Artificial Intelligence with Deep Learning Techniques

M. Anand Kumar[1](✉) and A. Suresh Kumar[2]

[1] Department of Computer Applications, Graphic Era Deemed to be University, Dehradun, India
anandkumar@geu.ac.in

[2] Department of Computer Science and Engineering, Graphic Era Deemed to be University, Dehradun, India
sureshkumar.cse@geu.ac.in

Abstract. Stroke is the second largest disease after heart disease that leads to death. Stroke-related diseases need immediate attention from health care experts. With rapid growth and advancements in the field of medical technologies across the world, there is a huge demand for the latest wireless communication technology, especially for the continuous monitoring of patients. Body area network is one among the promising technology which uses special-purpose sensor networks with design principles to function independently across various platforms to connect several medical sensors and related applications. The Body area network was applied in most of the medical and its related applications starting from basic patient monitoring systems to advanced critical disease diagnosis applications which provides a high degree of health care services not only to patients but also to health care professionals. The main objective of this research work was to propose a new medical approach for earlier disease diagnoses for stroke and its related disease that require immediate treatments. This work Integrates the body area network with artificial intelligence techniques which enables health care workers to speed up the diagnosis process that needs immediate attention. The experimental results show that the proposed approach obtained better results with an accuracy of 88.47% when compared to other existing models and also identified that this model is best suited for disease diagnosis, especially for stroke-related issues.

Keywords: Artificial intelligence body area networks · Deep learning · Sensors · Treatments · Wireless networks

1 Introduction

Health care systems play a vital role in the current pandemic scenario where more unknown diseases across the globe make it difficult for the health care experts for early disease diagnosis [1]. Early and accurate diagnosis is crucial which influences not only

M. Singh et al. (Eds.): ICACDS 2022, CCIS 1613, pp. 243–256, 2022.
https://doi.org/10.1007/978-3-031-12638-3_21

the efficiency of treatment but also affects the infected patients. For instance, infectious diseases like Covid, undiagnosed patients may transmit the disease to other persons. Clinical outcomes will be affected seriously by some of the diseases like pulmonary diseases and strokes which need immediate treatments as a late diagnosis will lead to more critical conditions for patients [2]. Most of the health care industries trying to incorporate the latest ICT (Information and Communication Technology) to effectively monitor the delivery of health care services. This type of technology, not only provides services for patients and medical experts inside the health centers; but also to their workplaces most efficiently with minimum cost and better quality. Some of the recent challenges in providing quality health care services by medical practitioners and service providers paved the way for adopting new technologies like Artificial intelligence, the Internet of Things(IoT), Deep learning algorithms, and data analytics for better patient outcomes.

According to the latest medical survey reports, IoT was the third most technology implemented in health care sectors. New healthcare-related challenges have opened the door for healthcare stakeholders to successfully deliver high-quality healthcare services using IoT technologies. IoT provides several benefits such as real-time monitoring systems, efficient methods to collect patient data and analyze data for the following outcomes: a) to better assess the patient medical conditions and complexity; b) to design modern treatment approaches; c) to implement decision support system to assess health care experts and mechanisms for better disease diagnosis. Machine learning techniques provide a sophisticated and automatic approach for analyzing high-dimensional biomedical data that improves medical diagnosis significantly. A well-trained machine learning algorithm can perform the given task with high reproducibility or accuracy, which is vital for clinical decision-making decisions, especially in disease diagnosis [3].

Recently Wireless Body Area Network (WBAN) started deployment in medical industries with huge expectations. It is an emerging and promising technology that will change the existing health care methodology to a higher level. One of the notable technologies as far as WBAN was concerned is wearable health care devices. When compared to traditional health care systems, wearable sensor devices are cost-effective and easy to deploy. It helps health care industries to reduce infrastructure as well as deployment costs.

2 Related Works

Stroke-related diseases became one of the major disabilities across the globe. The latest predictions show that by the year 2030, more than 70 million patients will be affected by stroke and 200 million disability patients related to stroke [4]. It was one of the serious issues in developed countries like India and it was noted that stroke-related diseases are rapidly increasing in developing countries [5]. So there is a huge need for classification algorithms and methods for the early prediction of diseases. Various health factors and data analysis has been carried out to classify stroke-related diseases. Due to the lack of labeled data in hospitals and health care centers, prediction methods with machine learning are usually not available or reliable. There was a huge need to analyze the factors that affect the growth rate of these cases.

The recent work presented a model which had been developed with Artificial Neural Network (ANN) for stroke detection. In this work, the Cardiovascular Health Study (CHS) dataset was used to compile their findings. Three datasets were created, each with 212 strokes (all three) and 52, 69, and 79 for non-stroke patients. The final data includes 357 characteristics and 1824units, as well as 212 stroke rates. The C4.5 decision tree approach had been utilized for feature selection, and Principle Component Analysis (PCA) was applied for length reduction. They employed the Back Propagation Learning approach to implementing ANN. For the three datasets, the accuracy was 95 percent, 95.2 percent, and 97.7 percent, respectively [6].

For stroke classification, Decision Trees, Bayesian Classifiers, and Neural networks are employed [7]. They used thousand patient data in their dataset. For dimensionality reduction, the PCA technique was applied. They achieved the highest accuracy in ten rounds of each algorithm, with 92 percent, for the neural network model, 91 percent for the Naive Bayes classifier, and 94 percent for Decision tree algorithms, respectively [8].

Several deep learning approaches which use segmentation and prediction of stroke-related diseases are based on the imaging concept. This research also presented the importance of deep neural network approaches for stroke diseases, particularly image processing such as Conventional Neural Network (CNN) and Fully Convolution Network (FCN) (FCN). It suggests that various deep learning models could be developed for improved outcomes in detecting stroke patients. This assessment also includes information on upcoming trends and advancements in stroke detection [9].

From the literature survey, this research work identified that stroke diseases are rapidly increasing worldwide due to the environmental changes and lifestyle of people. Most of the related work uses trusted machine learning classifiers such as Decision Trees, Bayesian Classifiers, and Neural Networks. Due to the advancements in healthcare technologies, the dataset plays a vital role. Most of the research was conducted with the available datasets and EMR patient records. Currently, deep learning models play a vital role alongside Body Area Network.

This research mainly focuses on two shortcomings of previous works. The first contribution was to use an existing dataset with the real dataset with the help of Wireless BAN. This initiative provides a better quality of dataset which is the main parameter for critical diseases. The second contribution was to use Deep Learning models and classifiers to predict diseases. The selection of neural-based learning models was purely based on the suitability of the environment and working principles of the models. Some models provide a better result for imaging and some for non-imaging clinical solutions.

3 Research Methodology

The dataset used in this research work was obtained from Kaggle dataset and real-time data from hospitals. Two different datasets were used for analysis. The first dataset contains 5000 records (Kaggle Dataset) and the second dataset (Real Dataset) was collected from Uttarakhand health centers and hospitals with the help of personal servers that are attached to different health centers. The second dataset contains 1025 records in which the readings are taken from WBAN Wearable devices. From the second dataset, three attributes such as BMI, Body temperature, and Blood Pressure are measured with WBAN wearable devices from real-time patients. The first dataset contains 11 attributes whereas the second dataset contains 10 attributes that are collected in real-time. The differences in attribute selection were purely based on a real-time collection set using WBAN devices. The following tables represent the attributes and their description (Tables 1 and 2).

Table 1. Dataset (One) parameters.

Attributes	Description	Values
Gender	Male or female	1 = male 0 = female
Age	Age denoted in years	Incessant
Hypertension	Yes or No	1 = Yes 0 = No
Heart disease	Yes, or No	1 = Yes 0 = No
RBP	Resting blood pressure	Incessant value in mm hg
Heart rate	Heart rate achieved	Incessant value
Avg. glucose level	The average glucose level in blood	Floating point number
BMI	Patient's body mass index	Floating point number
Smoking status	Smoking status of the patient	String literal
DEP	Exercise-induced depression when compared to rest	Incessant value
Stoke	Output	1 = Stroke 0 = No Stroke

Table 2. Dataset 2 parameters

Attributes	Description	Values
Gender	Male or female	1 = male 0 = female
Age	Age denoted in years	Incessant
Hypertension	Yes or No	1 = Yes 0 = No
Heart disease	Yes, or No	1 = Yes 0 = No
RBP	Resting blood pressure	Incessant value in mm hg

(continued)

Table 2. (*continued*)

Attributes	Description	Values
Heart rate	Heart rate achieved	Incessant value
FBS	Fasting blood sugar	1 ≥ 120 mg/dl 0 ≤ 120 mg/dl
Avg. glucose level	The average glucose level in blood	Floating point number
BMI	Patient's body mass index	Floating point number
Smoking status	Smoking status of the patient	String literal
DEP	Exercise-induced depression when compared to rest	Incessant value
Stoke	Output	1 = Stroke 0 = No Stroke

3.1 Data Preprocessing

Before building the machine learning model, data processing is the main requirement where unwanted redundancy and noise should be removed. Redundant data will lead the model to perform with less efficiency. As far as complex medical applications are concerned, the model should have good efficiency for better predictions which is one of the essential requirements from any machine learning model. The first dataset has eleven attributes that begin with the attribute Patient-ID which was not required by the model. Label encoding was done in the dataset to convert string literals to integer values which will be interpreted by the computer easily. Then the dataset will be inspected for Missing values. It was also identified that the quality of the dataset is very important for medical models which are critical in nature and need immediate attention. One of the better solutions for the missing value problem is to use better imputation methods. KNN imputation method is best suited for medical models. It selects K complete samples. The missing value will be filled by a median of the corresponding identifier in K complete samples. The Shortest Euclidean distance method will be used for filling values.

$$\text{where, } \begin{array}{c} d_{xy} = \sqrt{weight * squared\ distance\ from\ present\ coordinates} \\ weight = \frac{Total\ number\ of\ coordinates}{Number\ of\ present\ coordinates} \end{array} \quad (1)$$

3.2 Proposed Model

This Research work applied Convolutional Neural Networks, Generative Adversarial Networks, Multilayer Perceptron, Radial Basis Function Networks, and Deep Belief Networks (Fig. 1).

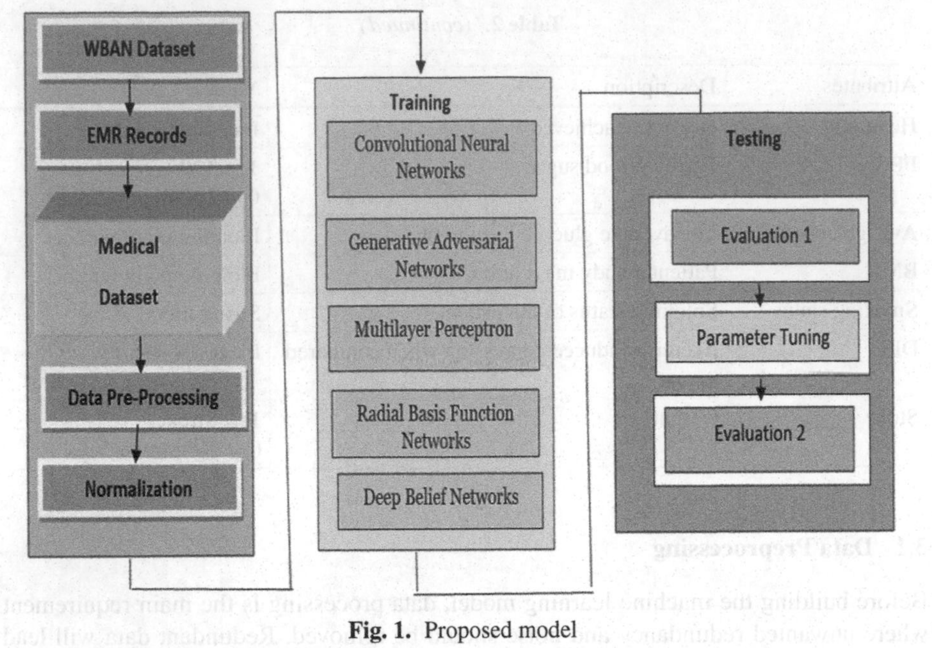

Fig. 1. Proposed model

Algorithm

Input: Patient record from Dataset 1

1. Begin
2. For each Record(Dataset)
3. The Preprocessing of Stroke Dataset 1 using preprocessing methods
4. Feature selection is employed to the processed data from Step 3
5. Train the Deep Learning classifiers using Training Dataset(TR1)
6. Validate the data using Testing Dataset(TS1)
7. Compute Performance Evaluation Metrics that are described in the Performance metrics section
8. Perform Parameter tuning if required based on Step 7
9. Output (1=Stroke; 0= No Stroke;)
10. Input Dataset 2 (EMR Dataset)
11. Follow Step 2 to 8 for Dataset 2
12. End

---

4 Deep Learning Models

4.1 Convolutional Neural Network (CNN)

The task of binary classification relates to the challenge of predicting stroke-related patients. In supervised contexts, neural networks have been shown to be an excellent parametric classifier under specific limitations [10–12]. With the rapid growth of structured data, deep learning networks incorporate a huge number of application-oriented hidden layers which significant improvements in several areas, including time series, prediction, image processing, and speech processing [13–15].

4.2 Radial Basic Function(RBF)

This type of network provides different architectural models when compared to other neural network models [16]. Most of the models contain several layers that introduce nonlinearity by applying nonlinear activation functions. But RBF network has only an input layer, hidden layer, and output layer. The input layer gets the input data and feed it to a hidden layer. The computation method is totally different from other neural network models [17–19]. This computation method makes this model very powerful where prediction features such as classification and regression. This method will be one of the best-suited models for healthcare application, especially for stroke-related disease diagnosis. The computation formula for hidden can be represented as

$$y(\mathrm{x}) = \sum_{i=1}^{N} w_i \, \phi(||\mathrm{x} - \mathrm{x}_i||), \tag{2}$$

4.3 Multilayer Perceptron (MLPs)

Multilayer Perceptrons [20, 21] are a feed-forward deep neural network with several layered perceptron as activation functions. MLPs are made up of two fully connected layers: an input layer and an output layer. It will have multiple hidden layers with the same amount of input and output layers that can be used to create pictures, voice, and machine translation software [22, 23].

4.4 Deep Belief Networks (DBNs)

Deep belief networks are the generative models of deep neural networks which consist of multi-layer variables like stochastic and latent. These variables are binary values that are hidden units. These are a stack of Boltzmann machines that consists of connections between layers and each RBM layer communicates with both preceding and following layers. The application of DBNs is video recognition, image processing, and motion capture data [24, 25].

5 Experiments

5.1 Experimental Setup

Computer simulation is a way for examining a wide range of Machine learning models that represent real-world systems using specific simulation application software designed to emulate some of the system's essential attributes. Microsoft's ML.NET Open-Source and Cross-Platform machine learning techniques were used in this study. The deep learning algorithms were implemented using this method. It also adds Artificial Inelegancy to any application by creating trained deep learning models for medical applications of any kind.

5.2 Evaluation Metrics

Deep learning algorithms' overall performance is measured using performance metrics. It was also used to track how well machine learning models are implemented and performed on a given dataset under different conditions. Selecting the appropriate metric is crucial for comprehending the model's behavior and making the necessary changes to improve it. The following table indicates the metrics (Table 3).

Table 3. Performance metrics

S.No	Metrics	Description
1	TP (True Positive)	It is the number of occurrences of stroke disease that are classified as true and are true
2	TN (True Negative)	It's the number of cases of stroke disease that are classified as fake and are, in fact, false
3	FN (False Negative)	It is the number of cases of stroke disease that are labeled as false but are true
4	FP (False Positive)	It refers to the number of cases of stroke disease that are labeled as true but are false

5.3 Mathematical Model

When it comes to medical data, a mathematical model is one of the most significant aspects to consider because these are vital data that must be properly assessed for better disease detection. This section explains the mathematical model that was used to evaluate this outcome.

- Accuracy: Accuracy is calculated as the number of right forecasts made by the model over all types of forecasts given in classification problems [18]. The proper forecasts (True positives and True negatives) are in the numerator, while the type of all algorithmic predictions (Right and Wrong) is in the denominator, and it's stated as:

$$Accuracy = \frac{TP + TN}{(TP + FP + FN + TN)} \quad (3)$$

- Precision: Precision [19] is a measure that determines what % of stroke patients have actually had a stroke. The projected positives (those who are expected to have a stroke, abbreviated as TP and FP) and the ones who have had a stroke, abbreviated as TP, are as follows:

$$Precision = \frac{TP}{TP + FP} \quad (4)$$

- Recall or Sensitivity: Recall is a measure that shows what percentage of stroke patients the algorithm correctly diagnosed. The true positives (those who have strokes are TP and FN) and the patient diagnosed with strokes by the model are both TP. FN is included since the Person experienced a stroke despite the model's prediction, and it is given as:

$$Sensitivity = \frac{TP}{TP + FN} \quad (5)$$

- F1 Score: When creating a model to solve a classification problem, it is not necessary to carry both Precision and Recall. So if it can receive a single score that represents both Precision (P) and Recall(R), that would be ideal (R). Taking their arithmetic mean is one approach to do this. (P + R)/2, with P denoting precision and R denoting recall. However, in other circumstances, this is not a good thing. It can be written as:

$$F1\,Score = \frac{2 * Precision * Recall}{Precision + Recall} \quad (6)$$

6 Results and Discussion

This work presents the observations of stroke diseases based on parameters, because it's one of the main sources of evidence for critical medical applications. Only the latest deep learning models will not be a better solution for critical stroke-related diseases. From this research, it was identified that only EGC imaging and the models that use only medical imaging are not suitable for accurate predictions of critical like strike-related issues. Even in this work, some of the classification algorithms are used only for medical image classifications are used with certain modifications. Some of the recent works related to stroke diseases use specific deep learning models that support only ECG images. But if these models are used for all criteria then it will be a better solution for critical applications.

6.1 Accuracy Results Based on Dataset

The following figure shows the accuracy results of Dataset Selection. Two datasets are used such as One Dataset from EMR and one dataset from WBAN. Even though the dataset contains 3000 records, only 250 records are selected for initial evaluations. The figure clearly shows that WBAN Dataset provides better accuracy than that of the EMR dataset. As far as critical medical applications such as strokes related diseases are concerned Wireless body area networks will give better solutions for the patients (Fig. 2).

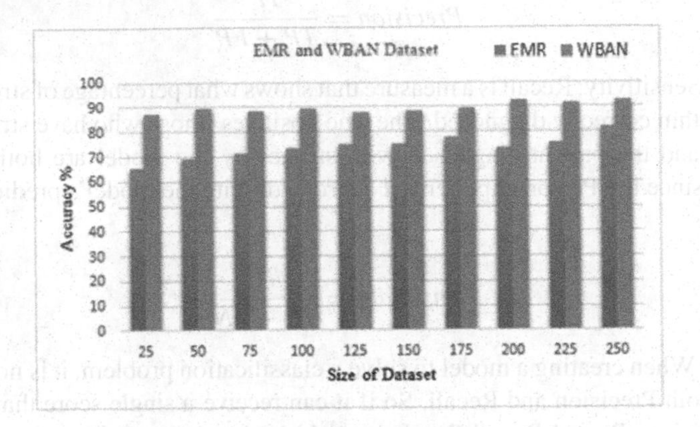

Fig. 2. Accuracy between the EMR and WBAN

6.2 Accuracy Results Based on Deep Learning Models

This section presents the various results obtained during the experiments based on the mathematical model that was presented in the previous section. The following Figure shows the results and accuracies of the implemented deep learning models such as Convolutional Neural Networks, Generative Adversarial Networks, Multilayer Perceptron, Radial Basis Function Networks, and Deep Belief Networks. The Experimental results clearly show that the Deep Belief Network model gives a better result with an accuracy of 93.78% when compared to other models. It also indicated that the Deep Belief Networks models were better suitable for stroke-related disease prediction mechanisms. It was also identified that Convolutional Neural Networks with accuracy with 90.25%, Generative Adversarial Networks with 80.19%, Multilayer Perceptron with 81.02, and Radial Basis Function Networks with 79.37% accuracy. Figure a) presents the accuracy results based on dataset selection and figure b) presents the precision results based on deep learning models. In both cases, WBAN provides better results (Fig. 3).

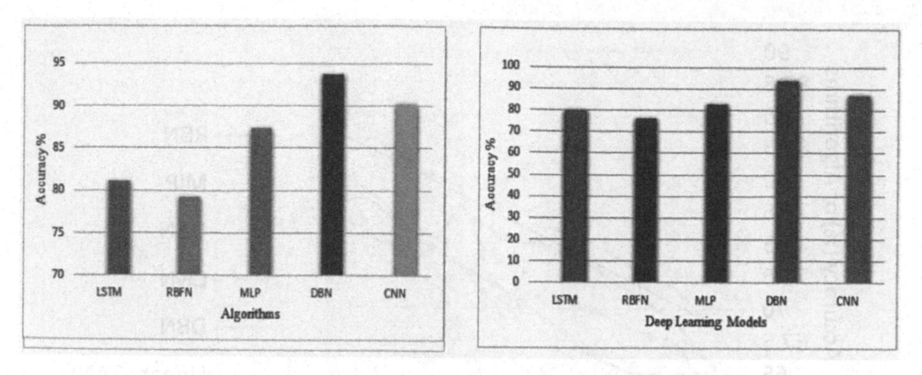

Fig. 3. a) Accuracy between the EMR and WBAN b) Precision value (%)

The following figures (a and b) present the results based on the Recall score and F1 Score. In both cases, DBN provides better results when compared to other models, which is a positive sign as far as medical applications are concerned (Fig. 4).

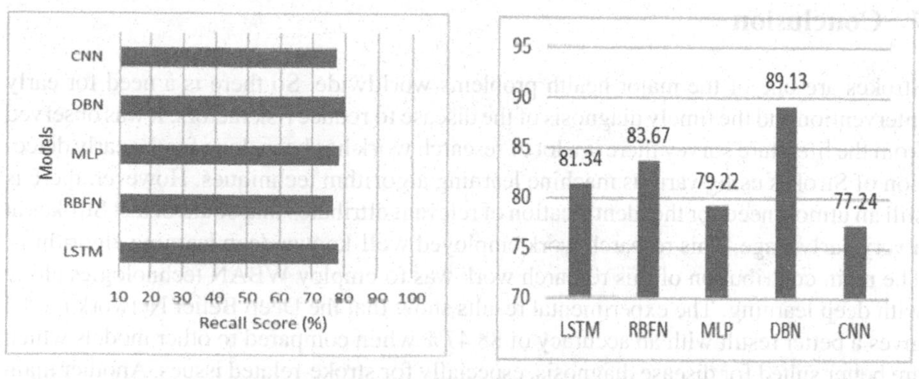

Fig. 4. a) Recall score (%) b) F1 score (%)

6.3 Accuracy Results Based Training Data Set

Figure 5 shows the results based on the selection of different data-set sizes ranging from 45 to 90. The highest value is attained for all the used algorithms in the 75-percent to an 80-percent range of data training, according to the analysis. Only SVM exhibits an increase in accuracy as the training size grows. The remaining algorithms, on the other hand, provide outstanding results and peak accuracy values of 80% in the training size realm. The best size range for forecasting the disease and getting the accuracy, precision, recall, and F1 score is 75-percent to 80-percent train data size, according to this research. It was also indicated that there is a larger possibility of under-fitting in the 50-percent Train data set. Over-fitting is possible for 90% of the Train data set. As a result, a data set with a 75 to 80-percent training rate is excellent for any disease diagnosis application.

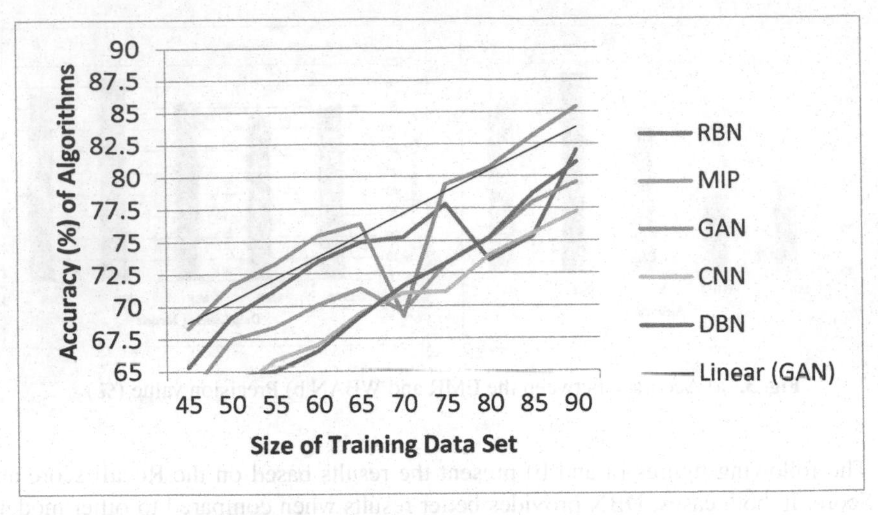

Fig. 5. Accuracy based on dataset size.

7 Conclusion

Strokes are one of the major health problems worldwide. So there is a need for early intervention and the timely diagnosis of the disease to reduce risk factors. It was observed from the literature survey there is a lot of research work has been done for the early detection of Strokes using various machine learning algorithm techniques. However, there is still an utmost need for the identification of relevant attributes that could detect Strokes at a very early stage. This research work employed well-known deep learning algorithms. The main contribution of this research work was to employ WBAN technologies along with deep learning. The experimental results show that the Deep Belief Network model gives a better result with an accuracy of 88.47% when compared to other models which are better suited for disease diagnosis, especially for stroke-related issues. Another main observation from this research work was that 80% training data set is ideal for accurate results as well as for good performances.

References

1. Liu, Q., Mkongwa, K.G., Zhang, C.: Performance issues in wireless body area networks for the healthcare application. A survey and future prospects. SN Appl. Sci. 155–158 (2021)
2. Salam, H.A., Khan, B.M.: Use of wireless system in healthcare for developing countries. Digit. Commun. Netw. **2**(1), 35–46 (2016)
3. Movassaghi, S., Abolhasan, M., Lipman, J., Smith, D., Pour, A.J.: Wireless body area networks. A survey. IEEE Commun. Surv. Tutor. **16**(3), 1658–1686 (2014)
4. Fang, G., Xu, P., Liu, W.: Automated ischemic stroke subtyping based on machine learning approach. IEEE Access **8**, 118426–118432 (2020)
5. Kim, A.S., Cahill, E., Cheng, N.T.: Global stroke belt. Geographic variation in stroke burden worldwide. Stroke **46**(12), 3564–3570 (2015)

6. Xu, H., Pang, J., Zhang, W., Li, X., Li, M., Zhao, D.: Predicting recurrence for patients with ischemic cerebrovascular events based on process discovery and transfer learning. IEEE J. Biomed. Health Inform. **25**(7), 2445–2453 (2021)
7. Rajora, M., Rathod, M., Naik, N.S.: Stroke prediction using machine learning in a distributed environment. In: Goswami, D., Hoang, T.A. (eds.) ICDCIT 2021. LNCS, vol. 12582, pp. 238–252. Springer, Cham (2021). https://doi.org/10.1007/978-3-030-65621-8_15
8. Chun, M., et al.: The china kadoorie biobank collaborative group. Stroke risk prediction using machine learning. A prospective cohort study of 0.5 million Chinese adults. J. Am. Med. Inform. Assoc. **28**(8), 1719–1727 (2021)
9. Choi, Y.-A., et al.: Deep learning-based stroke disease prediction system using real-time bio signals. Sensors **21**(13), 4269 (2021). https://doi.org/10.3390/s21134269
10. Tahmid, M., Alam, S., Akram, M.K.: Comparative analysis of generative adversarial networks and their variants. In: 23rd International Conference on Computer and Information Technology-ICCIT, pp. 1–6 (2020)
11. Bodyanskiy, Y., Pirus, A., Deineko, A.: Multilayer radial-basis function network and its learning. In: IEEE 15th International Conference on Computer Sciences and Information Technologies-CSIT, pp. 92–95 (2020)
12. Alsmadi, M.K., Omar, K.B., Noah, S.A., Almarashdah, I.: Performance comparison of multi-layer perceptron (back propagation, delta rule and perceptron) algorithms in neural networks. In: IEEE International Advance Computing Conference, pp. 296–299 (2009)
13. Tran, S.N., d'Avila Garcez, A.S.: Deep logic networks: inserting and extracting knowledge from deep belief networks. In: IEEE Transactions on Neural Networks and Learning Systems, vol. 29, no. 2, pp.246–258 (2018)
14. Roth, G.A., et al.: Global burden of cardiovascular diseases and risk factors. Update from the GBD 2019 study. J. Am. Coll. Cardiol. **76**, 2982–3021 (2020)
15. O'Shea, A., Lightbody, G., Boylan, G., Temko, A.: Neonatal seizure detection from raw multi-channel EEG using a fully convolutional architecture. Neural Netw. **12**(25) (2020)
16. Adhi, H.A., Wijaya, S.K., Badri, C., Rezal, M.: Automatic detection of ischemic stroke based on scaling exponent electroencephalogram using extreme learning machine. J. Phys. Conf. Ser. **820**, 12005–12013 (2017)
17. Kwon, Y.-H., Shin, S.-B., Kim, S.-D.: Electroencephalography based fusion two-dimensional (2d)-convolution neural networks (CNN) model for emotion recognition system. Sensors **18**(5), 1383 (2018). https://doi.org/10.3390/s18051383
18. Thara, D.K., PremaSudha, B.G., Xiong, F.: Epileptic seizure detection and prediction using stacked bidirectional long short term memory. Pattern Recognit. Lett. **128**, 529–535 (2019)
19. Shoily, T.I., Islam, T., Jannat, S., Tanna, S.A., Alif, T.M., Ema, R.R.: Detection of stroke disease using machine learning algorithms. In: 10th International Conference on Computing, Communication and Networking Technologies-ICCCNT, pp. 1–6 (2019)
20. Krishna, V., Sasi Kiran, J., Prasada Rao, P., Charles Babu, G., John Babu, G.: Early detection of brain stroke using machine learning techniques. In: 2nd International Conference on Smart Electronics and Communication–ICOSEC, pp. 1489–1495 (2021)
21. Emon, M.U., Keya, M.S., Meghla, T.I., Rahman, M.M., Manun, S.M., Kaiser, M.S.: Performance analysis of machine learning approaches in stroke prediction. In: 4th International Conference on Electronics, Communication and Aerospace Technology-ICECA, pp. 1464–1469 (2020)
22. Pande, A., Manchanda, M., Bhat, H.R., Bairy, P.S., Kumar, N., Gahtori, P.: Molecular insights into a mechanism of resveratrol action using hybrid computational docking/CoMFA and machine learning approach. J. Biomol. Struct. Dyn. 1–15 (2021)

23. Gupta, A., Lohani, M.C., Manchanda, M.: Financial fraud detection using naive bayes algorithm in highly imbalance data set. J. Discret. Math. Sci. Crypt. **24**(5), 1559–1572 (2021)
24. Singh, N., Singh, D.P., Pant, B.: ACOCA: ant colony optimization based clustering algorithm for big data preprocessing. Int. J. Math. Eng. Manag. Sci **4**, 1239–1250 (2019)
25. Singh, N., Singh, D.P., Pant, B., Tiwari, U.K.: μBIGMSA-microservice-based model for big data knowledge discovery: thinking beyond the monoliths. Wireless Pers. Commun. **116**(4), 2819–2833 (2020). https://doi.org/10.1007/s11277-020-07822-0

Accelerating the Performance of Sequence Classification Using GPU Based Ensemble Learning with Extreme Gradient Boosting

Karamjeet Kaur[1(✉)], Anil Kumar Sagar[1], Sudeshna Chakraborty[2], and Manoj Kumar Gupta[3]

[1] School of Engineering and Technology, Sharda University, Greater Noida, U.P., India
karam_7378@yahoo.com, aksagar22@gmail.com
[2] Lloyd Institute of Engineering and Technology, Greater Noida, U.P., India
sudeshna2529@gmail.com
[3] Shri Mata Vaishno Devi University, Katra, Jammu and Kashmir, India
manoj.gupta@smvdu.ac.in

Abstract. The classification of biological sequences is a very difficult and challenging task in the development of the bioinformatics field. The classification can be done by extracting the features from biological sequences. Recently machine learning in bioinformatics is attracting more researchers for its better results. This paper proposes an ensemble-based approach for DNA sequence classification and compares it with other machine learning models. This model is trained for the classification task under CPU and GPU configuration separately and evaluated using execution time and accuracy. The results show that XGBoost based model has better performance as compared to other classifiers if its parameters are chosen correctly and may help in more accurately classifying the large and variable size biological sequences. This is based on supervised learning in which gradient boosted decision trees are designed for enhancing speed and performance. Different models have experimented and it has been found that gradient boosting is a highly scalable and parallelizable method that outperforms other methods. This proposed model achieves an accuracy of 93% with reduced time as compared to other machine learning models.

Keywords: Machine learning · Biological sequence · Ensemble learning · Gradient boosting · Classification · GPU

1 Introduction

Proteins play a very important role to perform many functions in the living organism to sustain its life. A protein sequence is made of twenty different amino acids when arranged in a particular order. A sequence defines the primary structure of protein and it has structural and genetic information to infer the function of the protein. DNA sequence consists of only four nucleotides A, C, G, and T e.g., TACGTACGTA. Within an organism, protein plays different functions and has different structures [1]. A huge number

M. Singh et al. (Eds.): ICACDS 2022, CCIS 1613, pp. 257–268, 2022.
https://doi.org/10.1007/978-3-031-12638-3_22

of DNA, RNA, protein sequences are added in repositories daily which are located worldwide. It is necessary to classify sequences so that we can find similar sequences which have a similar function. Whenever new sequences arrive then this new sequence is compared with existing database sequences to predict the group of new sequences [2]. The traditional methods of pairwise sequence alignment and multiple sequence alignment are used to find similarities between two or more sequences based on computing scores that represent matches or mismatches between sequences. After that many sequence alignment tools are developed e.g. BLAST [3], FASTA [4], ClustalW [5], and MUSCLE [6] but it is found that it is a very time-consuming process and its complexity increases as the length of sequences increases.

In comparison to these traditional alignment algorithms and tools, recently this sequence data is considered big data, and machine learning is found to be appropriate for sequence classification. The input and output are already known to machine learning techniques and we have to prepare data according to the input acceptable by machine learning. The major challenge in sequence classification is that most classifiers take input as vector features but sequence data do not have explicit features. If sequence data is transformed using appropriate feature selection methods, then it leads to the problem of the high dimensionality of feature space which does not reduce the computation time. So, it is difficult to design a classifier for a feature-based sequence classification task. Examples of feature-based classification are decision trees and neural networks.

Sometimes distance-based methods are used to find the similarity between two sequences such as K nearest neighbor and support vector machine in which performance of classifier depends on choosing distance measure. Some deep learning architectures like convolutional neural networks are successfully applied to images and text data. Recurrent neural networks are used for speech recognition, natural language processing, and sentiment analysis. Machine learning is successfully applied in the medical field [7] and in real-time applications [8].

If sequence data is represented in a numeric format and the dimensionality of this sequence data is reduced then the machine learning technique can be applied to extract clear features from numeric data and used to analyze large-scale data with more accuracy and with less time.

2 Related Work

As the sequence data have high dimensionality so the problem of extracting the protein function from highly dimensional sequence data is a challenge [1] due to which classification becomes inefficient. The multi-label linear discriminant analysis approach is combined with Laplacian embedding for incorporating multiple biological network data into consideration to handle classification problems. When network data is combined with sequence data then it further improves the accuracy of prediction.

The real-world object has multiple labels instead of a single instance so multi-label learning [9] uses a set of labels and there is more progress toward multilabel learning. Such type of learning has solved many problems like multimedia content, image, email spam detection, web mining, tag recommendation, bioinformatics, and information retrieval. Parallel encoding and serial decoding are also used for multi-label text classification [10] for designing CNN-based models. In this model, a hierarchical decoder is designed to generate and decode label sequences. This model is very much competitive and robust as compared to RNN based model which can handle high label datasets.

The neural network is further improved by embedding label co-occurrence information [11]. During initialization, for each pattern of label co-occurrence, some neurons from the final hidden layer are assigned dedicatedly. These neurons then connect to stronger weights than others and improved the classification accuracy. Due to the exponential growth of label combinations, it is difficult to classify multi-label text as compared to multi-class text. Then CNN and RNN are combined [12] which capture local and global textual semantics to model high-order correlations for labels and can handle high dimensional data. The word-vector-based CNN is responsible for extracting features from text and RNN for multi-label task prediction.

One more approach is proposed [13] in which sequence data is considered as text and the positional information of sequence data remains the same after that classification is done by extracting features from input data by the convolutional neural network.

CNN model is combined with hybrid models [14]. For classification, CNN, CNN-LSTM, Bidirectional CNN-LSTM based architectures are used after label and K-mer encoding. Label encoding generates unique index values by preserving positional values. K-mer encoding converts the sequence into words after that any text classification can be applied to the sequence classification task. It has been found that the accuracy of k-mer encoding is better than label encoding.

Machine learning and statistical techniques like ANN, SVM, CRT, CHAID, QUEST, and C5.0 are also used for classification tasks [15]. The datasets are taken from UniprotKB and the data is divided into seven classes. This model also solved the high dimensionality problem by using SPSS. Some of the classes are balanced and some are unbalanced which affects the performance of the model also. It is proven that SVM and C5.0 give better results than other techniques. SVM supports sequence data whose nature is non-linear and it performs better for protein function prediction [16] and finding DNA binding sites [17].

A reference-based classification framework is designed [18] which is an improvement over pattern-based sequence classification [19]. This is a general platform and can be used as a tool for developing new classification algorithms. This is proved to have high classification accuracy.

A computational model named iDTi-CSsmoteB is developed [20] whose purpose is to identify an interaction between drug chemical structure and protein sequence. The datasets imbalance problem is dealt with by the over-sampling SMOTE technique and the XGBoost algorithm is used to predict drug-target prediction. This model is highly effective due to a good feature extraction technique named PSSM-Bigram, DP-PseAAC, and AMPseACC.

The main problem in the classification of the sequences is that sequences do not have clear features so the feature selection process is difficult. Although many machine learning techniques are available to complete this classification process but cannot be applied to sequence data. If we represent sequence data in a general way then it causes the high-dimensionality problem. So, it is very necessary to represent sequence features and at the same time reduce the dimension of data so that machine learning fits such representation. It is also very important to choose a suitable coding method and reduce the time taken for classification.

3 Methodology

The detailed methodology for sequence classification is represented in Fig. 1. An iterative approach has been used to classify DNA sequences which involves the conversion of sequence data into numeric data. The biological sequence data is in the form of characters. e.g. ACGTCGATACAC. As these sequences are not appropriate and ready to use for machine learning techniques so it needs pre-processing. The preprocessing of sequence data includes various steps:

1) These character sequences are converted into small case and divided into a list of words in which the number of characters in each word is fixed. This transformation helps in representing optimal features and then variable-length sequences are obtained.
2) The tokenization process is used to form tokens according to words that are generated by step 1.
3) Since the sequences are of variable length but for applying machine learning the dimension of data should be the same. So pre padding process is applied to generate an equal no. of tokens for each sequence.
4) One hot encoding technique is used to represent these tokens into feature vectors.
5) Now preprocessed data is ready. Ensemble-based XGBoost classifier is designed.
6) XGBoost classifier uses parallelized depth-first tree boosting and cross-validation process which increases its accuracy.

This ensemble learning comes under the category of machine learning methods that produce single superior output. This model classified all the sequences accurately. XGBoost classifier avoids overfitting, handles missing data, and has in-built cross-validation capability. All of these characteristics helped the classification task. XGBoost is a machine learning method that uses parallel tree boosting. It is highly scalable and all the decision trees are gradient boosted. It is used for supervised learning. It can do parallel computation on a single machine.

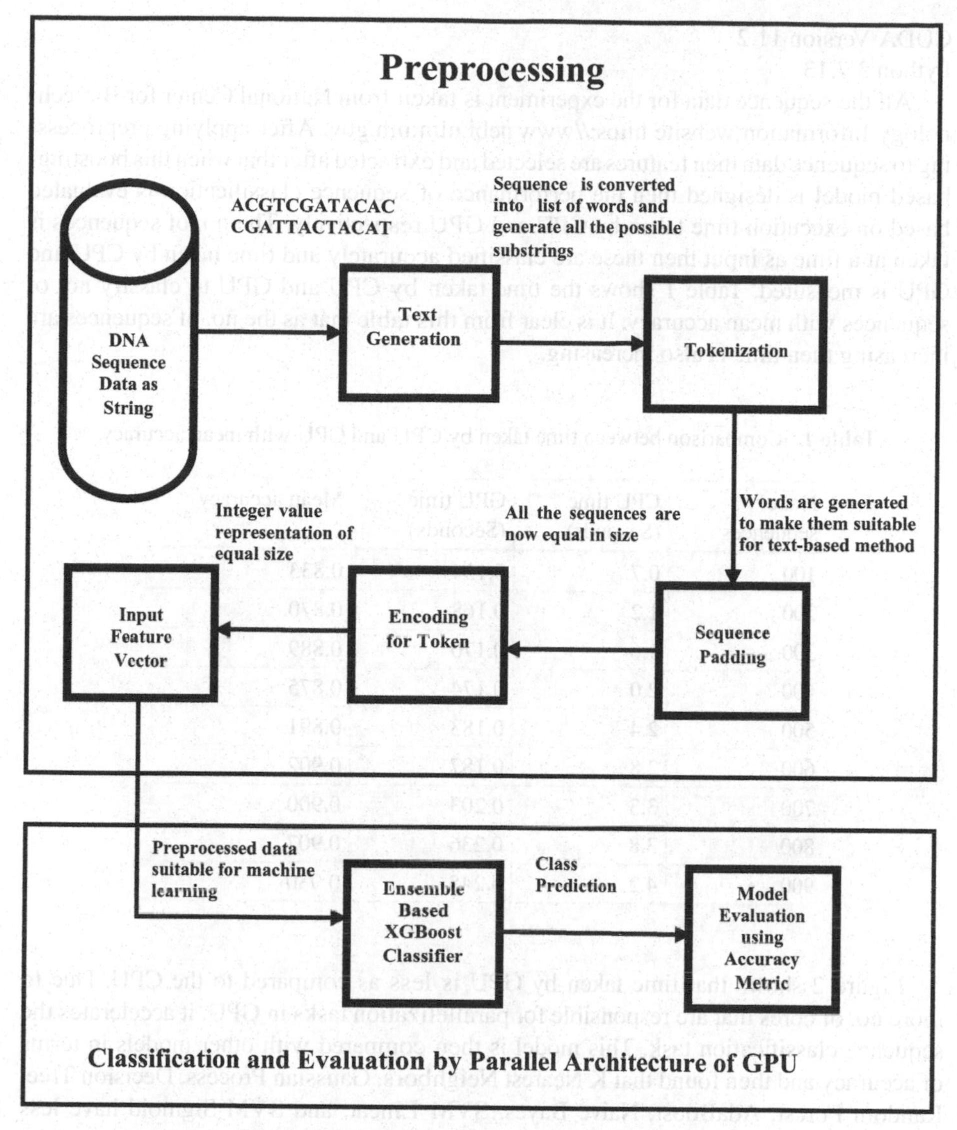

Fig. 1. Methodology for sequence classification.

4 Result and Discussion

The implementation of this paper is done under CPU and GPU configuration. The hardware and software configuration are as follows:
CPU Intel Core i7 9750H 2.60 GHz, 16 GB RAM, 1 TB SDD
GPU Tesla K80 which has 2 Kepler GK 210 GPUs, 24 GB GDDR5 memory
4992 CUDA Cores, 480 GB/sec memory bandwidth
NVIDIA Driver Version 460.32.03

CUDA Version 11.2
Python 3.7.13

All the sequence data for the experiment is taken from National Center for Biotechnology Information website https://www.ncbi.nlm.nih.gov. After applying preprocessing to sequence data then features are selected and extracted after that when this boosting-based model is designed then the performance of sequence classification is evaluated based on execution time taken by CPU and GPU respectively. The no. of sequences is taken at a time as input then these are classified accurately and time taken by CPU and GPU is measured. Table 1 shows the time taken by CPU and GPU to classify no. of sequences with mean accuracy. It is clear from this table that as the no. of sequences are increasing then time is also increasing.

Table 1. Comparison between time taken by CPU and GPU with mean accuracy.

No. of sequences	CPU time (Seconds)	GPU time (Seconds)	Mean accuracy
100	0.7	0.154	0.833
200	1.2	0.168	0.870
300	1.6	0.170	0.889
400	2.0	0.174	0.875
500	2.4	0.183	0.891
600	2.8	0.187	0.902
700	3.3	0.203	0.900
800	3.8	0.236	0.907
900	4.2	0.245	0.930

Figure 2 shows that time taken by GPU is less as compared to the CPU. Due to more no. of cores that are responsible for parallelization tasks in GPU, it accelerates the sequence classification task. This model is then compared with other models in terms of accuracy and then found that K Nearest Neighbors, Gaussian Process, Decision Tree, Random Forest, AdaBoost, Naive Bayes, SVM Linear, and SVM Sigmoid have less accuracy as compared XGBoost based classifier. This designed model has the highest accuracy of 93% and hence performs best. After selecting deep learning architecture, appropriate hyperparameters are set by experimentation. There are different parameters like no. of trees, depth of trees, learning rate, sub-sample ratio, and column ratio per tree which influence the results. The performance of the model depends on feature selection. All these parameters are represented by Figs. 3, 4, 5, 6 and 7 which affect the overall accuracy of the model.

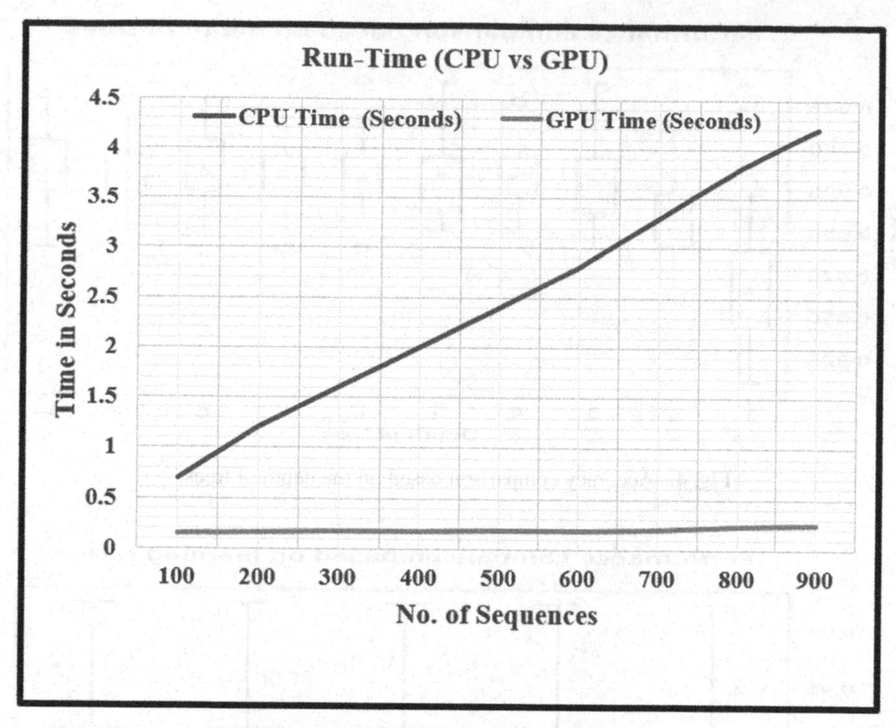

Fig. 2. CPU and GPU run-time comparison.

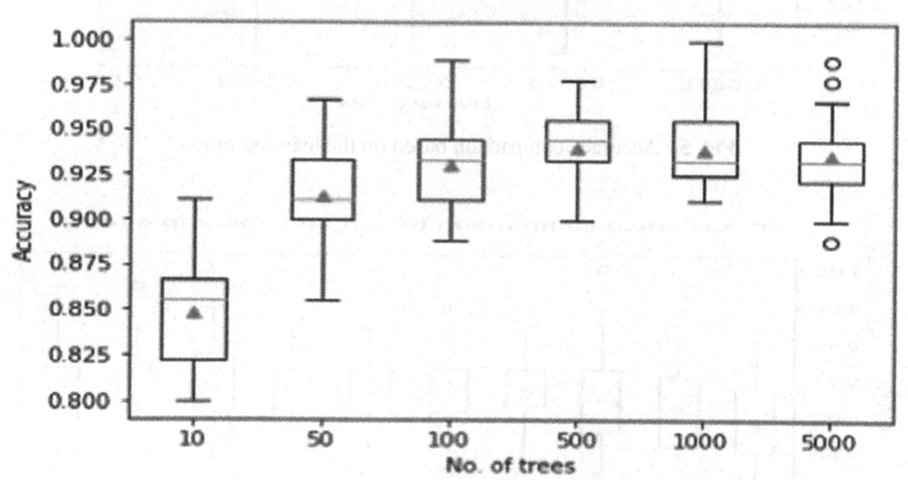

Fig. 3. Accuracy comparison based on no. of trees.

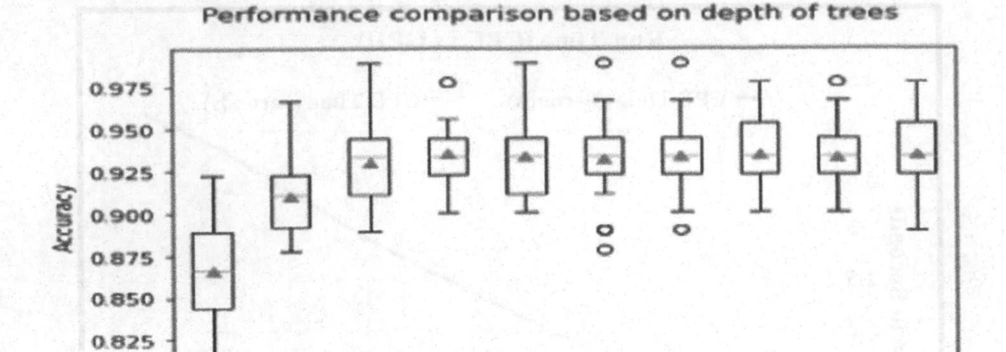

Fig. 4. Accuracy comparison based on the depth of trees.

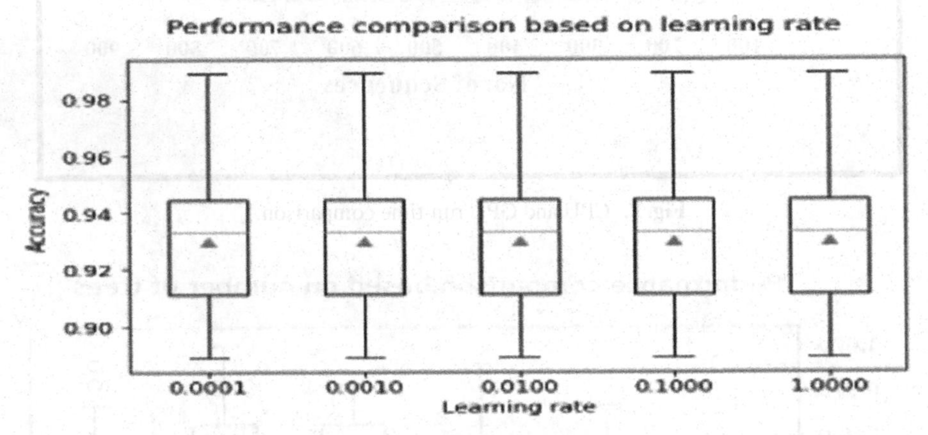

Fig. 5. Accuracy comparison based on the learning ratio.

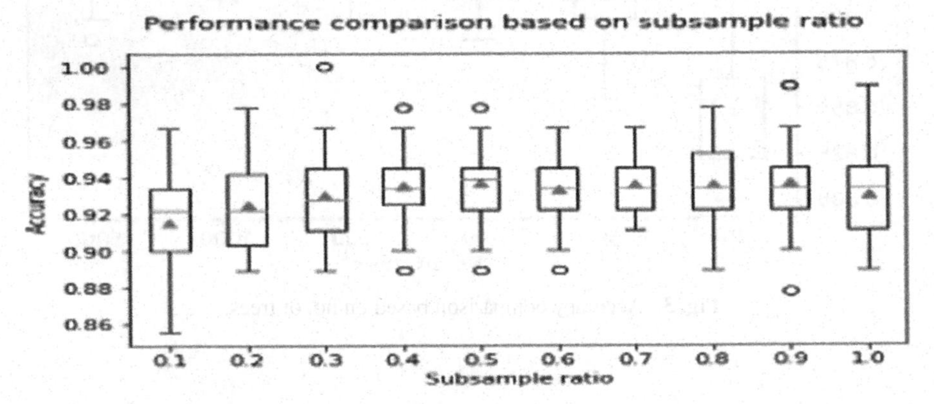

Fig. 6. Accuracy comparison based on subsample ratio.

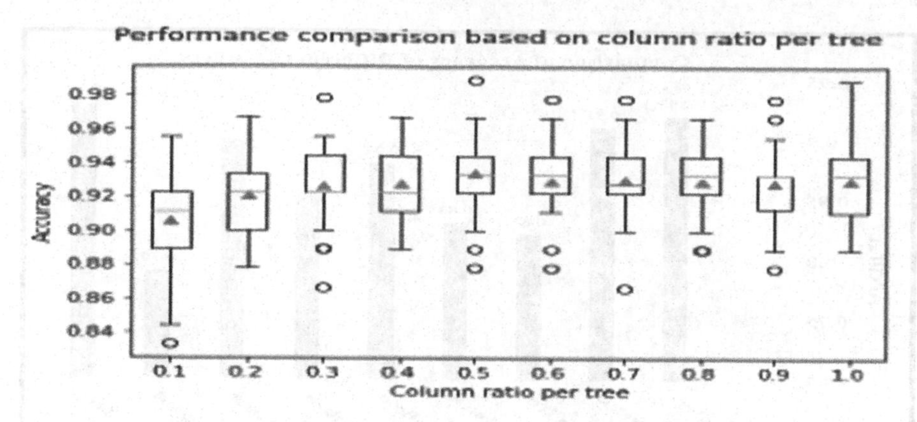

Fig. 7. Accuracy comparison based on column ratio per tree.

Table 2. Comparison between the accuracy of different types of classifiers with our proposed model.

Name of classifier	Mean accuracy
K Nearest neighbors	0.92
Gaussian process	0.91
Decision tree	0.82
Random forest	0.83
AdaBoost	0.88
Naive bayes	0.86
SVM linear	0.90
SVM sigmoid	0.79
XGBoost	0.93

Table 2 shows the accuracy score for different types of classifiers applied on the same datasets. The comparison of accuracy between different models is shown in Fig. 8. Our proposed model has best performance and can be used to classify sequences that predict the function of the protein.

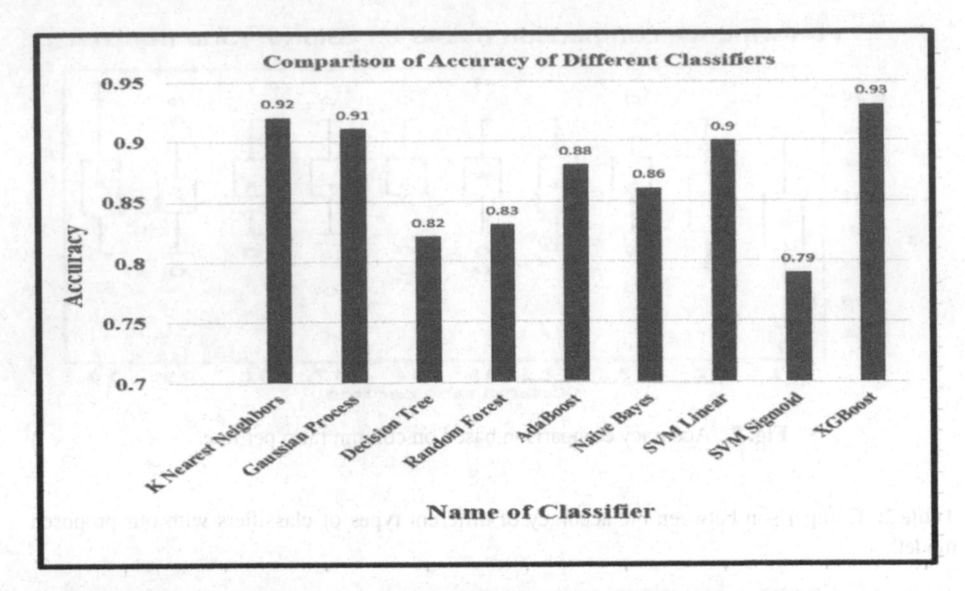

Fig. 8. Comparison between the accuracy of different models with proposed model.

This type of classification can reduce the computational complexity more effectively. For performance evaluation, 10-fold cross-validation is used, which means sequences are divided into 10 groups randomly and each group is equal in size. One group is unannotated and nine groups are annotated. The results are predicted using an unannotated group which is repeated ten times in which out of ten groups, one group becomes unannotated randomly. The final result is the average of all predictions. The accuracy remains same whether we use 5-fold or 10-fold cross-validation.

Conclusion and Future Scope

As new sequences are added in the biological databases at exponential rate then protein sequence classification method is required to identify the class of new sequence to reflect the function of the protein and how much it is similar to the existing sequence or it is entirely a new sequence. The sequence data is considered as big data, but this type of data is not suitable for applying machine learning techniques directly. So, to prepare this data, preprocessing is applied in which word embedding is implemented to convert protein sequences to numeric vectors. It is concluded that word embedding is working well for protein sequence representation and when implemented with boosting algorithm gives more accurate results in terms of accuracy which is 93%. Then proposed model is evaluated in terms of time taken by CPU and GPU to classify the sequences and our model is compared with other models. It is found that the time taken by GPU is less as compared to CPU hence it is accelerating the performance of sequence classification task. As the length of sequences is increasing it is giving more accurate results for balanced datasets. This model can be extended for handling imbalanced datasets.

References

1. Wang, H., Yan, L., Huang, H., Ding, C.: From Protein Sequence to Protein Function via Multi-Label Linear Discriminant Analysis. IEEE/ACM Trans. Comput. Biol. Bioinform. **14**(3), 503–513 (2017). https://doi.org/10.1109/TCBB.2016.2591529. PMID: 27429445
2. Alhalem, S., et al.: DNA Sequences Classification with Deep Learning: A Survey. Menoufia J. Electron. Eng. Res. **30**(1), 41–51 (2021). https://doi.org/10.21608/mjeer.2021.146090
3. Altschul, S.F., Gish, W., Miller, W., Myers, E.W., Lipman, D.J.: Basic local alignment search tool. J. Mol. Biol. **215**(3), 403–410 (1990). http://doi.org/10.1016/S0022-2836(05)80360-2. PMID: 2231712
4. Pearson, W.R.: BLAST and FASTA similarity searching for multiple sequence alignment. Methods Mol. Biol. **1079**, 75–101 (2014). https://doi.org/10.1007/978-1-62703-646-75. PMID: 24170396
5. Thompson, J.D., Higgins, D.G., Gibson, T.J.: CLUSTAL W: Improving the sensitivity of progressive multiple sequence alignment through sequence weighting, position-specific gap penalties and weight matrix choice. Nucleic Acids Res. **22**(22), 4673–4680 (1994). https://doi.org/10.1093/nar/22.22.4673. PMID: 7984417, PMCID: PMC308517
6. Edgar, R.C.: MUSCLE: Multiple sequence alignment with high accuracy and high throughput. Nucleic Acids Res. **32**(5), 1792–1797 (2004). https://doi.org/10.1093/nar/gkh340
7. Upreti, K., Kumar, N., Alam, M.S., Verma, A., Nandan, M., Gupta, A.K.: Machine Learning-based Congestion Control Routing Strategy for Healthcare IoT Enabled Wireless Sensor Networks. In: Fourth International Conference on Electrical, Computer and Communication Technologies (ICECCT), pp. 1–6 (2021). https://doi.org/10.1109/ICECCT52121.2021.9616864
8. Alam, M., Jalil, S.Z.A., Upreti, K.: Analyzing recognition of EEG based human attention and emotion using Machine learning. Mater. Today, Proc. **56**, 3349–3354 (2021). https://doi.org/10.1016/j.matpr.2021.10.190
9. Zhang, M.-L., Zhou, Z.-H.: A Review on Multi-Label Learning Algorithms. IEEE Trans. Knowl. Data Eng. **26**, 1819–1837 (2014). https://doi.org/10.1109/TKDE.2013.39
10. Yang, Z., Liu, G.: Hierarchical sequence-to-sequence model for multi-label text classification. IEEE Access **7**, 153012–153020 (2019). https://doi.org/10.1109/ACCESS.2019.2948855
11. Kurata, G., Xiang, B., Zhou, B.: Improved Neural Network-based Multi-label Classification with Better Initialization Leveraging Label Co-occurrence, pp. 521–526 (2016). https://doi.org/10.18653/v1/N16-1063
12. Chen, G., Ye, D., Xing, Z., Chen, J., Cambria, E.: Ensemble application of convolutional and recurrent neural networks for multi-label text categorization. In: International Joint Conference on Neural Networks (IJCNN), pp. 2377–2383 (2017). https://doi.org/10.1109/IJCNN.2017.7966144
13. Ngoc Giang, N., Tran, et al.: DNA Sequence Classification by Convolutional Neural Network. J. Biomed. Sci. Eng. **09**, 280–286 (2016). https://doi.org/10.4236/jbise.2016.95021
14. Gunasekaran, H., Ramalakshmi, K., Arokiaraj, A.R.M., Kanmani, S.D., Venkatesan, C., Dhas, C.S.G.: Analysis of DNA sequence classification using CNN and hybrid models. Comput. Math. Methods Med. 1–12 (2021). https://doi.org/10.1155/2021/1835056
15. Gupta, C.L.P., Bihari, A., Tripathi, S.: Human Protein Sequence Classification using Machine Learning and Statistical Classification Techniques. Int. J. Recent Technol. Eng. **8**, 3591–3599 (2019). https://doi.org/10.35940/ijrte.B3224.078219
16. Amidi, A., Amidi, S., Vlachakis, D., Paragios, N., Zacharaki, E.I.: A machine learning methodology for enzyme functional classification combining structural and protein sequence descriptors. In: Ortuño, F., Rojas, I. (eds.) IWBBIO 2016. LNCS, vol. 9656, pp. 728–738. Springer, Cham (2016). https://doi.org/10.1007/978-3-319-31744-1_63

17. Chowdhury, S., Shatabda, S., Dehzangi, I.: iDNAProt-ES: Identification of DNA-binding Proteins Using Evolutionary and Structural Features. Sci. Rep. **7** (2017). https://doi.org/10.1038/s41598-017-14945-1
18. He, Z., Xu, G., Sheng, C., Xu, B., Zou, Q.: Reference-Based Sequence Classification. IEEE Access **8**, 218199–218214 (2020). https://doi.org/10.1109/ACCESS.2020.3042757
19. Zhou, C., Cule, B., Goethals, B.: Pattern Based Sequence Classification. IEEE Trans. Knowl. Data Eng. **28**(5), 1285–1298 (2016). https://doi.org/10.1109/TKDE.2015.2510010
20. Mahmud, S.M.H., Chen, W., Jahan, H., Liu, Y., Sujan, N.I., Ahmed, S.: iDTi-CSsmoteB: Identification of Drug–Target Interaction Based on Drug Chemical Structure and Protein Sequence Using XGBoost With Over-Sampling Technique SMOTE. IEEE Access **7**, 48699–48714 (2019). https://doi.org/10.1109/ACCESS.2019.2910277

Automated Vehicle Number Recognition Scheme Using Neural Networks

T. P. Anish[1(✉)] and P. M. Joe Prathap[2]

[1] Department of Computer Science and Engineering, R.M.K College of Engineering and Technology, Chennai, India
anishcse@rmkcet.ac.in
[2] Department of Information Technology, R.M.D Engineering College, Tiruvallur 601206, India
drjoeprathap@rmd.ac.in

Abstract. The Autonomous License Plate Recognition method is a system of mass surveillance that reads and recognizes vehicle number plates using numerous Image Processing techniques and Optical Character Recognizing on imagery. Several conventional automatic number plate detection and recognition (NPDR) is carried out in a confined space, with photos acquired directly with high brightness, sharpness, and familiar characters. NPDR was always not accessible due to automobile shadowing, brightness variations, and non-uniform number plate character. This study offers a method for detecting vehicle plate numbers using a deep neural network model, a subset of Deep Learning. To overcome issues such as undesired luminance and tilt, which decrease categorization and, as a result, affect detection performance. This study uses a variety of techniques in each area, from license plate identification to character segmentation, to improve the overall system performance to the greatest extent possible with the least amount of effort and computational resources.

Keywords: Digital image processing · Convolutional neural network (CNN) · Optical character recognition · Deep learning

1 Introduction

As a consequence of an increasing number of passenger automobiles, one using its unique license number, in their use of in criminal activity, among several other things [1]. Therefore, it became necessary to build an automatic, automated process for vehicle identification based on the recognition of automobile license plates. This computer-controlled automated process, also known as fully automated NPDR is an innovation that reads vehicle license plates—using Optical Character Recognition (OCR) on image data to generate vehicle location information. It uses surveillance camera television, highway regulation camera systems, or webcams specially developed for the job [2].

Registration plates vary from one country to the next as shown in Fig. 1. Automobile license plates are subject to certain restrictions and regulations. (1) Two characters (which reference towards the province in the country in which the automobile was initially registered), (2) Two numerals (the year it became granted), and (3) Three letters

M. Singh et al. (Eds.): ICACDS 2022, CCIS 1613, pp. 269–280, 2022.
https://doi.org/10.1007/978-3-031-12638-3_23

are chosen randomly. Sizes, patterns, and characters of registration plates are a few background knowledge concerning automobile registration plates.

Fig. 1. Types of number plates in India

NPDR is a technology that combines image detection, text categorization, and recognition to identify automobiles based on their number plates. This method does not require any extra equipment on automobiles because it just uses number plate identification [3]. The use of LPR systems is advancing in prominence, particularly in surveillance and transportation control mechanisms. Accessibility monitoring in workplaces and driveways, police agencies, car theft tracking, adaptive cruise control, electronic toll payment, and sales forecasting are all common uses for LPRS [4].

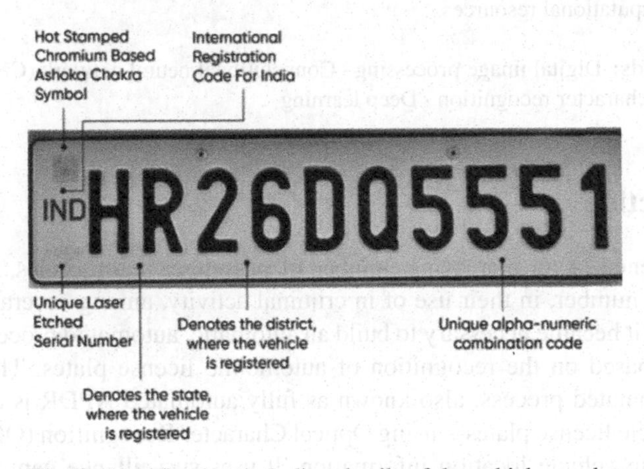

Fig. 2. Vehicle plate in India - details of the vehicle number

For number plate identification, LPR apps use image analysis and categorization procedures as shown in Fig. 2. Then each process takes a long time to complete. Governmental rules standards used in number plates can dramatically reduce computational complexity while improving accuracy [5]. Because the shape, layout, and alignment of number plate information might differ significantly in different photos, restrictions include a range of possible values rather than precise dimensions. The accuracy of automatic identification systems, such as the detection techniques utilized, and the reliability of imaging technology, comprising sensor and illumination, are the two most important aspects of vehicle detection systems [6]. Highest identification accuracy, quicker processing speed, managing quite so many varieties of plates as possible, managing the broadest selection of quality images, and highest data input distortions sensitivity are all factors to take into account.

2 Lıterature Survey

In image sequences of the registration plate image, machine learning and text categorization and techniques for vehicle detection play a significant role. A sensor, picture capture, and specifically created application intended for image recognition, investigation, and authentication put together the whole system for self-directed automobile LPRD. In the last few years, automobile recognition has been a hot subject of following a line of investigation. Several studies have been carried out to settle on the nature of automobiles, including cars, lorry, and motorcycles.

Aniruddh et al. [1] pattern matching was used on license plates derived from motionless photos, and an accuracy rate of 80.8 percent was attained. By setting the polarizing filter to acquire the whole shot and employing multiple pairs of machine learning, this reliability may be considerably enhanced. Utilizing cross genetic algorithms, the suggested service's architecture may be enhanced to recognize the license plates of several automobiles in an image representation frame.

Swati et al. [2] devised a framework for locating license plates for automobiles in India, segmenting the numerals so that each digit could be identified individually. We usually focus on the two stages: first, finding the license number, and subsequently, segmenting all of the letters and numbers to recognize individual numbers independently. A technique for identifying car number plates on multidimensional images is proposed that is more efficient and faster demanding. The Edge detection model is used to find boundaries and fill the holes fewer than 8 pixels wide. We eliminate related components just under 1000 pixels to remove the license plate. Our proposed method is primarily based on the Indian vehicle registration system. For poor light source images, license plate recognition reliability may well be improved.

Divya et al. [3] Car identification is captured on this paper and afterward utilized to retrieve vehicle owner information for the objectives mentioned above. Two-level decisions are utilized to segment vehicle numbers utilizing geometrical procedures and border characteristics. On a considerable set of photos, an efficient method is suggested and evaluated, demonstrating the effectiveness of the suggested technique. Apurva et al. [4] proposed an ideal features-based number plate recognition method. The suggested method combines a discrete wavelet technique with a BP neural network model. The method for recognizing license plates was created using the MATLAB7 program, which included CNN capabilities as well as programming files and modules programming. In addition, certain traditional license plates, such as provincial and traditional Indian number plates, were used in the qualitative research.

Atikuzzaman et al. [5] suggested an approach for detecting and recognizing serial numbers that is specifically made in the direction of operating on top of camera-captured films. Plate recognition, classification letter separation, and recognizing are the three main aspects of this unique approach. Employing a HAAR Feature-based Classier to recognize number plates, a class letter extractor with a suggested technique, and a CNN to recognize category characters completes those stages. In our obtained dataset, our proposed technique produced enthralling findings. With around 30 fps, their collection of 5500 registration plates produced an adequate recognition accuracy of 91.38 percent. With 390 sample photos, we evaluated the effectiveness of our Number plate Recognition system and found that it achieves 96.92 percent accuracy, while Class Letter Classification achieves 94.61 percent with much the same amount of data.

Shahnaj et al. [6] this research paper provides a brief discussion of automobile plate number identification and tracking strategies used for successful traffic control and assessing the systems' dependability. Vehicle plate detection and an identification system play a vital part in building an intelligent transport system. Although changing illumination, brightness, quasi kind of number plate, varied patterns, and based methods in the surroundings have made it harder to identify car number plates. It will always be a challenging task. To detect plate number characters, changing motion imagery, numerical systems, tilted or side-view photos. Recognition may employ specific computer vision techniques combined with neural network models. The vehicle plate recognition and detection techniques were categorized in this research depending on their reliability.

Emina et al. [7] developed an approach for extracting Nigeria plate numbers that demonstrated to be highly efficient and effective. Its license plate positioning accuracy was 100 percent, while its read accuracy was 90 percent. All of the photographs that were examined were of a stopped vehicle. Vehicles in motion were not evaluated. Each of these variables is used to create an image underneath various lighting circumstances.

V.V.K. Raju et al. [8] presents a technique for detecting and identifying automobile plate numbers that will aid in the identification of approved and unauthorized automobile license plates. This project presents a method based on the morphological operations that are basic yet effective, as well as the Edge detection method. Using the locality preserving technique, this technique simplifies segmenting all of the numbers and letters on the registration number. Following segmenting the digits and characters on the license number, the numbers and characters are recognized using a text detection methodology. The focus is on adequately locating the license plates region to divide all letters and numbers so each amount can be identified independently.

Olamilekan et al. [9] aims to employ data analysis to detect serious offenders based on their registration plates. It has an infrared sensor that detects the automobile. Throughout the evaluation, the detector was given a certain period to identify the item that was then registered by the microcontroller. The webcam was activated when the timer reached zero, capturing the number plate and saving the picture to the Raspberry Pi. The digits on the captured image are accessed through an Internet address on a website page. If deployed, the system can be utilized to enhance traffic safety and adaptive cruise control in upcoming smart cities.

Ravi Kiran Varma's et al. [10] article describes a novel wavelet transformation for Indian LPDR that can handle cheap, cross-angled, noisy, and quasi typeface plate numbers. Numerous image enhancement techniques are used in the well before stage, including geometric conversion, Gaussian smoothness, and edge detection. Following that, outlines are given by boundary tracking, and outlines are audit functions on feature quantities and spatial localization for registration plate segmentation. Following screening and de-skewing the bounding box, the K- nearest neighbor method is utilized for feature extraction. The strategies provided yielded good results.

Bagade's et al. [11] suggested idea proposes an automatic parking toll collecting system that relies on the vehicle's registration plates and the length of time it is kept in the parking space. This paper addresses an issue in AI systems, object tracking, and machine learning in developing an ANPR. This issue involves mathematical concepts and procedures that guarantee a variety of techniques for completing the product's stages. The images are taken with any camera capable of producing high-quality images.

This study focuses on the localization of license plates that use the contours tracking approach, as well as border identification and refinement utilizing Canny's image segmentation method. Furthermore, this paper applies a new approach based on neural networks for plate number text detection. This study uses a variety of techniques in each area, including plate number recognition to character segmentation, to improve the study's validity to the greatest extent possible only with the least amount of work and processing resources.

3 System Model

This article is about automatic license plate recognition using an enhanced neural network (ENN) to locate a plate number section out of imagery without a human. This study describes a new ENN approach with six phases of processing. It provides innovative tactics and strategies suitable for dealing with various possible hazards on every level [12]. The suggested ENN was developed and validated on 312 registration plate photos with various dangerous situations, including nighttime, interior, daytime, fuzzy, misty, wet, and slanted LP. The overall probability of identification is 95%. It also contrasts the results of the experiment to the available ENN approaches. In terms of probability of detection and operating time, the suggested ENN approach surpasses two current ENN algorithms as shown in Fig. 3.

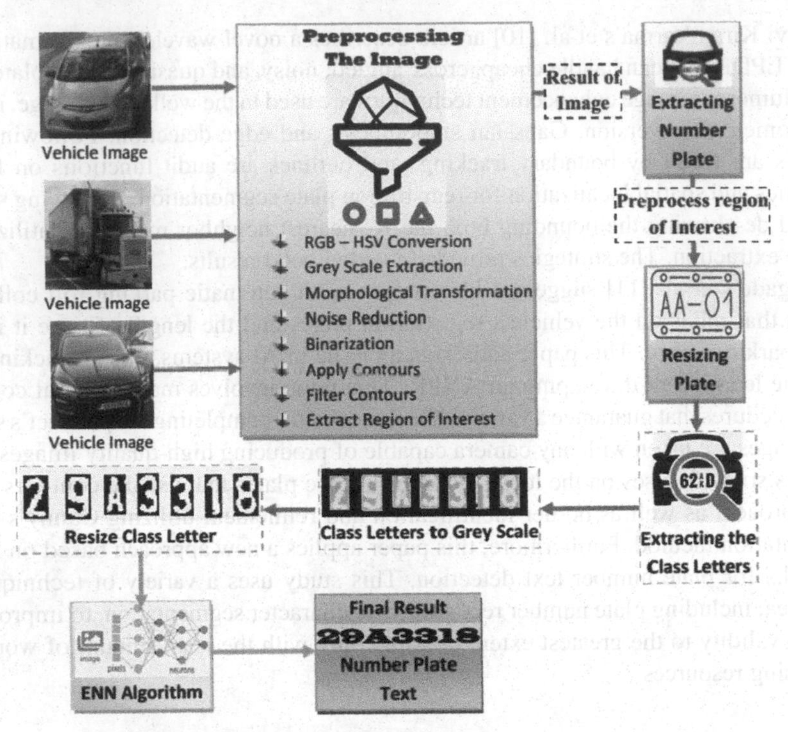

Fig. 3. Proposed automated vehicle number plate recognition system architecture

The rainfall reduction technique is only used on the monsoon automobile photos in the proposed ENN. There is no such phase to decide whether a digital image is rain-affected and what is not. It is entirely focused on the rapid recognition of target plate from an automobile picture without considering the imaging techniques and identification phases of an ALPR system. Different dangerous visual circumstances are the focus of attention. A classifier for rainfall and quasi images might be used to activate the rain reduction feature [13–16]. In the future, we hope to improve the suggested ENN to tackle these issues.

Furthermore, the ENN approach can be durable by integrating additional screening parameters and associating other concerns such as snowy weather and vertically inclined LP [17]. An LPR system uses computer vision techniques to assist in identifying vehicles based on their number plates. A procedure in which the number plate region is first localized inside an automobile photograph provided through one or several sensors. Afterward, the characters on the plate are detected by a character gesture recognition, known as vehicle detection [18–22]. In this study, an algorithm has been developed for recognizing plates utilizing photographs taken at multiple angles, locations, and times of the day or, including, under varying lighting circumstances.

Otsu's adaptive threshold approach and the plate's characteristics are used to detect and recognize the plate. Feature image is segmented using transverse and longitudinal scatter plots. Finally, Recursive Neural Networks are used to recognize characters.

The simulation results are included, as well as performance evaluations [23]. Throughout the simulations, Matlab software is employed. The process entails gray-level suppression, screening, and edge detection. The purpose of reducing input images to the grayscale image is to remove any unnecessary information. The computing power is greatly increased in this manner. Eventually, a Recursive Neural Network is used to classify the acquired texts.

4 Experimental Results

The proposal aims to identify and analyze texts on a license plate. Therefore, the photos are classified into three groups: educate, verify, and authentic (Table 1).

Table 1. Experimental settings

CPU	Intel(R) Core i5- 2.30 GHz
Software	Windows 7 64-bit
Memory	16.00 GB
Coding language	MATLAB
Data base	MYSQL
Image collections	5000

The photos are educated in a 75:25 ratio, with 75 percent of the pictures being used to construct the system and 25 percent that has been used to evaluate its correctness. Acknowledging the input from the user starts the procedure. The plate number imagery and identified text will then be presented at the end as shown in Figs. 4–6.

Fig. 4. Individualized Intelligent Vehicle License Plate Detection

Fig. 5. Computerized Detection of Vehicle License Plates – Group

Fig. 6. Computerized Detection of Vehicle Number Plates – Group (Hidden)

As a dataset, a total of 5000 plate number images are collected. The textual photos in the collection are initially utilized as a training dataset, and the Machine Learning technique would then be trained using this set. Simultaneous plates and recognition systems are used to assess the computational results [24]. The disadvantage of this system is that even if only one character is incorrect, the plate is not detected appropriately (Table 2).

Table 2. Template matching results

Actual plate	Predicted plate	Mismatched characters	Accuracy
KA14N2125	KA14N2125	0	100%
KA91WK6646	KA91MK6846	2	80%
KA51D9523	KA51D9523	0	100%
TN76AZ1124	TN78AZ1124	1	90%
KA51WC3803	KA51WC3803	0	100%
TN04AY9939	TN04AY9939	0	100%
TN63DB4367	TN88O84367	4	60%

Characters are configured to the standard required. License plates are created using several typeface layout sources and backdrop extraction from those other photos. The resulting image's dimensions are resized to the needed size.

Table 3. Accuracy of the proposed system

Work	Accuracy
KNN approach	92%
[1]	80.8%
[4]	89.39%
[7]	90%
[13]	85%

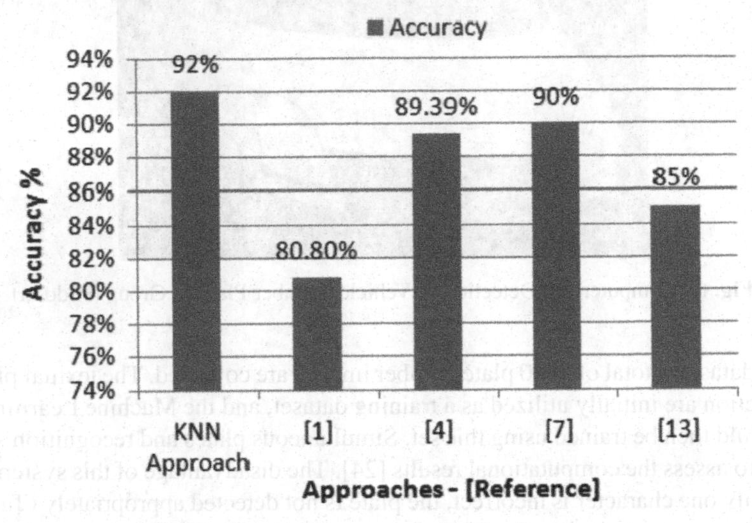

Fig. 7. Pictorial assessment - levels of accuracy

The effectiveness of plate number recognition and detection attained by the methods suggested in this paper is shown in Table 3. The results are then compared towards other number plate identification and recognition research. Figure 7 illustrates a pictorial assessment of the license plate recognition system's levels of accuracy with those of other related works.

5 Conclusion

Vehicle license plate recognition is a broad topic that can be approached utilizing various methods and methods. The preparatory procedures, including RGB to monochrome transformation and visual quantization, are performed first using our methodological approach. After that, the number plate is taken out. The characters then are fragmented, which is used as inputs to the CNN in need to identify the characters correctly. Our approach became more dependable and effective in determining after being trained with 5000 photos captured by us, resulting in an average accuracy of 92 percent. However, more research in this field is urgently required to account for all possible difficult instances and slight rotation and skew. As a result, a declining natural system was designed.

References

1. Puranic, A. et al.: Vehicle number plate recognition system: a literature review and implementation using template matching. Int. J. Comp. Appl. **134**(1) (Jan 2016)
2. Bhandari, S.: Indian vehicle number plate detection using image processing. Int. Res. J. Eng. Technol. **04**(04) (Apr 2017)
3. Divya, K.N., Ajit, D.: Recognition of vehicle number plates and retrieval of vehicle owner's registration details. Int. J. Innovat. Res. Technol. Sci. **2**(3), pp. 61–66
4. Apurva, B.: An enhancement of number plate recognition based on artificial neural network. Int. J. Comp. Sci. Info. Technol. **7**(3), 1291–1295 (2016)
5. Atikuzzaman: Vehicle Number Plate Detection and Categorization Using CNNs. In: 2019 International Conference on Sustainable Technologies for Industry 4.0 (STI), pp. 24–25. December, Dhaka
6. Parvin, S., Rozario, L.J., Islam, E.: Vehicle number plate detection and recognition techniques: a review. Adv. Sci. Technol. Eng. Sys. J. **6**(2), 423–438 (2021)
7. Etomi, E.E.: Automated number plate recognition system. Tropical J. Sci. Technol. **2**(1), 38–48 (2021)
8. Raju, V.V.K.: Vehicle number plate detection for traffic control. JASC: J. Appl. Sci. Computa. **VI**(V) (May 2019)
9. Shobayo, O., Olajube, A., Ohere, N., Odusami, M., Okoyeigbo, O.: Development of smart plate number recognition system for fast cars with web application. Appl. Computa. Intelli. Soft Compu. **2020**, 7 (2020). Article ID 8535861
10. Ravi Kiran Varma, P., Ganta, S., Hari Krishna, B., Svsrk, P.: A novel method for indian vehicle registration number plate detection and recognition using image processing techniques. Procedia Computer Science **167**, 2623–2633 (2020). ISSN 1877–0509, https://doi.org/10.1016/j.procs.2020.03.324
11. Bagade, J.V.: Automatic number plate recognition system: machine learning approach. IOSR J. Comp. Eng. (IOSR-JCE), 34–39. ISSN: 2278–0661, ISBN: 2278–8727.
12. Dhinakaran, D., Joe Prathap, P.M.: Preserving data confidentiality in association rule mining using data share allocator algorithm. Intelli. Auto. Soft Comput. **33**(3), pp. 1877–1892 (2022). https://doi.org/10.32604/iasc.2022.024509
13. Selvi, M., Joe Prathap, P.M.: WSN data aggregation of dynamic routing by QoS analysis. J. Adv. Res. Dyn. Control Syst. **9**(18), 2900–2908 (2017)
14. Prathap, P.M.J., Vasudevan, V.: Revised variable length interval batch rekeying with balanced key tree management for secure multicast communications. IJCSNS Int. J. Comput. Sci. Net. Secur. **8**, 232–241 (2008)
15. Premalatha, J., Prathap, P.M.J.: A survey on underwater wireless sensor networks: progresses applications and challenges. In: MATEC Web of Conferences, vol. 57, p. 02007 (2016)
16. Aruna Jasmine, J., Nisha Jenipher, V., Richard Jimreeves, J.S., Ravindran, K., Dhinakaran, D.: A traceability set up using digitalization of data and accessibility. In: International Conference on Intelligent Sustainable Systems (ICISS), pp. 907–910. IEEE Xplore, Tirunelveli, India (2021). https://doi.org/10.1109/ICISS49785.2020.9315938
17. Rabbani, G., Islam, M.A., Azim, M.A., Islam, M.K., Rahman, M.M.: A Bangladeshi license plate detection and recognition with morphological operation and convolution neural network. In: 2018 21st International Conference of Computer and Information Technology (ICCIT), pp. 1-5. IEEE (2018)

18. Dhinakaran, D., Joe Prathap, P.M., Selvaraj, D., Arul Kumar, D., Murugeshwari, B.: Mining privacy-preserving association rules based on parallel processing in cloud computing. Int. J. Eng. Trends and Technol. **70**(3), 284–294 (2022). https://doi.org/10.14445/22315381/IJETT-V70I3P232.
19. Sanaj, M.S., Joe Prathap, P.M.: An efficient approach to the map-reduce framework and genetic algorithm based whale optimization algorithm for task scheduling in cloud computing environment. Mater Today Proc **37**(2), 3199–3208 (2020)
20. Shidore, M.M., Narote, S.P.: Number plate recognition for indian vehicles. Int. J. Comp. Sci. Netw. Secu. **11**(2), 143–146 (2011)
21. Dhinakaran, D., Joe Prathap, P.M.: Ensuring privacy of data and mined results of data possessor in collaborative ARM. Pervasive Computing and Social Networking. Lecture Notes in Networks and Systems, vol. 317, pp. 431–444. Springer, Singapore (2022). https://doi.org/10.1007/978-981-16-5640-8_34
22. Selvi, M., Joe Prathap, P.M.: Analysis & classification of secure data aggregation in wireless sensor networks. Int. J. Eng. Advan. Technol. (IJEAT) **8**(4), 1404–1407 (2019)
23. Dhinakaran, D., Kumar, D.A., Dinesh, S., Selvaraj, D., Srikanth, K.: Recommendation system for research studies based on GCR. In: 2022 International Mobile and Embedded Technology Conference (MECON), pp. 61-65. Noida, India (2022). https://doi.org/10.1109/MECON53876.2022.9751920
24. Miyata, S., Oka, K.: Automated license plate detection using a support vector machine. In: 2016 14th International Conference on Control, Automation, Robotics and Vision, ICARCV 2016, pp. 13–15 (2016). https://doi.org/10.1109/ICARCV.2016.7838653

Noise Prediction Using LIDAR 3D Point Data - Determination of Terrain Parameters for Modelling

Shruti Bharadwaj[1](✉), Kumari Deepika[2], Rakesh Dubey[1], and Susham Biswas[1]

[1] Rajiv Gandhi Institute of Petroleum Technology, Jais, Amethi 229305, India
shruti01bharadwaj@gmail.com, {pgi19001,susham}@rgipt.ac.in
[2] Symbiosis Institute of Computer Studies and Research, Symbiosis International (Deemed University), Atur Centre, Model Colony, Pune 411016, India
kumari.deepika@sicsr.ac.in

Abstract. Noise prediction at a location necessitates topographical data (positional information about the building, road, etc.), noise data (of sources), and a prediction model to forecast noise levels. This is important for setting up urban planning for noise pollution management. Variability in outdoor conditions of grounds along with the locations of noise sources and non-sources (i.e., noise receivers) make it difficult to predict noise levels accurately. Good noise prediction models need to determine related factors accurately, i.e., how noise can transmit from noise source to receiver positions- directly or indirectly. Existing techniques for noise modelling suffer due to approximation in use of technique for generation of terrain parameters along with the use of inadequate qualities of terrain data. Use of highly accurate and dense LIDAR data is planned to overcome deficiencies in data quality. Detail source to receiver noise propagation paths is tried to be determined. While propagating from noise source, it can directly reach to the receiver or reach indirectly after diffracting around barriers and buildings. Further, it can reflect from the ground and wall before reaching to receiver location. In the paper a technique is described using LIDAR data to generate terrain parameters. LIDAR 3D point cloud data are used as terrain data. An algorithm is proposed to determine all the possible paths in 3D from the noise source to the receiver. A point-to-point rigorous routing system capable of operating in 3D is created, specifically designed to suit the sonic propagation concept (unlike well know shortest path determination algorithms used primarily in 2D). Once the detailed path information is obtained, the detailed terrain parameters (distance, difference, reflection contribution, ground absorption etc.) are determined. These are then applied over semi-empirical noise model to determine the noise levels at receivers (or unknown locations). From the LIDAR data of RGIPT campus, building coordinates are extracted and then by applying the new routing algorithm it was possible to identify the paths, and extract terrain parameters. These are then applied over noise model to predict the noise map for RGIPT campus. Accurate prediction is ensured through use of accurate terrain data and efficient extraction of terrain parameters from detailed paths in 3D (unlike the conventional 2D shortest routing algorithms). Accurate management schemes. prediction scheme can help in developing efficient noise.

M. Singh et al. (Eds.): ICACDS 2022, CCIS 1613, pp. 281–295, 2022.
https://doi.org/10.1007/978-3-031-12638-3_24

Keywords: Noise modeling · Noise mapping · Noise prediction · Sound propagation · Traffic noise · LIDAR

1 Introduction

Noise pollution is one of the most important and rising environmental pollution issues affecting human health [1]. Noise prediction is required for noise pollution management. To estimate noise levels surrounding distinct noise sources, significant terrain data (positional information about the building, road, etc.) is required, as well as noise data (of sources) and a prediction model. Noise propagates from the source to the receiver's position in the event of outside propagation. While moving, it can directly reach to a receiver or reach indirectly after diffracting around barriers and buildings. It can also be reflecting from the ground or can be absorbed to different surface before reaching a receiver location. It may be thought that the best way of finding noise levels is noise modelling and that requires possible path determination. Determination of the transmission is a big challenge in the real world. Terrain data is required for it. Terrain data usually comes in a different format, i.e., as satellite images, aerial photo, or ground points surveyed through Total Station, GPS or through LIDAR survey.

Fig. 1. LIDAR survey

LIDAR stands for Light Detection and Ranging (Fig. 1.). It is a remote sensing technology that can calculate the distance and other properties of targets by illuminating it with laser light and considering the backscattered light [2]. It has evolved into an effective tool for generating rich and precise elevation data across water bodies, landscapes, and project sites. These approaches are well-known in the surveying and engineering professions for their ability to generate exceptionally high accuracies [3] and point densities, allowing for the creation of exact, three-dimensional representations of highways, trains, buildings, and bridges [3].

Terrain data in the form of a raster, vector, or point cloud has its own set of challenges in terms of processing, determining distinct pathways, and extracting terrain characteristics for noise modelling. LIDAR point data are unique in that they are thick clouds (in X, Y, and Z) with very high precision. It is difficult to employ for terrain parameter generation.

2 Literature Review

With the growing concerns regarding the mitigation of noise pollution, it is necessary to estimate noise levels. All forecasts require an understanding of noise propagation over terrain. GIS is a sophisticated collection of tools for storing and retrieving spatial data from the actual world, as well as modifying and displaying it [4]. To trace the data alteration at each stage of the process, a GIS data management system might be employed. Alteration in input data, data simplification, interpolation, calculation, etc. that could affect the accuracy of outputs are all examples of data manipulation. There are a variety of noise prediction methods now available. They require a platform on which to construct a model from several components. Roadways, railways, etc. can be examples of linear noise sources, while industry sites, airports, etc. can be examples of area noise sources. Various noise prediction techniques employed various concepts to define these values, one of which is Lima, which used the Line segmentation principle to define these parameters. It is accomplished by using the projection approach; various portions are created from the source to the destination solely to examine the route variation for obstacle screening using "rubber-band logic" [5].

SoundPlan calculates the amount of noise generated by roads, trains, and industry. The model includes a separate toolbox with GIS-style data entry and manipulation devices. SoundPlan works on the idea that the user must import the attributes by setting heights, distances, and other parameters. Some programs can create DEMs from height data. For noise propagation modeling, the user must manually enter the information for sound sources, destinations, and barriers.

For numerous other urban applications and software that are not tailored for noise prediction, other techniques [6–8] are employed to discover the shortest routes. These algorithms' function using graph theory, with the root node as the source and the goal node as the destination. A graph, in this sense, is made up of vertices (also known as nodes) linked by edges (also called links). The Greedy algorithm, which is a type of algorithm that finds the maximum value at every node in the graph to find the shortest route, is important in this area. Even if it finds a node with a very large value throughout the search, the greedy algorithm will take that route because it can't return. As a result, it takes some time to find an appropriate solution. The Dijkstra algorithm, on the other hand, seeks to determine the quickest route from one location to another based on the least weight. This can only be used for ages that aren't negative. The Bellman ford algorithm is similar to Dijkstra's, with the exception that it can also be used for negative weight. However, it takes longer to search. There is another algorithm known as the A* algorithm, which performs the search for all existing nodes. This algorithm is the combination of the greedy and Dijkstra algorithms [9]. The problem of finding an optimal route has always been challenging in computer science, mathematics, and geoinformatics. The optimal route problem is also referred to as an optimization problem [10] attempted to provide solutions for both efficiency and accuracy. As we discussed earlier, many algorithms are designed to solve optimization problems [11]. A genetic algorithm is an evolutionary optimization technique that has been used to solve the challenge of finding the shortest path [12, 13] offered a genetic algorithm-based solution to the optimum route problem [14–17].

3 Research Methodology

In this work, the LIDAR data is used to determine terrain parameters through extraction of different paths between source and receiver positions. The building points, building corners, ground information etc. are first tried to be ascertained. Different paths, i.e., how noise (sound) propagates from various source positions (car, bus, train, etc.) to reach different receiver locations after, reflecting over different surfaces. An attempt has been made to determine all the possible paths in 3D from the noise source to the desired destination(s) using plane cutting technique. LIDAR data has been acquired for RGIPT campus. DSM, BEM are determined, along with buildings and building corners. Once building corners are determined the possible paths between source points and receiver points are tried to be determined. For RGIPT campus railway line is considered to be the linear source and within campus ground, building, and other objects are considered receiver points. Paths between source and receiver points are tried to be determined.

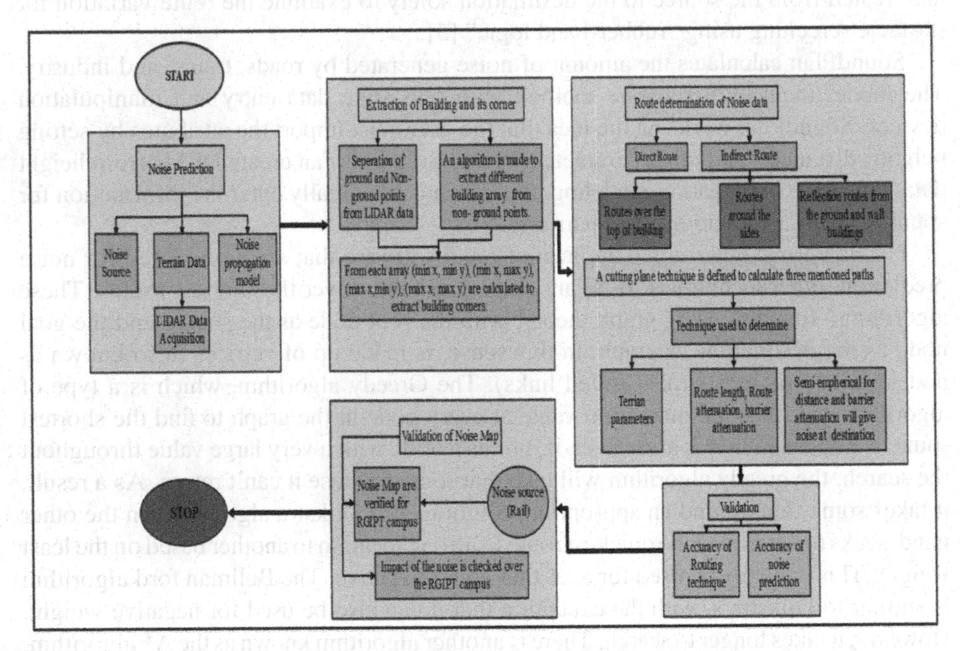

Fig. 2. Flowchart for the proposed research methodology.

First, the route that noise takes over the top of a structure as it travels from a source to a destination is attempted to be identified. Second, pathways around the sides of buildings or sideways paths are attempted to be established by taking into account a cutting plane between each pair of source and receiver and inter visibility between them.

Further, reflective paths are tried to be determined between each pair of source and receiver (destination) by checking availability of potential reflection points where angle of incidence becomes close to angle of reflection. Once the paths are determined, the terrain parameters such as distance, difference, reflection contribution are determined for each pair of source and receiver and put inside the noise model for prediction and mapping. This is the proposed technique which is shown in Fig. 2.

4 Result and Discussion

The detailed data processing worked on various stages such as LIDAR data acquisition, building extraction, Building corner estimation, Path Determination (Top way, Side way, and Reflection s). All these paths are used to determine terrain parameters and Noise Mapping. Step wise processing and results are given below.

4.1 LIDAR Data Acquisition

Kindly assure Part of RGIPT campus near academic area consisting of buildings and grounds is used for LIDAR data generation and determination of various paths.
Step 1: - Area of RGIPT campus is taken from google map (Fig. 3(a)) that consists of academic block 1 and academic block 2 and administrative area.
Step 2: - For the area taken in step 1 LIDAR data is generated (Fig. 3(b)) that is in the form of point cloud and consist of x, y, and z information.

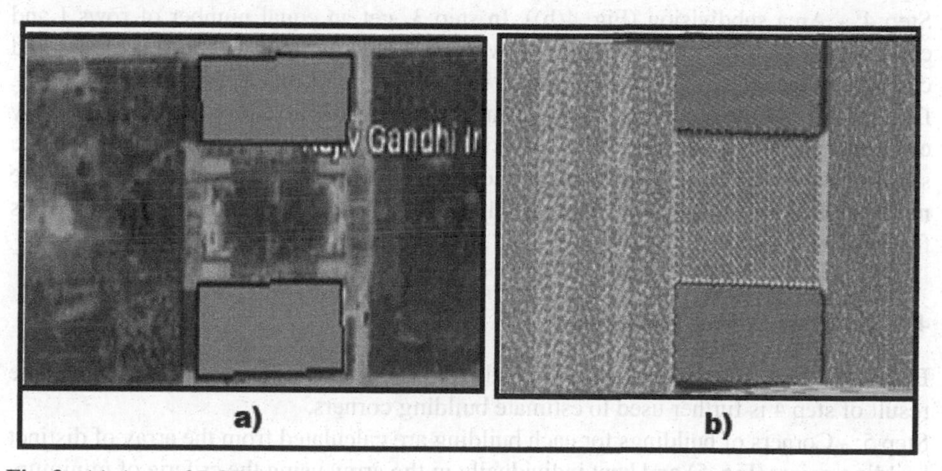

Fig. 3. (a) Area of RGIPT . taken from Google map, (b) The plot shows LIDAR point data for the area in step 1

4.2 Building Extraction

Part of RGIPT campus near academic area consisting of buildings and grounds is used for LIDAR data generation and determination of various paths.

Step 3: - In the step 2, we got the LIDAR plot of an area, for that plot digital elevation [18] and digital surface model are created. Basically, after the creation of two, this step gives the output of the subtraction of the BEM from the DSM to get elevated data of RGIPT area. The pink portion of the output (Fig. 4(a).) is the area contained elevated points of the building. This is done only for extracting the buildings in the area.

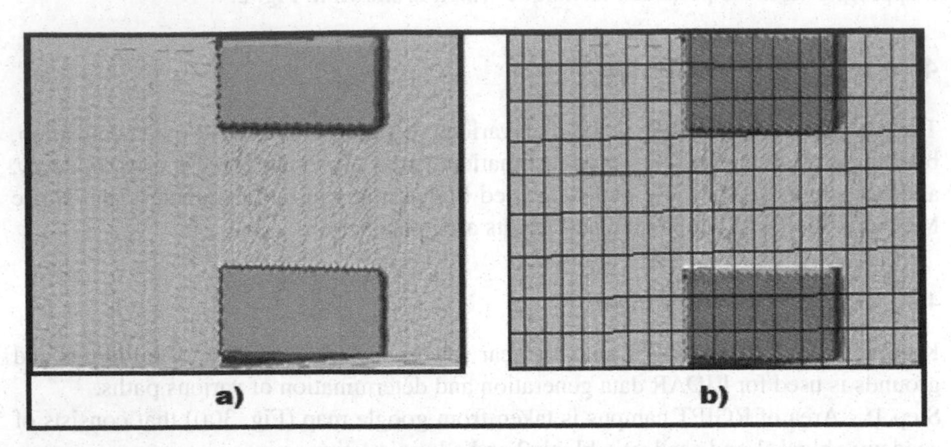

Fig. 4. (a) Elevation data after Subtraction of BEM from DSM, (b) Cell partition of an area in step 4.

Step 4: - Area subdivision (Fig. 4(b)). In step 3, get an equal number of rows I and columns (j), then start moving cell (i,j) with incrementing (i) row and (j) column and check if it includes point or not, and if it contains point, name that point buiding1. Determines the threshold for distinguishing one construction point from another by examining neighboring cells of the cell housing the point. If a neighboring cell has the same raised value, assign that point to the same building; otherwise, allocate it to the new building. Again, verify each point till the finish, and then create an array of points for each building.

4.3 Building Corner Estimation

Building edges and corners are estimated from the LIDAR 3D point cloud data. The result of step 4 is further used to estimate building corners.

Step 5: - Corners of buildings for each building are calculated from the array of distinct building points (Fig. 5) and kept individually in the array using the criteria of minimum and maximum of x and y (min x, min y), (min x, max y), (max x, min y) (max x, max y).

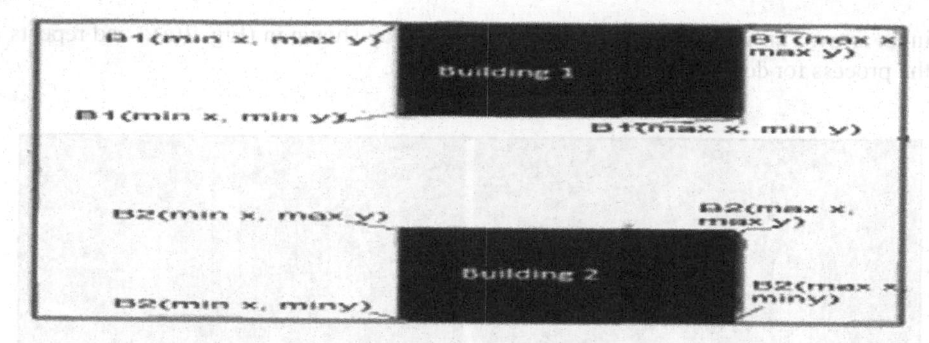

Fig. 5. Building corners details.

4.4 Path Determination

For the modelling different paths are determined, that describes how noise propagates from various sources (car, bus, train, etc.) to reach different locations through direct or indirect transmission after diffracting (around building), reflecting (from wall) over different surfaces.
Step 6: - When there is no blockage between the source and the receiver, the estimated path is the direct path. Paths that experience the diffracting or reflecting points are called as Indirect path which are given below.

Top Way Path. These are the path that are followed by noise to propagate from a source to destination over the top of buildings. So, these paths are called as Top Way Path.

Step 7: Examine the intersecting buildings between a source and destination line and determine the highest structure; buildings on the right side of the tallest building are regarded upward, while those on the left are considered below. (Fig. 6(a)).

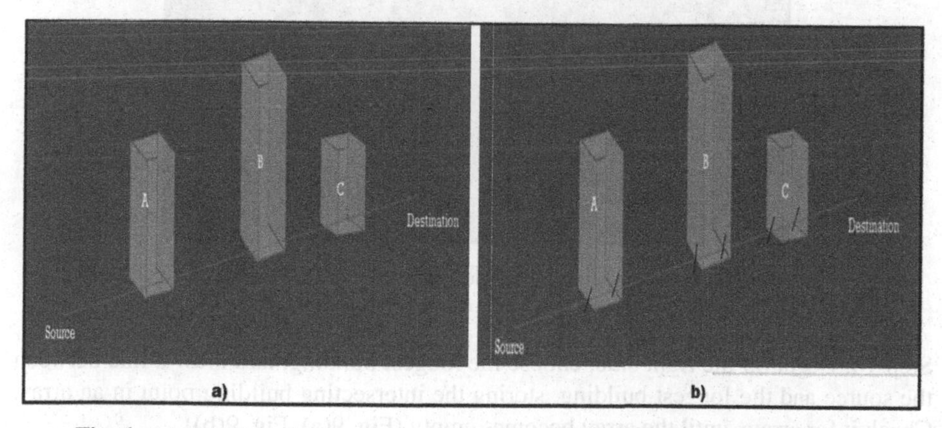

Fig. 6. (a) Buildings for the top way determination, (b) Working process of step 7.

Step 8: - Firstly it is done for upward, by creating a 3D line from source to tallest building and check is any building intersect or not (Fig. 6(b)). And store the intersected building in an array. This is done for determining upward way shown in (Fig. 7(a).) and repeats the process for downward way shown in (Fig. 7(b)).

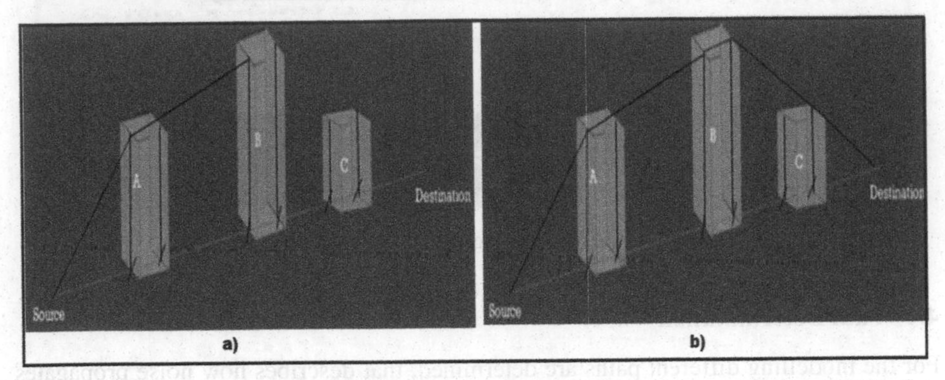

Fig. 7. (a) Working process for upward path determination, (b) The complete Top Way Path.

Side Way Path. Path followed by noise through the sides or corners of the buildings. So, these paths are called as Side Way Path.

Step 9: Setting up a buffer region in order to remove the outliers, by dividing the whole parts in two sides, one in right and one in left (Fig. 8.).

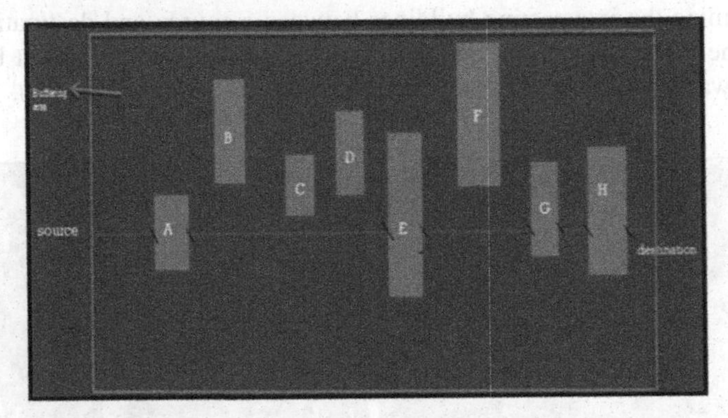

Fig. 8. Building setup for the determination of side way paths.

Step 10: Begin on the right side, choose the longest building, and draw a line between the source and the longest building, storing the intersecting building point in an array. Check it for arrays until the array becomes empty (Fig. 9(a), Fig. 9(b)).

Step 11: Determine the maximum length point of the non-intersecting building with the source and destination lines, then repeat the preceding step (Fig. 10(a)). A line from F to the destination is formed and checks the intersecting building and intersecting point is stored in an array (Fig. 10(b)). This is the complete right-side (Fig. 11(a)) and repeats the same for the left side (Fig. 11(b)).

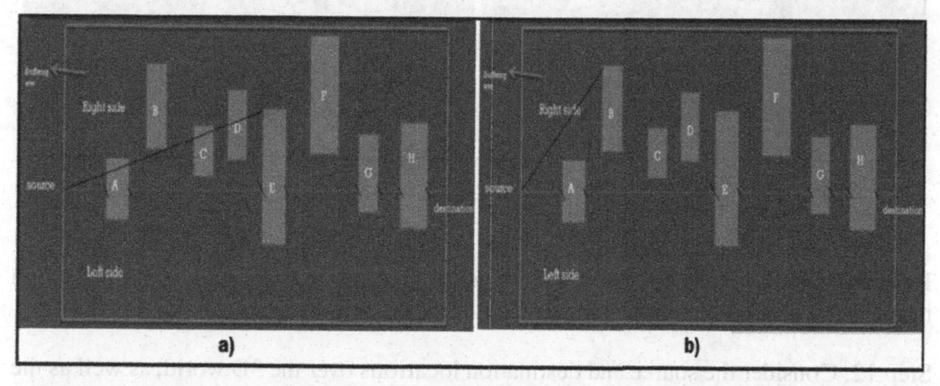

Fig. 9. (a) Working process of step 10 in Side way, (b) Process till the array become empty.

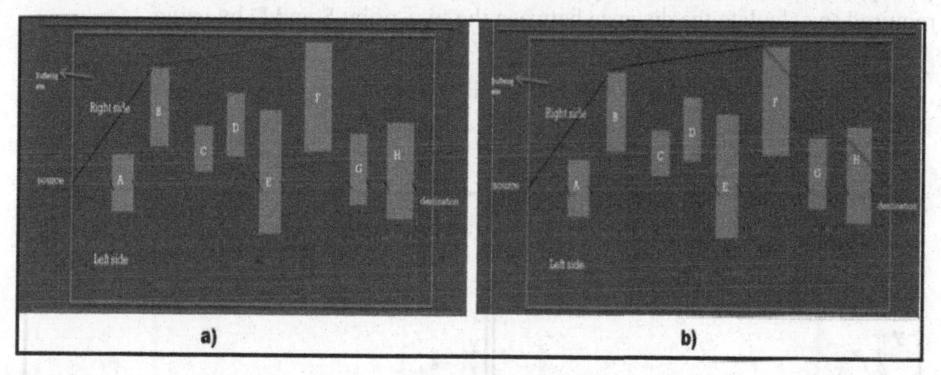

Fig. 10. (a) Working process of step 11 in Side way, (b) Working process for the downward array.

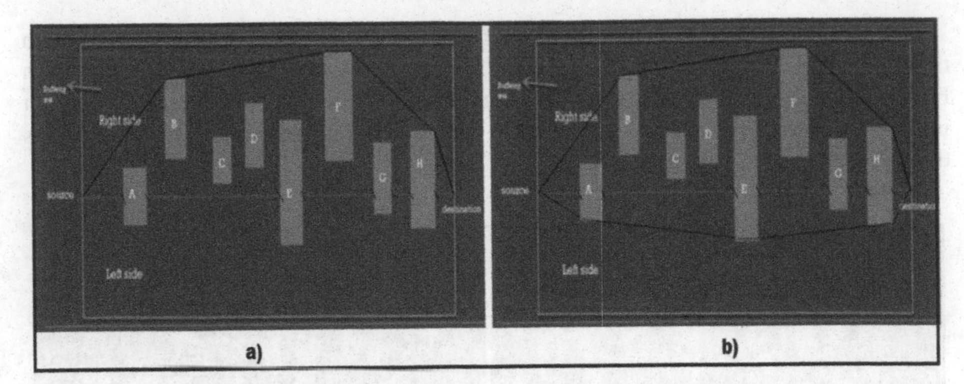

Fig. 11. (a) Complete the Left-side path, (b) Complete the Right-side path.

Reflection Path. Paths followed by the noise due to wall reflection and reflection through the ground. So, these paths are called as Reflection Path.

Step 12: Consider the source and destination locations over the 3D world, as well as the region, first for the Reflection. Make a connection between source S and destination S (×1, y1, 0) and the destination point D (×2, y2, 0) (Fig. 12(a)), after the creation, it is required to calculate the distance between the two-point S and D by using-

$$\text{Distance} = \sqrt[2]{\left((x2 - x1)^2 + (y2 - y1)^2 + (z2 - z1)^2\right)}$$

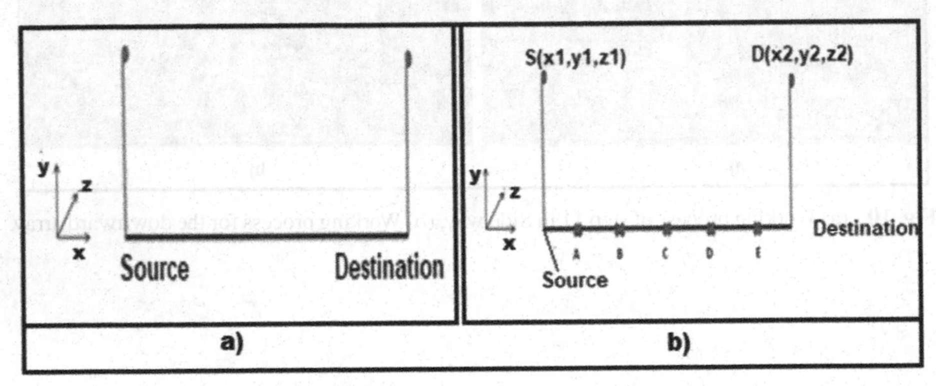

Fig. 12. (a) A . Source and destination for Reflection, (b) Working process for determination of potential intermediate ground reflection points

Step 13: According to Snell's law of reflection, the angle formed by the incident ray equals the angle formed by the reflected ray, sin(i) = sin(i) (r). For the figure indicated in Step 16, the points located on the line are A, B, C, D, and E. (Fig. 12(b)) [19]. The angle from S and angle from D are now measured for each point, say 'A.' The angle difference (d) is derived after computing angles from S and D. This is done for all of

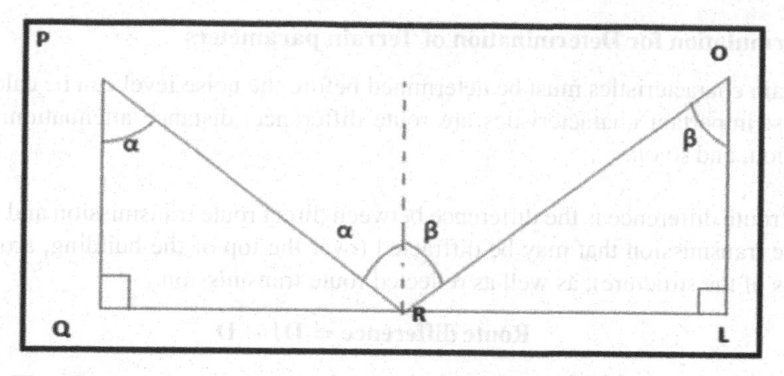

Fig. 13. Determination angles for reflecting point using Snell's law process.

the points discovered in Step 17 to determine the optimal point of reflection (Fig. 13, Fig. 14(a)).

Snell's law for reflection, sin(α) = sin(β)

$$(\alpha) = cos - 1(QR/PR$$

$$(\beta) = cos - 1(RL/OR)$$

$$d = |\sin(\alpha) - \sin(\beta)|$$

Step 14: - The likely reflection point is derived after calculating the angle difference for each point (Fig. 14(b)).

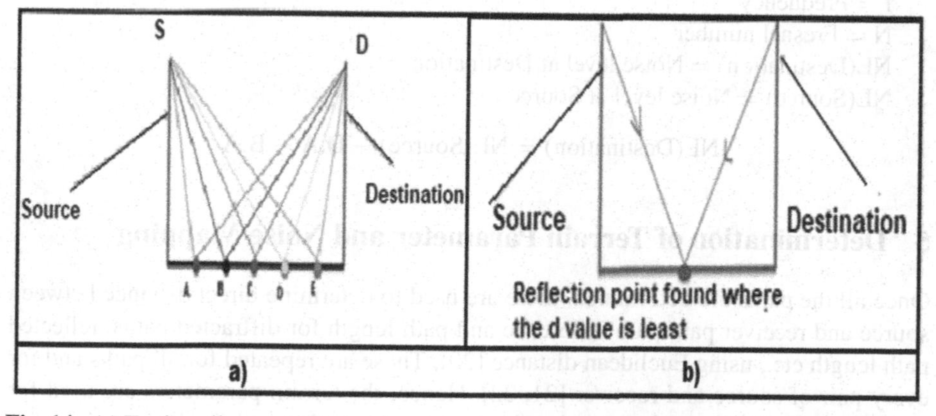

Fig. 14. (a) Testing all potential points between . source and destination to find the best reflection point, (b) Reflection point, the point at which d is minimum

4.5 Formulation for Determination of Terrain parameters

The terrain characteristics must be determined before the noise level can be calculated. The most important characteristics are route difference, distance attenuation, barrier attenuation, and so on.

1. The route difference is the difference between direct route transmission and indirect route transmission that may be diffracted (over the top of the building, around the sides of the structure), as well as reflected route transmission.

$$\textbf{Route difference} = \textbf{D1} - \textbf{D}$$

   where D1= the Indirect transmission route
   D = the Direct transmission route
2. Signal attenuation is determined by the distance and barrier between the source and destination locations. This is determined for each individual.

$$\text{Distance Attenuation} = 20\log_{10} D + 11$$

$$\text{B.A} = 5.65 + 66\text{N} + 244\text{N}^2 + 287\text{N}^3 \quad \textit{if } -0.30 \le N \le -0.02$$

$$\text{B.A} = 5.02 + 21.1\text{N} - 19.9\text{N}^2 + 6.69\text{N}^3 \quad \textit{if } -0.02 \le N \le 1.0$$

$$\text{B.A} = 10\log_{10} N + 18 \quad \textit{if } 1.0 \le N \le 18.0$$

$$\text{B.A} = 25 \quad \textit{if } N \ge 18$$

$$\text{N} = \frac{\text{Route difference}}{\frac{\lambda}{2}} \qquad \lambda = \frac{\text{c}}{\text{f}}$$

where D = Direct transmission route
B.A = attenuation due to barrier
λ = wavelength
c = Speed of light
f = Frequency
N = Fresnel number
NL(Destination) = Noise level at Destination
NL(Source) = Noise level at Source

$$\text{NL(Destination)} = \text{NL(Source)} - \text{D.A} - \text{B.A}$$

5 Determination of Terrain Parameter and Noise Mapping

Once all the paths are determined, these are used to determine direct distance between source and receiver pair, path difference and path length for diffracted paths, reflected path length etc., using Euclidean distance [20]. These are repeated for all paths and for every pair of source and receiver [21, 22]. Hence, the terrain parameters are used for noise mapping using noise prediction model. Noise mapping is done to show the noise propagation in the particular area. 25 noise sources at railway line near RGIPT campus are used to determine all the paths for all the receiver locations at the campus as shown in Fig. 15(a). These are then used to map the noise level for the RGIPT campus due to railway noise sources (Fig. 15(b)). Predicted noise levels are plotted in GIS environment using ArcGIS software [23].

Fig. 15. A (a) RGIPT campus with 25 noise sources on railway line, (b) Predicted Noise Level map for the RGIPT campus.

6 Conclusion

GIS based applications determines the routes primarily in 2D, however the acoustic pressure waves transmit in 3D, requiring extraction of route in 3D. Authors understood the need of establishment of solution in this direction. Author proposed the algorithm to overcome the data quality limitation and the technique to extract the terrain parameters from the LIDAR data for noise modelling. Use of very high resolution and dense LIDAR point offered a better terrain data for the terrain parameter extraction. A point-to-point rigorous routing approach for working in 3D is created utilizing LIDAR data, specifically designed for the sonic propagation concept (unlike well-known shortest path determination algorithms primarily used in 2D). The approach allows for the determination of direct, path over top, pathways around building sides, and reflected paths. These are then used to determine detailed terrain parameters, and comprehensive noise map.

References

1. Biswas, S., Lohani, B.: Development of high resolution 3D sound propagation model using LIDAR data and air photo. In: The International Archives of the Photogrammetry, pp. 1735–1740 (2008)
2. Lohani, B., Ghosh, S.: Airborne LiDAR technology: a review of data collection and processing systems. Proc. Natl. Acad. Sci. India Sect. A - Phys. Sci. **87**, 567–579 (2017) https://doi.org/10.1007/s40010-017-0435-9
3. Iordan, D., Popescu, G.: The accuracy of LiDAR measurement for the different land cover categories. Sci. Pap. Ser. E.l. Reclamation, Earth Obs. Surveyying, Environ. Eng. **4**, 158–164 (2015)
4. Choi, J., Zhang, B., Oh, K.: The shortest path from shortest distance on a polygon mesh. J. Theor. Appl. Inf. Technol. **95**, 4446–4454 (2017)
5. Tandel, B., Sonaviya, D.: A Quick review on Noise propagation models and software. 6 (2018)

6. Kogut, J.P., Pilecka, E.: Application of the terrestrial laser scanner in the monitoring of earth structures. Open Geosci. **12**, 503–517 (2020). https://doi.org/10.1515/geo-2020-0033
7. Veronese, L.D.P., Ismail, A., Narayan, V., Schulze, M.: An accurate and computational efficient system for detecting and classifying ego and sides lanes using LiDAR. In: IEEE Intell. Veh. Symp. Proc. 2018-June, pp. 1476–1483 (2018). https://doi.org/10.1109/IVS.2018.8500434
8. Asal, F.F.F.: Comparative analysis of the digital terrain models extracted from airborne LiDAR point clouds using different filtering approaches in residential landscapes. Adv. Remote Sens. **08**, 51–75 (2019). https://doi.org/10.4236/ars.2019.82004
9. Madkour, A., Aref, W.G., Rehman, F.U., Rahman, M.A., Basalamah, S.: A Survey of Shortest-Path Algorithms (2017)
10. Zarrinpanjeh, N., Dadrass Javan, F., Naji, A., Azadi, H., De Maeyer, P., Witlox, F.: Optimum path determination to facilitate fire station rescue missions using ant colony optimization algorithms (case study: City of Karaj). Int. Arch. Photogramm. Remote Sens. Spat. Inf. Sci. - ISPRS Arch. **43**, 1285–1291 (2020). https://doi.org/10.5194/isprs-archives-XLIII-B3-2020-1285-2020
11. Alasadi, H.A.A., Aziz, M.T., Dhiya, M., Abdulmajed, A.: A network analysis for finding the shortest path in hospital information system with GIS and GPS. J. Netw. Comput. Appl. **5**, 10–23 (2020). https://doi.org/10.23977/jnca.2020.050103
12. Canali, C., Lancellotti, R.: GASP: genetic algorithms for service placement in fog computing systems. Algorithms. **12**, 1–19 (2019). https://doi.org/10.3390/a12100201
13. Mittal, H., Okorn, B., Jangid, A., Held, D.: Self-Supervised Point Cloud Completion via Inpainting (2021)
14. Dubey, R., Bharadwaj, S.: Collaborative air quality mapping of different metropolitan collaborative air quality mapping of different metropolitan. In: ISPRS - Int. Arch. Photogramm. Remote Sens. Spat. Inf. Sci. XLIII-B4–2, pp. 87–94 (2021). https://doi.org/10.5194/isprs-archives-XLIII-B4-2021-87-2021
15. Bharadwaj, S., Dubey, R.: Raster data based automated noise data integration for noise raster data based automated noise data integration for noise. In: ISPRS - Int. Arch. Photogramm. Remote Sens. Spat. Inf. Sci. XLIII-B4-2, pp. 159–166 (2021). https://doi.org/10.5194/isprs-archives-XLIII-B4-2021-159-2021
16. Joshi, G., Pal, B., Zafar, I., Bharadwaj, S., Biswas, S.: Developing intelligent fire alarm system and need of UAV. Lect. Notes Civ. Eng. **51**, 403–414 (2020). https://doi.org/10.1007/978-3-030-37393-1_33
17. Bharadwaj, S., Dubey, R., Zafar, M.I., Srivastava, A., Bhushan Sharma, V., Biswas, S.: Determination of optimal location for setting up cell phone tower in city environment using lidar data. In: ISPRS - Int. Arch. Photogramm. Remote Sens. Spat. Inf. Sci. XLIII-B4-2, pp. 647–654 (2020). https://doi.org/10.5194/isprs-archives-xliii-b4-2020-647-2020
18. Liu, X., Zhang, Z., Peterson, J., Chandra, S.: Large area DEM generation using airborne LiDAR data and quality control. In: Spatial Accuracy Assessment in Natural Resources and Environmental Sciences, pp. 79–85 (2008)
19. Gaol, F.L.: Bresenham algorithm: implementation and analysis in raster shape. J. Comput. **8**, 69–78 (2013). https://doi.org/10.4304/jcp.8.1.69-78
20. Bharadwaj, S., Dubey, R., Zafar, M.I., Tiwary, S.K., Faridi, R.A., Biswas, S.: A novel method to determine the optimal location for a cellular tower by using LiDAR data. Appl. Syst. Innov. **5**, 30 (2022). https://doi.org/10.3390/asi5020030
21. Dubey, R., Bharadwaj, S., Zafar, M.I., Mahajan, V., Srivastava, A., Biswas, S.: GIS mapping of short-term noisy event of diwali night in lucknow city. ISPRS Int. J. Geo-Information. **11**, 25 (2021). https://doi.org/10.3390/ijgi11010025

22. Sharma, V.B., et al.: Review of structural health monitoring techniques in pipeline and wind turbine industries. Appl. Syst. Innov. **4**(3), 59 (2021). https://doi.org/10.3390/asi4030059
23. Maguire, D.J.: ArcGIS: general-purpose GIS software. Encycl. GIS. 1–8 (2016). https://doi.org/10.1007/978-3-319-23519-6_68-2

Block Based Resumption Techniques for Efficient Handling of Unsuccessful Loads in Data Warehouse

N. Mohammed Muddasir[1](✉) and K. Raghuveer[2]

[1] Department of IS&E, VVCE, Mysuru 570002, India
mohammed.muddasir@vvce.ac.in
[2] Department of IS&E, NIE, Mysuru 570008, India

Abstract. ETL is the acronym for extract transform and load, it's a process to extract, transform and load data into the data warehouse from sources that could be other transactional database, text files, logs etc. This source could be heterogeneous or homogeneous. Any of the steps in ETL could fail and measures are required to resume the process. There could be several reasons for failure like network break down, hard ware crash, database unavailable, over loaded systems, etc. The issue of ETL failure is a serious one because it's a time consuming processes. In case of failure of ETL what should be the strategy to resume the process so that the focus of resumption is on the data that failed to load rather than on the data that is already in the data warehouse. In this paper a block based approach is used to load the data that failed during load of ETL. Empirical results show the block based approach performs better in terms of resumption time as compared to SQL EXCEPT.

Keywords: Near real time ETL · Block based resumption techniques · Data warehouse

1 Introduction

Data ware house is according to Bill Inmon is subject oriented, integrated, time variant, non-volatile collection of data. Data in data ware house is never subject to deletion and it always growing in size, variety. Source of data or information for the data warehouse are the transactional processing systems. The process of ETL is used to extract data from source, transform to suit the data warehouse environment [1] and finally load into the data warehouse. The ETL was done previously during off peak hours, but due to the requirements for real time [2] analysis the ETL is also subject to real time loading. As the transactional database cannot be in synch at a real time basis because the data has to be extracted, transformed and loaded, the terminology for real time ETL is called as near real time ETL [3]. The processes guarantee the data available for analysis is as fresh as possible and the results of analysis are very accurate. The ETL and near real time ETL deals with data in huge size, variety and volume of data arrival is also fast. Hence the

M. Singh et al. (Eds.): ICACDS 2022, CCIS 1613, pp. 296–308, 2022.
https://doi.org/10.1007/978-3-031-12638-3_25

ETL developers have to come up with solution to make sure the ETL process completes in time and without any interrupt or failure.

In case of failure of any of the steps in the ETL processes, the situation is tense as how to handle the current load that failed as well as the new data that is arriving at a faster rate. Handling failure of ETL is a major concern for all the stake holders. There are solutions and algorithms that handle this scenario and try to resume the ETL, as discussed in related work of this paper. The ETL resumption solutions could be categorized into one with less over head or more over head during the normal operation and also the one that does resumption fast or slow. A detailed comparison of categories of solutions is discussed in [4].

The focus of any work on ETL resumption should be with less over head during the normal operation and load fast during resumption in case there is a failure of ETL. The solution this paper discusses is called as block based approach to resume failed ETL data. The data in the source is made into blocks by taking a range of primary key values. The overhead part in this work is to keep a separate table having all the blocks for a particular table or a set of tables after a equi-join is performed on those tables. This additional blocks table helps during resumption of failed ETL processes. The idea is based on comparison and this work reduces the number of comparisons required to identify the data that which is not loaded due to failure. To illustrate in case of loading 100 rows if a failure happens after 80 rows loaded, then to load the remaining 20 rows one has to compare or processes all the 100 rows to know which 80 rows were loaded and which 20 were not loaded. In case of blocks approach the comparison is based on which block was loaded and which one did not. If the 100 rows are made into 10 blocks and then the maximum number of comparison in case of failure is 10 thus reducing the amount of processing required for the resumption of the ETL process.

2 Related Work

The most cited work on resumption of failed data load is in this [4] paper. The authors have devised efficient re-load of failed data loads based on certain simple properties of the workflow. The technique proposed in their work does not have any kind of overhead of storing snap shots or save points to reload failed data loads. The idea is to look at the amount of data or tuples that were successfully loaded before a failure of load happens viz. if out of 100 rows 80 were loaded successfully before a failure the need is to load only the remaining 20 and not the entire 100. So the crux of their solution is to identify those missing rows. To do this they have developed a frame work for loading and reloading. Within this frame work they define properties for efficient resumption. The algorithm is called as design resume algorithm. The approach is getting all records in same order or different order and subset of records that are in different order or suffix of records (subset of records in same order). Same order here means the order in which the records were processed prior to load failure. They make use of key attribute and analyze the input tuples that are contributing to the warehouse tuples or output tuples. The input tuples after transformation may be contributing to one or more warehouse tuples. Identifying those input tuples that have already processed and contributed to the output tuples to avoid those tuples from going through the transformation and load is the solution they provide. To identify such tuples they have formulated properties like map to one, suffix safe, in-det-out and set to sequence. Using these properties they identify the tuples that contribute to the warehouse tuples on one to one basis or a prefix set basis or sequence basis. The design phase of the algorithm helps to identify tuples to be discarded based on the properties in filtering sub set of tuples or prefix of tuples. After this phase the resume phase shall re extract the missing tuples and load into the warehouse.

The improvement to design-resume algorithm was proposed in hybrid design-resume algorithm [5]. The authors have claimed to improve design resume algorithm by allowing simultaneous loading with the help of staging techniques. They identified the limitation of design resume as not able to work in case of distributed database environment and hence their first improvement is to identify loads that have already finished in case of distributed loads. They introduce new property called as sameSuffix to filter out tuples that are performing the same operation but simultaneously through various paths. One more enhancement was to provide a hint subset not recommended to design-resume algorithm that helps to avoid processing huge amounts of data; this data can be filtered out at a later phase. The processed data at each stage are stored in disk as stage data; if complete data is available the subsequent transformations are performed from staged data rather than from the original source. Further the design-resume logic is applied on the stage data. Although the approach can be applied on a distributed environment it has the overhead of additional storage space.

Checkpoint technique is used in database management systems to make sure data until the check point are present in the database in case of failure while inserting the data. Periodic check points are created in the system logs during transaction processing and if in case of failure the resumption of failed loads is done after the check point, because any updates before the checkpoint are already saved in the database. The authors of [6] have extended check point based resumption for ETL. They combine check point with design resume [4] to create a more efficient design resume algorithm. Checkpoint techniques could increase the processing delay and reduce efficiency, but the authors have managed to create check points in way to increase the efficiency. This is done by building a graph where the nodes are the inserter; transformer and extractor each of these nodes are identified as check point feasible based on certain conditions. Conditions are as follows inserter is check point feasible if it holds suffix safe and map to one property. Transformers related to check point feasible are themselves check point feasible transformers; similarly extractors are check point feasible if their corresponding transformers are check point feasible. Additional properties were defined for filtering same set, set to sequence and same suffix.

A control table technique is proposed by Hitachi in pentaho data integration (PDI) technique restart ability [7]. The idea is to maintain a separate table containing the details of the files to be loaded. If the file load is successful then the control table entry in updated to success for that file. If in case of failure the control table entry is updated as failure for that file entry. Only the failed status file in the control table is reloaded.

SQL supports many techniques for reloading the data in case of failed ETL [8]. They are inserting based on time stamp ordering, Insert where not exist, insert except, merge and left join etc. Each of the techniques is supported by many available commercial data base management systems.

Some of the other work on ETL in near real time known as stream ETL [9–11] explore on accuracy and efficiency of the development of the process the data in the fashion of streams. Another study is using of grammar and machine learning to perform the analysis of incoming queries and predict query processing time [12].

The above discussed works have given us the thought of having resumption techniques that are only working on part of data not yet loaded in case of failure of ETL loads. We developed a block based approach where we divide the given data into various blocks. The criteria for data to be present in the blocks is inspired from [13] a blog where the author makes blocks of data. He takes data of 3 million rows and creates 300 blocks each of size 10,000 rows. The novelty here is the blocks are made up using the domain of the data i.e. Each block is made of range of primary key values. The block one is made from primary key value one till primary key value 10,000 as the bounds. Similarly next block from primary key value 10,001 till 20,000 and so on. We assume that each table that which is very important in the database would have a primary key column or set of columns making primary key. If the data is partitioned based on primary key values it could be applied to any general purpose database.

We used this concept of making blocks of data but extended it to tables that are joined for the purpose of transformation. One of the most common forms of transformation in ETL is that of joins [14, 15] apart from filtering and grouping. Since the focus of this work is on resumption we have applied blocks based partition concept on joined data and track each block if it is loaded successfully or failed. In case of failure of loading only the particular block is reloaded. Further this work uses a status variable with an initial value of 3 for checking the possible outcome of every block. After every stage of ETL, if extract is successful status is decremented by one similarly if transform is successful status is decremented by one and finally if load is successful the status is decremented by one. Finally if status is 0 then the ETL for the block is successful. If not depending on the value of status only the relevant phase is re executed for that particular block.

3 Block Based Solution

In this section we explain the block based approach with an illustration. If the major chuck of data is transferred before a failure to calculate the remaining rows we have to process a huge amount of data. If 80–90 % of data is transferred to know the remaining 10–20 % of data that was not transferred, one had to process the already transferred data according to design-resume algorithm [4, 1]. As pointed out earlier it's not possible to know the remaining rows to be transferred if zero rows are present at the data warehouse. Failure can detect remaining rows only if we have some data at the data warehouse. Also this solution could require huge amount of processing if the majority of data is present at the data warehouse and then a failure happens. Hence we go for a block based approach where we have to work only on blocks of data in case of failure and not required to look up on the entire database.

We need to make the blocks of data for all the tables that are involved in the ETL process. Example if two tables are to be joined to get the data for loading viz. EMP and DEPT. EMP contains 100 rows and DEPT contains 10 rows. The scenario is each employee belongs to utmost one department. The natural join operation of EMP and DEPT would result in 100 rows.

Now if one has to load these 100 rows in to the data warehouse as one full chunk by normal ETL process and in case of failure knowing which record failed is a challenge. The challenge lies if 80 rows of the chuck is loaded successfully but only 20 rows remains to be loaded processing 80 rows is mandatory to know which 20 rows failed to load. To avoid reprocessing this huge amount of data to know what was not loaded the approach in this work is to make the data into blocks viz. blocks of size 10 i.e. all the data in employee now are considered as 10 blocks each containing 10 records equal to the original 100 records of EMP table. Now if major chunk of data is loaded, again taking the example of 80–90% of data being loaded to know the remaining data to be loaded one has to only know which block of data failed to load. In this way reprocessing huge amount of data to know which part of data failed is skipped.

The scenario is explained with the below Fig. 1, Fig. 2 and Fig. 3 with two table EMP and DEPT. Due to shortage of space we have taken on 5 records for DEPT table and 10 records for EMP table. These tables were taken from [16]. The blocks are made using ranges for primary key values. In the present scenario the DNUM is the primary key. The DEPT table is made into two blocks. In blocks concept range of primary keys are created and used while joining two to more tables. The detailed can be seen in below two tables DEPT and EMP. DEPT has two blocks primary key ranging from 100 and 200. Another block of primary key ranging from 300–500. Based on these blocks the EMP table also gets two blocks as shown in the EMP_DEPT table.

The block1 range from DNUM 100 to 200 and the block2 range from DNUM 300 to 500. If we have a transformation to join EMP and DEPT table on DNO in EMP and DNUM in DEPT then the resulting no. of rows would be 10 in the joined tables. But if we take the block wise approach there would be only 2 blocks. In case of failure for loading after the transformation EMP_DEPT we had to load all the 10 rows again. But if we maintain the block concept then we need to load which ever block failed. If for example block2 failed then we require to load only 6 rows as compared to the previous case of loading 10 rows. If block1 failed then we require to load only 4 rows. In any case the number of rows to be loaded is less as compared to loading the entire table. If block based approach is not used and 8 rows are loaded before a failure one had to reprocess the entire 8 rows to know which 2 rows were not loaded. But with block based approach this huge percent of reprocessing data is avoided.

A separate blocks table is maintained after creating the blocks from the base tables. Example is table with a name BLOCKS is created with column as block number, block size, starting primary key vaue and ending primary key value as shown below. The relevant columns are filled with the sample data from table DEPT.

The blocks are not same as partioning the table in any general purpose database management system. Partitition is usually done on single table but blocks in this paper are create after the join operation performed on two or more tables. The idea of making the tables in to blocks is different than partition concept of DBMS. In partition the single table is segregated according to the partitioning criteria like list key or hash based partition. Partition-wise join [17] in available in DBMS like oracle that combines partition and indexes to join but those are single row joins.

EMP

SSN	FNAME	LNAME	SAL	DNO
00000000	John	Smith	20000	100
11111111	Raju	Kumar	25000	100
22222222	Mohammed	Aleem	30000	200
33333333	Khalid	Imran	35000	300
44444444	Adil	Rashid	22000	300
55555555	Ravi	Kumar	21000	300
66666666	Raghu	Jain	26000	400
77777777	Anoop	Kumar	31000	400
88888888	Sudhir	Kumar	33000	100
99999999	Guru	Prasad	25000	500

Fig. 1. Emp table showing sample data

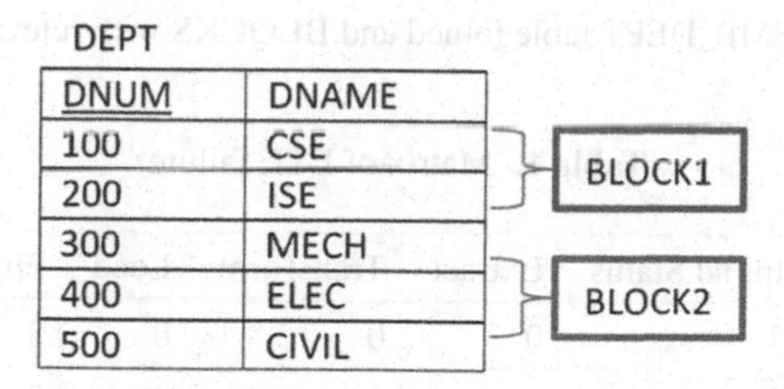

DEPT

DNUM	DNAME
100	CSE
200	ISE
300	MECH
400	ELEC
500	CIVIL

Fig. 2. Dept table showing block concept

This work also has a control table concept to maintain the status of each block then keeping track of failed block is easier as compared to keeping track of all the rows in the entire table. Control table contains information about blocks that are successful or not successful. We start by assigning flags to each of the steps in ETL process. The flag values are 1 for success and 0 for failure of a step. If the initial step for Extract fails we have to simply repeat the entire cycle there is no short cut here. If in case the transformation step fails then we could avoid repeating the extract step and continue with re-executing the transformation step only. Similarly other phases as shown in Table 1;

In the above table the column with the name blocks is the number of blocks to be processed. The extract column with status 0 is the extract for this block is a failure and need to be processed. Similarly status 1 represent success for that column be if transform of load for this particular block. The status column has four values 0, 1, 2, and 3. Initial value of status for all the blocks is 3. Once a ETL step is completed successfully the status is decremented by 1 and hence it's automatically known which step have to work upon for each block. If after processing the block if the final status value remains 3 indicate the ETL for this block in failure are all the 3 steps extract, transform and load have to repeat for that particular block. Hence there is a need to repeat the entire ETL for a particular block. Final status value 2 means extract was successful but need to work on

EMP_DEPT

SSN	FNAME	LNAME	SAL	DNO	DNAME
00000000	John	Smith	20000	100	CSE
11111111	Raju	Kumar	25000	100	CSE
22222222	Mohammed	Aleem	30000	200	ISE
33333333	Khalid	Imran	35000	300	MECH
44444444	Adil	Rashid	22000	300	MECH
55555555	Ravi	Kumar	21000	300	MECH
66666666	Raghu	Jain	26000	400	ELEC
77777777	Anoop	Kumar	31000	400	ELEC
88888888	Sudhir	Kumar	33000	100	CSE
99999999	Guru	Prasad	25000	500	CIVIL

BLOCK1

BLOCK2

BLOCKS

BLOCKNO	BLOCKSIZE	START_PK_VALUE	END_PK_VALUE
1	2	100	200
2	3	300	500

Fig. 3. EMP_DEPT table joined and BLOCKS with relevant data.

Table 1. Matrix of ETL failure

Blocks	Initial Status	Extract	Transform	Load	Final status
1	3	0	0	0	3
2	3	1	0	0	2
3	3	1	1	0	1
4	3	1	1	1	0

2 more steps for this block i.e. transform and load. Status value 1 means need to work on 1 step i.e. loading. Lastly if final status values 0 means need not work on any part of ETL for this block and every stage of ETL was successfully completed.

4 Implementation Methodology and Results

Block based approach allowed us to track individual blocks. Initially we created a table by the name blocks and this table is populated with the blockno, block size, starting primary key value and ending primary key value for each block. The simulation for failure was through delete operation of the percentage of data. Initially both the source and ware house tables were in sync, having the exact amount of rows. To simulate 10% reload the ware house tables was subjected to deletion of 10% of rows. Later these 10% data or rows were reloaded from the source to the ware house. To reload the rows two techniques were used on based on DBMS approach where the SQL EXCEPT feature was used as given in [5]; second the block based approach was used. The experimental results shows block based approach performs better than the SQL EXCEPT feature.

The experiment was carried out on TPC-DS bench [18] mark data set. The store sales table was considered as the source on which the blocks of data were created. The store sales table had composite primary key that comprises of ss_item_sk and ss_ticket_number. These two primary key ranges were used for creating the blocks. To simulate the ware house, same table with exact same column was created and loading would happen from store sales table to its replica in the data warehouse. The intention was to show that resumption works well with block based approach hence not much transformation part was taken in to consideration while implementing. The system used for the experimentation have the following specification Intel core i5 2.5 GHz CPU with 4 GB RAM runs 64bit windows10 operating system.

The data size was in rows. Initial source table had 1000 rows and blocks of size 100 were created. This was loaded completely into the warehouse table. The simulation of failure was done through deletion as said previously, so the percentage of failure was taken from [4]. In this work the percentage of failure was starting from 10%, then 25%, 50%, 75% and lastly 90%. Figure 4 shows the resumption time for each percentage failure of reload. The comparison is with SQL EXCEPT and proposed block based approach. The X-axis is the percentage failure and Y-axis is the reload or resumption time. The time is recorded in mille seconds. The criteria was showing which of the two approaches is better is based on resumption time. The one with a less resumption time is better.

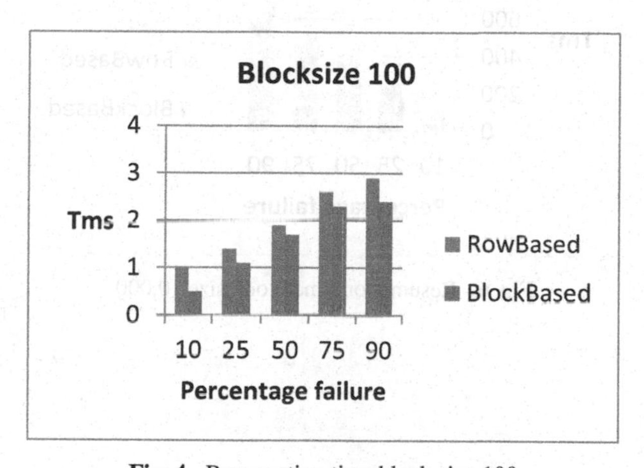

Fig. 4. Resumption time block size 100

Similar experiments were done on data of size 10,000 rows where the block size was 1000 and also on data of size 100,000 rows where the block size was 10,000. Figure 5 and Fig. 6, show the result. The block based approach performs better in all cased. The above experimentation were conducted like a proportionality test where in the initial data size and the block size were related like 100 rows blocks size was 10 rows and 1000 rows blocks size was 100 rows etc. The results show that as the rows increase the time to reload also increasing naturally.

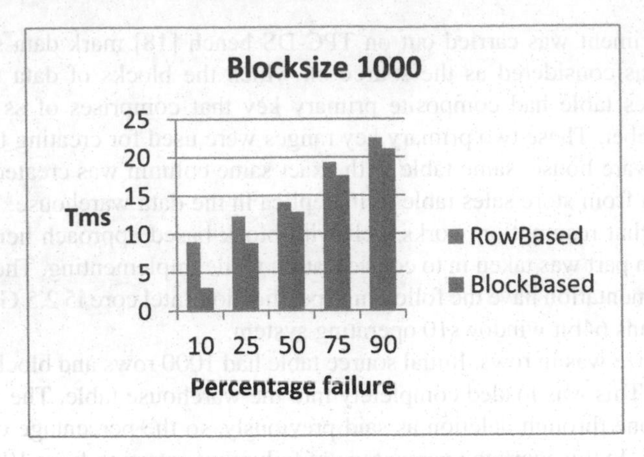

Fig. 5. Resumption time block size 1000

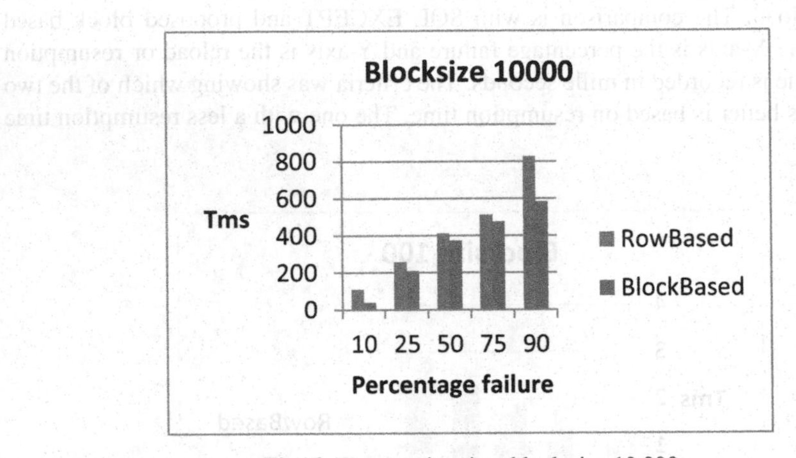

Fig. 6. Resumption time block size 10,000

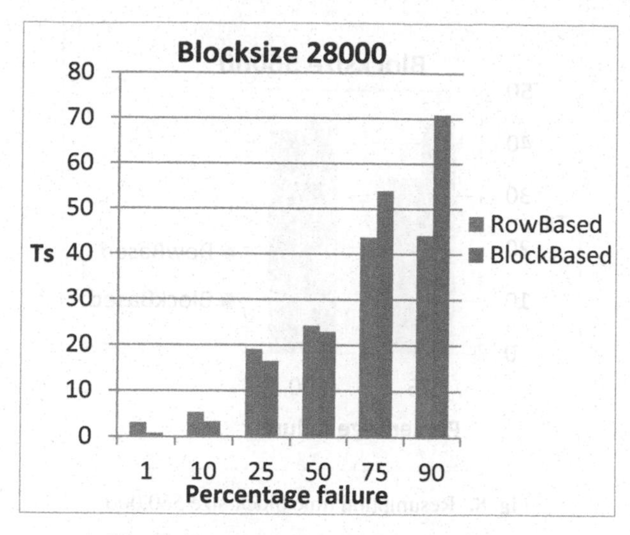

Fig. 7. Resumption time block size 28,000

The next phase was to have block size that are variable. In this case we took 2,800,000 (28 lac) rows and created blocks of 28,000 each so we got around 100 blocks, and not only repeated the same reload percentage as above but also additional 1% failure was tested. The time in this case was recorded in seconds. Figure 7 above shows the result and the resumption based on block based approach is better in all cases expect for 75% and 90% reload. The 1% reload is the most important as block based approach out performs SQL EXCEPT approach by almost 4 times, better in terms of resumption time, block based is 0.7 s and SQL EXCEPT is 3.01 s. This is very useful because of the huge percentage of data that has to be re looked. The 28 lac rows failed at just the end. i.e. about 1% failure the resumption based on block approach is better. This was one of the most significant observation of block based approach.

Another important observation is that the block approach was not performing well with 75% and 90% reload with the block size of 28,000 rows. The results in this study matches with the claim in [8] where SQL EXCEPT performs best of huge data. Hence the empirical study to take up different block size was considered to know whether the block approach fails for huge data. The data of 28 lac rows was made in to 50 blocks each of size 56,000 rows each, then 25 blocks of size 112,000 rows and finally 10 blocks of 280,000 rows. Since the blocks based approach failed for 75% and 90%, only these percentage resumption were experimented. The Fig. 8, Fig. 9 and Fig. 10 show the result for blocks size 25, 50 and 10 respectively. The results below show that for each of the blocks size considered for experimentation every time blocks based approach performs better than SQL EXCEPT. Lastly a plot of different blocks size is shown in Fig. 11. The X-axis is the blocks size in number of rows per block and Y-axis is the time in seconds. The empirical study leads to the conclusion that a bigger block size performs better for resumption of failed loads.

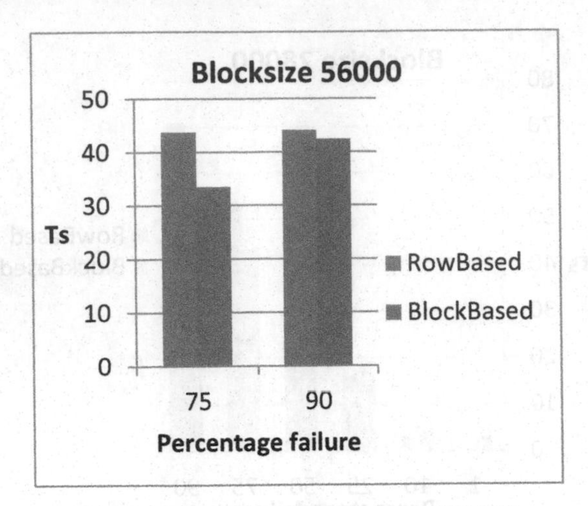

Fig. 8. Resumption time block size 560,000

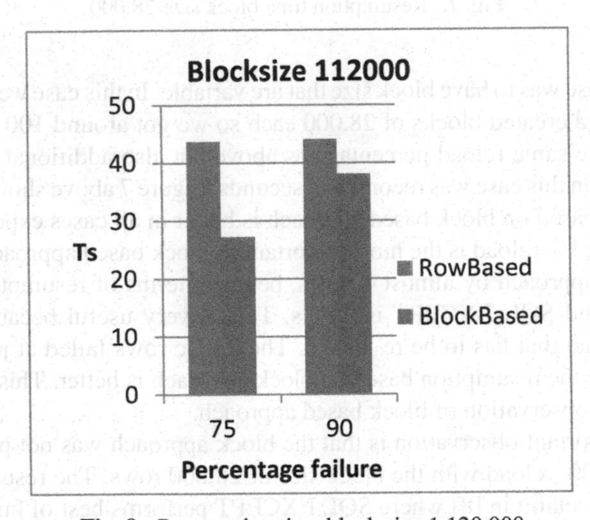

Fig. 9. Resumption time block size 1,120,000

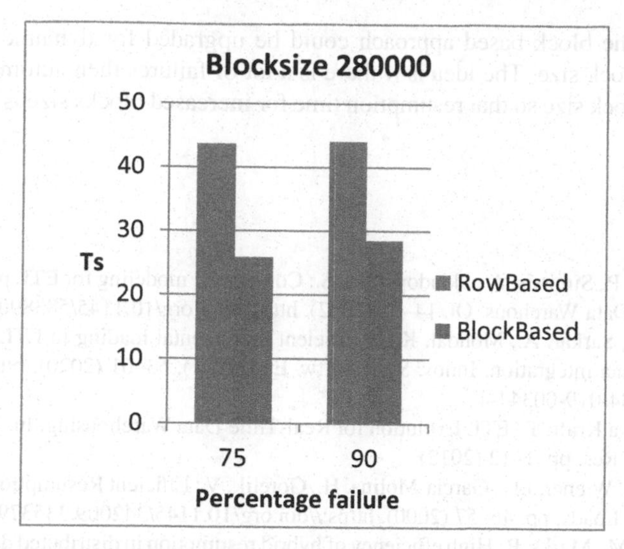

Fig. 10. Resumption time block size 280,000

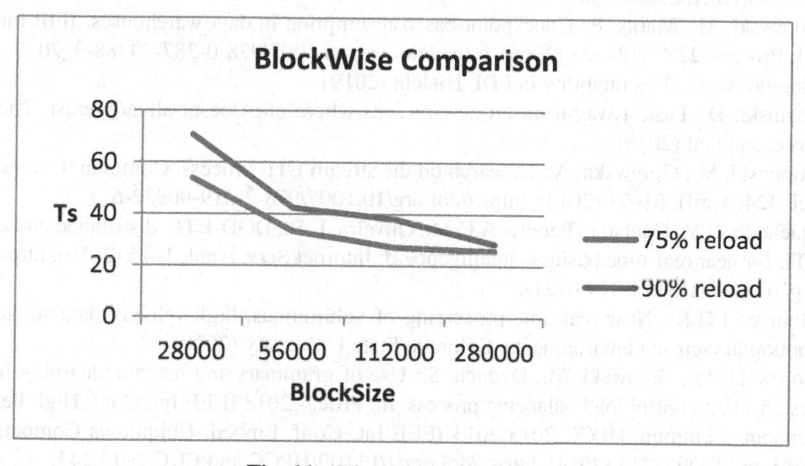

Fig. 11. Block wise comparison

5 Conclusion and Future Enhancement

The approach to resume failed ETL load have a significant importance in the age of real time data analytics. The work in this paper contributes towards reducing the time of resumption using the block based approach. The overhead in block based approach is less and hence it's feasible to implement in an existing ETL environment. The most noted reduction in time is where block based approach performed better in case of 1% load failure while loading a 28 lac rows. The block based approach was almost reduced the resumption time by a factor of 4 as compared with the resumption time using SQL EXCEPT.

In future the block based approach could be upgraded for dynamic tracking and changing of block size. The idea is if there is a lot of failures then automatically try to increase the block size so that resumption time for increased blocks size is less as shown in this work.

References

1. Vassiliadis, P., Simitsis, A., Skiadopoulos, S.: Conceptual modeling for ETL processes. ACM Int. Work. Data Warehous. Ol. 14–21 (2002). https://doi.org/10.1145/583890.583893
2. Biswas, N., Sarkar, A., Mondal, K.C.: Efficient incremental loading in ETL processing for real-time data integration. Innov. Syst. Softw. Eng. **16**(1), 53–61 (2020). https://doi.org/10.1007/s11334-019-00344-4
3. Kakish, K., a Kraft, T.: ETL Evolution for Real-Time Data Warehousing. In: Proc. Conf. Inf. Syst. Appl. Res. pp. 1–12 (2012)
4. Labio, W.J., Wiener, J.L., Garcia-Molina, H., Gorelik, V.: Efficient Resumption of İnterrupted Warehouse Loads, pp. 46–57 (2000). https://doi.org/10.1145/342009.335379
5. Gorawski, M., Marks, P.: High efficiency of hybrid resumption in distributed data warehouses. Proc. - Int. Work. Database Expert Syst. Appl. DEXA **2006**, 323–327 (2005). https://doi.org/10.1109/DEXA.2005.108
6. Gorawski, M., Marks, P.: Checkpoint-based resumption in data warehouses. IFIP Int. Fed. Inf. Process. **227**, 313–323 (2006). https://doi.org/10.1007/978-0-387-39388-9_30
7. Morehouse, C.: Restratability in PDI. Hitachi (2019)
8. Lozinski, D.: Fastest-way-to-insert-new-records-where-one-doesnt-already-exist. The curious consultant (2015)
9. Gorawski, M., Gorawska, A.: Research on the stream ETL process. Commun. Comput. Inf. Sci. **424**(April), 61–71 (2014). https://doi.org/10.1007/978-3-319-06932-6_7
10. Machado, G.V., Cunha, Í., Pereira, A.C.M., Oliveira, L.B.: DOD-ETL: distributed on-demand ETL for near real-time business intelligence. J. Internet Serv. Appl. 1–15 (2019). https://doi.org/10.1186/s13174-019-0121-z
11. Munige, T.H.R.: Near real-tıme processıng of volumınous, hıgh-velocıty data streams for contınuous sensıng envıronments. Colorado State University (2020)
12. Gorawski, M., Gorawski, M., Dyduch, S.: Use of grammars and machine learning in ETL systems that control load balancing process. In: Proc. - 2013 IEEE Int. Conf. High Perform. Comput. Commun. HPCC 2013 2013 IEEE Int. Conf. Embed. Ubiquitous Comput. EUC 2013, pp. 1709–1714 (2014). https://doi.org/10.1109/HPCC.and.EUC.2013.243
13. Vanlıghtly, J.: Building-synkronizr-a-sql-server-data-synchronizer-tool-part-1. RabbitMQ (2016). [Online]. Available: https://jack-vanlightly.com/blog/2016/11/12/building-synkronizr-a-sql-server-data-synchronizer-tool-part-1
14. El-Sappagh, S.H.A., Hendawi, A.M.A., El Bastawissy, A.H.: A proposed model for data warehouse ETL processes. J. King Saud Univ. - Comput. Inf. Sci. **23**(2), 91–104 (2011). https://doi.org/10.1016/j.jksuci.2011.05.005
15. Stitchdata: ETL Transforms. Talend (2019)
16. Navathe, R.E.S.B.: Database Systems (2016)
17. Oracle: Partition wise Join. Oracle (2010)
18. Transaction Processing Council (2020). [Online]. Available: http://www.tpc.org/tpc_documents_current_versions/pdf/tpc-ds_v2.13.0.pdf

Augmentations: An Insight into Their Effectiveness on Convolution Neural Networks

Sabeesh Ethiraj[1] and Bharath Kumar Bolla[2(✉)]

[1] Liverpool John Moores University, Liverpool, UK
[2] Salesforce, Hyderabad, India
bolla111@gmail.com

Abstract. Augmentations are the key factor in determining the performance of any neural network as they provide a model with a critical edge in boosting its performance. Their ability to boost a model's robustness depends on two factors, viz-a-viz, the model architecture, and the type of augmentations. Augmentations are very specific to a dataset, and it is not imperative that all kinds of augmentation would necessarily produce a positive effect on a model's performance. Hence there is a need to identify augmentations that perform consistently well across a variety of datasets and also remain invariant to the type of architecture, convolutions, and the number of parameters used. This paper evaluates the effect of parameters using 3×3 and depth-wise separable convolutions on different augmentation techniques on MNIST, FMNIST, and CIFAR10 datasets. Statistical Evidence shows that techniques such as Cutouts and Random horizontal flip were consistent on both parametrically low and high architectures. Depth-wise separable convolutions outperformed 3×3 convolutions at higher parameters due to their ability to create deeper networks. Augmentations resulted in bridging the accuracy gap between the 3×3 and depth-wise separable convolutions, thus establishing their role in model generalization. At higher number augmentations did not produce a significant change in performance. The synergistic effect of multiple augmentations at higher parameters, with antagonistic effect at lower parameters, was also evaluated. The work proves that a delicate balance between architectural supremacy and augmentations needs to be achieved to enhance a model's performance in any given deep learning task.

Keywords: Deep learning · Depth-wise separable convolutions · Global average pooling · Cutouts · Mixup · Augmentations · Augmentation paradox

1 Introduction

Data augmentations have become a crucial step in the model building of any deep learning algorithm due to their ability to give a distinctive edge to a model's performance and robustness [1]. The efficiency of these augmentations largely depends on the type and the number of augmentations used in each scenario. The effectiveness of augmentations in improving a model's performance depends on model capacity, the number of training parameters, type of convolutions used, the model architecture, and the dataset

M. Singh et al. (Eds.): ICACDS 2022, CCIS 1613, pp. 309–322, 2022.
https://doi.org/10.1007/978-3-031-12638-3_26

used. Augmentations are very specific to the factors mentioned herewith, and it is not imperative that all kinds of augmentation would necessarily produce a positive effect on a model's performance. Furthermore, very few studies have evaluated the relationship between augmentations, model capacity, and types of convolutions used.

Data augmentations can range from simple techniques such as Rotation or random flip to complex techniques such as Mixup and Cutouts. The efficiency of these techniques in improving a model's performance and robustness is critical in the case of smaller architectures as they are ideal for deployment on Edge/Mobile devices due to the lesser model size and decreased training parameters. Not every deep learning task requires an architecture optimized on generic datasets like Imagenet. Hence, there is a need to build custom-made lightweight models that perform as efficiently as an over-parameterized architecture. One such technique to help achieve reduced model size is the utilization of Depth-wise Separable convolutions [2]. In this paper, experiments have been designed to answer the primary objectives mentioned below.

- To evaluate the relationship between model capacity and augmentation
- To evaluate the effect of model capacity on multiple augmentations
- To evaluate the effect of depth wise separable convolution on augmentation

2 Literature Review

Lately, work on augmentations and architectural efficiency has been the hallmark of research, making models more efficient and robust. The research methods implemented in the paper have been implemented in various works summarized below.

2.1 Depth-Wise Separable Convolutions and GAP

Depth-wise convolutions were first established in the Xception network [3] and were later incorporated in the MobileNet [2] architecture to build lighter models. Due to the mathematical efficiency, these convolutions help reduce the number of training parameters in contrast to a conventional 3×3 kernel. Traditional CNNs linearize the learned parameters by creating a single dimension vector at the end of all convolutional blocks. However, it was shown that this compromises a network's ability to localize the features extracted by the preceding convolutional blocks [4]. It was also shown that [5] that earlier layers capture only low-level features while the higher layers capture task-specific features, which need to be preserved to retain much of the information. This was made possible by the concept of Global Average Pooling, wherein the information learned by the convolutions is condensed into a single dimension vector. The work done by [6] has also stressed the importance of Depth-wise convolutions and Global average pooling layer wherein deep neural networks were trained using these techniques, and a significant improvement in model performance was observed despite the reduction in the number of parameters. The techniques have been widely used in industrial setting as in the case of detecting minor faults in case of gear box anomalies [7], where in retaining special dimension of features is very important, considering the minor nature of these faults and a high probability of missing them during regular convolutions. A similar study in the classification of teeth category was done where in the results of max pooling was compared with that of average pooling [8].

2.2 Data Augmentation

Work on Regularization functions was done as early as 1995 to make models more robust, such as the radial and hyper basis functions [9]. These focus on a better approximation of the losses. Bayesian regularized ANNs [10] were more robust than conventional regularization techniques, which work on the mathematical principle of converting non-Linear regression into ridge regression, eliminating the need for lengthy cross-validation. State-of-the-art results were obtained on CIFAR-10 and CIFAR-100 datasets using Cutouts [11] to make models more robust and generalizable. Similarly, [12] the concept of mixup wherein combining the input and target variables resulting in a completely new virtual example resulted in higher model performance due to increased generalization. Even conventional baseline augmentations [1] have been shown to perform complex augmentation techniques such as GANs in augmenting the training samples. Combining self-supervision learning and transfer learning has also been shown to boost model performance [13] when no label is provided in some instances to enable powerful learning of the feature representations that are not biased. However, the effects of augmentations are not simply restricted to the improvement of model performance alone, as seen in the work done by [14], where other properties of augmentation such as test-time, resolution impact, and curriculum learning were studied. Though augmentations generally improve model performance most of the time, augmentation may not necessarily and always positively affect a neural network. They do have their shortcomings, as described in work done by [15].

3 Research Methodology

The methodology focuses on studying the effect of augmentations on different architectures (varying parameters) across three different datasets (CIFAR10, FMNIST and MNIST consisting of 50000 training and 10000 validation samples) using 3×3 and depth wise separable convolutions. Augmentation techniques have been chosen judiciously to perform a wide variety of transformations to the input images so that the effect of these techniques would be more explainable from the research perspective. The augmentations have been performed using Pytorch and the Mosaic ML library [16].

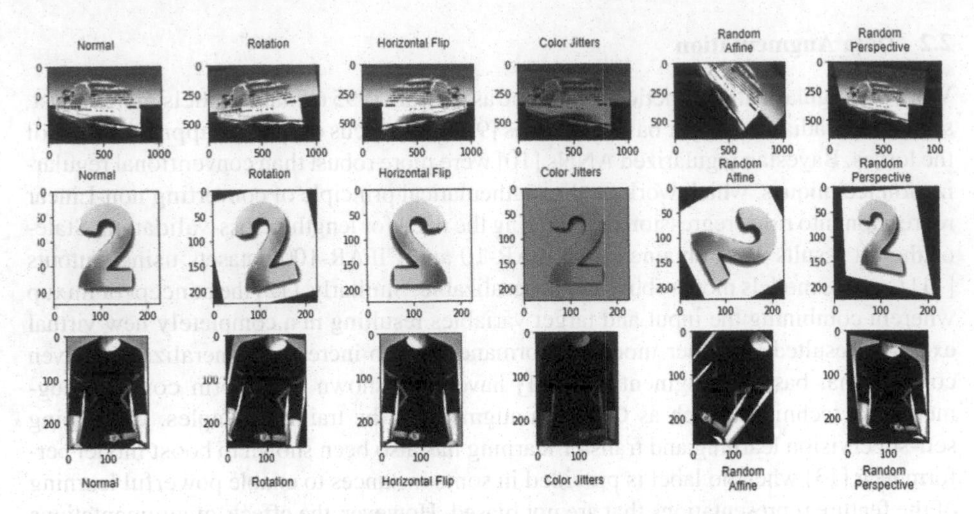

Fig. 1. A representation of various augmentation techniques

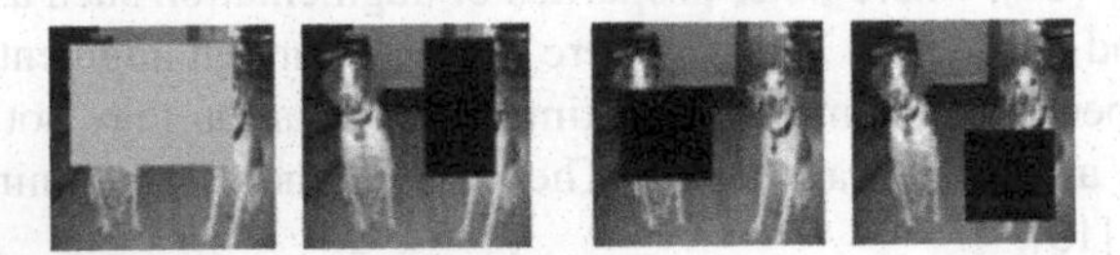

Fig. 2. Cutout

Fig. 3. Mixup

3.1 Depth-Wise Separable Convolutions

The mathematical intuition behind the depth wise convolutions is reducing the number of training parameters for the same number of features extracted by a 3×3 kernel. This is achieved by combining depth-wise channel separation and a 3×3 kernel followed by a point-wise convolution to summate the separated features learned in the previous step.

3.2 Global Average Pooling

Global average Pooling (GAP) is performed by taking the average of all the neurons/pixels in each output channel of the last convolutional layer resulting in a linear vector.

3.3 Random Rotation and Random Horizontal Flip

Random Rotation of 10° (Fig. 1) was chosen and applied uniformly on all the datasets to keep the variation constant across datasets. The rotation was kept minimal at 10° to avoid any significant distortion of the original distribution. In case of random horizontal flip, the augmentation is applied with a probability of 0.5, where there is a lateral rotation of the images.

3.4 Random Affine and Random Perspective

Random Affine (Fig. 1) is a combination of rotation and a random amount of translation along the width and the height of the image as defined by the model's hyperparameters. Random perspective performs a random perspective transformation of the input image along all the three axes.

3.5 Cutout

The Cutout is a regularisation or augmentation technique in which pixels from an input image are clipped. Random 8 × 8 masks (Fig. 2 – Sample) have been clipped in the experiments to avoid any extreme influence of both smaller and larger Cutouts which can affect model performance significantly.

3.6 Mixup

Mixup is a regularization technique in different input samples and their target labels to create a different set of virtual training examples, as shown in Fig. 3. A hyperparameter δ controls the mixup. The mathematical formula for mixup is shown in Eq. 1 where $\hat{x}$ and $\hat{y}$ are new virtual distributions created from the original distribution.

Virtual distribution.

$$\begin{aligned} \hat{x} &= \delta x_i + (1-\delta)x_j \\ \hat{y} &= \delta y_i + (1-\delta)y_j \end{aligned} \tag{1}$$

3.7 Loss Function

The loss function used here is the cross-entropy loss as this is a multiclass classification.

Cross-Entropy Loss.

$$CE = -\sum_{t}^{c} t_i \log s_i \tag{2}$$

3.8 Evaluation Metrics

Validation Accuracy and the percentage of accuracy change from the baseline model for every architecture are used as evaluation metrics.

3.9 Model Architecture

Table 1. Model architectures

Architectures incorporating Global Average Pooling across datasets				
MNIST	Fashion MNIST		CIFAR-10	
1.5K – 1560	**1.5K DW**	**−1560**	5.8K NDW	−5,886
	5.7K NDW	−5722	7.9K NDW	−7,92
	5.6K DW	**−5626**	25K NDW	−25,298
	7.8K NDW	−7,777	**25K DW**	**−24,788**
	7.6K DW	**−7,621**	140K NDW	−143,218
	25K NDW	−25,154	**140K DW**	**−143,406**
	25K DW	**−24,644**	340K NDW	−340,010
	140K NDW	−142,930	**340K DW**	**−344,508**
	140K DW	**−143,118**	600K NDW	−590,378
	600K NDW	−600,575	**600K DW**	**−599,913**
	600K DW	**−599,625**	1M NDW	−1,181,970
			1M DW	**−1,159,474**

(*NDW – non depth wise (3 × 3)/DW – depth-wise separable convolutions)

Architectures (Table 1) have been built by sequentially reducing the number of parameters using the concepts of Depth Wise Convolutions and Global Average Pooling followed by application of various augmentations (Fig. 4).

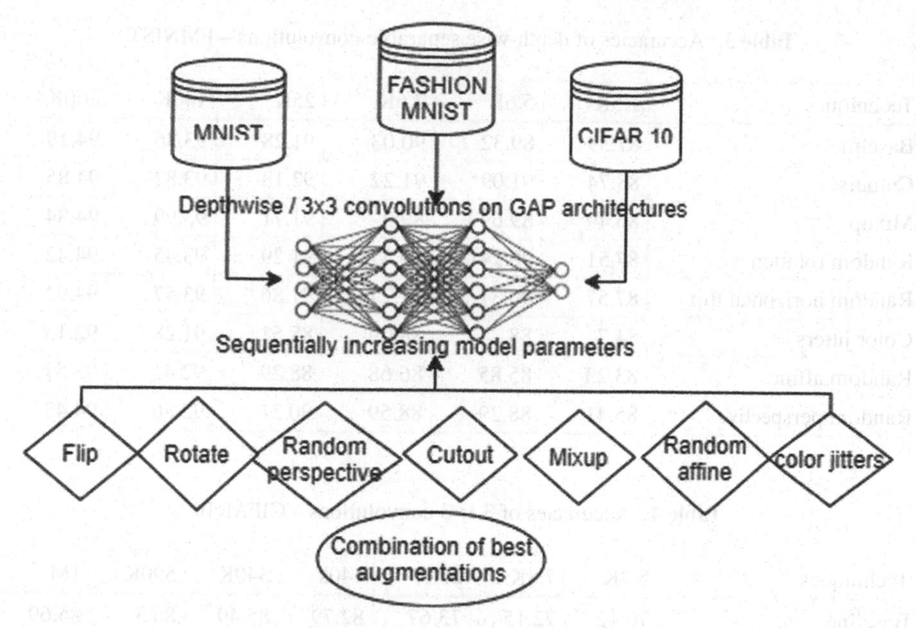

Fig. 4. Proposed methodology

4 Results

The validation accuracies of different architectures (parameters) have been summarized in Tables 2, 3, 4, and 5 on both the FMNIST and CIFAR 10 datasets.

Table 2. Accuracies of 3×3 convolutions – FMNIST

Techniques	5.7K	7.8K	25K	140K	600K
Baseline	90.12	90.6	91.81	93.55	94.19
Cutouts	91	91.51	92.45	93.87	94.7
Mixup	90.03	90.29	91.65	94.17	94.7
Random rotation	90.17	90.34	91.64	93.57	94.39
Random horizontal flip	91.07	91.23	92.17	93.81	94.72
Color jitters	88.04	88.69	88.61	91.38	92.17
Random affine	86.38	86.52	89.12	92.36	93.51
Random perspective	88.72	89.17	90.89	92.83	94.06

Table 3. Accuracies of depth-wise separable convolutions – FMNIST

Techniques	1.5K	5.6K	7.6K	25K	140K	600K
Baseline	87.39	89.32	90.03	91.28	93.36	94.19
Cutouts	88.74	91.09	91.22	92.18	93.82	94.85
Mixup	86.49	89.69	89.9	91.71	93.99	94.84
Random rotation	87.51	90.24	90.47	91.29	93.45	94.42
Random horizontal flip	87.57	90.37	91.27	91.86	93.57	94.92
Color jitters	84.7	88	88.09	88.51	91.28	92.13
Random affine	83.23	85.85	86.68	88.29	92.41	93.57
Random perspective	85.41	88.29	88.59	90.31	92.86	94.43

Table 4. Accuracies of 3 × 3 convolutions – CIFAR10

Techniques	5.7K	7.8K	25k	140k	340K	590K	1M
Baseline	70.12	72.15	73.67	82.79	85.49	86.3	85.69
Cutouts	70.89	72.06	75.91	84.53	87.04	87.6	87.49
Mixup	71.62	73.33	76.33	84.81	87.62	87.54	88.21
Random rotation	70.35	72.11	74.97	83.02	85.84	86.45	86.16
Random horizontal flip	71.58	73.32	77.41	84.9	87.76	88.53	88.12
Color jitters	69.2	68.79	71.96	79.91	82.92	83.93	83.31
Random affine	59.93	60.61	68.39	81.13	84.76	85.55	85.17
Random perspective	67.31	67.85	74.55	82.7	86.71	87.66	86.83

Table 5. Accuracies of depth-wise separable convolutions – CIFAR10

Techniques	25k	140K	340K	590K	1M
Baseline	70.75	81.06	83.7	86.7	85.96
Cutouts	72.49	82.56	85.36	87.85	87.89
Mixup	72.85	83.45	86.14	88.86	88.55
Random rotation	72.1	81.34	83.81	87.62	86.71
Random horizontal flip	74.66	84.04	86.72	89.12	89.01
Color jitters	70.89	78.98	81.09	84.2	84.17
Random affine	61.8	79.43	82.58	86.74	86.3
Random perspective	69.95	81.34	84.77	88.38	86.38

4.1 Augmentations on Depth-Wise Separable Convolutions

Equivocal Performances on Architectural Saturation. On the FMNIST dataset, baseline architectures with fewer parameters and 3 × 3 convolutions performed better than depth wise separable convolutions due to a 3 × 3 convolution's improved feature extraction. However, on the incorporation of augmentation techniques, the difference in validation accuracies between the depth-wise Conv and 3 × 3 Conv architectures diminishes considerably ("*Diminishing differences*") with **equivocal performances** by both types of convolutions in most cases (Table 6 and Table 7) as the architecture approaches **saturation**, beyond which there is no significant improvement in accuracy even with augmentations.

Table 6. The difference in accuracies (3 × 3 conv – depth wise conv) – FMNIST

Techniques	5.5K	7.8K	25K	140K	600K
Baseline	0.8	0.57	0.53	0.19	0
Cutouts	-0.09	0.29	0.27	0.05	-0.15
Mixup	0.34	0.39	-0.06	0.18	-0.14
Random Rotation	-0.07	-0.13	0.35	0.12	-0.03
Random horizontal flip	0.7	-0.04	0.31	0.24	-0.2
Colour jitters	0.04	0.6	0.1	0.1	0.04
Random affine	0.53	-0.16	0.83	-0.05	-0.06
Random perspective	0.43	0.58	0.58	-0.03	-0.37

Table 7. Difference in accuracies (3 × 3 conv – depth wise conv) – CIFAR10

Techniques	25K	140K	340K	600K	1M
Baseline	2.92	1.73	1.79	-0.4	-0.27
Cutouts	3.42	1.97	1.68	-0.25	-0.4
Mixup	3.48	1.36	1.48	-1.32	-0.34
Random Rotation	2.87	1.68	2.03	-1.17	-0.55
Random horizontal flip	2.75	0.86	1.04	-0.59	-0.89
Colour jitters	1.07	0.93	1.83	-0.27	-0.86
Random affine	6.59	1.7	2.18	-1.19	-1.13
Random perspective	4.6	1.36	1.94	-0.72	0.45

Additive Effect of Model Capacity in Depth Wise Convolutions. The above phenomenon of diminishing differences is not seen in the CIFAR-10 experiments suggesting that there is scope for improvement in model performance by fine-tuning the layers. However, with **a higher number of parameters (>600K), depth wise convolutions perform better than 3 × 3 convolutions as they enable a neural network to go deeper in terms of the number of convolutional layers.**

4.2 Consistency of Augmentations on Architectural Diversity

Augmentation techniques have been applied to different architectures and a relative ranking score was given to each of these techniques based on the average change in accuracy and standard deviation. A higher ranking was given to augmentations with the least standard deviation and higher gain in accuracy (Table 8). It was observed that Cutouts and a simple technique such as Random Horizontal Flip performed consistently superior to other techniques. It remained invariant to change in model capacity, architectural depth, and convolutions as evident from the least standard deviation and highest change in accuracy. On the CIFAR-10 dataset, mixup achieved the highest accuracy which is attributed to the wide distribution of classes. At the same time, the same technique and random horizontal flip decreased model performance on the MNIST dataset. Random affine, colour jitters, and random perspective negatively impacted (Augmentation paradox) on the accuracy on both the datasets.

Table 8. Change in accuracies/change in standard deviation – augmentation

Aug	FMNIST			CIFAR 10			MNIST
	STD	Change in Acc	Rank	STD	Change in Acc	Rank	Change in Acc
Cutouts	**0.47**	**0.96**	**1**	0.76	1.78	2	**0.1**
Mixup	0.53	0.15	4	**0.59**	**2.59**	**1**	−0.5
Rand rotate	0.34	0.17	3	0.61	0.65	3	0.46
Rand HFlip	0.38	0.67	2	**1.10**	**3.21**	**1**	−0.81
Color jitters	0.54	−2.41	5	1.13	−2.56	5	0.74
Rand affine	1.52	−2.81	6	5.87	−4.80	6	−2
Rand perspective	0.69	−1.04	7	2.35	−0.14	4	−0.99

4.3 The Superiority of Parameters over Augmentations

The difference in accuracy gains/losses across the various augmentation strategies reduces as the number of model parameters increases. This tendency is particularly pronounced in augmentation strategies that degrade model performance. This is seen statistically (Fig. 5), where higher architectures have lower standard deviations of accuracies. This phenomenon is indicative of the following hypothesis. Positive augmentations such as Cutouts, Mixup, Random rotation, and random horizontal flip have no meaningful influence on accuracy at larger parameter counts (low SD). In contrast, the performance of negative augmentations such as Color Jitters, Random affine, and Random perspective improves with the number of parameters. The neurons' enhanced learning capacity mitigates the effect of negative augmentation strategies at higher settings. In contrast, at lesser number of parameters, the effect of negative augmentation is exaggerated on both datasets as evident by the increased SD.

4.4 Combining Augmentations

Augmentations that positively affect the model performance in terms of accuracy were combined in varying combinations. At higher number of parameters combining different augmentations resulted in a synergistic effect. However, an antagonistic effect was observed at a lesser number of parameters. This can be attributed to models' relatively lesser learning capability with fewer parameters. The observations are summarized in Tables 9, 10, 11, and 12.

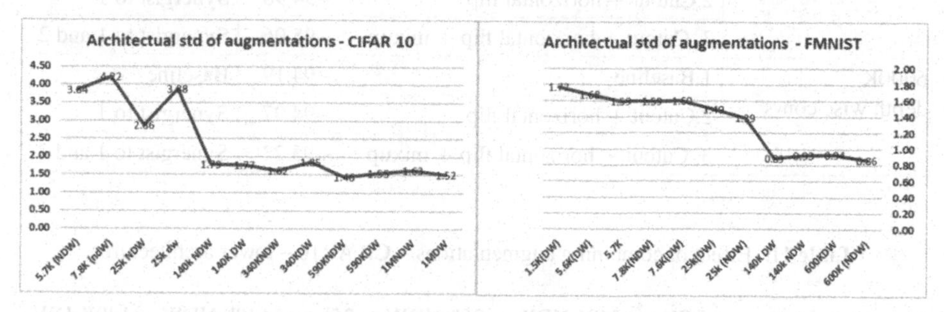

Fig. 5. Standard deviation of changes in accuracy for various techniques

Table 9. Effect of combining augmentations – MNIST

Models (1.5K params) – MNIST	Accuracy	Effect
1. Baseline	98.35	Baseline
2. Cutout + random rotation	98.73	Synergist to 1
3. Cutout + random rotation + colour jitters	98.23	Antagonist to 1 & 2
4. Cutout + colour jitters	98.9	Synergist to 1, 2 & 3

Table 10. Effect of combining augmentations – FMNIST

Architecture	Accuracy w.r.t augmentations – FMNIST		Effect
140k 3 × 3 convolutions	1.Baseline	–93.55	Baseline
	2.Cutout + horizontal flip	–94.55	Synergist to 1
	3. Cutout + horizontal flip + mixup	–94.22	Antagonist to 2

(*continued*)

Table 10. (*continued*)

Architecture	Accuracy w.r.t augmentations – FMNIST		Effect
140k depth wise convs	1. Baseline	−93.36	Baseline
	2. Cutout + horizontal flip	−94.48	Synergist to 1
	3. Cutout + horizontal flip + mixup	−93.71	Antagonist to 2
600K 3 × 3 convolutions	1.Baseline	−94.19	Baseline
	2.Cutout + horizontal flip	−94.98	Synergist to 1
	3. Cutout + horizontal flip + mixup	−95.06	Synergist to 1 and 2
6000K depth wise convs	1.Bascline	−94.19	Baseline
	2.Cutout + horizontal flip	−94.97	Synergist to 1
	3. Cutout + horizontal flip + mixup	−95.22	Synergist to 1 and 2

Table 11. Effect of combining augmentations – CIFAR10 – lower architectures

Sl	Techniques	5.7K NDW	7.8K NDW	25k NDW	25k DW	140k NDW	140K DW
1	Baseline	70.12	72.15	73.67	70.75	82.79	81.06
2	MU + RHP	72.51	73.88	78.05	74.71	86.55	85.65
3	MU + RHP + CO	71.46	73.93	77.59	75.02	86.64	86.64
4	Effect of 3 w.r.t 2	**Ant**	Syn	**Ant**	Syn	Syn	**Ant**

Table 12. Effect of combining augmentations – CIFAR10 – higher architectures

Sl No	Techniques	340K NDW	340 K DW	590 K NDW	590 K DW	1M NDW	1M DW
1	Baseline	85.49	83.7	86.3	86.7	85.69	85.96
2	MU + RHP	89.72	88.51	90.05	91.88	89.76	90.58
3	MU + RHP + CO	89.92	88.75	90.69	91.98	90.2	90.98
4	Effect of 3 w.r.t 2	**Syn**	**Syn**	**Syn**	**Syn**	**Syn**	**Syn**

5 Conclusion

The focus area of research in this paper has primarily been evaluating various augmentation techniques and arriving at an understanding of how model capacity and depth wise convolutions affect the outcome of an augmentation. The work has identified a new direction in appreciating those consistently invariant techniques and would apply them across a wide variety of datasets. Furthermore, this is the first study of its kind to unravel the relationship that exists between depth-wise convolutions, model capacity, and augmentations across a wide variety of standard datasets. The conclusions of the experiments are summarized below.

Augmentations such as Cutout, Random Horizontal flip, and Random Rotation performed consistently across all architectures. Considering the trade-off among training time, mathematical computational time, and model accuracy, it is suggested that a simple technique such as random horizontal flip, which performs equally well, may be used as a baseline augmentation. Further, these techniques were invariant to the number of parameters and the type of convolutions used, hence making them ideal for deployment on other real-life datasets. Combining augmentations worked well on over-parameterized architectures with the synergistic effect seen in all the cases. Depth wise separable Convolutions were effective on a higher number of parameters as they gave the ability of a model to go deeper and hence outperformed models with lesser parameters. Though on lesser parameterized architectures, 3×3 performed better, the application of augmentations bridged the accuracy gap between these architectures.

References

1. Mikołajczyk, A., Grochowski, M.: Data augmentation for improving deep learning in image classification problem. In: 2018 International Interdisciplinary PhD Workshop, IIPhDW 2018, Jun 2018, pp. 117–122. https://doi.org/10.1109/IIPHDW.2018.8388338
2. Falconi, L.G., Perez, M., Aguilar, W.G.: Transfer learning in breast mammogram abnormalities classification with Mobilenet and Nasnet. In: International Conference on Systems, Signals, and Image Processing, vol. 2019 June, pp. 109–114, Jun 2019. https://doi.org/10.1109/IWSSIP.2019.8787295
3. Chollet, F.: Xception: Deep Learning with Depthwise Separable Convolutions (2016)
4. Lin, M., Chen, Q., Yan, S.: Network in network. arXiv:1312.4400 [cs], Mar 2014. Accessed 17 Jan 2022 [Online]. Available http://arxiv.org/abs/1312.4400
5. Zhou, B., Khosla, A., Lapedriza, A., Oliva, A., Torralba, A.: Learning deep features for discriminative localization. In: 2016 IEEE Conference on Computer Vision and Pattern Recognition (CVPR), pp. 2921–2929 (2016). https://doi.org/10.1109/CVPR.2016.319
6. Ethiraj, S., Bolla, B.K.: Classification of astronomical bodies by efficient layer fine-tuning of deep neural networks. In: 2021 5th Conference on Information and Communication Technology (CICT), Kurnool, India, pp. 1–6, Dec 2021. https://doi.org/10.1109/CICT53865.2020.9672430
7. Li, Y., Wang, K.: Modified convolutional neural network with global average pooling for intelligent fault diagnosis of industrial gearbox. EiN **22**(1), 63–72 (2019). https://doi.org/10.17531/ein.2020.1.8
8. Li, Z., Wang, S., Fan, R., Cao, G., Zhang, Y., Guo, T.: Teeth category classification via seven-layer deep convolutional neural network with max pooling and global average pooling. Int. J. Imaging Syst. Technol. **29**(4), 577–583 (2019). https://doi.org/10.1002/ima.22337
9. Girosi, F., Jones, M., Poggio, T.: Regularization theory and neural networks architectures. Neural Comput. **7**(2), 219–269 (1995). https://doi.org/10.1162/neco.1995.7.2.219

10. Burden, F., Winkler, D.: Bayesian regularization of neural networks. In: Livingstone, D.J. (ed) Artificial Neural Networks, vol. 458, pp. 23–42. Humana Press, Totowa, NJ (2008). https://doi.org/10.1007/978-1-60327-101-1_3
11. DeVries, T., Taylor, G.W.: Improved regularization of convolutional neural networks with cutout. arXiv:1708.04552 [cs], Nov 2017. [Online]. Available http://arxiv.org/abs/1708.04552. Accessed 17 Jan 2022
12. Zhang, H., Cisse, M., Dauphin, Y.N., Lopez-Paz, D.: mixup: beyond empirical risk minimization. arXiv:1710.09412 [cs, stat], Apr 2018 [Online]. Available http://arxiv.org/abs/1710.09412. Accessed 17 Jan 2022
13. He, X., et al.: Sample-efficient deep learning for COVID-19 diagnosis based on CT scans. Health Inform. preprint, Apr 2020. https://doi.org/10.1101/2020.04.13.20063941
14. Shorten, C., Khoshgoftaar, T.M.: A survey on image data augmentation for deep learning. J. Big Data **6**(1), 1–48 (2019). https://doi.org/10.1186/s40537-019-0197-0
15. Wang, J., Perez, L.: The effectiveness of data augmentation in image classification using deep learning (2017)
16. T.M.M. Team: composer [Online] (2021). Available https://github.com/mosaicml/composer/

A Comprehensive Analysis of Linear Programming in Image Processing

Ankit Shrivastava(✉), S. Poonkuntran, and V. Muneeeswaran

Vellore Institute of Technology (VIT), Bhopal, India
ankit15121992@gmail.com

Abstract. This paper provides a basic background consisting of mathematical concepts of linear programming (LP) and implementation to process the images. This analysis consists of a basic introduction to optimization techniques used for and image processing. Image processing consists of various problem domains like segmentation, color-based segmentation, multiple objects tracking problems in video streams, image registration and image de-noising etc. Various algorithmic approaches devised to address these challenges using LP. The use of linear programming facilitates the designing of digital filters, utilization of prior knowledge for occlusion detection and efficient data fitting for image restoration along with the enhancement of the image quality. Reconstructed image exhibits optimal results then counterparts. LP helps in minimizing errors using estimators, extended dynamic programming for object tracking. It is also capable to separate meaningful regions, or interactive colors and with the integration of correlation clustering for segmentation of images. This analysis gives a insights versatility of the linear programming to implement, integrate and solving any problem. The analysis helps to conclude the future scopes and ideas to target and modify the problem.

Keywords: Optimization · Linear programming · Image processing · Objective function · Constraints · Segmentation · Restoration · Tone mapping · Object tracking

1 Introduction

Image processing is an enormous region which incorporates various examination subfields. Image processing basically alludes to a huge assortment of strategies and calculations which among different objectives endeavor to extricate helpful data from images, observe focuses and areas of interest inside images, convert images to more productive portrayals, and further develop a representation of data. To be more explicit well-known subfields of image processing incorporate example recognition, object coordinating, image blurring, image pressure, edge identification, and image reclamation [1–4]. For example, the use of image processing is the processing of pictures got from satellites which could be to some extent harmed and might be missing data [5].

As a general rule images can be simple or computerized. The latest applications include the utilization of advanced images. Computerized images can be considered

M. Singh et al. (Eds.): ICACDS 2022, CCIS 1613, pp. 323–335, 2022.
https://doi.org/10.1007/978-3-031-12638-3_27

rectangular clusters comprising of various qualities, and every area in the exhibit is known as a pixel [6].

As the name recommends improvement is a method for taking care of an issue by turning a bunch of boundaries to accomplish an ideal arrangement towards a characterized objective. We overall have restricted resources and time, and we really want to exploit them [7]. Everything utilizes improvement from utilizing your time gainfully to tackling inventory network issues for your organization. It is a particularly intriguing and significant point in image processing [8].

For instance, one might be keen on tracking down the most extreme or the base worth of a capacity. The capacity could be related to a genuine issue and the potential applications are perpetual. For example, the capacity could address the benefit of an organization which one might want to boost or it could address the costs which should be limited. The reason for utilizing enhancement methods is to tune sure boundaries on which the benefit and the costs depend with the goal that the capacity amplification or minimization is conceivable [9].

Linear programming (LP) is probably the easiest method for performing advancement. It assists him with tackling some exceptionally mind-boggling advancement topics by creating a twosome improve on beliefs. It will undoubtedly go over the applications and difficulties that Linear Programming can solve as an expert [10].

1.1 Optimization Through Linear Programming

Linear programming is a basic procedure that portrays complicated networks through linear functions and tracks down the ideal places afterwards. The preceding sentences crucial word is shown. Although open connections may be more puzzling, linear networks may be worked with [11].

Linear programming is used in a variety of ways all over the world. At discrete and expert levels, one uses linear programming. When he rides from home to work and must take the shortest route one employs linear programming. When person has a task transport, on the other hand he takes steps to ensure that cooperates effectively in order to complete the assignment on time [12, 13].

1.2 The Following Is an Example for Linear Programming Problem

Let courier person has to deliver six parcels in a single day. Stop A is where the distribution centre is located. Six transport termination points are U, V, W, X, Y & Z. Space between urban communities has been depicted by the data on the lines. To save money and time the person in charge of transportation must pick the shortest route possible (Fig. 1).

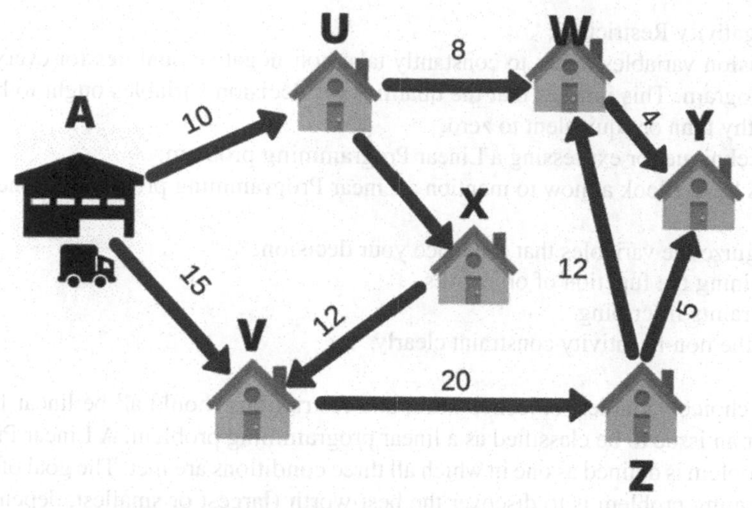

Fig. 1. Example of linear programming

As a result, the person in charge of transportation will calculate various routes to each of the delivery points & find the shortest route. Linear programming will be applied to find shortest route. The goal of the delivery person is to deliver all the parcels on time. Operation-Research refers most typical method of selecting the optimal course.

Linear programming is used to find the best solution to any problem with constraints. Scientists use linear programming to turn image processing problems into numerical systems. It includes work-line discrepancies that are subject to limits.

1.3 Linear Programming Terminology that Are Commonly Used

Let's utilize the following example to define some Linear Programming terminology.
Variables in Decision Making:
The decision variables are those that will determine the research outcome. They deal with the last details of the research. To address any problem, we must first determine the decision variables. A & B represented by X and Y are the research choice variables in aforementioned model.

Function of the objective:
It is defined as an aim that is simply conclusive. In the aforementioned scenario, the system is required to construct all of Z's support. As a result, the proposed work has an advantage.

Constraints:
Constraints are restriction or obstacles imposed on the choice factors. They, for the most part, limit the value of choice variables. The accessibility cutoffs for assets Milk and Choco in the above model are research restrictions.

Non-Negativity Restriction:
The decision variables ought to constantly take non-negative qualities for every single linear program. This implies that the qualities for decision variables ought to be more noteworthy than or equivalent to zero.

A Technique for expressing a Linear Programming problem:

Let's have a look at how to mention a Linear Programming problem in general.

1. Recognize the variables that influence your decision.
2. Explaining the function of objectives
3. Constraints Inscribing
4. State the non-negativity constraint clearly.

The choice variables, objective limit, and restrictions should all be linear limits in order for an issue to be classified as a linear programming problem. A Linear Programming Problem is defined as one in which all three conditions are met. The goal of a linear programming problem is to discover the best worth (largest or smallest, depending on the problem) of the linear expression (known as the objective work)

$$f = c_1x_q + \ldots\ldots\ldots\ldots\ldots\ldots + c_nx_n$$

under the influence of a set of restrictions stated as inequalities:

$$a_{11}x_1 + \ldots\ldots\ldots\ldots\ldots\ldots + a_{1n}x_n \leq b_1$$

$$a_{m1}x_1 + \ldots\ldots\ldots\ldots\ldots\ldots + a_{mn}x_n \leq b_m \; with \; \forall x_i \geq 0$$

2 Related Works

This research provides a comprehensive survey of the available linear programming approaches used for different image processing tasks. The below sections will cover those methods equipped with a linear programming approach.

2.1 Digital Filter for Edge Enhancement and Deblurring

Linear programming can process the image so that the edges of the objects in an image can be enhanced using a filter designed with the help of linear programming. Edge finding is an essential process in several applications of image processing, e.g. frame analysis and recognition of characters. Edges were depicted as important and rapid variations in the brightness of image. The typical point of edge identification is to find edges having a place within the limits of objects of interest. A methodology that is regularly utilized before distinguishing edges is to upgrade these edges by edge improvement channels. The deblurring of an image obscured because of environmental disturbance. The environmental disturbance point spread capacity has the type of a Gaussian capacity. This capacity is a consequence of quite a while averaging the image of a point source going through the atmosphere [1]. A linear solidity constraint has been planned and is revealed to produce steady recursive digital filters in a subclass of likely, steady designs.

2.2 Image Restoration

Wide surveys have been conducted on the challenge of restoration for linearly deteriorated images. It discovered that if no spectral components are present and signal noise which detached, logical extension may used to restore signal from low passband data for images with a limited scope. Spatial freq. domain extrapolation shows improvement for spatial resolution in spatial domain. Linear programming (LP) techniques are used in the ideal data fitting phase [2].

2.3 Motion Detection

The use of linear programming to recover high-request movement models from highlight line correspondences is described. Highlight line correspondences are robust because, unlike conventional highlight point correspondences, they are generally unfeeling toward gap impacts and T-intersections. Highlight line correspondences can estimate different estimations, too. The linear programming solver needs to bother with underlying speculation or a noise gauge, which are likely for recursive weighted least-square calculations. Examining estimators depend on L1 variable versus powerful LMedS estimators reveals that while the L1 estimator is slightly low powerful in compare to LMedS estimator, L1 calculation is significantly fast. Processing a model-based arrangement between progressive edges forms the basis of the movement examination. The arrangement cycle is divided into two stages: (I) estimating correspondences and addressing them as emphasize line interactions, and (II) converting the arrangement issue into a linear programme and settling [3].

2.4 Image De-noising

Extending an input image to the subspace of permitted pictures created, for example, by PCA, is a common approach of image de-noising. In any case, a fundamental disadvantage of this method is that when a few pixels are undermined by noise or obstruction, the projection will refresh all pixels. Another technique is to use a '1-standard penalty to distinguish the outraged pixels and to update the identified pixels in particular. The recognisable proof and refreshing of boisterous pixels are formed as one linear program which can be tackled productively. Notably, one can apply the v-stunt to determine the negligible portion of pixels to be recreated straightforwardly. In light of linear programming, here's another image de-noising technique. Our main idea is to portray sparsity by minimally isolating the arrangement from the complicated. Our technique's on-complex organisation is linked to existing strong, measurable procedures. Surprisingly, our method can handle block noise while maintaining the improvement problem's convexity (each linear programme is edged) [4].

2.5 Object Tracking

Many vision applications, such as visual navigation and object activity identification, need the simultaneous tracking of many objects. Even if each object can be tracked independently, if objects have interrelationships, tracking them all together is required for good results. Another strategy that can be employed for more efficient object tracking search is linear programming (LP). Using 0-1 Integers to optimise object tracks Radar data association has been examined using programming (Fig. 2).

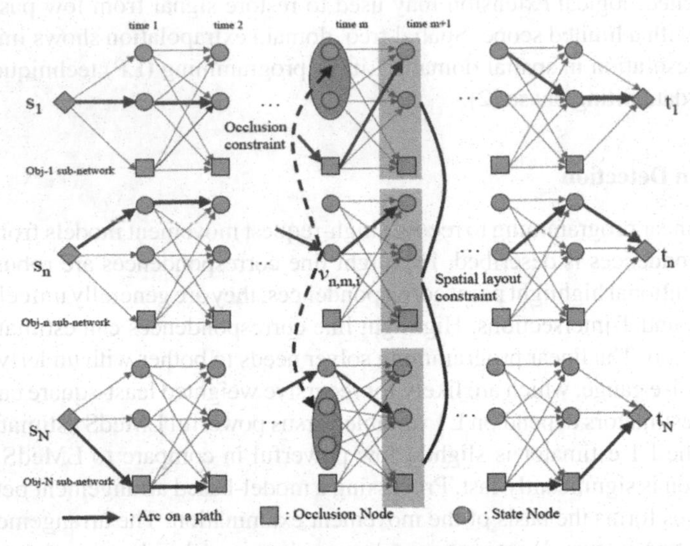

Fig. 2. Multiple object tracking [5]

This design differs from the predicted scheme in that each possible path is assigned a variable, and item tracking is treated as a single pressing issue. This scheme models track collaboration like the spatial design imperative and common occlusion. Tests reveal that the suggested system is effective at tracking objects in long video sequences with complicated relationships [5].

Multi-target tracking is described as a distinct process. A number of stream variables that can be coupled to give overall directions is significantly easier to implement with Linear Programming, a consistent form of it. The number of followed items will usually fluctuate with time means that certain articles will be seen in tracking area while other will exit. As a result, the framework's absolute mass varies and permit streams to be in & out the zone. The quantity of variables of our enhancement issue is exceptionally high, and addressing it straightforwardly is just down to earth for modestly estimated frameworks. Two simple Trimming the Chart and Batch processing solutions can be used to get around this restriction. Instead of presenting mass preservation as a prerequisite, the improvement might substantially low expensive by conducting on subspace directly spread across maps which save mass. Different strategies in light of various levelled dividing of the diagram may likewise give elective approaches to decrease the computational expense altogether [6].

New study represented capacity for universal progress of monitoring in crowded settings where synchronization issue has been addressed for all tracks simultaneously. Multi-target tracking, in this view, is a required minimum-cost stream enhancement concern that can be resolved using LP to regard as the global ideal. In LP ways to deal with tracking, worldwide streamlining is done over a succession of edges utilising static elements, for example, appearance-likeness or distance between discoveries (Fig. 3).

Fig. 3. Ground plane grid used for pedestrian detection in [6]

A novel strategy for adding motion modelling in to a low-cost network flow tracker is to add edges to the tracking graph to show linkages between far distant detections. Costs of the connections have been estimated using Novel energy function which includes velocity [7] (Fig. 4)

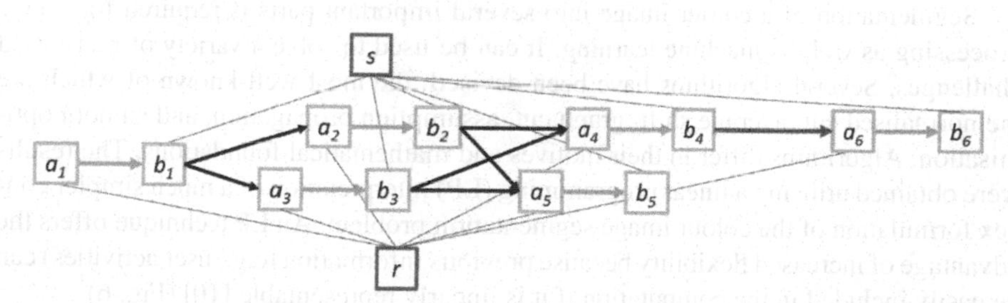

Fig. 4. Tracking problem in [7] was represented using this graph

2.6 Segmentation of Images

One critical challenge is Image segmentation for image analysis & computer vision. Image is being segmented in to foreground area matching item of interest and back ground areas rather than uniform colour regions which is the challenge with object-respecting colour picture segmentation. The colour segmentation problem should be rephrased as a problem of energy minimization. It's basically a 0-1 integer programming issue as variables be minimised 0-1 labels of pixels (1 means foreground and 0 means background). There are various effective strategies for discovering approximate optimum solutions if one opts for an approx. sub optimal solution rather genuine universal maximal solution. Linear Programming Relaxation [8] is an efficient algorithm (Fig. 5).

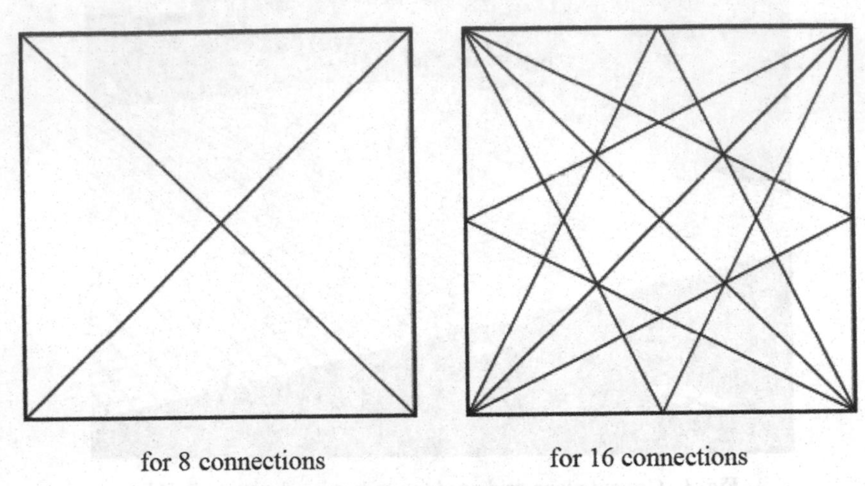

Fig. 5. Selected primary regions in [9]

A relaxed type of area segmentation can be solved in the best feasible way. The main concept to be put the problem for region segmentation curvature consistency into integer linear programme. A typical implementation of the dual simplex algorithm was employed to compute the solutions, and satisfactory running times were discovered [9].

Segmentation of a colour image into several important parts is required for image processing as well as machine learning. It can be used to solve a variety of real-world challenges. Several algorithms have been devised, the most well-known of which are the normalised cut, average shift, graph cut, assumption propagation, and smooth optimisation. Algorithms differ in their motives and mathematical foundations. The results were obtained utilising a linear programming (LP) interpretation of a much simpler convex formulation of the colour image segmentation problem. An LP technique offers the advantage of increased flexibility because previous information (e.g., user activities) can be easily included in the computation if it is linearly representable [10] (Fig. 6).

Rather than thinking about how to improve segmentation methods. Another technique is to improve performance by merging segmentations from multiple algorithms on the same image. It is logical to assume that better segmentation results can be obtained by auto merging several segmentations of an input image into a mono segmentation that, in some ways, is the average of all the segmentations, similar to the concept of 'wisdom of the crowd' in social choice theory [11].

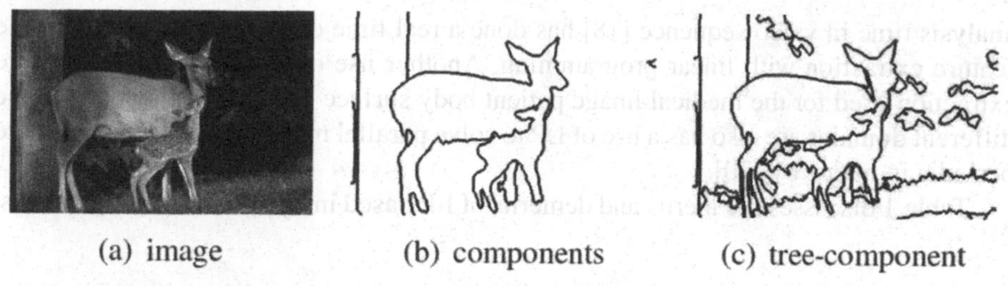

(a) image (b) components (c) tree-component

Fig. 6. The ILP problem has been broken down into small sub problems [11]

2.7 Image Registration

Clinical image examination is a laid-out space in computational, numerical and organic sciences. The capacity to contrast or break data across subjects and starting points of various modalities is a basic and essential part of PC supported conclusion. The term utilised frequently to communicate this need is enlistment. The enrolment issue regularly includes three angles: (a) a model of change, (b) a resemblance rule, and (c) a technique for advancement. Enrolment may be done globally or locally. Parametric models are typically used to handle universal enlistment with limited no. of degree of opportunity such as inflexible or nearby. Such a course will present earlier information in the enrolment cycle and will make the improvement step more proficient [12].

2.8 Tone Mapping

The difference of a crude image can be not so significant because of different causes such as helpless enlightenment conditions, inferior quality economical imaging sensors, client activity mistakes, media disintegration (e.g., old blurred prints and movies) and so forth contrast upgrade methods can be grouped into two methodologies: setting delicate (point wise administrators) & setting free (point administrators). In setting delicate methodology the difference is characterised by the pace of progress in power between adjoining pixels. Based on pixel-by-pixel, difference is increasing by simply changing neighbourhood wave-form. The counter proportion of tone bending accomplishes the harmony between tone nuance and difference upgrade. Utilising a linear programming system to tackle the fundamental obliged advancement issue and arrangement can build image contrast while safeguarding tone progression, two clashing quality standards [13].

2.9 Classification

Classification of the images content has been implemented by [15] using convex linear programming with similarity metrics. On the other hand [16] has used the feature of incremental learning classifier to enable deep learning models. For the deep neural networks a multi label classification with integration of LP has been done by [17].

2.10 Feature Extraction

Feature extraction is a method to facilitates the various classification and recognition tasks to train the various machine learning data base and networks. To reduce the visual analysis time in video sequence [18] has done a real time distributed approach for the feature extraction with linear programming. Another use of Integer LP based feature extraction used for the medical image patient body surface point clouds by [19]. Some different domains are also has a use of LP to solve parallel machine problems which are periodic in nature by [20].

Table 1 discusses the merits and demerits of LP based image processing algorithms.

Table 1. Comparison of literature review over merits and demerits

Ref. Work	Merit	Demerit
[1]	Recursive digital filter	Less linear stability
[2]		Limited solution area
[3]	LMedS estimators has robust performance	Time complexity against non-deterministic and exponential time complexity
[4]	Complex image reconstruction	Limited capability
[5]	Specific object tracking	Not adaptive
[6]	Multiple object tracking	Requirement of pedestrian grid
[7]	Multi target tracking	Failure scenarios with occlusions
[8]	Conceptually-meaningful foreground and background segmentation	Not suitable for video surveillance
[9]	Simplex method approach	Limited to region defined segmentation
[10]	Segmentation is based on colour only	Modification required for video sequences
[11]	Several into one segmentation	Modification needed for global
[12]	Gradient-free method	Prior knowledge of features are required
[13]	Constraint optimization approach	Contrast while preserving tone continuity, two conflicting quality
[14]	Correlation clustering for exact global solution	Not accurate in some sub pixel regions
[15]	Hybrid iterative method for optimization	Problem in selection parameter consistency
[16]	Incremental learning classifier	Not adaptable for new data
[17]	Multi-Label classification	Security of deep neural network
[18]	Works in real time	Lifetime of network is finite
[19]	Has precision over 90%	Surface point cloud needs to evaluated first
[20]	Parameter tuning is not required	Designed to specific problem domain

3 Problem Identification

After getting into the literature, some problems have been identified, which is either not targeted by the author or unable to resolve up to the mark. Such issues are discussed below domain wise.

Image Restoration/Reconstruction

- The half-plane recursive filter designed in [1] has less linear stability, which needs to be addressed and removed in future works.
- The image restoration defined in [2] has a limited solution area and is limited to impulsive noise-affected images. After tweaking some of the restrictions and decision variables, it can also outperform other types of noise.
- The image reconstruction approach followed in [4] has limited capability due to convex optimisation. Convex optimisation can be improved by modifying the structure of the objective function.
- The approach followed in the [12] will work more efficiently using prior knowledge of expected deformation to get proper re-registration. The method for tuning two conflicting parameters has been handled in [13] a very balanced manner to get good tone and contrast results.

Object Tracking

- The given LMedS estimators in [3] have robust performance but show some time complexity against non-deterministic and exponential time complexity.
- Object tracking has been a great algorithm to get the specific object tracked in [5], and the approach can search and track multiple objects in video streams. The approach can do some recognition tasks when integrated with CNN or ANN models.
- The work done in the [6] has limitations for the multiple object tracking is the use of the pedestrian grid, limiting the practical use of the approach or designed system. Another tracking approach mentioned in [7] has some failure scenarios with occlusions, so it needs to resolve in further optimisation.

Image Segmentation

- To achieve colour image segmentation in [8] has done a very good job to isolate the object programmatically; the linear programming function performed well and has scope for the modification for the video surveillance applications.
- A similar problem has been targeted using the simplex method approach based on the region defined in [9] and needs to be tweaked for another segmentation needs.
- For segmentation based on colour has another approach too defined in [10], where the SDP method is adopted and will be used for video sequence cut operations with some changes.
- In [11], the ensemble method with integer linear programming and global usage steps would be modified with an m-step.

Image Classification

- Selection parameter consistency is there in [15] to achieve and target the desired classification problem.
- In [16] the proposed model needs to updated for the new data.

- There is security issue with DNNs in [17] for classification.

Image Feature Extraction
- Feature extraction is good in [18] but limited lifetime of video sensor network needs to be resolved.
- In [19] has targeting the surface point clouds for feature extraction but needs preprocessing to get those points.
- An optimal feature extraction approach is developed in [20] but currently it is focused for one problem specific.

4 Conclusion

This paper has performed a comprehensive analysis of linear programming for image processing. Implementing linear programming to achieve segmentation, tracking of objects, motion detection, de-blurring, classification, feature extraction, tone and contrast mapping has done an excellent job to achieve the results. This is very helpful in pedestrian detection in automaton. These challenges have been seem more accessible using linear programming. From the study it can be said that every processing tasks have a different set of objective functions to design to get the optimised results. The decision variables and objective functions can be modified by integrating with other mathematical optimisation in post or pre stage with linear programming to better fit the output restrictions of the results.

References

1. Chottera, A., Jullien, G.: Design of two-dimensional recursive digital filters using linear programming. IEEE Trans. Circuits Syst. **29**(12), 817–826 (1982). https://doi.org/10.1109/TCS.1982.1085107
2. Mammone, R., Eichmann, G.: Super resolving image restoration using linear programming. Appl. Opt. **21**(3), 496 (1982). https://doi.org/10.1364/AO.21.000496
3. BenEzra, M., Peleg, S., Werman, M.: Real-time motion analysis with linear programming. Comput. Vis. Image Underst. **78**(1), 32–52 (2000). https://doi.org/10.1006/cviu.1999.0826
4. Tsuda, K., Ratsch, G.: Image reconstruction by linear programming. IEEE Trans. Image Process. **14**(6), 737–744 (2005). https://doi.org/10.1109/TIP.2005.846029
5. Jiang, H., Fels, S., Little, J.J.: A linear programming approach for multiple object tracking. In: 2007 IEEE Conference on Computer Vision and Pattern Recognition, Minneapolis, MN, USA, Jun 2007, pp. 1–8. https://doi.org/10.1109/CVPR.2007.383180
6. Berclaz, J., Fleuret, F., Fua, P.: Multiple object tracking using flow linear programming. In: 2009 Twelfth IEEE International Workshop on Performance Evaluation of Tracking and Surveillance, Snowbird, UT, Dec 2009, pp. 1–8. https://doi.org/10.1109/PETSWINTER.2009.5399488
7. McLaughlin, N., Rincon, J.M.D., Miller, P.: Enhancing linear programming with motion modeling for multitarget tracking. In: 2015 IEEE Winter Conference on Applications of Computer Vision, Waikoloa, HI, Jan 2015, pp. 71–77. https://doi.org/10.1109/WACV.2015.17
8. Li, H., Shen, C.: Object respecting color image segmentation. In: 2007 IEEE International Conference on Image Processing, San Antonio, TX, USA, 2007, pp. II-257–II-260. https://doi.org/10.1109/ICIP.2007.4379141

9. Schoenemann, T., Kahl, F., Cremers, D.: Curvature regularity for region based image segmentation and inpainting: a linear programming relaxation. In: 2009 IEEE 12th International Conference on Computer Vision, Kyoto, Sep 2009, pp. 17–23. https://doi.org/10.1109/ICCV.2009.5459209
10. Li, H., Shen, C.: Interactive color image segmentation with linear programming. Mach. Vis. Appl. **21**(4), 403–412 (2010). https://doi.org/10.1007/s00138-008-0171-x
11. Alush, A., Goldberger, J.: Ensemble segmentation using efficient integer linear programming. IEEE Trans. Pattern Anal. Mach. Intell. **34**(10), 1966–1977 (2012). https://doi.org/10.1109/TPAMI.2011.280
12. Glocker, B., Komodakis, N., Tziritas, G., Navab, N., Paragios, N.: Dense image registration through MRFs and efficient linear programming. Med. Image Anal. **12**(6), 731–741 (2008). https://doi.org/10.1016/j.media.2008.03.006
13. Wu, X.: A linear programming approach for optimal contrast-tone mapping. IEEE Trans. Image Process. **20**(5), 1262–1272 (2011). https://doi.org/10.1109/TIP.2010.2092438
14. Alush, A., Goldberger, J.: Hierarchical image segmentation using correlation clustering. IEEE Trans. Neural Netw. Learning Syst. **27**(6), 1358–1367 (2016). https://doi.org/10.1109/TNNLS.2015.2505181
15. Ni, K., Phelps, E., Bouman, K.L., Bliss, N.: Training image classifiers with similarity metrics, linear programming, and minimal supervision. In: 2012 Conference Record of the Forty Sixth Asilomar Conference on Signals, Systems and Computers (ASILOMAR), Nov 2012, pp. 1979–1983. https://doi.org/10.1109/ACSSC.2012.6489386
16. Bai, J., et al.: Class incremental learning with few-shots based on linear programming for hyperspectral image classification. IEEE Trans. Cybernet. **52**, 1–12 (2020). https://doi.org/10.1109/TCYB.2020.3032958
17. Zhou, N., Luo, W., Lin, X., Xu, P., Zhang, Z.: Generating multi-label adversarial examples by linear programming. In: 2020 International Joint Conference on Neural Networks (IJCNN), Jul 2020, pp. 1–8. https://doi.org/10.1109/IJCNN48605.2020.9206614
18. Eriksson, E., Dán, G., Fodor, V.: Real-time distributed visual feature extraction from video in sensor networks. In: 2014 IEEE International Conference on Distributed Computing in Sensor Systems, May 2014, pp. 152–161. https://doi.org/10.1109/DCOSS.2014.30
19. Chen, J., et al.: 3D rigid registration of patient body surface point clouds by integer linear programming. In: 2019 International Conference on Image and Vision Computing New Zealand (IVCNZ), Dec 2019, pp. 1–6. https://doi.org/10.1109/IVCNZ48456.2019.8960993
20. Pang, J., Tsai, Y.-C., Chou, F.-D.: Feature-extraction-based iterated algorithms to solve the unrelated parallel machine problem with periodic maintenance activities. IEEE Access **9**, 139089–139108 (2021). https://doi.org/10.1109/ACCESS.2021.3118986

Machine Learning Based Approaches to Detect Loan Defaulters

Nishanth Ramesha(✉)

PESIT South Campus, Bengaluru, Karnataka, India
nishanth.ramesh1@gmail.com

Abstract. Consumers acquire many loans from banks when they need money, and banks provide many low-interest rates offers to entice people to take out loans. However, if consumers do not pay their loans on time, the bank may incur a significant loss. The problem statement seeks to categorize whether people can repay their debt, preventing banks from incurring substantial losses. Defaulters can bankrupt banks due to large loan non-payment, resulting in a financial crisis in the country or for any bank that provides the loan. Before issuing a loan to a person, a comprehensive check is performed on their profile to ensure that they do not default, but it is still difficult to determine who will default and who will not. Because the number of individuals taking out loans is increasing year after year, a system to identify and handle this rising problem is urgently needed to find a solution. As the number of people taking out loans increases, so will the number of defaulters. There are a variety of classification machine learning techniques and deep learning approaches that may be used to solve the difficulties. The study's primary goal is to compare and contrast the Random Forest, Logistic Regression, and XGBoost models to see which one performs and provides the best accuracy.

Keywords: Loan default prediction · XGBoost · Random Forest · Logistic Regression · Machine learning · Ensemble techniques

1 Introduction

Individuals all across the globe count on banks that can lend them money for a variety of reasons, including to assist them in overcoming financial limits and achieving personal goals. The action of getting a loan has become unavoidable due to the banking or finance industry's evolving and rising competitiveness. Furthermore, both large and small banks or enterprises rely on the scheme of issuing loans to earn a high level of profit so that they can manage and operate it correctly during hard times or during a recession, which enables them to survive in this competitive market. For a big proportion of the assets that the bank provides, the value or interest that is collected on the loans is issued [1].

Traditionally, the person had to offer an asset to the bank as collateral for the loan, but now because profit is the main requirement of the bank, it is issuing

M. Singh et al. (Eds.): ICACDS 2022, CCIS 1613, pp. 336–347, 2022.
https://doi.org/10.1007/978-3-031-12638-3_28

loans only on credit, with no collateral required. A credit score is a number that banks assign to customers based on financial taxation and the risk of lending money. To determine if the individual is deserving of the loan, an algorithm is required. However, researchers and banking authorities have recently chosen to train classifiers based on various machine learning and deep learning algorithms to automatically predict an applicant's credit score based on credit history and other historical data, making the process of selecting qualified applicants much easier before the loan is approved.

2 Related Work

The primary reason that loan prediction is so contentious in the financial sector is the competitive market that has formed as the business is booming and the potential for expansion is enormous. Furthermore, the field has attracted the interest of the research community. In the field of artificial intelligence. In recent years, it has drawn more attention to studies on loan prediction and credit risk assessment. Due to the current high demand for loans, there is a massive increase in the need for additional improvements in credit scoring and loan prediction models. A range of methods have been used to give credit scores to individuals, and considerable study has been done on the subject throughout the years. With years of research, many different types of algorithms were used, such as Random Forest, Logistic Regression model, Gradient Boosting, Naïve Bayes, etc.

According to other researchers, they have found different accuracy levels from these algorithms. Lin Zhu et al. [2] concluded from their research on Loan default prediction that the Random Forest Algorithm results in 98% accuracy, which is better than logistic regression, decision tree, and SVM, which gives us accuracy levels of 73%, 85%, and 75%. In [3], the authors use CatBoost, to detect loan defaulters and also added a verification module for better detection of the same. Techniques such as Federated Learning [4] and Evolutionary Game Analysis [5] were among others used apart from the approaches that dealt with Data Mining techniques [6].

A researcher named Pidikiti Supriya worked on the Decision Tree model and achieved an accuracy level of 81% [7]. Now research in 2021 showed work on Lending Club. Here the dataset was acquired from Kaggle. They needed to educate themselves with the Lending Club dataset before moving on with machine learning modeling. Importing the essential libraries and data files for the model was the first and most crucial step. Here, the leading club has classified borrowers into 6 Homeownership types. They got 73% accuracy results for the decision tree [8]. Another research by the School of Global management showed that the bank's management's goal to boost profits for its shareholders comes at the expense of greater risk. Interest rate risk, market risk, credit risk, off-balance-sheet risk, based on technological risk, foreign exchange risk, national or state risk, liquidity risk, liquidity risk, and collapse risk are only some of the hazards that banks face. The same machine learning techniques that are powering developments in search engines and self-driving cars may be used in the banking sector [9].

Several technological advancements have enabled the financial industry to investigate and mine a vast data infrastructure that includes a wide range of unstructured financial data on markets and consumers. Despite machine learning's limits in determining causation, economists are increasingly using it with other tools and knowledge to examine complicated relationships. Machine learning's adoption has been sparked by the promise of cost savings, increased productivity, and better risk management. New laws have forced banks to automate with the requirement for efficient regulatory compliance [10].

Researcher Stefano Piersanti in [11] found few methods for predicting default loans in his studies. Here he took the Balance sheet as data set. The algorithms used here were Decision Tree, Naïve Bayes, Gradient Boost, Adaboost, Logistic Regression, and Multinomial Bayesian Classifier. For further procedures, it can be listed in short forms as Predicting business failure is a critical problem for economic analysis and the smooth operation of the financial system. Furthermore, due to the recent financial crisis, the gravity of this problem has arisen. Many recent studies have used various machine-learning approaches to forecast the collapse of businesses.

In [12], the authors use a real dataset from a company in China to test their approach which dealt with undersampling and oversampling methods. Other research works in this field include using option pricing theory [13], considering psychological factors during loan repayment [14] and, in [15], they explore how various attributes come into the picture while providing loans.

3 Methodology

3.1 Data Preparation

The dataset has been acquired from different rural banks. The dataset consists of 5 features with about 6432 entries that have been received from the banks collected by the author personally. The features in the dataset are index, employed, bankbalance, annualsalary, and defaulted. The primary use of making and managing a dataset rather than taking a dataset that is already available is one of the novel things and also helps the model to get trained and appropriately tested on actual data, and the received output would be a reasonable accuracy.

3.2 Data Preprocessing

Before making any conclusions or training ML models, the dataset is analyzed (or preprocessed), otherwise some undesirable data will be trained, which will influence our ML model and its accuracy, which is not preferable. Dataset has five numerical features in which features like 'employed' and 'defaulted' are categorical. Categorical characteristics are those whose explanations or values are selected from a specified set of options. Absolute values include house colors, animal species, and months of the year. True/false; positive, negative, neutral, etc.; positive, negative, neutral, etc.; positive, negative, neutral, etc. Numerical

	Index	Employed	Bank Balance	Annual Salary	Defaulted?
0	1	1	8754.36	532339.56	0
1	2	0	9806.16	145273.56	0
2	3	1	12882.60	381205.68	0
3	4	1	6351.00	428453.88	0
4	5	1	9427.92	461562.00	0

Fig. 1. Real collected data

characteristics are those that have values that are either continuous or integer. The dataset has no null values. The feature index will be removed as it is not required. Categorical features are encoded to turn string data into integer data types. The 'defaulted' feature has approximately 3% data, indicating that a person can pay their loan and the rest can't. According to the dataset, around 30% of people are unemployed (Fig. 1).

3.3 Feature Visualization

The author has tried to find out the percentage of employed people who can pay their loans—distribution of numerical values among the dataset of the defaulted feature. Annual salary has a bimodal distribution, and the number of people with the defaulted loan decreases after USD 250,000, which can be seen in Fig. 2.

From Fig. 3, the author can see that the Bank balance has skewed values. For bank balances higher than USD 10,000, the number of non-defaulted loans decreases, and for bank balances lesser than USD 18,000, the number of defaulted loans decreases. Finally, Fig. 4 and 5 represents the respective scatter & box plot of annual salary and bank balance based on default status. Figure 6 expresses an idea about the importance of which feature is the most from the features in the dataset and which feature needs to be more addressed or focused rather than the others.

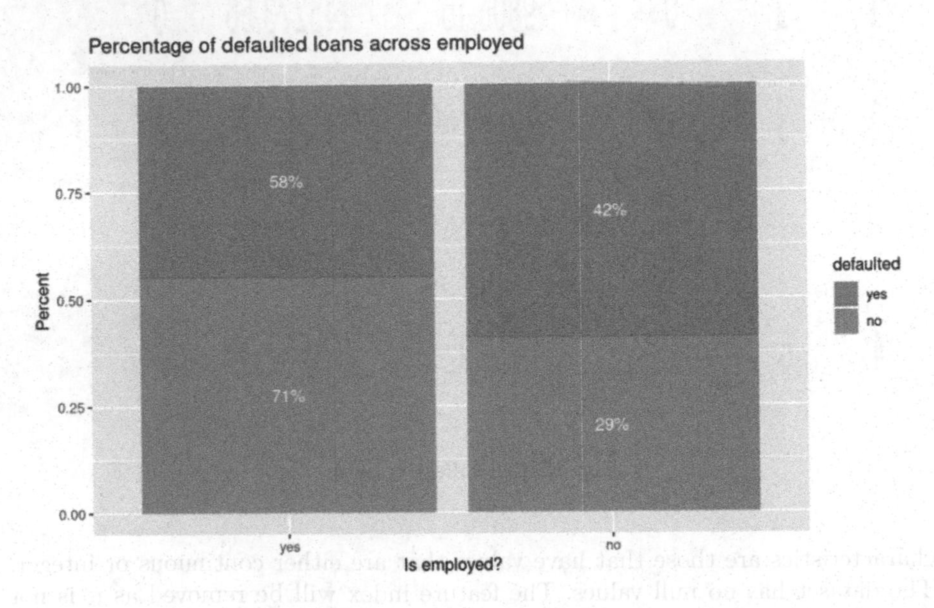

Fig. 2. Percentage of defaulted loans

3.4 Model Architecture

For classification and regression, Random Forest is a supervised machine-learning method. It's a decision tree-based ensemble classifier. Random Forest is a bagging strategy in which it picks a sample of data from a training dataset with replacement, utilizing feature sampling and row sampling, and feeds it to the decision tree classifiers each time the model is trained. For the final prediction label during testing, a majority vote is employed. The average of all predicted brands in the decision tree for the final prediction was used. For the vast majority of data, Random Forests are excellent.

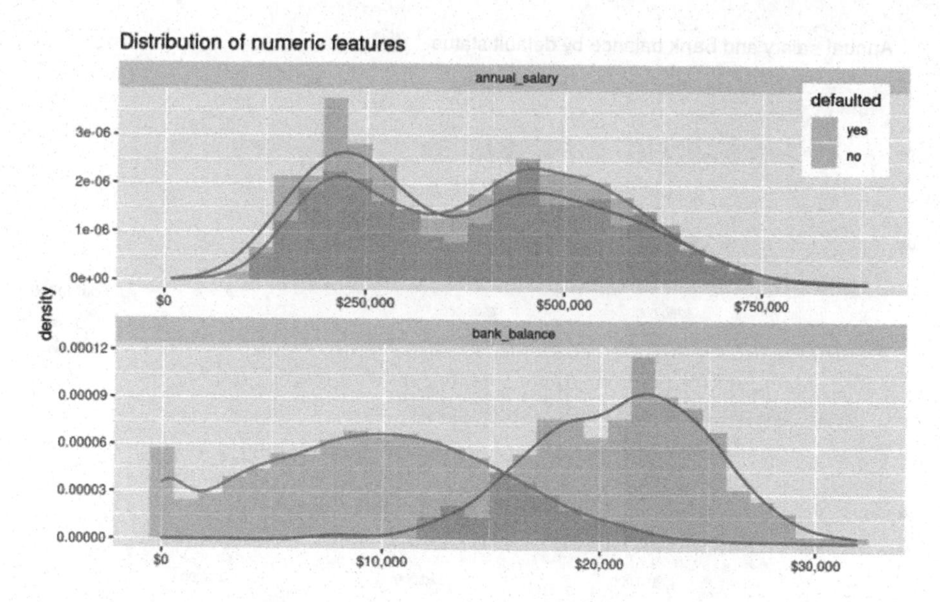

Fig. 3. Distribution of numerical features

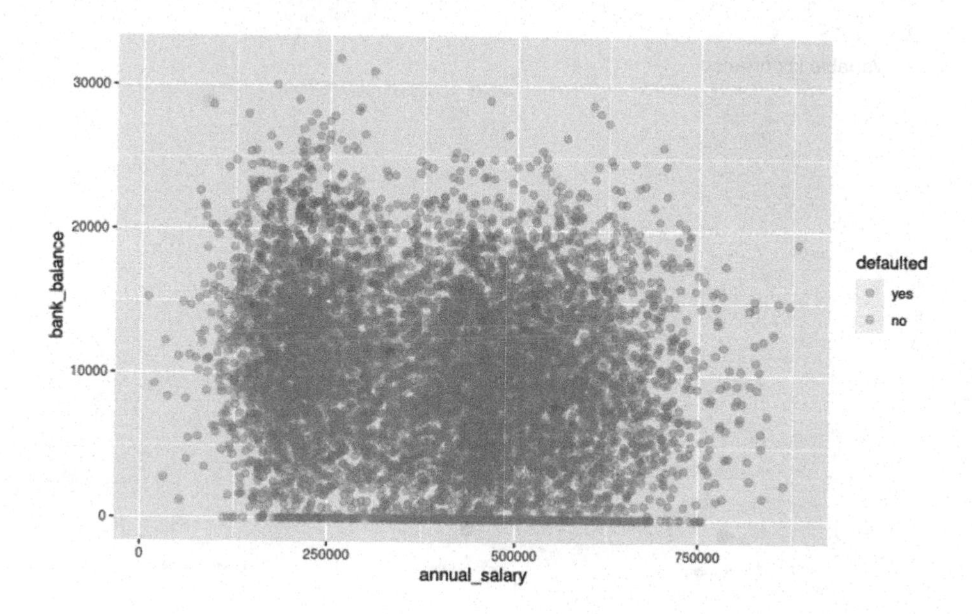

Fig. 4. Scatter plot

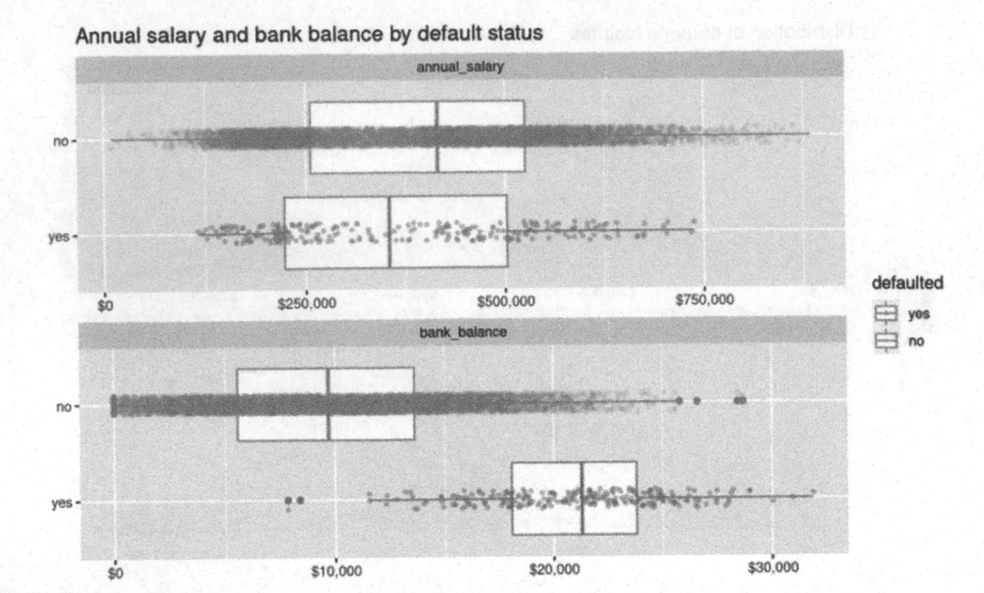

Fig. 5. Box plot

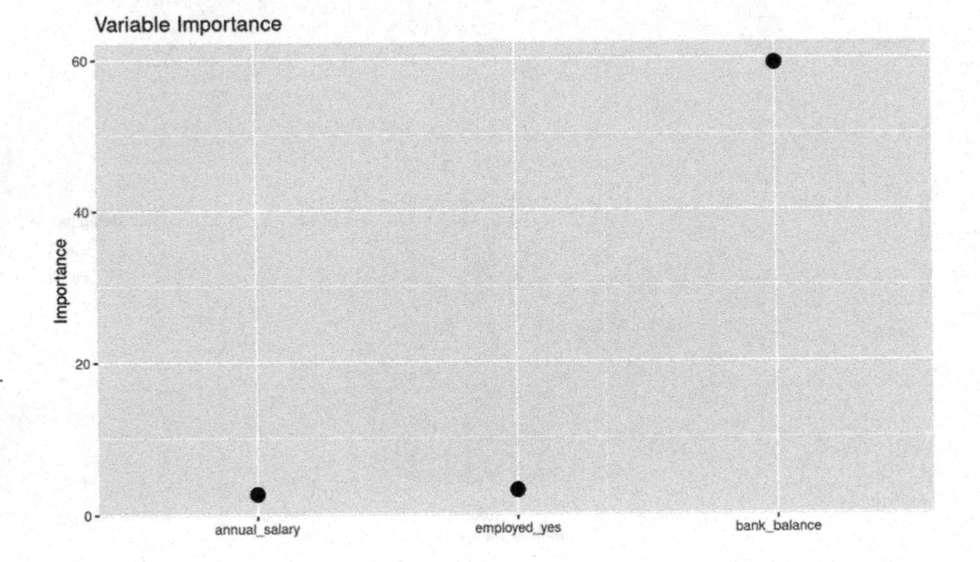

Fig. 6. Feature importance

Multiple decision trees reduce the high variance by lowering the low bias and high variance of single decision trees. After performing Random Forest, the author moved towards the XGBoost, part of the ensemble method. Generally, the XGBoost gives a systematic solution used to combine the power of the model. It is one of the powerful ways to build a supervised regressive model, which is why the author uses this on the supervised problem statement. The function consists of the loss function and regularization, generally used to tell the difference between the predicted value and the actual one. Below Fig. 7 shows the equation used for XGBoost. Finally, Logistic Regression uses the same binary approach for a supervised problem.

$$\mathcal{L}^{(t)} = \sum_{i=1}^{n} l(y_i, \hat{y}_i{}^{(t-1)} + f_t(\mathbf{x}_i)) + \Omega(f_t)$$

Fig. 7. XGBoost equation used

4 Results

The author has performed all three models: Random Forest, Logistic Regression, and XGBoost. After collecting unique data from the different banks and preprocessing well to optimize the data for getting the best results. After the model has been performed, the author visualized the five-fold validation for the training and the testing data, which can be seen in Figs. 8 and 9. The accuracy obtained in the training data was higher for Random Forest rather than logistic regression and XGBoost. A similar thing occurred in the results of the testing data, and hence the Random Forest has better accuracy. Figure 10 gives a glimpse of the confusion matrix happening in Random Forest. Below Table 1 represents the accuracy obtained for each model.

Table 1. Accuracy metrics

Model name	Training accuracy (in percentage)	Testing accuracy (in percentage)
Random Forest	88	90
XGBoost	84	85
Logisitic Regression	86	88

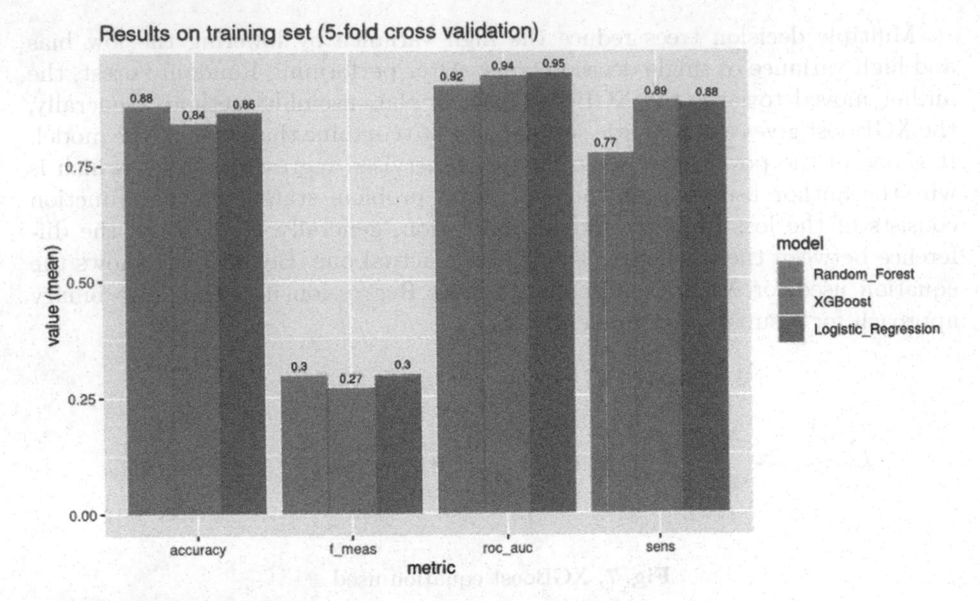

Fig. 8. 5 fold validation for training data

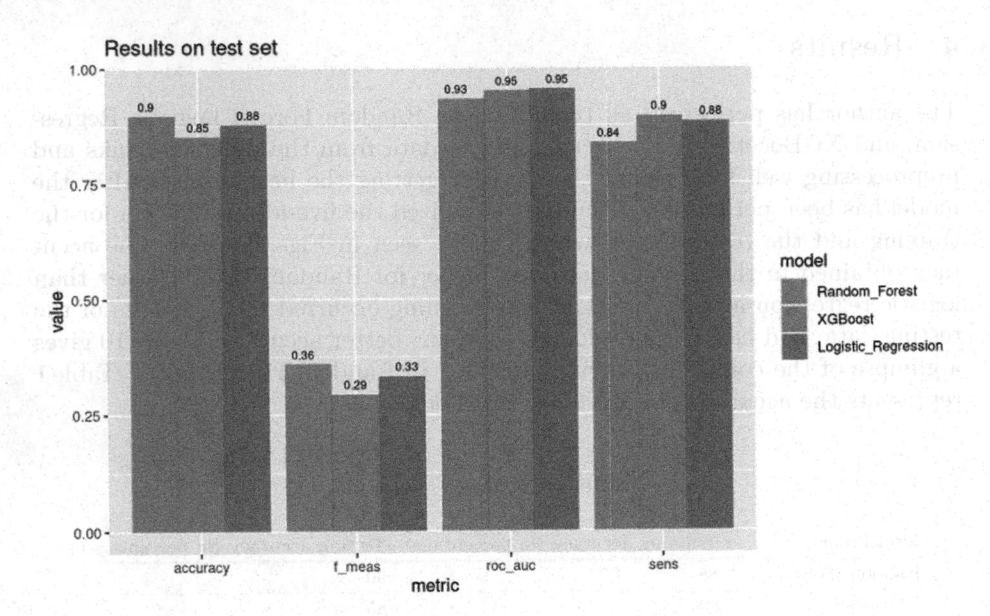

Fig. 9. 5 fold validation for testing data

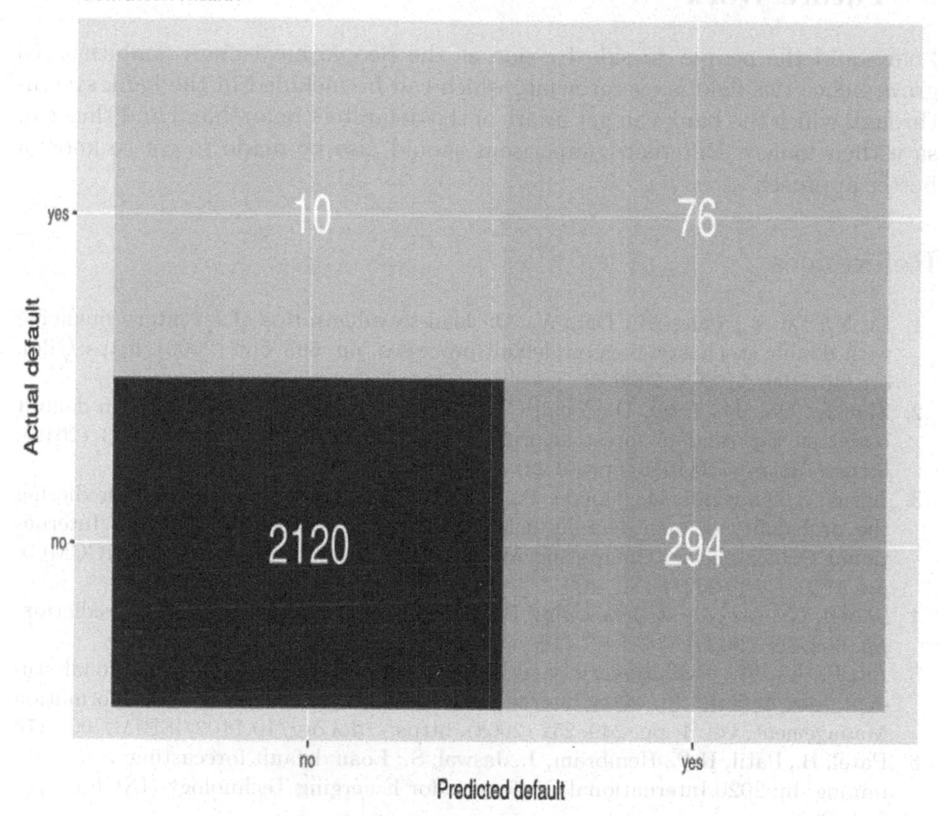

Fig. 10. Confusion matrix

5 Conclusion

The banking sector is the backbone of the economy for any country as it helps in functioning the businesses and other stuff that help in the country's development. As people take loans and some don't repay, such a loan default prediction system is required. Hence, the author worked on the problem statement and compared three models: Random Forest, Logistic Regression, and XGBoost. After comparing the performance of the models, the author concluded that the Random Forest approach works better than the Logistic Regression and XGBoost approach. The system developed by the author will help the banking sector grow at its max by stopping the defaulters.

6 Future Work

Loans and the people go side by side as the people have many ambitions to grow; hence this field is overgrowing, which can be included in the bank system through which the bank can get aware of the defaulters beforehand and thus can save their money. Different comparisons should also be made to get to know a better approach.

References

1. Li, Y., Xu, Y., Feng, G., Dai, W.: On loan-to-value ratios of inventory financing with doubly stochastic poisson default processes, pp. 663–666 (2006). https://doi.org/10.1109/APSCC.2006.73
2. Zhu, L., Qiu, D., Ergu, D., Ying, C., Liu, K.: A study on predicting loan default based on the random forest algorithm. Procedia Comput. Sci. 503–513 (2019). https://doi.org/10.1016/j.procs.2019.12.017
3. Barua, S., Gavandi, D., Sangle, P., Shinde, L., Ramteke, J.: Swindle: predicting the probability of loan defaults using catboost algorithm. In: 2021 5th International Conference on Computing Methodologies and Communication (ICCMC), pp. 1710–1715 (2021)
4. Shingi, G.: A Federated Learning Based Approach for Loan Defaults Prediction, pp. 362–368 (2020)
5. Lu, F., Lu, M.: Evolutionary game analysis on the phenomenon of national student loan default. In: 2008 International Seminar on Business and Information Management, vol. 1, pp. 249–252 (2008). https://doi.org/10.1109/ISBIM.2008.176
6. Patel, B., Patil, H.P., Hembram, J., Jaswal, S.: Loan default forecasting using data mining. In: 2020 International Conference for Emerging Technology (INCET), pp. 1–4 (2020)
7. Supriya, P.U., Pavani, M., Saisushma, N., Kumari, N.V., Vikas, K.: Loan prediction by using machine learning models. Int. J. Eng. Tech. **5**, 144–148 (2019)
8. Madaan, M., Kumar, A., Keshri, C., Jain, R., Nagrath, P.: Loan default prediction using decision trees and random forest: a comparative study. IOP Conf. Ser.: Mater. Sci. Eng. **1022**, 012042 (2021). https://doi.org/10.1088/1757-899X/1022/1/012042
9. Shaheen, S.K., ElFakharany, E.: Predictive analytics for loan default in banking sector using machine learning techniques. In: 2018 28th International Conference on Computer Theory and Applications (ICCTA), pp. 66–71 (2018). https://doi.org/10.1109/ICCTA45985.2018.9499147
10. Leo, M., Sharma, S., Maddulety, K.: Machine learning in banking risk management: A literature review. Risks, 29 (2019). https://doi.org/10.3390/risks7010029
11. Aliaj, T., Anagnostopoulos, A., Piersanti, S.: Firms default prediction with machine learning. In: Mining Data for Financial Applications: 4th ECML PKDD Workshop, MIDAS 2019, W urzburg, Germany, vol. 11985, p. 47 (2020)
12. Chen, Y.-Q., Zhang, J., Ng, W.: Loan default prediction using diversified sensitivity undersampling, pp. 240–245 (2018). https://doi.org/10.1109/ICMLC.2018.8526936
13. Wenqin, L., Shen, L.: Mortgage loan pricing model under the default risk. In: International Conference on E-Business and E-Government (ICEE), pp. 1-4 (2011). https://doi.org/10.1109/ICEBEG.2011.5884506

14. Yaohua, S.: Research on the theoretical model of psychological factors of borrowers with residential mortgage loan. In: Second International Conference on Business Computing and Global Informatization, pp. 175–178 (2012). https://doi.org/10.1109/BCGIN.2012.52
15. Sheikh, M.A., Goel, A.K., Kumar, T.: An approach for prediction of loan approval using machine learning algorithm. In: International Conference on Electronics and Sustainable Communication Systems (ICESC), pp. 490–494 (2020). https://doi.org/10.1109/ICESC48915.2020.9155614

Quantifying Nodes Trustworthiness Using Hybrid Approach for Secure Routing in Mobile Ad Hoc Networks

M. Venkata Krishna Reddy[1,2](✉), P. V. S. Srinivas[3], and M. Chandra Mohan[1]

[1] Department of CSE, Jawaharlal Nehru Technological University Hyderabad, Hyderabad, Telangana, India

[2] Department of CSE, Chaitanya Bharathi Institute of Technology(A), Hyderabad, Telangana, India

krishnareddy_cse@cbit.ac.in

[3] Department of CSE, Vignana Bharathi Institute of Technology, Hyderabad, Telangana, India

Abstract. Mobile Adhoc Networks (MANET) is a group of moving nodes where these nodes communicate each other through wireless links thus forming adhoc network. Each node works as router and forward packets from source to destination. Multiple wireless devices that connects on the fly in any situation are being gained importance today. These devices are flexible in nature, ad hoc and they can be temporarily setup at any point of time, in any place. These networks have lesser infrastructure costs due to decentralized administration. Message routing in those networks has become more difficult because of their innate dynamic character combined with restrictions like limited energy power, less bandwidth, wireless communication transmit nature and intervention of signals. Due to infrastructure less dynamic topology, distribution of bandwidth and limitations on resource usage among mobile nodes, the secure data transmission between source and destination is always a challenge. Security is always a critical concern in providing quality of service (QoS) and secure routing in MANET's since the presence of malicious nodes in the network pose all possible threats to MANETs. Many mechanisms exist and proposed to solve security issues in routing. But all those are complex and not properly addressing the elimination of malicious/selfish nodes. In this article, a new approach collaborative Trust Based Approach (CTBA) is proposed to provide node authentication using trust factor by combining direct and neighbor observations to form resultant collaborative trust before initiating route discovery process for performing data transmission in MANET's. Simulation results has proved that proposed CTBA approach performs well and fine compared to the routing without nodes trust calculation.

Keywords: Security · Direct observation · Neighbor observation · Collaborative trust · Secure routing

1 Introduction

MANET's are the networks that are formed with bunch of moving nodes which communicates each other through wireless paths [1]. There is no need of any infrastructure

M. Singh et al. (Eds.): ICACDS 2022, CCIS 1613, pp. 348–360, 2022.
https://doi.org/10.1007/978-3-031-12638-3_29

or centralized admin support. At any moment, any node can enter or exit the network. Every node behaves as a router in the network and forwards the packets. The nodes are free to move arbitrarily at any moment. The movement of the nodes can be distinguished from nearly stagnant mode to nodes which moves constantly [2]. So network structure and interlinks between the nodes can change in faster mode and unpredictably. Routing always involves in sending packets from source to destination. Sometimes destination may not be in the range of source, so source has to depend on the behaviour of intermediate nodes to establish route for forwarding packets to the destination. This ensures that every node in the network has to cooperate with each other to forward the data. Nodes in MANET's have energy constraints and limited resources. The biggest challenge arises when nodes hesitate to dissipate energy to forward others data packets. Hence they may behave maliciously or selfishly and drop the packets. Therefore, node security becomes one of the major issues of MANET. MANET's are frequently affected by attacks than fixed structure networks because of their characteristics. The major properties of any secured network are integrity, confidentiality, authentication, non-repudiation and availability [3]. So node authenticity is always a critical concern in MANET's.

Trust can play an crucial role in evaluating nodes authenticity before involving it in the routing in MANET's. To provide secure routing with the involvement of the nodes, quantifying of node's trustworthiness is essential and fair move. Several cryptographic, signature based and token based mechanisms are existing and proposed for MANET security. These mechanisms are not suitable to prevent and even can't detect security threats to the network due to random behaviours [4]. All the approaches provided for security of MANET's are stressing the factor of cryptography and very few approaches concentrated around trust factor. Existing cryptography mechanisms fails in isolating the malicious nodes. In existing and proposed trust based approaches, computation of trust factor was not properly addressed. Even in some approaches, it was calculated without any relation to the network parameters and nodes behaviour in en routing the packets. However Routing involving Trust calculation is an efficient measure to counter all security challenges posed by misbehaving/selfish nodes through detection and isolation of unfaith nodes in the network. More emphasis should be on the computation of trust factor for isolation of the malicious nodes by considering the nodes forwarding behaviour.

In this article, a novel trust involved security model where nodes reputation is computed by aggregating Direct trust values along with Neighbor trust values is proposed. The direct trust value is evaluated through a quantifiable approach by taking into consideration network parameters. Neighbor trust is calculated by taking into consideration the weights assigned to the neighbors depending on the distance. The proposed scheme Collaborative Trust Approach CTBA combines direct observations of the node and recommendations from various neighbors collected for calculating final resultant trust treated as hybrid, collaborative trust. The proposed solution CTBA depends on the trust factor for providing secure routing.

The remaining sections of the paper is arranged as below. Findings and observations of related literature work is presented in Sect. 2. A hybrid collaborative trust approach is proposed in Sect. 3 and Sect. 4 shows results obtained through the simulation and shows the performance of the proposed method CTBA. Finally, Sect. 5 concludes the paper.

2 Literature Survey

A variety of approaches are proposed for node discovery using trust factor. The issues of supporting each other between nodes in MANET's has gained many researchers attention over several years.

Authors in [5], proposed a new routing method which considers all the nodes that are trustworthy, but due to lack of energy, they behaves as selfish nodes. By default, all the selfish nodes may be malicious, and they behave selfishly because of low energy levels. This concludes that nodes with low power should not be included in routing as they drops packets due to energy consumption and act as malicious or selfish. A monitoring-based cooperation technique in MANET's to find and resolve unexpected routing behavior is proposed in [6]. Results obtained from simulation prove that proposed scheme performs well in packet delivery and throughput. But, this scheme is not fulfilling the task of identifying malicious nodes fully but works better when compared with other schemes. Authors of [7] elevated a trust approach depending on vector trust based model (VT). The main goal of this approach is to find misbehaving nodes and later mitigating them. This method is also implemented using routing protocols: dynamic source routing (DSR) [8] and ad hoc on-demand distance vector (AODV) [8].

The scheme in [9] proposes a trustcentric model for predicting malicious node's characters with the help of a reputation system and a supervision technique. A statistical theory on which trust depends is proposed as security solution in [10]. This method supports owner's decision policy, results based on intrusion detection, observations of other nodes and their experiences if packet forwarding. In [11], a general methodology based on node authentication to find collaboration among nodes in order to mitigate packet dropping nature is proposed.

In [12], authors proposed a adaptive approach that points about nodes trust is its belief on its neighboring node. Various trust values are defined in for providing security accordingly. This approach is efficient in terms of energy saving, secured communication. In [13], a trust secure routing solution is presented for finding trust that has proven very efficient against the presence of misbehaving nodes collaborating together. Authors of [14] proposed a method to identify malicious nodes. Nodes drop packets in order to save their own energy. Those nodes which drops, will subjected to a scrutiny for their selfish nature. Simulated results show this model as a good one. A small-payment model for two-hop wireless sensor networks is proposed in the article [15] to support multiple nodes participation in packet transmission by allowing all the users get benefited from depending on others' packets. Many schemes are also proposed for identifying and providing rewards for the cooperation and support on the other hand detecting and penalizing malicious behaviour. The proposed scheme in [16] presents an efficient token approach to simulate the node support in MANETs by establishing communication between source and destination. It has also highlighted that nodes should help each other in forwarding packets for other nodes in the network otherwise, establishing adhoc network may not be possible [17].

Authors, of [18] presented a method to separate homogeneous peers from selfish ones purely depending on "ad hoc on-demand distance vector (AODV)" routing protocol. In [19], authors discussed security issues of MANET which are associated with "Internet of Things (IoT)". The method concentrated on security challenges and vulnerabilities for pervasive computing including IoT in connection with Mobile ad hoc networks. An algorithm is presented in [16] where trust is calculated to isolate misbehaving nodes. It was observed that the projected method was helpful in routing data securely in Mobile Ad hoc Networks. In [20], a technique is proposed to isolate malicious nodes and a path discovery method that helps all the routing protocols in avoiding such nodes while forwarding any packet. All the above approaches discussed from the literature survey carried are based on either cryptographic mechanisms or trust centric mechanisms. But most of the trust centric mechanisms in existence are not exposing the computation of trust factor depending on the network parameters. It is opined that computation of trust factor based on nodes behaviour and strengthening through the two tier observations may lead to effective isolation of the malicious/selfish nodes and secure routing from source to destination. The proposed method, CTBA is aimed to utilize the above feature and quantifies the trust factor depending on nodes packet forwarding behaviour and two tier observations thus strengthening the computation of trust factor for efficient isolation of misbehaving nodes and effective secure routing.

3 Proposed Method (CTBA)

The proposed CTBA method quantifies the trustworthiness of the nodes by considering hybrid approach using both direct and indirect observations on the node in the network before involving them in the routing in order to ensure secure routing thus enhancing the performance of MANET's. Several Network parameters that can have impact on the network in routing are taken into consideration for calculation of resultant trust using both the ways of trust factor: direct and neighbour observations.

3.1 Process Flow

Figure 1 explains the process flow of trust quantification, nodes classification and Routing Decision. Nodes are classified as good or bad, malicious/selfish or efficient based on their trust value evaluated using the proposed approach. Later Routing is performed only by involving those trusted nodes.

In a network, source node wants to deliver packets to the destination which is located at out of its range, and then it depends on other nodes to form the route and to forward packets. Source Node initiates the quantification of trust for each node in order to classify them and involve in routing path. For this, source node quantifies its own observations on the neighbour node and also takes into consideration others observations on that neighbour node which should be included in routing and for which trustworthiness is required to calculate. This is performed in step wise fashion.

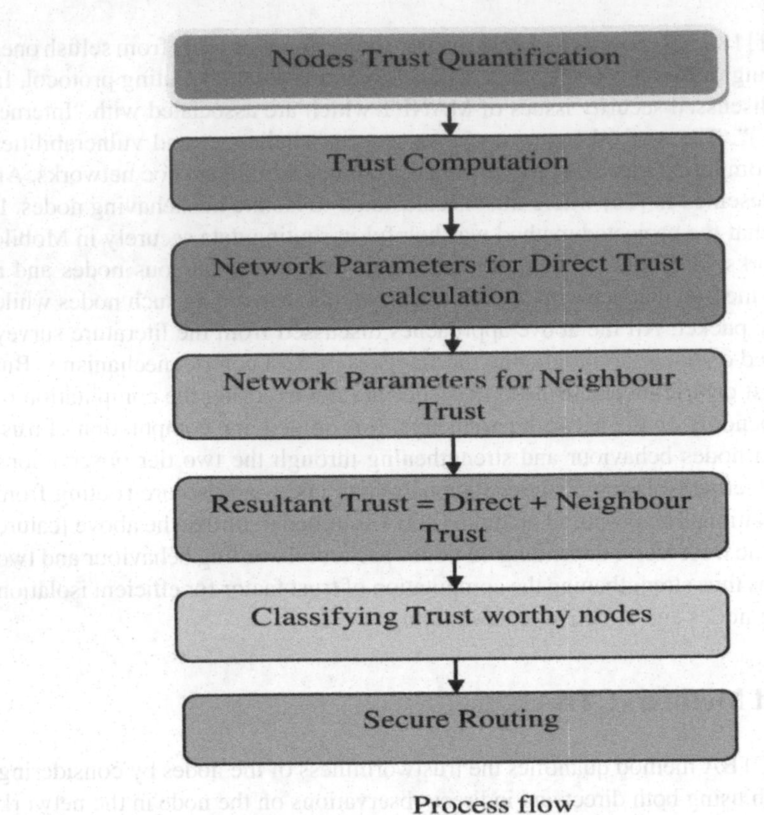

Fig. 1. Process flow of nodes classification

3.2 Algorithm

Collaborative Trust CTBA(D, N, R_T, S_T)
{
1. Initially Nodes 'n' trust quantification is initiated by source node
2. Direct trust D is evaluated based on Data Packets αand Control packets β,
$D_n = t1 * \alpha + t2 * \beta$
3. Neighbor trust N is evaluated based on the other nodes observation and their Weights, N_n= m1*NBN1 + m2 * NBN2 + m3 * NBN3…. + mn * NBNn
4. Hybrid, collaborative trust R_T is calculated as combination of direct and neighbor observations. $R_T = w_1 \cdot$(Direct Trust) $D_n + w_2 \cdot$(Neighbour Trust) N_n
5. Repeat the steps 2,3,4 for all the nodes in the network.
6. If node's resultant trust is above the Threshold S_T, classify as trusted and involve in routing. ***Trusted:*** **if $R_T \geq S_T$,**
Untrusted: if $R_T < S_T$
7. **Else** classify as untrusted and isolate.
Endif }

3.3 Direct Trust Evolution

Source node observes the behaviour of its neighbor nodes to notice their trust. This is because of node's mobility in Mobile ad hoc networks. A node can enter or exit at any point of time in MANETs. So a node can observe its nearest neighbor node to notice its behavior while communicating directly in a passive manner. The following are the network parameters taken into consideration for quantifying the direct observations D_n of source node on its neighbour node.

a. *For Data Packets*

1. No of data packets delivered at the node = D_d
2. No of data packets transmitted correctly by the node = D_c
3. No of data packets dropped by the node = D_r

b. *Control Packets Forwarded*

- Total no of Route Request packets passed to the Node = R_{cp}
- Total no of Route Reply packets received at the Node = R_{cr}
- Total no of Route Request packets sent by the Node = R_{cs}

Source node keeps the track of above parameters and collects the information from its observations of the traffic en routed through that neighbour node. With the help of above information, two types of ratios are calculated: Data forwarding ratio and Control forwarding ratio can be calculated by using Eq. 1 and Eq. 2.

Data ratio, α

$$\alpha = x * (D_c/D_d) + y * (D_r/D_d) \tag{1}$$

where x, y are the proportionate weights assigned and x + y = 1.
Control ratio, β

$$\beta = x * \left(R_{cr}/R_{cp}\right) + y * \left(R_{cs}/R_{cp}\right) \tag{2}$$

where x, y are the proportionate weights assigned and x + y = 1.

Then Direct Observations are

$$D_n = t1 * \alpha + t2 * \beta \tag{3}$$

t_1, t_2 are the weights assigned for Data ratio and Control ratio, Where t1 > t2.

3.4 Neighbor Trust Evaluation

Neighbour Trust is evaluated with above mentioned parameter by the other nodes observations on the node under consideration for trust evolution.

$$N_n = m1 * NBN1 + m2 * NBN2 + m3 * NBN3 \ldots .. + mn * NBNn \quad (4)$$

where m1, m2, m3.....mn is weights given to the neighbour other nodes based on their distance in the network from the node under trust evaluation.

NBN1, NBN2, NBN3......NBNn are neighbour node observations in the network.

3.5 Collaborative Trust Evaluation CTBA

Final hybrid, resultant collaborative trust calculation of a node is done based on Direct observations evaluated in (3) and neighbour observations calculated in (4).

$$R_T = w_{1*}(\text{Direct Trust})D_n + w_{2*}(\text{Neighbour Trust})N_n \quad (5)$$

where $w_1 > w_2$ and w_1, w_2 are the weights given to the direct and neighbour observations accordingly.

Once the 'Hybrid, resultant collaborative trust' R_T is determined for the Node, it is propagated to all the nodes present in the network.

3.6 Node Classification

Based upon the Resultant Trust/Hybrid Trust R_T calculated, the nodes are categorized by comparing to the static threshold value S_T.

$$\textit{Trusted} : \text{if } R_T \geq S_T$$
$$\textit{Untrusted} : \text{if } R_T < S_T$$

Static threshold value S_T is considered on basic conditions of the network and fixed as 0.6 which represents average static trust threshold. Source nodes involve only those nodes quantified as trusted ones in the routing path. Thus establishing the secure communication in the Mobile adhoc networks.

4 Simulation

The Simulation is carried out in a 700 × 500 m^2 network area which resembles a small conference gathering and IEEE 802.11 MAC for 600 s with 80-nodes are used for simulation as they represent the minimum gathering in any real time scenario. In the simulation part, it is observed that fair nodes drop the data packets because of the network problems which likely to be nodes mobilization, packet collisions etc. whereas the malicious/selfish nodes drop the data packets deliberately. Simulation parameters are given in Table 1.

Table 1. Parameters used for simulation

Simulation parameter	Value
Simulator	NS2.34
No of nodes	80
Network area	700 × 500
Packet size	512 bytes
No. of malicious nodes	16
Traffic type	CBR/UDP
Mobility	4–25 m/s
Pause time	5 s
Simulation time	600 s

5 Results and Discussion

The efficacy of the proposed CTBA approach is shown by comparing it to the existing routing protocol where intermediate nodes are involved without any prior trust calculations. The proposed scheme of this paper calculates the resultant trust value of a node by observing its data transmission nature and also others observations. Trust is calculated periodically due to the dynamic behaviour of MANET's. Nodes are classified based on their good and bad characterisations by comparing with the Trust threshold.

Performance Evolution
It is detected that almost 18% of the data packets are found lost because of the packet collision and network problems. Performance Metrics used for evolution of the proposed scheme are Packet Delivery Ratio (PDR), Packet Drop Ratio (PDR), Detection of False Positives and Throughput are taken into consideration. Calculations carried and results obtained results using CTBA approach after simulation are tabulated in Table 2.

Table 2. Calculations carried out for computing resultant collaborative trust using proposed CTBA approach

Node	Direct trust calculation	Neighbor calculation	Node collaborative trust
0	0.86	0.423	0.76341
1	0.85	0.379	0.73372
2	0.02	0.4782	0.21386
3	0.45	0.6257	0.35726
4	0.43	0.6984	0.71863
5	0.56	0.28534	0.18864

(*continued*)

Table 2. (*continued*)

Node	Direct trust calculation	Neighbor calculation	Node collaborative trust
6	0.68	0.23185	0.79324
7	0.05	0.435	0.32014
8	0.63	0.5791	0.65323
9	0.79	0.24986	0.713436
10	0.81	0.312	0.68463

Isolation of malicious nodes depending on the collaborative trust calculated using CTBA approach is shown in Table 3.

Table 3. Classification of trustworthy nodes using collaborative trust based approach CTBA

Node	Node collaborative Trust	Static trust threshold	Decision
0	0.76341	0.6	Trustworthy
1	0.73372	0.6	Trustworthy
2	0.21386	0.6	Malicious
3	0.35726	0.6	Malicious
4	0.71863	0.6	Trustworthy
5	0.18864	0.6	Malicious
6	0.79324	0.6	Trustworthy
7	0.32014	0.6	Malicious
8	0.65323	0.6	Trustworthy
9	0.713436	0.6	Trustworthy
10	0.68463	0.6	Trustworthy

5.1 Packet Delivery Ratio

In Fig. 2, packet delivery ratio and its impact due to the presence of malicious nodes is shown. The graph exhibits the comparison between the existing routing without trust and the proposed hybrid trust method. It clearly projects that Packet Delivery Ratio is high. In case of 5% of malicious nodes presence, packet delivery ratio for the existing routing protocol without trust calculation is 71 packets per second and 82 packets per second for the proposed scheme where trust is calculated in case of 100 packets scenario. It is proved that the proposed scheme performs well in case of packet delivery.

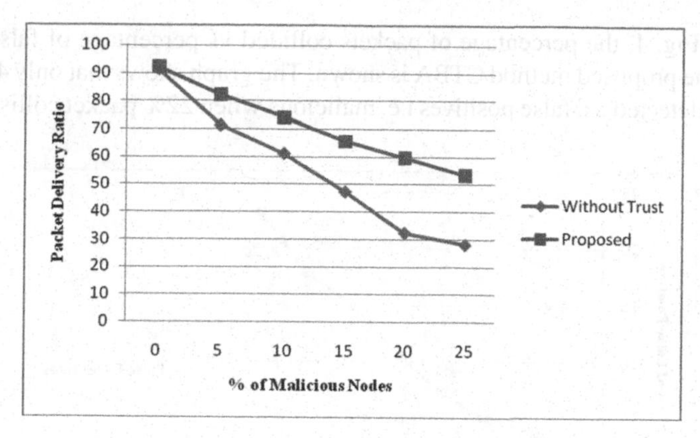

Fig. 2. Analyzing of packet delivery ratio

5.2 Packet Drop Ratio

Packet drop ratio can be defined as no of packets dropped out of total no of packets received by the node for forwarding. In Fig. 3, packet drop ratio for the both cases is mapped. The graph shows clearly that the proposed hybrid trust method outperforms when compared to the without trust strategy in terms of packet drop ratio. In case of the proposed hybrid scheme only 4% of packets drops are recognized for a total of 100 packets where 5% of malicious nodes are present and whereas 9% of packets drop are recognized for 100 packets where 5% of malicious nodes are present in case of routing without any trust calculation.

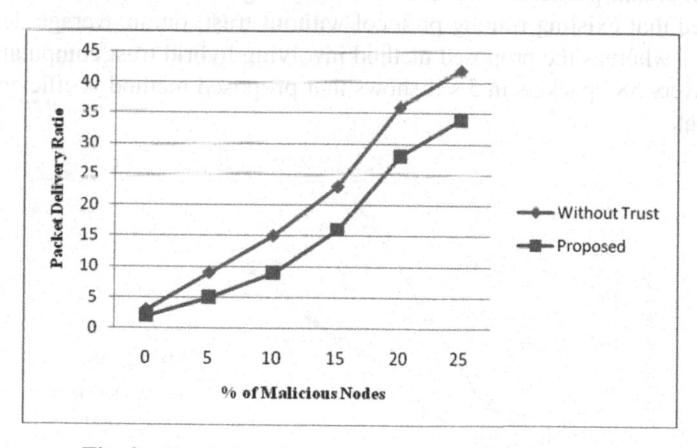

Fig. 3. Graph showing the results for packet drop ratio

5.3 False Positives Detection Ratio

The ratio defined as the available number of fair behaving nodes wrongly identified as misbehaving/selfish to the total available count of nodes present in the network, is 'False positive'. In Fig. 4, the percentage of packets collided vs percentage of false positives detected of the proposed method CTBA is shown. The graph shows that only 4.7% nodes are wrongly detected as. false positives i.e. malicious when 22% packet collisions occur.

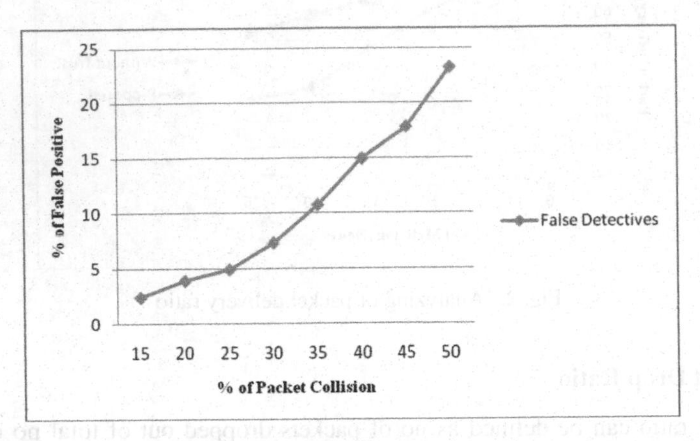

Fig. 4. False positive detection ratio

5.4 Throughput

Figure 5 shows the throughput comparison between existing routing protocol without trust computation and proposed hybrid method with resultant trust calculation. Throughput is number of data packets delivered successfully in given time. From the below figure, it is observed that existing routing protocol without trust, on an average delivers 489 packets in 5 s whereas the proposed method involving hybrid trust computation, on an average delivers 587 packets in 5 s It shows that proposed method is efficient in terms of throughput.

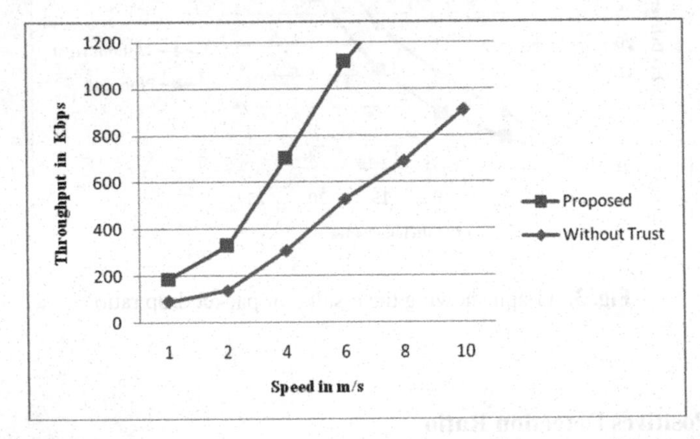

Fig. 5. Performance analysis of the proposed method in terms of throughput

6 Conclusion

In this article, a quantitative hybrid trust computation model CTBA is discussed which calculates the hybrid collaborative trust of every node as part of trust quantification and identifying nodes trustworthiness for secure and efficient routing by taking into consideration various network parameters of both categories: data and control packets. This involves calculating direct observations along with indirect observations of all the nodes in 1-hop distance. Simulation results shows that the method outperforms with good efficacy in terms of performance metrics: packet delivery ratio, false positive detection, packet drop ratio and throughput. Trusted Nodes can be easily identified using this method and malicious nodes are easily isolated. So the path from source to destination is formed only using the trusted nodes identified thus leading to secure and efficient routing in mobile adhoc networks.

This method can be enhanced by considering the nodes even in 2-hop distance. Their observations may also be quantified as part of trust evaluation. This may increase the efficiency of the routing. The proposed method can be applied in collaboration with adaptive trust mechanism and further can be extended with more sophisticated observations for computation of trust factor.

Acknowledgements. We would like to thank and acknowledge for all the members for providing their continuous support.

References

1. Alnumay, W., Ghosh, U., Chatterjee, P.: A Trust-based predictive model for mobile ad hoc network in internet of things. Sensors **19**(6), 1467 (Jan 2019)
2. Wu, B., Chen, J., Wu, J., Cardei, M.: A survey of attacks and countermeasures in mobile ad hoc networks. In: Wireless network security, pp. 103–135. Springer, Boston, MA (2007)
3. Alnumay, W.S., Chatterjee, P., Ghosh, U.: A trusted framework for secure routing in wireless ad hoc networks. In: 2015 International Conference on Collaboration Technologies and Systems (CTS), pp. 190–195. IEEE (1 Jun 2015)
4. Sumathi, K., Priyadharshini, A.: Energy optimization in manets using on-demand routing protocol. Procedia Computer Science. **1**(47), 460–70 (2015). Jan
5. Hinge, R., Dubey, J.: Opinion based trusted AODV routing protocol for MANET. In: Proceedings of the Second International Conference on Information and Communication Technology for Competitive Strategies, pp. 1–5 (4 Mar 2016)
6. Kumar, S.G.: A novel routing strategy for ad hoc networks with selfish nodes. J. Telecommun. **3**, 23–8 (2010)
7. Bansal, S., Baker, M.: Observation-based cooperation enforcement in ad hoc networks (4 Jul 2003). arXiv preprint cs/0307012
8. Gong, W., You, Z., Chen, D., Zhao, X., Gu, M., Lam, K.Y.: Trust based routing for misbehavior detection in ad hoc networks. Journal of Networks. **5**(5), 551 (2010). May 1
9. Perkins, C.E., Royer, E.M.: Ad-hoc on-demand distance vector routing. In: Proceedings WMCSA'99. Second IEEE Workshop on Mobile Computing Systems and Applications, pp. 90–100. IEEE (25 Feb 1999)
10. Li, N., Das, S.K.: A trust-based framework for data forwarding in opportunistic networks. Ad Hoc Networks. **11**(4), 1497–509 (2013). Jun 1

11. Sirisala, N.R., Bindu, C.S.: A novel QoS trust computation in MANETs using fuzzy petri nets. Int. J. Intelli. Eng. Sys. **10**(2), 116–25 (2017)
12. Michiardi, P., Molva, R.: Core: a collaborative reputation mechanism to enforce node cooperation in mobile ad hoc networks. In: Advanced communications and multimedia security, pp. 107–121. Springer, Boston, MA (2002)
13. Yan, Z., Zhang, P., Virtanen, T.: Trust evaluation based security solution in ad hoc networks. In: Proceedings of the Seventh Nordic Workshop on Secure IT Systems, vol. 14. Citeseer (15 Oct 2003)
14. Nekkanti, R.K., Lee, C.W.: Trust based adaptive on demand ad hoc routing protocol. In: Proceedings of the 42nd annual Southeast regional conference, pp. 88–93 (2 Apr 2004)
15. Guaya-Delgado, L., Pallarès-Segarra, E., Mezher, A.M., Forné, J.: A novel dynamic reputation-based source routing protocol for mobile ad hoc networks. EURASIP J. Wireless Comm. Netw. **2019**(1), 1–6 (2019). Dec
16. Gite, P., Kanellopoulos, D.N., Choukse, D.: An extended AODV routing protocol for secure MANETs based on node trust values. Int. J. Internet Technol. Secured Trans. **7**(3), 270–91 (2017)
17. Baras, J.S., Jiang, T.: Cooperative games, phase transitions on graphs and distributed trust in MANET. In: 2004 43rd IEEE Conference on Decision and Control (CDC) (IEEE Cat. No. 04CH37601), vol. 1, pp. 93–98. IEEE (14 Dec 2004)
18. Chiejina, E., Xiao, H., Christianson, B.A.: Proceedings of the 6th York Doctoral Symposium on Computer Science & Electronics, York, UK, 2013. Candour-based trust and reputation management system for mobile ad hoc networks. University of York, York (2013)
19. Wang, X., Liu, L., Su, J.: RLM: a general model for trust representation and aggregation. IEEE Trans. Ser. Comp. **5**(1), 131–43 (2010). Dec 23
20. Zafar, S., Soni, M.K.: Trust based QOS protocol (TBQP) using meta-heuristic genetic algorithm for optimizing and securing MANET. In: 2014 International Conference on Reliability Optimization and Information Technology (ICROIT), pp. 173–177. IEEE (6 Feb 2014)

Medical Ultrasound Image Segmentation Using U-Net Architecture

V. B. Shereena[1(✉)] and G. Raju[2]

[1] Department of Computer Applications, MES College, Marampally, Kochi, India
shereenavb@gmail.com

[2] SMIEEE, Department of Computer Science and Engineering, Faculty of Engineering, Christ (Deemed to Be University), Bengaluru, India

Abstract. This research article discusses the implementation aspects of a Deep Learning architecture based on U-Net for medical image segmentation. A base model of the U-Net architecture is extended and experimented. Unlike the existing model, the input images are enhanced by applying a Non-Local Means filter optimized using a metaheuristic Grey wolf optimization method. Further, the model parameters are modified to achieve better performance. Tests were performed using two benchmark B-mode Ultrasound image datasets of 200 Breast lesion images and 504 Skeletal images. Experimental results demonstrate that the modifications resulted in more accurate segmentation. The performance of the modified implementation is compared with the base model and a Bidirectional Convolutional LSTM architecture.

Keywords: U-Net · Deep learning · Ultrasound · Non-Local Means · Grey Wolf Optimizer · Segmentation

1 Introduction

Radiologists diagnose disorders in Ultrasound images by observing the textural characteristics of affected regions compared to neighborhood areas, which rely on their subjective expertise [1]. Computer-assisted systems that can provide accurate segmentation of regions of interest would support them in making an accurate assessment of the image [2]. In this direction, several works focused on the Ultrasound B-mode image segmentation have been published [3]. But the precise segmentation of Ultrasound images remains a challenge due to Speckle noise, low contrast, and intensity inhomogeneity [4]. In this article, a deep learning approach based on U-Net is explored for the segmentation of Ultrasound B-mode images.

Segmentation is the process of subdividing an image into disjoint, eloquent regions to locate prominent anatomical structures or anomalies such as tumors or cysts [5]. Even though a variety of classical algorithms are available for segmentation, the selection of an appropriate technique for a particular modality of medical images remains a challenging task owing to human biases or variances in images.

Deep learning offers a new endeavor with its revolutionary performance in segmentation and classification of medical images and associated computer vision tasks [6]. It

M. Singh et al. (Eds.): ICACDS 2022, CCIS 1613, pp. 361–372, 2022.
https://doi.org/10.1007/978-3-031-12638-3_30

extracts complex features directly from input data [7]. The highlight of the deep learning method is that they do not depend on hand-crafted features like conventional machine learning methods. Deep Learning Networks belong to the class of Artificial Neural Network (ANN) where multiple layers are used to build the network structure. The benefit of a Neural network is that any class of problems can be solved with a customized network architecture [8].

The most ideal and dominating algorithm for medical image analysis is Convolutional Neural Network (CNN). CNN application commenced with identifying handwritten characters [9] to acquire patterns from images with multiple layers stacked one above the other. AlexNet, a modified CNN model was developed in 2012 [10] which classified 1.2 million high-quality images to 1000 classes with an effective GPU implementation. To eliminate overfitting, this network used data augmentation and dropout concepts. Subsequently, numerous CNN architectures including ZFNet, VGG, GoogleNet, ResNet, and DenseNet were developed for medical image segmentation and classification [11].

Despite these developments, the deficiency of large volumes of labeled medical image datasets for training and testing of CNN hampered its widespread use. U-Net architecture is developed with a focus on image segmentation with a relatively small number of samples for model building [12, 13]. The structure of the U-Net resembles the letter U with a contracting and an expansive path. U-Net model came into the limelight with limited microscopic images of the ISBI 2015 challenge [12] and afterward used efficiently for brain tumor segmentation [14] evaluating on BRATS 2015 datasets. An improved performance was reported for a fully automated U-Net-based segmentation experimented using a Breast Ultrasound image dataset of 221 images [15]. Since the U-Net strategy learns directly from the input images and corresponding ground truths, there is no need for initiating a seed point in the specified region of interest. This article presents an implementation of the U-Net Segmentation method, by introducing changes to a base model.

It is a challenging task to detect and segment a small region of interest from a larger background. Such imbalance problems among the foreground and background classes (pixels) are reduced using a properly selected weight loss function. The Data Augmentation technique is incorporated to improve the quality of segmentation and avoid overfitting. For tackling the Speckle noise in Ultrasound images, they are Pre-processed using a Non-Local Means Filter optimized using a meta-heuristic called Grey Wolf Optimizer and denoised images are input to the U-Net architecture. Two publicly available Ultrasound image datasets are used for the experimentation. Further, for comparison, the experimentation is also conducted with baseline U-Net architecture [12] and Bidirectional Convolutional LSTM architecture [16]. The performance of the modified U-Net architecture is better than the base model considered.

The remainder of this article is organized as follows. Literature Review is presented in Sect. 2; Methodology in Sect. 3; Experimental Setup in Sect. 4; Results and Discussions in Sect. 5 respectively. Section 6 gives a Conclusion of the article.

2 Literature Review

Ronneberger et al., [12] proposed a U-Net strategy for the segmentation of microscopy images. The method incorporated data augmentation techniques to use the obtained labelled samples with more efficiency. Its architecture comprises a U-shaped contracting path and an expanding path. The method outperforms the preceding convolutional architecture for the segmentation of microscopic images. Experiments were conducted with training data of 30 images obtained from the ISBI 2012 challenge. The advantage of the architecture is that it performs faster segmentation in systems with a Graphical Processing Unit. But the experimentation is done for a limited dataset of microscopic images.

Dong et al., [14] proposed a reliable segmentation model using U-Net architecture for the extraction of tumors from brain MRI images. In this architecture, the accuracy of segmentation was enhanced using the data augmentation method and Soft Dice loss function. fivefold cross-validation is used to assess the proposed method which was implemented on BRATS 2015 datasets which showed that the method obtained efficient and promising segmentation for outlining the tumor regions. The limitations to this work include the lack of objective assessment using an independent testing dataset and the requirement of tuning numerous parameters. Also, it is less successful in segmentation of the enhanced tumor regions of low-grade tumor datasets.

Shi et al., [17] proposed a modified U-Net model with stacked architecture for medical image segmentation. The model proposed a zoom approach and a loss function for minimizing the overfitting risk and hybrid features for segmenting the affected regions. The proposed model was evaluated by experimenting on CT, histopathology, and digital colonoscopy images and found the improvement in performance without fine-tuning of parameters and at reduced computational complexity.

Du et al., [18] performed a systematic review on medical image segmentation and concluded that among convolutional neural networks, U-Net has the competent performance in the analysis of medical images. This article studied the application of U-Net in various imaging modalities including Computed Tomography, Magnetic Resonance Imaging, Ultrasound, X-Ray, Optical Coherence Tomography and Positron Emission Tomography. U-Net has the advantage of many feature channels in the up-sampling path to propagate related information to higher layers and the limitation of obtaining large labeled datasets and computation power.

Khoong, W.H., [19] proposed BUSU architecture based on U-Net based hybrid deep learning network for segmentation of Ultrasound images. Here a hybrid of two similar BCDU networks with different sizes improved the performance of the original model. The model employed bidirectional convolutional LSTMs with input data processed in forward and backward directions for decision making. Tests were performed on the DRIVE dataset of 40 retina images and found that the proposed method performs better than most of the models taken for comparison. The resource constraints put forth challenges to this research work for experimenting with more hybrid models.

In the research article presented by Ardhianto et al., [20], a comprehensive review of the deep learning challenges and trends regarding muscle Ultrasound images is conducted. Segmentation of Skeletal muscle is difficult because of the regular movement of muscles and resultant noise. In the classification of skeletal muscles, the cropping strategy overcomes the limitations. This study enhances knowledge of the deep learning strategies for handling muscle Ultrasound images. The first limitation of this study is the use of data augmentation or transfer learning which needs further investigation and the second is that the study is limited to two muscle types.

From the review, it is found that the segmentation performance using deep learning network architecture relies on the volume and variety of images used for training and testing as well as the quality of images. Lack of a large collection of labeled datasets results in overfitting. Also, if the hyperparameters are not properly tuned, it results in detecting false positive regions. Achieving faster convergence with reduced training time is yet another challenge in the field of deep learning. Deep Networks have the limitation of vanishing or exploding gradient problems and class imbalance issues. The survey emphasizes the merits of U-Net architecture in image segmentation. This article explores the possibility of improving U-Net architecture.

3 Methodology

In this research article, the implementation details of U-Net-based Segmentation are presented. The key steps are Image Pre-processing, Data Augmentation, and Image Segmentation using the U-Net model. The U-Net architecture model is given in Fig. 1.

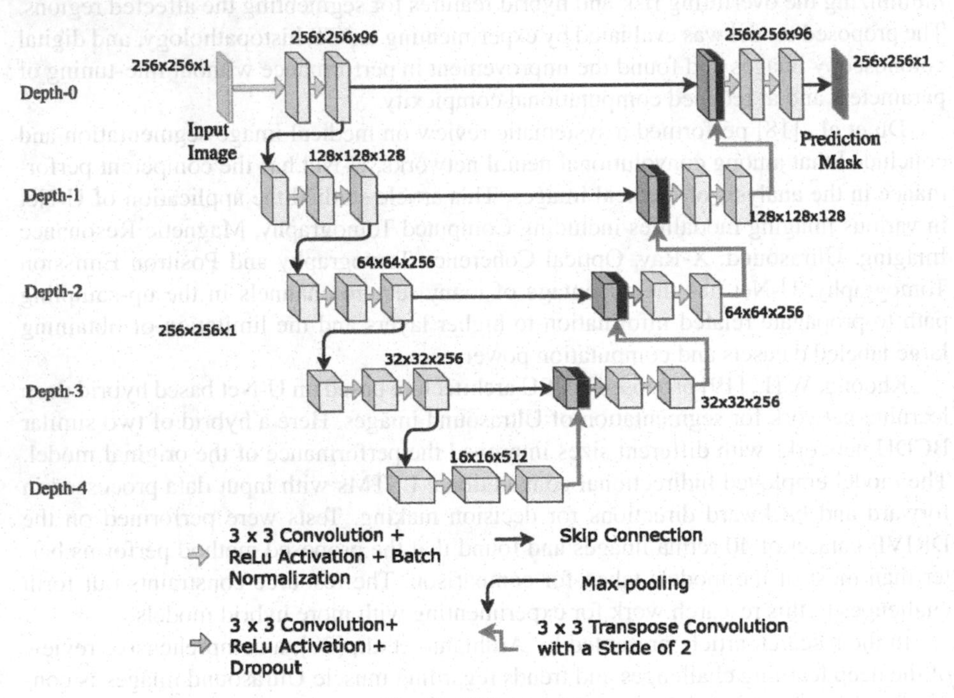

Fig. 1. U-Net model

3.1 Image Pre-Processing

Speckle noise present in Ultrasound images makes the segmentation process challenging [4]. To eliminate the Speckle noise and enhance the quality of segmentation, pre-processing is carried out with an Optimized Non-Local Means filter. Ultrasound lesion image is input to the Grey Wolf Optimization (GWO) method [21] to find the design parameters of the Non-Local Means (NLM) algorithm and then the optimized parameters are passed to the NLM filter.

3.2 Data Augmentation

A limitation of medical image datasets is that they are hard to obtain owing to ethical, privacy, and storage concerns. The need for expert annotation of the dataset increases the complexity of medical image database creation since labeling is time-consuming and expensive [14]. As a result, the open-access benchmark datasets available for research contain only limited number of images. Deep learning techniques require a large dataset for better performance. In this work, the Data Augmentation methods such as Rotation with Range 30, Zoom in the Range 0.8 and 0.9, Feature-wise Center, Feature-wise Standard Normalization, Sample-wise Center, and Sample-wise Standard Normalization are applied.

3.3 U-NET Model

U-Net model comprises of an encoder and decoder of depth-4, where the input image size is 256×256. Each depth consists of two convolutional layers with Rectified Linear Unit (ReLU) activation followed by batch normalization and Max-pooling. In depth-0 and depth-1, 96 and 128 filters are used in each convolution layer respectively. The number of filters in depth-2, depth-3, and depth-4 are 256, 256, and 512 respectively. As depth increases, the feature map size reduces with Max-pooling. To compensate for the feature reduction after each encoding depth, the number of filters is increased in the extreme depth levels.

In the decoding path, the up-convolution is achieved by a transposed convolutional operation and concatenation to the up sampled resultant feature space in each depth. To further reduce the overfitting, dropout is used at higher depths. Finally, a 1×1 convolution layer with sigmoid activation is performed at the last layer of the network. All convolution operations use a 3×3 kernel with the 'same' padding to get the resultant output image size same as that of the input. The final sigmoid layer results in producing output in the range 0 to 1. Each pixel value corresponds to a probability value which is further applied a threshold of 0.5 to convert to the required binary segmentation map.

Various hyperparameters, like the filter kernel size, the number of filters, depth, and dropout are selected based on experimentation.

4 Experimental Setup

Experiments were conducted using the Google Colab environment as well as on a workstation with Intel® Core [TM] i7-8550U CPU @ 1.99 GHz processor, 8 GB RAM, and an NVIDIA® GeForceMX130 graphics card. All Pre-processing stages are implemented using MATLAB®2019B and Deep Learning experiments done using a Python-based deep learning framework Keras with Tensorflow as the back end in Google Colab with the support of GPU.

4.1 Data Set

From the limited publicly available datasets, two heterogeneous datasets are identified for the experimentation, which carries a unique challenge of its own concerning the acquisition method and skill of the radiologist in ground truth labeling.

The dataset consists of the Skeletal Muscle and Breast Lesion images. The skeletal muscle dataset consists of 504 images verified at 25 Hz (AlokaSSD-5000 PHD, 7.5 MHz) from the calf muscles [22] and the ground truth of soleus muscle that lies beneath the gastrocnemius muscle. The breast Ultrasound dataset has 200 images which consisted of 104 malignant and 96 benign cases attained with Ultrasonix SonixTouch Research Ultrasound scanner [23] and the expert annotated affected regions to precisely specify the lesion area.

The images in the skeletal dataset are of size 436 × 528. The Breast dataset consists of images of size 1824 × 510. Before performing experiments, both the Ultrasound image datasets are converted to gray-scale, the images resized to 256 × 256 pixels, and the image data is normalized. From both the datasets, 80% of images are used for training, 20% for testing.

4.2 Training and Optimization

During training, the weights are initialized with the Glorot-uniform weight initializer. Glorot-uniform initializer provides for faster convergence. For updating the network weights, Adam Optimizer [24] is used. Adam optimizer is computationally efficient, requires less memory and minimal tuning of hyperparameters. The parameter values set in Adam optimizer are learning rate - 0.001, batch size - 8, beta1 - 0.9 and beta2 - 0.999. The model is trained for 200 epochs with early stopping patience of 50 epochs. If the epochs are increased further, training loss decreases, but validation loss increases. The number of epochs is selected by checking the accuracy and loss curves so that no overfitting is possible.

Proper selection of loss function is crucial in the success of segmentation models, particularly with imbalanced data. A linear Dice loss (DL) [25], which is the minimization of overlap between the prediction result and ground truth reduces the ill effects of class imbalance. But it gives equal weights to the false positives and false negatives while computing the loss and hence gives equal weightage to recall and precision. The Tversky Index (*TI*) [26], a generalization of the Dice coefficient is formulated as follows.

$$\mathrm{TI} = \mathrm{TP}/[\mathrm{TP} + \alpha\mathrm{FN} + \beta\mathrm{FP}] \tag{1}$$

TP, FP, and FN are true positives, false positives, and false negatives, respectively. α is the weight parameter for false negatives (FN), and β penalizes the false positives (FP). When $\alpha > 0.5$ and $\beta < 0.5$, more weightage will be given to false negatives and when α

< 0.5 and β > 0.5, more weightage will be given to false positives. Based on the features of the given dataset, an ideal trade-off between the precision and recall can be attained by fine-tuning the α and β parameters. Focal Tversky Loss (FTL) [26], a generalization of Tversky loss controls the loss behaviour at diverse values of the Tversky Index using the parameter γ and is selected as the loss function in this work.

$$Focal\ Tversky\ Loss = (1 - TI)^{[1/\gamma]} \tag{2}$$

K-fold Cross-validation is used for the evaluation of the model, where K is chosen [27] based size of the dataset. In this work, 5 - fold cross-validation is adopted to estimate the skill of the proposed machine learning model.

4.3 Performance Evaluation

The performance assessment of the U-Net segmentation model is done visually and quantitatively. For the quantitative analysis of the performance of the U-Net model, the metrics used are Dice Coefficient [28], Specificity [29], Precision, Recall [30], Accuracy [30], and Computation time.

Dice Coefficient (*DC*), is a measure of overlap between segmented image and ground truth image where the value ranges from 0 (no overlap) to 1 (perfect matching) [28].

$$Dice\ Coefficient = \frac{2|TPP|}{2|TPP| + |FPP| + |FNP|} \tag{3}$$

Specificity metric is a measure of the appropriately extracted region of interest by the segmentation model [29] and reaches its best at 1 or 100%.

$$\text{Specificity} = \frac{|\text{TNP}|}{|\text{TNP}| + |\text{FPP}|} \tag{4}$$

Precision is a metric that gives a measure of the predictive power of the algorithm to represent true positive pixels and samples incorrectly classified as positives (false positives). Precision is a numeric scalar value that ranges in the interval [0, 1] or [0,100] and reaches its best at 1 or 100%. The precision of the extracted region of interest by the proposed method can be computed as follows [30]:

$$Precision = \frac{|TPP|}{|TPP| + |FPP|} \tag{5}$$

Recall gives a measure of correctly segmented samples (true positives) and its misclassified samples (false negatives). The Recall is a numeric scalar value in the range [0, 1] or [0,100] and reaches its best at 1 or 100%. The recall value is computed as follows [30].

$$\text{Recall} = \text{Sensitivity} = \frac{|\text{TPP}|}{|\text{TPP}| + |\text{FNP}|} \tag{6}$$

Accuracy of the segmented image concerning ground truth image can be calculated as [30]:

$$\text{Accuracy} = \frac{\text{TPP} + \text{TNP}}{\text{TNP} + \text{FNP} + \text{TPP} + \text{FPP}} * 100\% \tag{7}$$

where TPP is the true positive pixels, TNP is the true negative pixels, FNP is the false negative pixels and FPP is the false positive pixels. For an efficient segmentation model, the value of accuracy will be close to 100%.

5 Results and Discussions

The performance of the modified U-Net model is assessed using two medical Ultrasound image datasets as explained in Sect. 4.1. The model implemented is compared with the base model [12]. In addition, a Bidirectional Convolutional LSTM architecture BCDU model [16] is also considered. The average values of the evaluation metrics obtained for the three approaches when applied on the Skeletal Muscle dataset are given in Table 1. The modified U-Net model performs better than the other two methods in accuracy, specificity, precision, recall, and Dice coefficient. The accuracy of skeletal datasets is 95.98% whereas that of Breast dataset is 95.18%, which is higher than that of other models. The accuracy can be increased further by acquiring more images for training and increasing the number of epochs. If the number of epochs is increased with lesser number of training images, overfitting will be the result. The performance metrics, viz, specificity, precision, recall and dice coefficient of modified U-Net have also higher values compared to the other two models. But the computational overhead of the modified method is higher than the base model. The BCDU model gives better performance than the base model of U-Net considered, at the cost of computation time.

Table 1. Average performance measures obtained by U-Net model and other selected models on skeletal muscle ultrasound dataset

Performance metric	Modified U-Net model	Classical U-Net model (Ronneberger et al., 2015)	BCDU model (Azad et al., 2019)
Accuracy	0.9598	0.9429	0.9469
Time (s)	23.569	7.865	34.746
Specificity	0.9641	0.9560	0.9594
Precision	0.9594	0.909	0.9178
Recall	0.9084	0.8868	0.8972
Dice coefficient	0.9270	0.8975	0.9062

A qualitative assessment of the performance of the three models is shown in Figs. 2 and 3 respectively for Skeletal Muscles and Breast Lesions images. The modified U-Net model extracts more regions similar to the ground truth image compared to the other models.

Table 2. Average performance measures obtained by U-Net model and other selected models on Breast Lesion Ultrasound dataset

Performance metric	U-Net model	Classical U-Net model (Ronneberger et al., 2015)	BCDU model (Azad et al., 2019)
Accuracy	0.9518	0.9442	0.9223
Time(s)	9.575	2.16	8.324
Specificity	0.9691	0.9685	0.9618
Precision	0.6331	0.5465	0.4059
Recall	0.5440	0.5115	0.5404
Dice Coefficient	0.5670	0.5276	0.5196

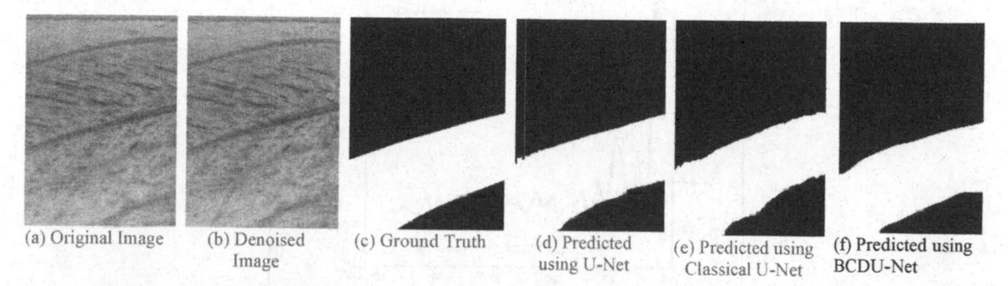

Fig. 2. Original Image, Denoised Image, Ground Truth and Predicted Outputs from U-Net and other selected methods using Skeletal Muscle Ultrasound image dataset

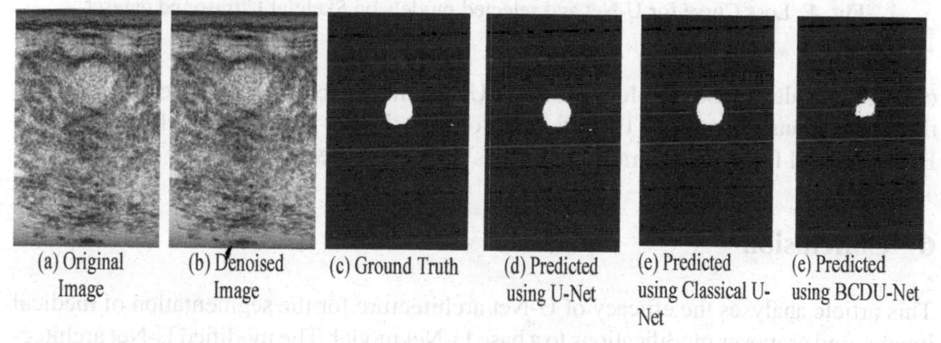

Fig. 3. Original Image, Denoised Image, Ground Truth and Predicted Outputs from U-Net and other selected methods using Breast Ultrasound image dataset

In Fig. 4, the training and validation loss during the model building for the Skeletal Ultrasound dataset is given. The loss curve gives a measure of the goodness of the model for accurate prediction. For all the models, the validation error approaches minimum with an increase in epochs.

From the quantitative and qualitative analysis of datasets using the U-Net model, it is observed that for both the datasets, modified U-Net model performs better than the

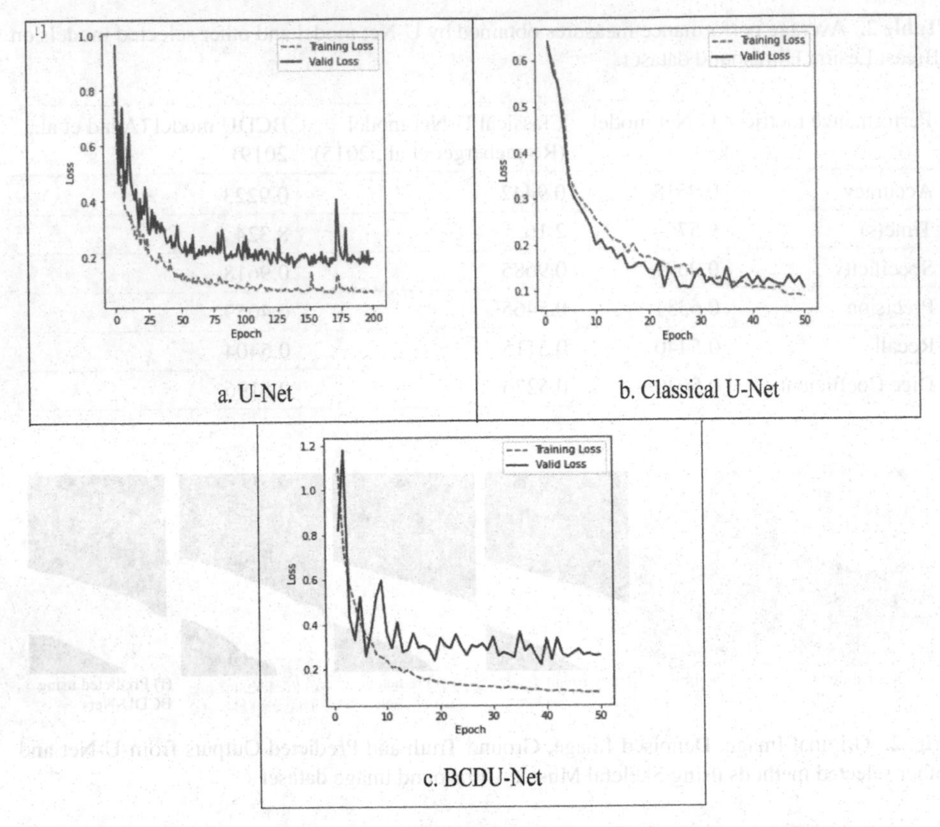

Fig. 4. Loss Curve for U-Net and selected models on Skeletal Ultrasound dataset

other two models. Also, it is found that the U-Net model achieves good performance for the Skeletal muscle dataset. The efficiency of Classical U-Net is higher than that of the BCDU model for Breast Ultrasound images but slightly less for Skeletal images.

6 Conclusion

This article analyses the efficacy of U-Net architecture for the segmentation of medical images and proposes modifications to a base U-Net model. The modified U-Net architecture is implemented and tested using two publicly available Ultrasound image datasets of Skeletal muscle fibers and Breast Lesions. The modifications introduced are the incorporation of a novel Pre-processing step and the use of the Tversky loss function. For the pre-processing, the NLM Filter optimized using the Grey Wolf Optimization method is adopted. The parameters/hyperparameters required for the model are selected through experimentation. Data augmentation is incorporated to increase the robustness of the model.

The performance of the proposal is compared with the base model as well as BCDU, a recent deep learning architecture. The proposed method is efficient in segmenting the Skeletal muscle images, based on the evaluation metrics used, and outperforms the other two models. For the breast image dataset, the performance of all three methods is inferior. The study brings out the significance of the U-Net model in medical image segmentation and the possibilities of improving the model for effective segmentation of different categories of medical images.

References

1. Lee, W.L.: An ensemble-based data fusion approach for characterizing ultrasonic liver tissue. Applied Soft Computing **13**(8), 3683–3692 (2013). https://doi.org/10.1016/j.asoc.2013.03.009
2. Kaltenbach, T.E.M., et al.: Prevalence of benign focal liver lesions: ultrasound investigation of 45,319 hospital patients. Abdominal Radiology **41**(1), 25–32 (2016). https://doi.org/10.1007/s00261-015-0605-7
3. Schindelin, J., Rueden, C.T., Hiner, M.C., Eliceiri, K.W.: The ImageJ ecosystem: an open platform for biomedical image analysis. Molecular Reproduction and Development **82**(7–8), 518–529 (2015). https://doi.org/10.1002/mrd.22489
4. Yuan, J., Wang, J.: Active contour based on local statistic information and an attractive force for ultrasound image segmentation. In: Proceedings of the 2017 2nd International Conference on Modelling, Simulation and Applied Mathematics (MSAM2017). Published (2017). https://doi.org/10.2991/msam-17.2017.23
5. Kumar, S.N., Lenin Fred, A., Muthukumar, S., Ajay Kumar, H., Sebastin Varghese, P.: A voyage on medical image segmentation algorithms. Biomedical Research (2018). https://doi.org/10.4066/biomedicalresearch.29-16-1785
6. Chen, Y., Jiang, H., Li, C., Jia, X., Ghamisi, P.: Deep feature extraction and classification of hyperspectral images based on convolutional neural networks. IEEE Trans. Geoscience and Remote Sensing **54**(10), 6232–6251 (2016). https://doi.org/10.1109/tgrs.2016.2584107
7. Liu, S., et al.: Deep learning in medical ultrasound analysis: a review. Engineering **5**(2), 261–275 (2019). https://doi.org/10.1016/j.eng.2018.11.020
8. Li, Q., Cai, W., Wang, X., Zhou, Y., Feng, D.D., Chen, M.: Medical image classification with convolutional neural network. In: 2014 13th International Conference on Control Automation Robotics & Vision (ICARCV) (2014). https://doi.org/10.1109/icarcv.2014.7064414
9. Lecun, Y., Bottou, L., Bengio, Y., Haffner, P.: Gradient-based learning applied to document recognition. Proceedings of the IEEE **86**(11), 2278–2324 (1998). https://doi.org/10.1109/5.726791
10. Krizhevsky, A., Sutskever, I., Hinton, G.E.: ImageNet classification with deep convolutional neural networks. In: Proceedings of NIPS, pp. 1106–1114 (2012). papers.nips.cc/paper/4824-imagenet-classification-with-deep-convolutionalneural-networks.pdf.
11. Cao, Z., Duan, L., Yang, G., Yue, T., Chen, Q.: An experimental study on breast lesion detection and classification from ultrasound images using deep learning architectures. BMC Medical Imaging, **19**(1) (2019). https://doi.org/10.1186/s12880-019-0349-x
12. Ronneberger, O., Fischer, P., Brox, T.: U-Net: convolutional networks for biomedical image segmentation. Lecture Notes in Computer Science, 234–241 (2015b). https://doi.org/10.1007/978-3-319-24574-4_28
13. Litjens, G., et al.: A survey on deep learning in medical image analysis. Medical Image Analysis **42**, 60–88 (2017). https://doi.org/10.1016/j.media.2017.07.005
14. Dong, H., Yang, G., Liu, F., Mo, Y., Guo, Y.: Automatic brain tumor detection and segmentation using U-Net based fully convolutional networks. Communications in Computer and Information Science, MIUA 2017 **723**, 506–517 (2017). https://doi.org/10.1007/978-3-319-60964-5_

15. Almajalid, R., Shan, J., Du, Y., Zhang, M.: Development of a deep-learning-based method for breast ultrasound image segmentation. In: 2018 17th IEEE International Conference on Machine Learning and Applications (ICMLA), pp. 1103–1108 (2018). https://doi.org/10.1109/icmla.2018.00179
16. Azad, R., Asadi-Aghbolaghi, M., Fathy, M., Escalera, S.: Bi-directional ConvLSTM U-Net with densley connected convolutions. In: 2019 IEEE/CVF International Conference on Computer Vision Workshop (ICCVW). Published (2019). https://doi.org/10.1109/iccvw.2019.00052
17. Shi, T., Jiang, H., Zheng, B.: A stacked generalization U-shape network based on zoom strategy and its application in biomedical image segmentation. Comp. Methods and Programs in Biomedicine **197**, 105678 (2020). https://doi.org/10.1016/j.cmpb.2020.105678
18. Du, G., Cao, X., Liang, J., Chen, X., Zhan, Y.: Medical image segmentation based on U-Net: a review. J. Imaging Sci. Technol **64**(2), 1–12 (2020). 20508–1. https://doi.org/10.2352/j.imagingsci.technol.2020.64.2.020508
19. Khoong, W.H.: BUSU-Net: An Ensemble U-Net Framework for Medical Image Segmentation, Image and Video Processing (2020). e-print: 2003.01581, arXiv:2003.01581
20. Ardhianto, P., et al.: A review of the challenges in deep learning for skeletal and smooth muscle ultrasound images. Applied Sciences **11**(9), 4021 (2021). https://doi.org/10.3390/app11094021
21. Shereena, V.B., Raju, G.: Modified non-local means model for speckle noise reduction in ultrasound images. In: Proceedings of 2nd Congress on Intelligent Systems, CIS 2021 (2021)
22. Cunningham, R., Sánchez, M., May, G., Loram, I.: Estimating full regional skeletal muscle fibre orientation from B-Mode ultrasound images using convolutional, residual, and deconvolutional neural networks. J. Imaging **4**(2), 29 (2018). https://doi.org/10.3390/jimaging4020029
23. Piotrzkowska-Wróblewska, H., Dobruch-Sobczak, K., Byra, M., Nowicki, A.: Open access database of raw ultrasonic signals acquired from malignant and benign breast lesions. Medical Physics **44**(11), 6105–6109 (2017). https://doi.org/10.1002/mp.12538
24. Kingma, D.P., Ba, J.: Adam: A Method for Stochastic Optimization. CoRR, abs/1412.6980. In: Proceedings of the International Conference on Learning Representations (ICLR) (2015). https://arxiv.org/abs/1412.6980
25. Wang, P., Chung, A.C.S.: Focal dice loss and image dilation for brain tumor segmentation. Deep Learning in Medical Image Analysis and Multimodal Learning for Clinical Decision Support, Lecture Notes in Computer Science **11045**, 119–127 (2018). https://doi.org/10.1007/978-3-030-00889-5_14
26. Abraham, N., Khan, N.M.: A novel focal tversky loss function with improved attention U-Net for Lesion segmentation. In: 2019 IEEE 16th International Symposium on Biomedical Imaging (ISBI 2019), pp. 683–687 (2019). https://doi.org/10.1109/isbi.2019.8759329
27. Refaeilzadeh, P., Tang, L., Liu, H.: Cross-Validation. Encyclopedia of Database Systems, 532–538 (2009). https://doi.org/10.1007/978-0-387-39940-9_565
28. Taha, A.A., Hanbury, A.: Metrics for evaluating 3D medical image segmentation: analysis, selection, and tool. BMC Medical Imaging **15**(1), (2015). https://doi.org/10.1186/s12880-015-0068-x
29. Kumar, S.N., Lenin Fred, A., Ajay Kumar, H., Sebastin Varghese, P.: Performance metric evaluation of segmentation algorithms for gold standard medical images. Advances in Intelligent Systems and Computing, 457–469 (2018b). https://doi.org/10.1007/978-981-10-8633-5_45
30. Sokolova, M., Japkowicz, N., Szpakowicz, S.: Beyond accuracy, F-Score and ROC: a family of discriminant measures for performance evaluation. Lecture Notes in Computer Science, 1015–1021 (2006). https://doi.org/10.1007/11941439_114

Rule Based Medicine Recommendation for Skin Diseases Using Ontology with Semantic Information

S. Subbulakshmi(✉), S. Sri Hari, and Devajith jyothi

Department of Computer Science and Applications, Amrita Vishwa Vidyapeetham, Amritapuri, India

subbulakshmis@am.amrita.edu

Abstract. Generally, skin diseases are taken seriously only after it gets aggravated. Many patients do not feel comfortable to consult physician for their skin problems and try to cure with home remedies. Identifying good dermatologist is more challenging. Disease gets cured with ease, if treated properly otherwise it becomes complicated. To handle this, knowledge-based decision support system is developed which provide recommendation for treatment. System realizes patient's symptoms with location, cause and recommends appropriate medicines. Ontology is used to elaborate data concepts and their relationships in skin disease, which permits sharing and reuse of domain knowledge. Semantics described using Web Ontology Language (OWL) with vocabularies, resources, logic's, and inference rules are queried for relevant information through SPARQL. System proved to render most valuable service to skin patients with ease and accuracy of its recommendation is high. It renders best treatment in COVID situation where people's movement is restricted to great extent.

Keywords: Knowledge information · Semantic web · Ontology · Rule-based system · Inference rules · Medicine recommendation

1 Introduction

Overall health of a person is directly proportional to cumulative functionalities of different organs in human body. Nature of skin reflects health condition of a person. Skin protects inner organisms from various environmental aspects like heat, cold, strong wind etc. It has direct contact with all real-world entities and thus, avoids dehydration and protects from dangerous microorganism. It protects from infection, injury, it provides sensation, balances fluid, regulates temperature of body and produces necessary vitamins. With certain variants of coronavirus, infected patients are at high risk of having skin problems that damage mucous membrane, which leads to chronic and acute dermatitis. Recently discovered infections namely, black fungus and white fungus are the main cause of deadly skin disease related to COVID infections. Fungus attacks patients with low immunity and patients suffering from cancer, diabetes, critical illness, and

M. Singh et al. (Eds.): ICACDS 2022, CCIS 1613, pp. 373–387, 2022.
https://doi.org/10.1007/978-3-031-12638-3_31

excess blood sugar. White fungus kicks off from tongue or private parts and from there spreads to another organs.

Acquiring proper skin treatment in this pandemic period is very difficult due to reasons that people are scared to go to hospitals, doctors also treat only patients with emergency needs, after confirming that they are not infected by virus. There is need to create an application to provide treatment without physical examination of patients, by going through some data like readings or images describing skin disease. Based on the data provided, doctors may prescribe medicines [1] after having minimal online interactions with patients. It ultimately ensures safety of both patients and doctors, as their exposure to virus is restricted.

Research renders medicine recommendation to patients when relevant information about disease is provided. Recommendation system helps in decision-making [2] based on the nature of disease. Patients provide details of skin problem like location and discomfort of skin like itching, dryness, cracks etc., which is categorized as symptoms. Location plays a vital role in skin disease as it is related to area of infection. Same symptoms and issues at different locations are categorized as different disease whose treatment process is also different. Having these aspects in mind, available medicines are also categorized for specific disease coupled with location details.

Main objective of this system is to find skin disease name, cause of disease, and appropriate medicine for that diseases. Medicine recommendation system uses [3] evidence-based knowledge retrieval process, which uses information stored in ontology which is extracted from different web resource. It aids [4] in decision-making process of selecting appropriate drugs for specific disease.

To populate knowledge-based information, system builds ontology with relationship between vocabularies in this domain relating to disease, symptoms, location, and medicines. It defines association rules [5] between entities and individuals. Rules and axioms defined in domain specific ontology is used to recommend medicines for a given disease. Then, mapping of ontology is implemented to interconnect and build relationship between one or more ontology which helps to retrieve linked information from ontology, based on requirements [6].

Basic task of ontology creation is done with Protege tool. It is a free, open-source ontology editor and knowledge-based framework based on Java concepts. Protege-Frame and Protege-OWL editors are used [7] for modeling ontologies. After creating ontologies, reasoners are used to identify conflicts in defined relationships, rules, and axioms. Identified conflicts are resolved by making required corrections in ontology definition. Once ontology creation is complete, SPARQL is used for retrieving knowledge information. Implementation of ontology-based system with domain specific details about skin disease, renders valuable knowledge-based information to both patients and medical practitioners as per their requirements with ease.

This paper is arranged as follows Sect. 2 related works explains summary of existing work in fields of knowledge-based system, Sect. 3 system architecture describes flow of the proposed work, Sect. 4 Knowledge Representation Techniques, illustrates methodology adopted to create this system. Section 5 Recommendation using SPARQL Query elaborates implementation of ontology for skin disease and it also explains execution of queries to render medicine recommendation. Section 6 conclusion and future work, summaries the work with future enhancements.

2 Related Work

Many researchers concentrated on advancement of ontologies to populate knowledge information associated with health care domain, Rich set of ontologies enables to create intelligent systems with machine independent operations. Alharb et.al. [8] have developed ontology for medicines comprising symptoms, diseases, and treatment related to specific disease. It is used for detection and treatment of diabetes. It uses patient information, symptoms with signs, risk factors, lab test results as input data and recommends a treatment plan defined by Clinical Practice Guidelines (CPG). It includes modelling and implementation of web ontology with resources in diabetes domain, for detection and treatments of diabetes in an early stage.

Ontology is used to represent relationships among domain concepts with description rules for sharing knowledge information among applications. Knowledge-based system [9] have developed ontology with the principles of Semantic Web Rule Language (SWRL) following OWL-DL standard. The system is executed with JESS inference engine to share details of disease and pharmacy ontology in order to recommend apt medicines for symptoms and issues related to heart diseases. Ontology holds information about varied types of heart disease with relevant mapping to symptoms, causes, etc. Similarly, pharmacy ontology holds medicines in two domains allopathy and ayurveda which is populated with medicines and diseases to which it could be used. Recommendation of medicines is implemented based on the inference rules defined in the ontology.

CDSS - Clinical decision support system [10] is a framework that supports ontological modelling with contents and rules which enables to recommend right information based on patient data. It uses subsumption graph representation of patient profiles to find decision solutions by combining historical patient's data from multiple sources. Results are provided in form of clinical case, like number of recommendations issued for a patient. It provides log details of patient's profile, which is used to understand and analyses disease model for making further recommendations. Nahhale et al. [11] created ontology for medical domain especially related to glaucoma disease. which is machine understandable. Ontology created includes a rich set of vocabulary as basic concepts and relations between them. They have created a reusable glaucoma disease knowledge base which could be used in machine independent intelligent systems.

Ontology-based clinical decision support system [12] mainly focused in creating semantic based data store of diabetic patients. Diabetic patient ontology, which is the core of system, analyses, examines, and suggests treatment procedure. Semantic profile encapsulated with patient details are used to successfully suggest appropriate solution which could be further used for similar patients. Tools are used to add patient details from existing medical records. The paper renders solutions only for diabetic patients. Ontology-based drug recommendations discovery [13] describes medical terminology and information as ontological expression and it combines them with clinical expertise of medical practitioners. Here, ontology creation is done using Galen-OWL approach and traditional business logic rule engine called Galen-Drools. Extraction of medical knowledge using queries helps to create modular ontology. Semantic web and Galen OWL deliver excellent match for medicine recommendation.

Mobile based disease ontology [14] focused to create insights for the vocabularies in traditional Persian medicine which is mapped to relevant English terminologies. Concepts related to 24 categories of diseases handled in Persian medical community are included in ontologies and each disease is enriched with sustainable disease data in English. It holds a rich set of diseases data which could be reused by medical practitioners to have in depth study of traditional disease data and it could be used in creating clinical decision support systems.

Sherimon, P.C., [15] demonstrated use of two ontologies disease ontology (DOID) and symptom ontology (SYMP). It uses algorithm to align these ontologies and creates a combined ontology (DSO) which holds knowledge data of disease and its symptoms. Ontology is created only for few sets of diseases and it must be extended to large set of diseases to have a complete dataset.

Most of research works focused on creation of knowledge information about disease and symptoms. It provides the base foundation to know about disease and its characteristics. Our research work aims to create knowledge-based clinical support system to recommend medicines based on the symptoms provided by the users. The system is developed by creating and mapping domain details of symptoms to disease, disease to medicine, and medicine to symptoms. Various combinations of propositional logic's are developed and derived using ontology descriptions to provide clinical assistance services with minimal information shared by the patients. When user provides information about his or her medical issues, a machine-independent system retrieves necessary data from the rich set of knowledge information and provides more appropriate medicine recommendation. The focus of this work is to create personalized clinical assistance service for skin related diseases.

3 System Architecture

Proposed system is designed to enable users for retrieval of domain specific knowledge information [16] related to skin disease without much difficulty. Users of the system retrieves relevant information by providing details of the health issue. The system helps user to identify cause for the disease and to get solution in-terms of medicine recommendation. Users comprises of patients or doctors. Patients use the system by directly specifying his ailments, and system recommend medicines, which is approved by doctors. Doctors use the system to acquire knowledge about disease, symptoms, causes and medicines available in the market. The knowledge gathered could be useful for him to provide best treatment for skin diseases. Thus, the system acts as an intelligent based clinical assistance.

Figure 1 depicts input given by users as symptoms and location, which is selected from list of symptoms and location retrieved from the ontology. Relationship between symptoms and location shown helps to identify name of disease. Above mentioned relationships are included in skin disease ontology. Two-way relationship exists between medicine and skin disease. Medicines are used to cure the disease and disease can be cured with medicines. First one is used to know about existing medicines and second one used for treating patients with specific disease. Every disease is due a reason or infection which is depicted with relationship between disease and causes in ontology.

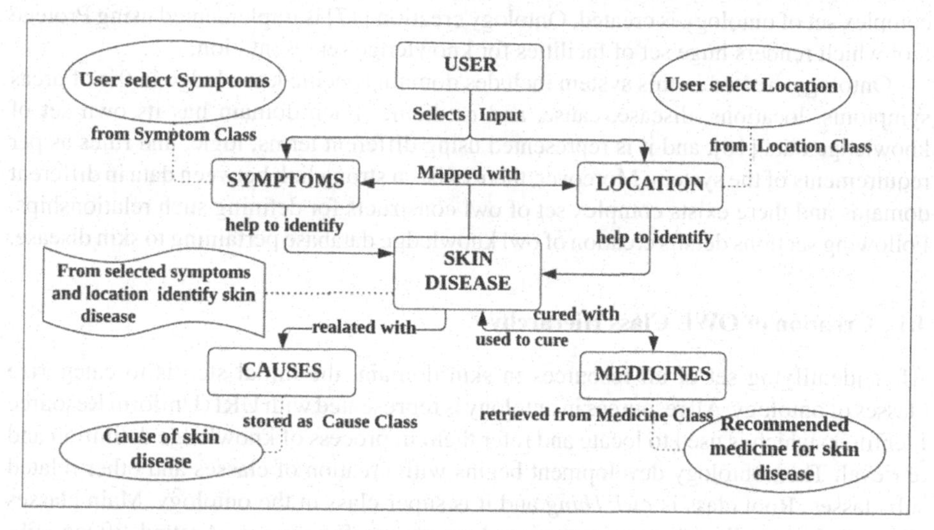

Fig. 1. System architecture

Generally, treatment for skin disease is given by practitioners after examining exact spot of skin and enquiring details about physical ailments experienced by patient. Since, our system eliminates direct contact of patient and practitioner, here user must select symptoms and location for their skin problem. To enhance patient to provide this information with ease, list of symptoms stored in repository is displayed, then based on symptom selected, location which are mapped with that symptoms are listed. On selection of location, exact name of skin disease is identified with the help of inference rules defined for semantic information in ontology of disease and symptoms.

As knowledge information of different disease is mapped with its related symptoms, causes and location, the system can make decision about nature of patient's disease in detail without any human interaction. On identification of disease and its cause, next step of finding appropriate treatment is initiated. Here comes the use of knowledge data of medicines which are linked with one to more diseases. Retrieval of medicines is executed based on inference rules and axioms defined to relate medicines and disease. Thus, most efficient medicines for given disease is realized by system with minimum effort of user and it is recommended for approval by medical practitioners. This will help patients to get proper treatment with minimal effort.

4 Knowledge Representation Techniques

The project aim is to retrieve knowledge from the system to be used by the patients affected with skin disease. Research requires a valid design to completely understand the procedure of creating knowledge repository with semantic information. Due to lack of existing ontology for skin disease, we create ontologies that are fabricated from numerous medical experts, books and websites. After identifying complete set of resource details in the form of classes, predicates, literals and other related set of rules and relationships, complex set of ontology is created. Ontology creation [17] is implemented using Protege tool which renders huge set of facilities for knowledge representation.

Ontology created in this system includes domain specific knowledge in the of areas symptoms, locations, disease, cause, and medicine. Each domain has its own set of knowledge data [18], and it is represented using different terms, logic, and rules as per requirements of the system. Moreover, there exists a strong link between data in different domains and there exists complex set of owl constructs for defining such relationships. Following sections details creation of owl knowledge database pertaining to skin disease.

4.1 Creation of OWL Class Hierarchy

After identifying set of all resources in skin domain, the initial step is to categorize classes of ontology. All resources in ontology is represented with URI (Uniform Resource Identifier) which is used to locate and refer them in process of knowledge definition and retrieval. Then ontology development begins with creation of classes and other related sub classes. Root class is *owl:Thing* and it is super class in the ontology. Main classes of ontology are Skin disease, Location, Causes and Symptoms. Around fifteen sub-classes are created for the class *skindisease* as shown Fig. 2. Similarly, class hierarchy is maintained for *symptoms, location, causes* and *medicines.*

Different aspects of knowledge details for classes are defined with OWL constructs *EquivalentTo, SubClassOf, Instance, DisjointWith, unionOf, intersectionOf, complemenfOf* and *restriction.* Restriction are used to impose condition that should be satisfied to form class with defined logic. Classes in proposed ontology are created with the aim of recommending medicines for skin disease.

4.2 Linking of Classes with Predicates

On defining resources as class hierarchy with proper logics, properties to classes and values are defined using predicates either as *object properties* or *data properties*. Object properties is used to define predicate to link one class or resource with another class, which results in linking of web data [19]. Linked data is the core aspect of system which enables to relate complex set of knowledge information of all domains with ease.

Predicates for classes is enriched with characteristics like *inverse property, symmetric property, functional property, transitive property* as per nature of properties related to domains symptoms, location, causes and disease. Moreover, restriction for property like *cardinality restriction, value restrictions* are used to define number of relations between classes and type of value that can be given.

4.3 Predicates as Object Property

Main purpose of object property is to link resources, to create a connected graph of information which enables machines to understand and retrieve knowledge information independently. Object property is a predicate which is defined as resource in semantic web and it includes two aspects domain and range. Property is related to one or more defined classes with OWL construct domain. Set of values that a property can be given is defined using OWL construct range. So, properties with *domain* and *range* describes how classes are related to each other instance, it connects an individual with another individual. Sample set of object property of the system shown Table 1. Similarly, object properties are defined to link all classes in the domain.

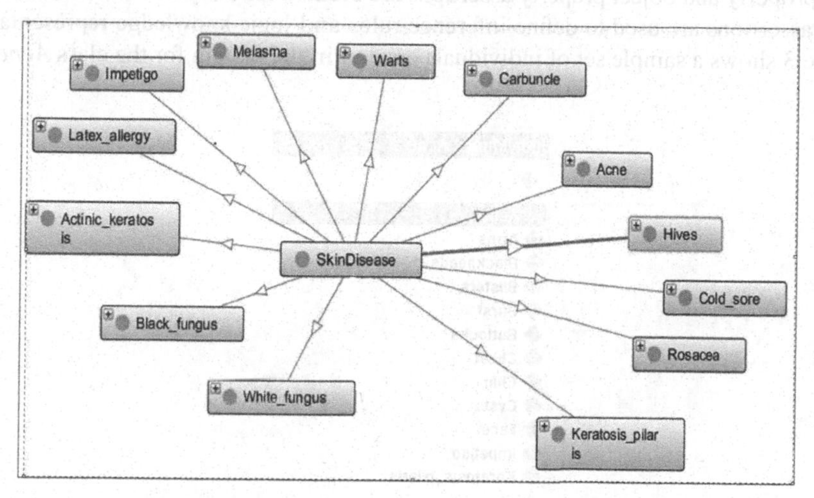

Fig. 2. Onto-graph of skin disease and its subclasses

Table 1. Sample object property with domain and range

Object property	Domain	Range
hasSymptoms	skin_disease	symptoms
haslocated	skin_disease	location
usedtoCure	medicine	Skin_disease

4.4 Predicate as Data Property

Certain predicates take literal values of any predefined datatype are defined using *dataProperty*. For which domain is Class and range being Datatype. It supports all xml schema datatypes. Here, data property is created as per requirements with value restrictions and sample set is shown below which specifies the value used for property *curedBy* which takes values of string data type.

- curedBy "Retin-A (tretinoin)" ^^xsd:String.
- curedBy "Differin (adapalene)' ^^xsd:String.

– curedBy "famciclovir (Famvir)"^^xsd:String.
– curedBy "acyclovir (Zovirax)"^^xsd:String.

4.5 Instances of Classes as Individuals

Class instances with respect to skin disease domains are created using owl: *Individuals*. Objects are referred as Individuals and they can be likely a member of more than one class. Data property and object property assertions are created for instances of the ontologies. Such assertions are used to define inference rules and logic knowledge representation. Figure 3 shows a sample set of individuals created in the system for the class *Acne*.

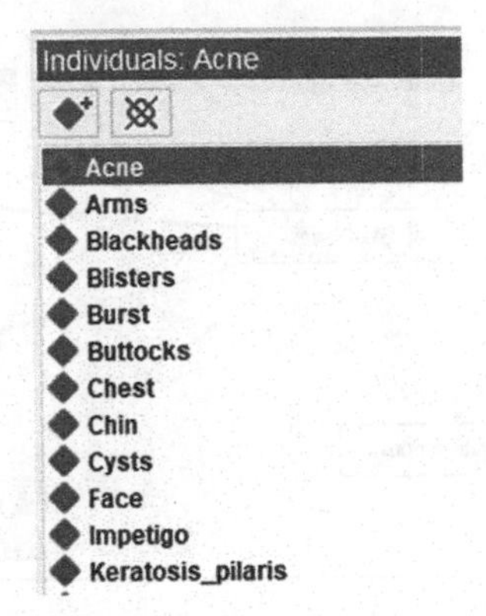

Fig. 3. Sample set of individual from Protégé

4.6 Linking Classes with Object Properties Using Restrictions

Owl construct *Restrictions* like *someValues, allValues*, *hasValues* are used to filter the set of values or objects for an object Property. Figures 4 and 5 show how *acne* and *cold_sore* is linked with its object properties with classes of *symptoms* and *location* using some restriction. All possible knowledge information relevant to domain of interest [20] gathered from different sources are bundled in web ontology with classes, relations, predicates, and individuals with rich set of restrictions and characteristics with the help of Protégé. Our system ontology is complete with details of skin disease, by including set of skin diseases as classes.

Individuals are created for those skin diseases to represent different variations of a disease. Each individual is linked with its symptoms, causes location and medicines. After populating ontology with contents and rules, HermiT OWL Reasoner is used to check for conflicts and errors. Conflicts reported by the reasoners are rectified by checking the definition of rules with respect to classes and properties. Then, ontology information is used for recommending appropriate medicines to specific skin problem mentioned by patient on executing proper queries written in SPARQL Protocol a Resource Description Framework Query Language.

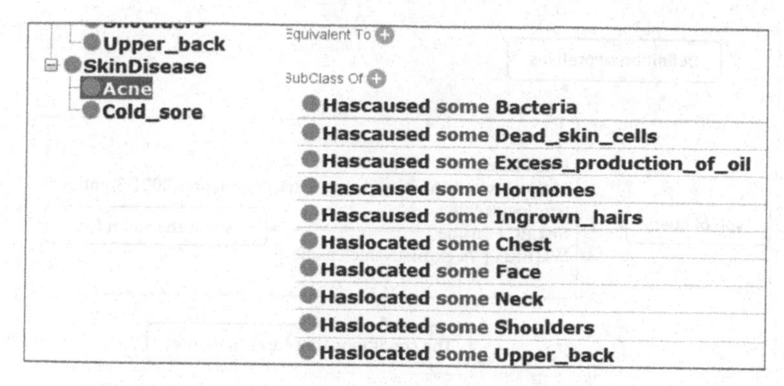

Fig. 4. Use of restrictions with object properties for class Acne.

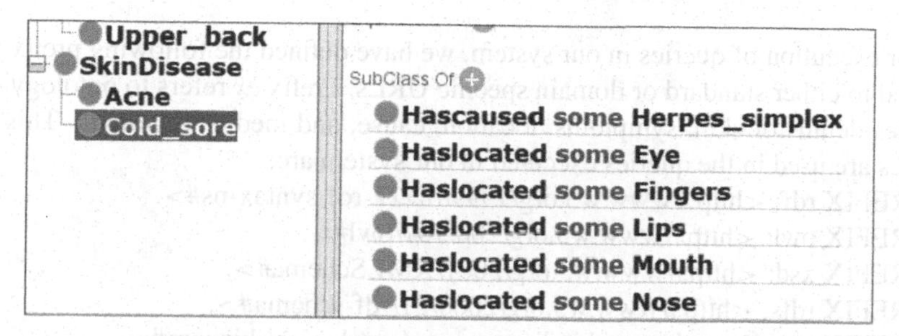

Fig. 5. Use of restrictions with object properties for class Cold_sore

5 Recommendation Using SPARQL Query

SPARQL is a semantic web query language which uses RDF data model in triples format - subject, predicate, object to retrieve data from ontology. It helps querying unknown relationships, providing valuable data which are extracted through inference rules from both structured and semi-structured data defined in ontology with relevant domain information. Structure of SPARQL query is shown in Fig. 6.

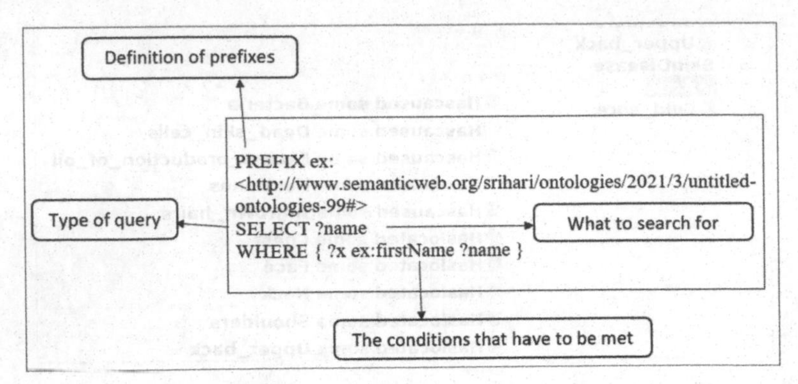

Fig. 6. Structure of SPARQL query

For execution of queries in our system, we have defined the following prefix which mapped to either standard or domain specific URLs. Prefix *ex* refers to ontology which includes details of skin, symptoms, location, cause, and medicine domain. This set of prefixes are used in the queries executed in our system are:

PREFIX rdf: <http://www.w3.org/1999/02/22-rdf-syntax-ns#>.
PREFIX owl: <http://www.w3.org/2002/07/owl#>.
PREFIX xsd: <http://www.w3.org/2001/XMLSchema#>.
PREFIX rdfs: <http://www.w3.org/2000/01/rdf-schema#>.
PREFIX ex: <http://www.skindisease.com/ontologies/skincare#>.

5.1 SPARQL Query for Displaying Symptoms

As the initial step of medicine recommendation, first we have to list, symptoms details so that user could select appropriate one from the given listing by using the predicate *ex:hassymptoms*. Following query used to retrieve symptoms from ontology which generates output as shown in Fig. 7.

SELECT DISTINCT? symptoms
WHERE { ?z ex:hassymptoms ?symptoms.}

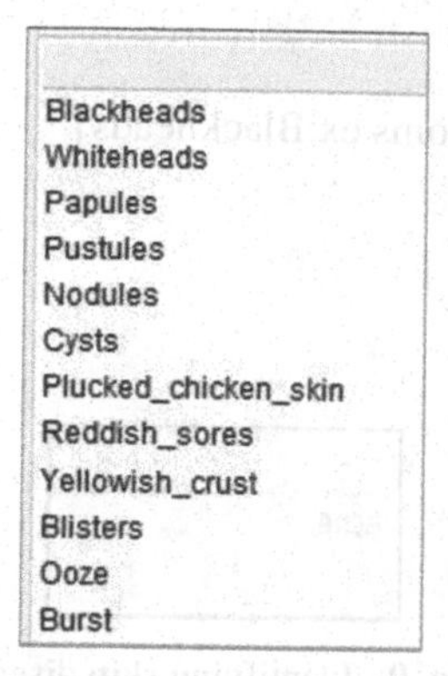

Fig. 7. Partial list of symptoms

5.2 SPARQL Query for Displaying Location

Location plays an important role in identifying exact skin disease, so after selecting symptoms, location must be mentioned. For this purpose, possible location of skin disease is listed by executing the given query with the predicate *ex:haslocated.*

```
SELECT DISTINCT ?location
WHERE { ?y ex:haslocated ?location.}
```

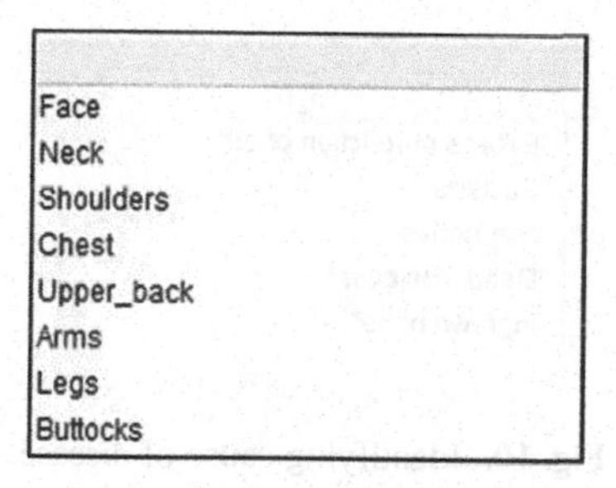

Fig. 8. Partial list of locations

It allows to select an option from the list generated as query output shown in Fig. 8.

5.3 SPARQL Query for Finding Skin Disease

On selection of symptoms and location details, system will try to find disease, so that further treatment process could be recommended. Query to identify distinct disease, uses union operation in where clause for predicates *ex:hassymptoms* and *ex:haslocated*. Sample query is executed when user have selected *Blackheads* as symptoms and *Neck* as location and its output is shown in Fig. 9.

```
SELECT DISTINCT *
WHERE { { ?z ex:hassymptoms ex:Blackheads}
UNION
{ ?z ex:haslocated ex:Neck} }
```

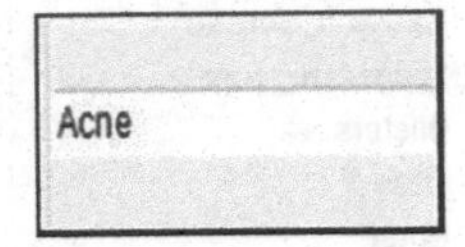

Fig. 9. Identifying skin disease

5.4 SPARQL Query for Finding Skin Disease

Causes for disease should be identified to give most appropriate treatments. Only based on that medicines are recommended for that disease. Following is sample query which list causes of disease *Acne* and its output is shown in Fig. 10.

```
SELECT DISTINCT (str(?cause) as ?skin

WHERE { ex:Acne ex:Caused ?x}.
```

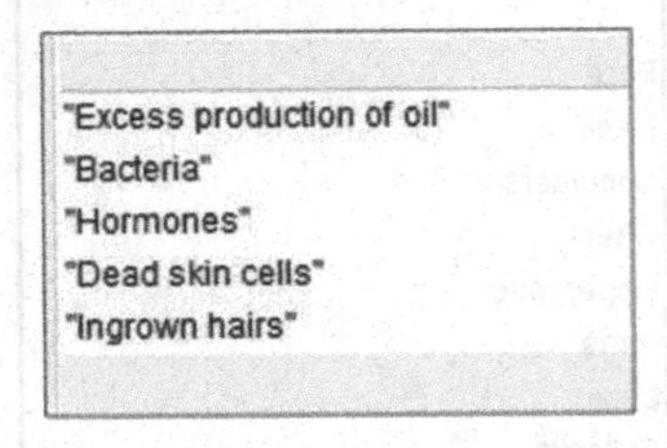

Fig. 10. Identifying cause of disease.

5.5 SPARQL Query to Find Medicine for the Skin Disease

After identifying disease based on relevant details selected by user, system retrieves treatment details as medicine, which is mapped to different diseases in ontology. Following query shows list of medicines that could be given for disease *Acne* which uses predicate *ex:TreatmentMedicine*. Result of query is shown in Fig. 11.

```
SELECT DISTINCT (str(?treatmedicine) as ?skin
WHERE { ex:Acne ex:TreatmentMedicine ?treatmedicine}
```

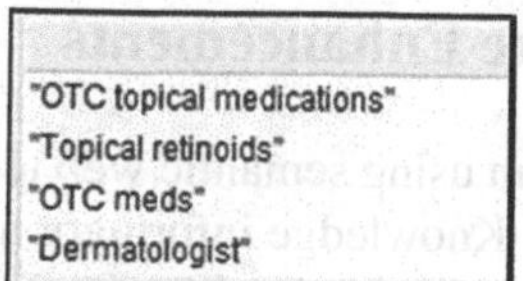

"OTC topical medications"
"Topical retinoids"
"OTC meds"
"Dermatologist"

Fig. 11. Recommended medicine list for specific disease.

Thus, recommendation system could assist practitioners to collect essential details like symptoms, location of the skin disease without meeting them in person, get suggestions from the system in form of medicines to be recommended using rule based semantic web technologies [21]. Finally, practitioners can approve the most appropriate medicines based on users age and other medical history. The system automatically tries to gather essential information like causes and diseases before retrieving medicines details for problems specified by the user.

Query execution is interfaced with java application which list essential details to user retrieved from SPARQL queries. On selection of such details, further queries are executed dynamically to suggest list of medicines to practitioners, which is recommended to patients on his approval.

Application is tested for different types of skin diseases related to acne and cold sores by specifying varied set of inputs. Input data is collected from WebMD website which provides the type of skin diseases with its subtypes and major causes and symptoms. We consulted local dermatologist for acquiring details of medicines and treatment for those skin diseases. Input are fed into the system in the form of parameters for execution of SPARQL queries. Query results are tested for the validity of the system with real time enquiry provided by dermatologist for treating skin diseases. Accuracy of system is assessed based on the predicted recommendation for the given query with disease details, against the actual medicines used for treatment. Query results with the set of all recommendations predicted during the implementation phases are categorized as correct, partial, wrong and omitted recommendations. Precision and recall are calculated for those recommendation results using the following formulas:

$$Precision = \frac{Correct + (0.5 * PartialCorrect)}{Correct + Wrong + PartialCorrect}. \quad (1)$$

$$Recall = \frac{Correct + (0.5 * PartialCorrect)}{Correct + Omitted + PartialCorrect}. \quad (2)$$

$$F1_{score} = 2 * \frac{Precision * Recall}{Precision + Recall}. \quad (3)$$

Around 100 possible queries are executed, and the results showed precision and recall values as 0.8 and 0.9 respectively. F1_score for the same is 0.84. Results shows a moderately good recommendation and further this score could be improved by populating rich set of knowledge information in the ontology.

6 Conclusion and Future Enhancements

Efficient decision support system using semantic web techniques is designed to retrieve knowledge from skin ontology. Knowledge information related to symptoms, location, cause, disease, and medicines are maintained in ontology which is coupled with relevant logic and rules to incorporate complex knowledge representation. The system helps patient suffering from skin disease to get treatment details without meeting medical practitioners in person, which is most promising feature during this COVID pandemic situation. It retrieves appropriate medicine list by giving patient symptoms and location. Symptoms will help in identifying type of skin disease and cause for that disease. Based on disease and cause, system retrieves list of medicines from ontology with logic, axioms and inference rules defined. Most appropriate medicines from the above list is approved by medical practitioners and those medicines are recommended to patient for treating his/her disease. In future, domain information could be gathered from linked web resources which will reflect most accurate, dynamic and updated knowledge base.

References

1. Drenovska, K., et al.: Covid-19 pandemic and the skin. Int. J. Dermatol. **59**(11), 1312–1319 (2020). https://doi.org/10.1111/ijd.15189
2. Alkhatib, B., Briman, D.: Building a herbal medicine ontology aligned with symptoms and diseases ontologies. J. Digit. Inf. Manag. **16**, 114 (2018). https://doi.org/10.6025/jdim/2018/16/3/114-126
3. Subbulakshmi, S., Krishnan, A., Sreereshmi, R.: Contextual aware dynamic healthcare service composition based on semantic web ontology. In: 2019 2nd International Conference on Intelligent Computing, Instrumentation and Control Technologies (ICICICT), pp. 1474–1479 (2019). https://doi.org/10.1109/ICICICT46008.2019.8993303
4. Muppavarapu, V., Gowtham, R., Gyrard, A., Noura, M.: Knowledge extraction using semantic similarity of concepts from Web of Things knowledge bases. Data Knowl. Eng. Elsevier. 2021 Aug 25:101923 (Impact Factor: 1.992, SCI Indexed) **135**, 101923 (2021)
5. Guefack, B., Lasbleiz, V.D., Bourde, J., Duvauferrier, R.A.: Creating an ontology driven rules base for an expert system for medical diagnosis. In: MIE, pp. 714–718 (2017)
6. Mohan, A.K., Venkataraman, D.: The forensic future of social media analysis using web ontology. In: International Conference on Advanced Computing and Communication Systems (ICACCS-2017) (2017)
7. Ahmad, A., Bandara, M., Fahmideh, M., Proper, H.A., Guizzardi, G., Soar, J.: An overview of ontologies and tool support for COVID-19 analytics. In: 2021 IEEE 25th International Enterprise Distributed Object Computing Workshop (EDOCW), pp. 1–8 (2021). https://doi.org/10.1109/EDOCW52865.2021.00026
8. Alharbi, R.B., Jawad Al-Masri, S.: Ontology based clinical decision support system for diabetes diagnostic (2015). https://doi.org/10.1109/SAI.2015.7237204
9. Subbulakshmi, S., Ramar, J.D., Hari, S.S.: Knowledge-based medicine recommendation using domain specific ontology. In: Karrupusamy, P., Balas, V.E., Shi, Y. (eds.) Sustainable Communication Networks and Application. LNDECT, vol. 93. Springer, Singapore (2022). https://doi.org/10.1007/978-981-16-6605-6_14
10. Galopin, A., Bouaud, J., Pereira, S., Seroussi, B.: An Ontology Based Clinical Decision Support System for the Management of Patients with Multiple Chronic Disorders. Studies in Health Technology and Informatics, vol. 216, pp. 275–279 (2015)

11. Nahhale, Y., Ammoun, H.: Construction of glaucoma disease ontology. In: Fakir, M., Baslam, M., El Ayachi, R. (eds.) CBI 2021. LNBIP, vol. 416, pp. 399–413. Springer, Cham (2021). https://doi.org/10.1007/978-3-030-76508-8_29
12. Sherimon, P.C., Krishnan, R.: OntoDiabetic: an ontology-based clinical decision support system for diabetic patients. Arab. J. Sci. Eng. **41**(3), 1145–1160 (2015). https://doi.org/10.1007/s13369-015-1959-4
13. Doulaverakis, C., Nikolaidis, G., Kleontas, A., Kompatsiaris, I.: GalenOWL: ontology-based drug recommendations discovery. J. Biomed. Semant. **3**, 14 (2012). https://doi.org/10.1186/2041-1480-3-14
14. Shojaee, H., Ayatollahi, H., Abdolahadi, A.: Developing a mobile based disease ontology for traditional Persian medicine. Inform. Med. Unlocked **20**, 100353 (2020). https://doi.org/10.1016/j.imu.2020.100353
15. Mohammed, O., Benlamri, R., Fong, S.: Building a diseases symptoms ontology for medical diagnosis: an integrative approach. In: International Conference on Future Generation Communication Technology (2012)
16. Venkataraman, D., Haritha, K.C.: Knowledge representation of university examination system ontology for semantic web. In: 4th International Conference on Advanced Computing and Communication Systems, ICACCS 2017 (2017)
17. Subbulakshmi, S., Ramar, K., Shaji, A., Prakash, P.: Web service recommendation based on semantic analysis of web service specification and enhanced collaborative filtering. In: Thampi, S.M., Mitra, S., Mukhopadhyay, J., Li, K.-C., James, A.P., Berretti, S. (eds.) ISTA 2017. AISC, vol. 683, pp. 54–65. Springer, Cham (2018). https://doi.org/10.1007/978-3-319-68385-0_5
18. Jaspers, M.W., Smeulers, M., Vermeulen, H., Peute, L.W.: Effects of clinical decision-support systems on practitioner performance and patient outcomes: a synthesis of high-quality systematic review findings. J. Am. Med. Inform. Assoc. **18**(3), 327–334 (2011)
19. Barton, A., Rosier, A., Burgun, A., Ethier, J.-F.: The cardiovascular disease ontology. Front. Artif. Intell. Appl. **267** (2014). https://doi.org/10.3233/978-1-61499-438-1-409
20. Fisher, H.M., Hoehndorf, R., Bazelato, B.S., et al.: DermO; an ontology for the description of dermatologic disease. J. Biomed. Semant. **7**, 38 (2016). https://doi.org/10.1186/s13326-016-0085-x
21. Schriml, L.M., Mitraka, E.: The disease ontology: fostering interoperability between biological and clinical human disease-related data. Mamm. Genome **26**(9–10), 584–589 (2015). https://doi.org/10.1007/s00335-015-9576-9

FRAMS: Facial Recognition Attendance Management System

Anagha Vaidya[1], Vipin Tyagi[2], and Sarika Sharma[1](✉)

[1] Symbiosis Institute of Computer Studies and Research, Symbiosis International (Deemed University), Model Colony, Pune 411016, India
sarika4@gmail.com

[2] Jaypee University of Engineering and Technology, Raghogarh-Guna 473226, Madhya Pradesh, India

Abstract. Student attendance monitoring is one of the most important activity in the education domain. In the traditional attendance system it is rollcall calling method which leads to human errors and chance of marking proxy attendance is very high. The teacher also spends a considerable amount of time on attendance task. This paper introduced new model Facial Recognition Attendance Management System (FRAMS) for marking students' attendance through facial recognition. The machine learning algorithms are developed and proposed by the authors. Two models are created with train images of each students. The Face Detection and Face Recognizer models are developed in this study by using open source software libraries on static images. The experiment results show 90% accuracy is recorded by the model for marking the student's attendance successfully. The model is comparing the images with different students and marking accordingly.

Keywords: Face recognition · RFID · Figure print · Learning Analytics · Educational data mining

1 Introduction

Student attendance monitoring system is integral part of any educational system. It is considered as one of the measurement criteria for students learning measurement in Learning analytics (LA) and educational data mining (EDA) process. The age-old attendance process of calling out a student's name and then marking their attendance on a sheet of paper is outdated, not to mention the high chances of giving out proxy or fake attendance. Currently there are multiple attendance management systems available for the universities and schools to use. Examples of such systems are:

1) Using RFID tags on Identity Cards.
2) Moodle Attendance Plugin.
3) Fingerprint Recognition
4) Barcode
5) Microsoft team

M. Singh et al. (Eds.): ICACDS 2022, CCIS 1613, pp. 388–398, 2022.
https://doi.org/10.1007/978-3-031-12638-3_32

However, these systems have certain drawbacks, owing to which the problem regarding fake attendance is not fully handled. The RFID tags, Barcode Scanner can be presented by one student for multiple students, giving them proxy attendance [1]. Learning management systems (LMS) such as Moodle attendance is basically the digitalization of the manual attendance process. Once student login then their attendance is marked, although the student may not attend the full session. Fingerprint Recognition is a good way to mark attendance since it requires the biometric of a student which cannot be faked easily. Nonetheless, the implementation of fingerprint recognition is tough. The scanner organization and completion of the process is time consuming; it is a costly solution.

Thus, in-order to overcome these drawbacks of existing attendance management systems, the study proposed solution of recording attendance of students based on facial recognition. When this system is implemented, during the lecture hours, the student to look at the camera for a few moments and get the image captured which will then be used to detect all the faces of the students and accordingly mark their attendance. Once it is recorded, the different analytical report can be generated from it.

The paper is arranged as follows: It starts with the introduction to the topic of the study followed by the review of literature of the related studies. The methodology part explains the method and the algorithm developed by the authors. The results obtained are discussed in the discussion section and the paper concludes with limitations and future research scope.

2 Literature Review

LA and EDA are two emerging fields in data analytics, specific to education domain. In LA the data is collected, analysis is done, and the result is displayed to the educational stakeholders for further decision making. The data is collected through different students' activities like student's exam, attendance etc. The student attendance data collection is one of the crucial activities as the chances of human error and proxy data collection are there.

In literature different techniques are proposed for the same [2]. Researchers [3] proposed system by combining face recognition and RFID technique. The basic pattern of face detection is nose, eyes, hair, ears, and sometimes it is based on skin tone. And then the different algorithm AdaBoost Algorithm for Face Detection, Viola Jones Face Detection Algorithm, SMQT Features and SNOW Classifier Method, Local Binary Pattern (LBP) are used for further analysis. All these algorithm has their own advantages and disadvantages [10]. Adaboost Algorithm for Face detection is an ensemble of learning algorithms. It takes a collection of classifiers called as weak learners or base learners. It combines them to produce a strong classifier. Adaboost's biggest advantage is that it has multi-class variant. The Viola-Jones face detector a cutting edge face detector [4]. It is based on the framework of Adaboost. The Viola-Jones Face Detecting Algorithm reported a detection speed of 0.07 s per frame of size ~ 300 × 300 pixels on a regular desktop in its original implementation. The deep learning approach is also used in recent research for face recognition [12,13].

In recent times, a number of algorithms for face recognition have been proposed [2]. Although most of these works deal with only a single image of the face at any given point of time, instead of continuously observing facial information. The problem of facial detection thus improving the accuracy of the recognized face. In recent times it is important to constantly observe the face rather than using a static image since the background subtraction and interframe subtraction process become easy and detection of the face becomes faster and more accurate.

[5] Proposed algorithm which provides small hint by using the predictions of which student generally sits on which bench and thus giving the recognition software a small hint as to who might be sitting there. They also make use of background subtraction and inter-frame subtraction to estimate if a student is sitting on a particular bench or not.

[6] Have presented their idea of Face recognition using algorithms Landmark Estimation and Convolution Neural Network. Another method is Local Binary Pattern (LBP) algorithm to record the student attendance. LBP is one among the various methods, and is popular as well as effective technique used for the image representation and classification [7]. Researchers proposed face recognition technique by using Eigenface values, Principle Component Analysis (PCA) and Convolutional Neural Network (CNN) [8].

Recently android based system are being prominently used. The QR code containing the complete course information can be generated and displayed at the front of classroom. The student only needed to capture his/her face image and displayed QR code using his/her smartphone. The image can then sent to server for attendance process [9].

[11] Presented a method where the face detection part was done by using viola-jones algorithm method while the face recognition part was carried on by using local binary pattern (LBP) method.

After going through relevant papers and research studies conducted by respected members of this community, the authors have developed a better understanding of facial recognition, the various algorithms used for it, and how to implement the same technology for the current study undertaken for facial detection for an attendance management system.

3 Methodology

This section presents the methodology which was adopted to develop the proposed Facial Recognition Attendance Management System.

3.1 The Web Portal for Recording Attendance:

The web application was developed by considering two actors student and faculty. For this development authors have used the high-definition camera which is necessary in the different classrooms and Python 3.6, OpenCV, XLSX module. The following algorithms are developed by the authors, considering two stakeholders i.e. student and faculty (Table 1).

Table 1. Algorithms for the attendance

Algorithm for faculty access	Algorithm for Student access
1. Faculty visits the FRAMS web portal 2. Node server connects with MongoDB Atlas using the URI in hosting environment variable 3. Faculty Login: Faculty credential will be checked and the Dashboard will be viewed to the respective faculty 4. Faculty selects the class to mark attendance 5. Faculty gets the option to upload class image 6. Faculty uploads image and clicks on mark attendance 7. Student list with image URLs of each student is fetched from the database for selected course and is forwarded to the flask server for facial recognition a. Face detection: Using Python and facial recognition library with face recognition API b. Update the status of the recognized students to "Present" and send list back to Node server and the Node server makes a new entry in attendance collection and sends back the attendance sheet to the client 8. Attendance marked 9. END	1. Student visits FRAMS web portal 2. Node server connects with MongoDB Atlas using the URI in hosting environment variable 3. Student accesses the Application with required credential 4. Credentials are verified from the Backend Database 5. Student Dashboard will display 6. Student Clicks on Subject attendance card to view the Detailed Attendance 7. Report on Respective Subjects 8. Student Has Read-Only Rights but can report any discrepancies in attendance 9. END

The following figures show some of the screen shots (Figs. 1, 2, 3):

Login Screen

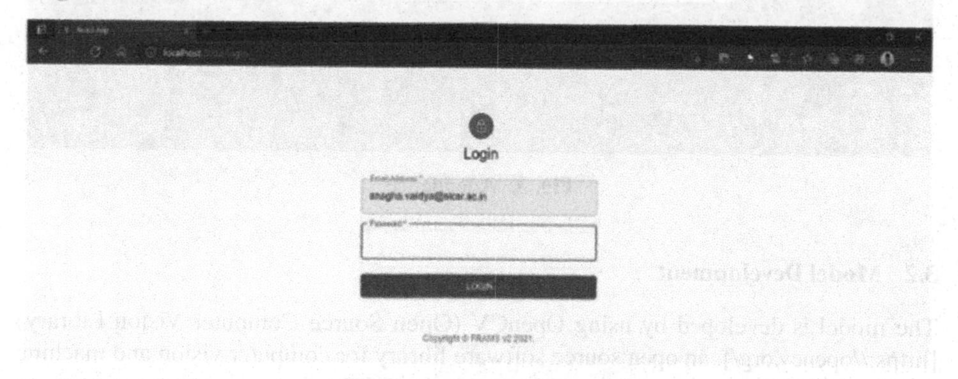

Fig. 1. Login screen

Dashboard

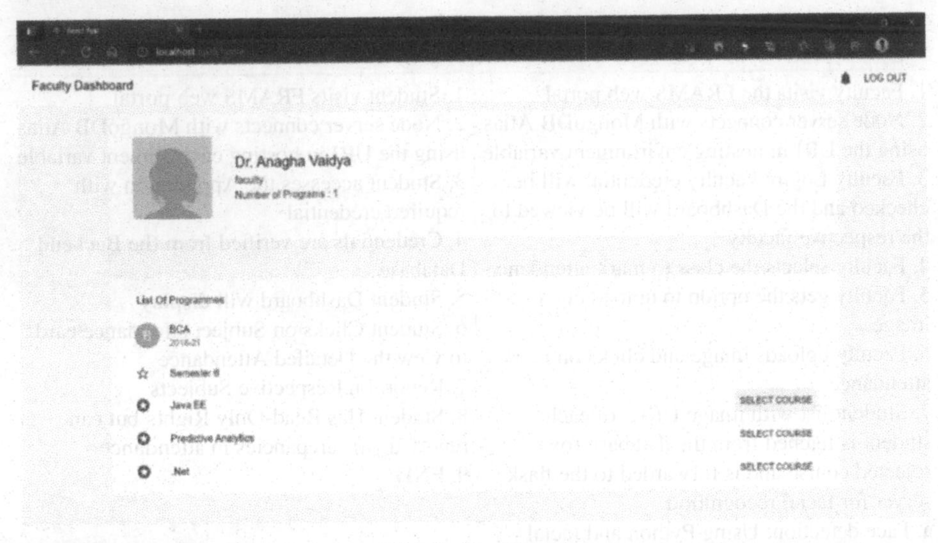

Fig. 2. Dashboard

Mark Attendance

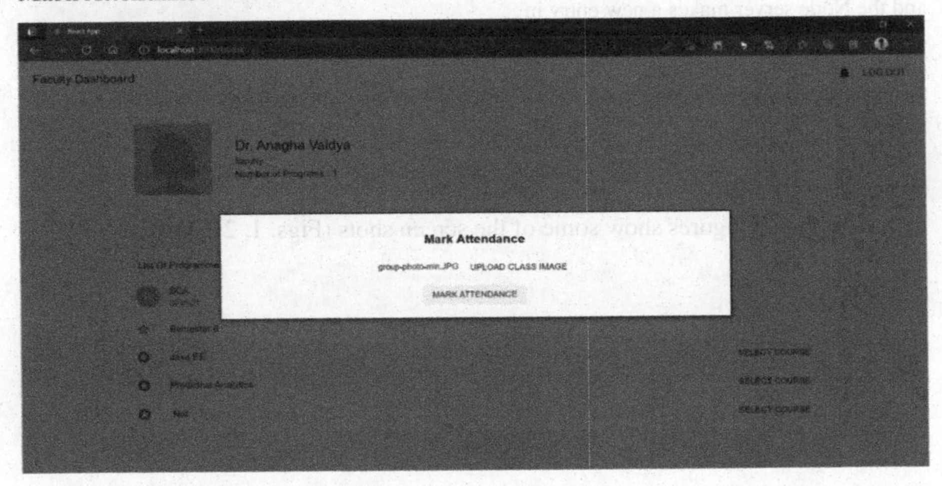

Fig. 3. Attendance

3.2 Model Development

The model is developed by using OpenCV (Open Source Computer Vision Library) [https://opencv.org/], an open source software library for computer vision and machine learning. This library is used for video analysis, CCTV footage analysis and image analysis. The image reading and processing is easily performed through cv2 module.

The cv2 module overcome the drawback of the traditional OpenCV algorithm for reading and writing images. For the face recognition model, these will be the steps:

Step 1: Prepare training data.

a) Import `openCv` and `numpy` library
b) The data folder contains the name of courses as each courses has different folder for each subject.
c) Every subject has three lists namely PRN, Name of students and vector of faces. The faces vector contains list of face samples of a person in different angles.

Step 2: Train Face Recognizer.

a) Load the training images from `dataSet` folder.
b) Capture the faces and Id from the training images: The `trainer()` function of `OpenCV` recognizer is used for training the images.
c) Put them in a list of name, PRNs and face samples and return it. The result will be in a `.yml` file that will be saved on a `trainer/` directory.

Step 3: Face Detection Algorithm.

`OpenCV` library comes with several pre-trained classifier that are trained to find different things like faces, eyes, smiles etc. For face detection is performed through "`HaarCascades`" classifier algorithm. The Haar-like features are input to the algorithm. The different `Haar` features are stored as `XML` and found in `Lib\site-packages\cv2\data`.

a) The face-cascaded object is created through `v2.CascadeClassifier` which is used to detect faces in the images along with '`detectMultiScale()`' function. The function takes the following parameters:
b) `scaleFactor`: Parameter specifying how much the image size is reduced at each image scale.
c) `minNeighbors`: Parameter specifying how many neighbors each candidate rectangle should have to retain it.
d) `minSize`: Minimum possible object size. Objects smaller than that are ignored.
e) `maxSize`: Maximum possible object size. Objects larger than that are ignored.
f) The method returns a list of rectangles of all detected objects with the tuples(x,y,w,h), where the x, y values represent the top-left coordinates of the rectangle, while the w, h values represent the width and height of the rectangle, respectively.
g) The `cv2.rectangle()` function is used to easily draw the rectangles where a face was detected. The faces are bigger or smaller therefore the images are resized several times so the face will end up with detectable size.
h) The `imshow()` method simply shows the past image in a window with the provided title.

Step 4: Face Recognition Algorithm.

a) For prediction of different images, `cv2.predict()` function is used, the function returns the label associate with the images along with the "confidence" of the recognition The `recognizer.predict()`, will take as a parameter a captured portion of the face to be analyzed and will return its probable owner, indicating its id and how much confidence the recognizer is in relation with this match. If the confidence index will return "zero" if it will be considered a perfect match.

b) The recognizer could predict a face, we put a text over the image with the probable id and how much is the "probability" in % that the match is correct. If not, an "unknown" label is put on the face.

Step 5: Update the attendance Sheet.

Read the identified label and add the attendance of the student in the respective subject folder. If unknown faces are detected, they are put into different list.

Step 6: Testing.

a) The web camera capture the images. The video streaming images are provided as an input. The `cv2.VideoCapture` class read these images.
b) The `read()` function reads individual image from the input stream. The function returns retval and image, The image retrieve frame and the return value '`True`' if the frame has been retrieved,else it return '`False`'.
c) Once the different images are stored into the face vector, then call the face recognizer function to predict the supplied image

Two models are created with 30 train images of each students. This number is accepted through experiment, as experiment gets good results when we compared the image with these images. The lesser stores of images were not able to recognize person very clearly. These images train the database. The model1 is developed on static images and the Face Detection and Face Recognizer is always developed on fly. The model2, the training dataset is created by using algorithm of Face Detection and Face Recognizer, when the testing images are provided it performs the comparison and stores the result (Figs. 4, 5, 6).

Fig. 4. Result after code 2 execution

Image Selector

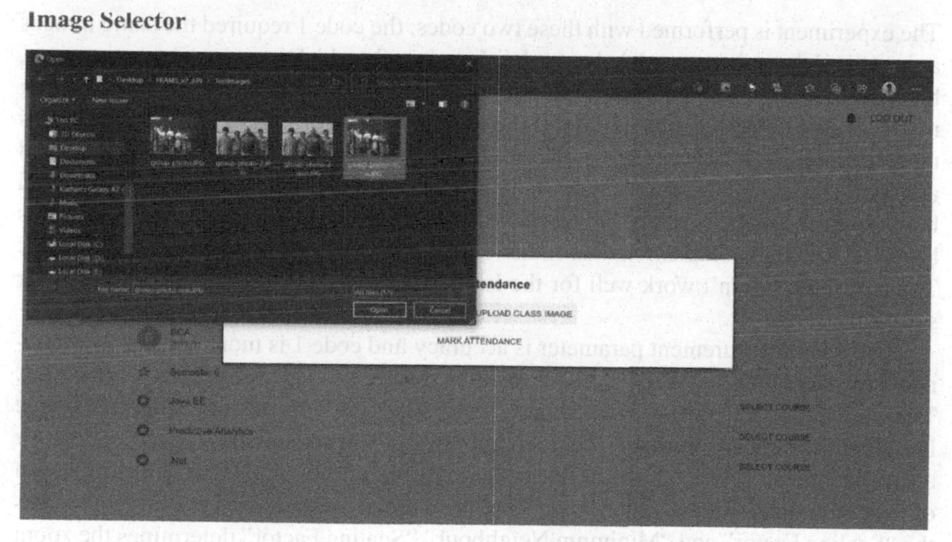

Fig. 5. Output after code 1 execution

Mark Attendance

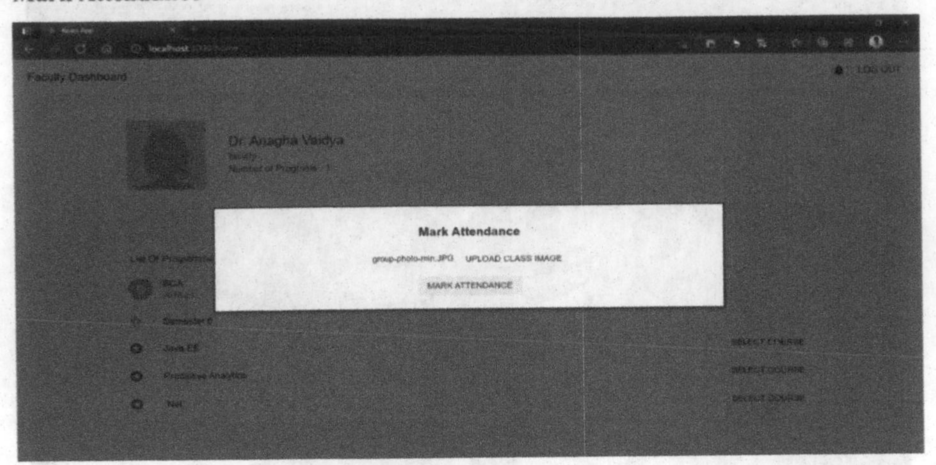

Fig. 6. After code1 execution

4 Result and Discussion

The experiment is performed with these two codes, the code 1 required the static images as input and the training model always developed on fly which increases execution time, but all different images are recognized very well. While code 2 used the existing training model of face detection which reduced the timing. The algorithm fail in face detections under different light conditions e.g. if the training image is clicked in a bright background and in the testing model the recognition of the same person under different lighting will be difficult. It has been observed by the developers that the face detection algorithm is based on different complexion than Indian skin. code2 is tested with different skin tone. The algorithm doesn't work well for the Indian students those are having darker color skins.

The other measurement parameter is accuracy and code 1 is more accurately recognized the faces then code2. By considering these result we made some changes in code1. The training model is being created and stored in a ".yml" format and Captures image live from the webcam and uses it for recognition. But still some issues are recorded, they are a) confidence of the code is low due to lack to training dataset. B) Due to the limitation of the currently developed algorithms, there is no pre-set value which can be provided as the "Scaling Factor" and "Minimum Neighbour" "Scaling Factor" determines the zoom in effect of the image being processed. However, there cannot be a constant value for this since the image will contain faces of people that are at different distances from the camera. This limits us in recognizing faces that are far away from the camera when the image is clicked, along with limiting us with the number of faces that we can recognize in a single image. By trial-and-error method, we have set the "Scaling Factor" to 1.25 and "Minimum Neighbour" to 5.

5 Conclusion, Limitations, and Future Scope

Facial detection and recognition are a technology that is still currently emerging and hasn't been explored to its complete potential. The algorithms and library modules used by us in this project were chosen carefully after conducting extensive research in the said domain. This fully automated system of facial recognition for attendance management system showed high accuracy with a very limited dataset. Another limitation is that we have not tested the algorithm for multiple cameras.

The model has certain limitations such as since the number of students may be high in a class, so the time consumption is will be high. There was a gradual fall in the accuracy as the dataset increased since the algorithms implemented by us are yet to be perfected. Furthermore, the images used to train the model were all taken under different conditions such as different lighting and variable distance from the camera which made the task of processing the images even harder as the scaling factor varied for each volunteer's picture.

If implemented correctly in the future, this system of using Facial detection and recognition for attendance management system will save time and effort on a very large scale. Such automated systems will never be as robust and innate as the human brain's capability to do the same, but this research study gave an insight as to what the future of facial detection and recognition holds. With further development in the algorithms used for face detection and recognition, it is possible that researchers would be able to accurately recognize multiple faces simultaneously. MATLAB and neural networks can be used to create a better face recognition software.

References

1. Rahman, S., Rahman, M., Rahman, M.M.: Automated student attendance system using fingerprint recognition. Edelweiss Appl. Sci. Technol. **2**(1), 90–94 (2018)
2. Bai, K.J.L., Sreemae, K., Sairam, K., Praveen Kumar, B., Saketh, K.: A survey on real-time automated attendance system. In: Mai, C.K., Kiranmayee, B.V., Favorskaya, M.N., Satapathy, S.C., Raju, K.S. (eds.) Proceedings of International Conference on Advances in Computer Engineering and Communication Systems. LAIS, vol. 20, pp. 473–480. Springer, Singapore (2021). https://doi.org/10.1007/978-981-15-9293-5_43
3. Akbar, M.S., Sarker, P., Mansoor, A.T., Al Ashray, A.M., Uddin, J.: Face recognition and RFID verified attendance system. In: 2018 International Conference on Computing, Electronics & Communications Engineering (iCCECE). IEEE, pp. 168–172 (2018)
4. Krishna, M.G., Srinivasulu, A.: Face detection system on AdaBoost algorithm using Haar classifiers. Int. J. Mod. Eng. Res. **2**(5), 3556–3560 (2012)
5. Charity, A., Okokpujie, K., Etinosa, N.O.: A bimodal biometric student attendance system. In: IEEE International Conference on Electro-Technology for National Development, pp. 464–471 (2017)
6. Patel, N., Sharma, N.: Face recognition using landmark estimation and convolution neural network. Int. Res. J. Eng. Technol. **6**(8), 1279–1284 (2019)
7. Preethi, K., Vodithala, S.: Automated smart attendance system using face recognition. In: 2021 IEEE 5th International Conference on Intelligent Computing and Control Systems (ICICCS), pp. 1552–1555 (2021)

8. Sawhney, S., Kacker, K., Jain, S., Singh, S.N., Garg, R.: Real-time smart attendance system using face recognition techniques. In: 2019 IEEE 9th International Conference on Cloud Computing, Data Science & Engineering (Confluence), pp. 522–525 (2019)
9. Sunaryono, D., Siswantoro, J., Anggoro, R.: An android based course attendance system using face recognition. J. King Saud Univ.-Comput. Inf. Sci. **33**(3), 304–312 (2021)
10. Othman, N.A., Aydin, I.: A smart school by using an embedded deep learning approach for preventing fake attendance. In: 2019 IEEE International Artificial Intelligence and Data Processing Symposium (IDAP), pp. 1–6 (2019)
11. Elias, S.J., et al.: Face recognition attendance system using Local Binary Pattern (LBP). Bull. Electr. Eng. Inform. **8**(1), 239–245 (2019)
12. Teoh, K.H., Ismail1, R.C., Naziri, S.Z.M., Hussin, R., Isa, M.N.M., Basir, M.S.S.M.: Face recognition and identification using deep learning approach. J. Phys. Conf. Ser. **1755**, 1–10 (2021)
13. AbdELminaam, D.S., Almansori, A.M., Taha, M., Badr, E.: A deep facial recognition system using computational intelligent algorithms. PLOS ONE **15**(12), 1–27 (2020)

American Sign Language Gesture Analysis Using TensorFlow and Integration in a Drive-Through

Abhineet Sharma, Anupam Chopra(✉), Mayank Singh(✉), and Adesh Pandey(✉)

Department of Information Technology, KIET Group of Institutions, Ghaziabad, India
abhineetsharma77@gmail.com, anupamchopra1607@gmail.com

Abstract. The algorithm described in this paper aims to achieve Gesture recognition in identifying American Sign Language Symbols. There are a total of 29 characters that were added to the image database for the model to classify. The Database includes 87,000 images and more than 1200 images per character. Since it is a multiclass classification problem, a CNN model is implemented. The primary objective of the algorithm is to correctly classify all the gestures that are mapped on the proposed ASL Alphabet matrix integrated with the Digital Signage. This is done using OpenCV Library which captures the real-time feed and is then converted to grayscale that is provided as an input to the Convolutional Neural Network for the classification. The CNN model uses an ADAM optimizer that mediates the learning rate by quickly reaching the global minima based on the *Vector Component* Precedence.

Keywords: Gesture recognition · Multiclass · CNN · OpenCV · ADAM

1 Introduction

The most important tool in human life is communication. It is a basic and effectual way to share thoughts, feelings, and ideas. However, the vast majority of people in the world do not have this ability. Many of them have hearing problems, speech problems, or both. Hearing loss is defined as hearing loss partially or completely in one or both ears. A deaf person, on the other hand, is a speech impediment that limits his ability to speak. The ability to learn a language is impaired when hearing/speech loss occurs in childhood, resulting in language impediment, also known as hearing mutism. These diseases are part of the most common disability in the world. A report on Statistics on children with physical disabilities revealed that over the past decade has seen an increase in the number of newborns with disabilities and hearing impairments and has created a barrier to communication between the world and them [1].

M. Singh et al. (Eds.): ICACDS 2022, CCIS 1613, pp. 399–414, 2022.
https://doi.org/10.1007/978-3-031-12638-3_33

For those with speech limitations and hearing impairments, one of the key modes of communication is sign language, which is a gesture-based language. Sign language has hundreds of symbols generated by diverse gestures and facial expressions, comparable to any spoken language [2]. The purpose is to learn to recognize American Sign Language signs. The conversion of these signals into words or characters of existing spoken languages is known as sign language recognition. As a result, using an algorithm or model to translate sign language into words can help to bridge the gap between those with hearing and speech issues and the rest of the world.

The field of vision-based hand gesture recognition applies to current research, especially computer-assisted visualization and machine learning. Many researchers are working on the subject of human-computer interaction (HCI) to make it easier and more natural to connect with computers without the use of additional devices. As a result, the primary goal of gesture recognition research is to develop systems that can recognize a person's distinctive gesture and utilize it to send information, for example. Vision-based hand gesture interfaces require rapid and reliable hand detection as well as the capacity to recognize movements in real-time to be useful. The hand gesture, which has various applications, is one of the most potent elements of human communication modal properties. In this context, we can recognize sign language, a form of communication used by the deaf community [3].

The ASL Detection System is a Convolutional Neural Network program. The basic idea of the project was to make any customer-based service more friendly to people with disabilities (especially deaf and mute). Being taught sign language as another common way of interacting with others to overcome barriers to communication, many of them face problems in daily life such as ordering food in Drive-thru or having a conversation with an attendant who is oblivious to their predicament. ASL Detection is based on Image Processing and Object Detection integrated with a computer webcam and any other video visual aids to assist both; customer and seller to have a better interaction while having a conversation.

The authors aim to distinguish ASL Gestures i.e., alphabet matrix and special characters defined in the database using image recognition. Images are first converted to grayscale and then normalized. The processed images are then added to make the Dataset more solid and generic. This helps the Neural Network become less prone to overfitting and improves the overall accuracy of the network itself. We applied one-hot coding for the class element of each sample, converting the integer into a 28-element binary vector. ASL Gestures are used as an ordering matrix on the drive-thru menu board which is known as a Digital Signage. Each food item will have an ASL **Gesture** associated with it. The CNN model classifies live feeds from a webcam based on Multivariate class labels that are ASL characters. The probabilities for each gesture are shown along with the gesture recognized by the model. The CNN model used in this project uses Adam as an optimizer that simulates the rate of learning by quickly reaching the global minima based on Vector component precedence. As the model uses 87,000 images, hyper-parameter optimization for a better classification is necessary along with adding dropout layers with the frequency of rate. The Dropout layer, that aims to reduce overfitting, converts input values to 0 at random with a rate of frequency for every training step. The non-zero inputs are upscaled by 1/ (1 - rate) so the sum remains constant across all inputs. Float between 0 and 1 for the rate. Drop in the percentage of input units [4].

2 Literature Review

Machine learning has been around for quite some time as it has been implemented in various fields. Many people are trying to use various ML Techniques to make the world better since the 1970s. [5] turn invariant stances that utilize limit histogram have been manifested [6, 7]. To secure the information picture, a camera was used. The bunching process uses the channel for skin shading discovery [8] to find the limit for each gathering in the gathered picture using a common form after calculation. The picture was separated into numerous networks once the limits were standardized. Harmony's size chain was used as a histogram by dividing the image into multiple areas N in an outspread structure, according to the explicit edge. [9], Multilayer Perceptron and dynamic programming, and DP coordinating were used for the classification process. Various analyses have been executed on multiple highlight positions nonetheless using distinctive harmony's size histogram and harmony's size FFT [10]. Convolution Neural Networks (CNN) [11, 12] have an entrenched role in the field of image recognition and it as has been distinctly exhibited by several researchers in past.

One of the most competitive areas of research is Gesture Recognition. It may involve the use of sensory hardware as one of the different ways to detect the gesture. However, employing hardware is both costly and difficult in practice. Researchers are striving to discover the best recognition accuracy using computer vision algorithms. Various glove-based and vision-based sign language detection systems were tested for effectiveness. Sruthi C. J. and A. Lijiya [13].

Many scientists and students are encouraged to apply the Transfer Learning approach or to enforce their Neural Network architecture. As research into sign language recognition progresses, an ASL database has become increasingly important. Certain databases have been made public around the world by some Computer Vision scientists and laboratories to develop new solutions [14].

CNN has an important role to play in areas such as image processing. It has had a huge impact on many areas. CNN is used for error detection and classification in nanotechnology such as making semiconductors. Researchers have been interested in digit recognition [15]. On the topic of Digit Recognition, a considerable number of papers and articles have been published recently. According to studies, Deep Learning methods like multilayer CNN using Keras and TensorFlow deliver substantially higher accuracy than machine learning algorithms like SVM, KNN, and RFC, which are also extensively used. Because of its great accuracy, the Convolutional Neural Network (CNN) is frequently utilized in image classification, video analysis, and other applications. Many academics attempt to include sentimental appreciation in their sentences. By adjusting different parameters, CNN is employed in natural language processing and emotive recognition [16].

Mohit Patil et al. have developed a grid-based recognition system for Indian Sign Language gestures and poses. Provides real-time accuracy and detection. many techniques such as object stabilization and skin color, and facial recognition, are utilised.

tracking and detection of hand gestures, segmentation is used. A system that can effectively distinguish all 33 sign languages in ISL (Indian Sign Language) has been described in the paper. the program on this page uses techniques including skin colour removal, object stabilization, hand removal, and facial removal to reliably detect the person's hand movements. 96.4% accuracy. in addition, the system correctly recognises 10 gestures with an accuracy rate of 98.23%. every sign is classified using the Hidden Markov Model chain method, and each hand posture is classified using a KNN model. The results show that the system can accurately and in real-time detects hand positions and the signs present in Indian Sign Language. The method given is simple and can be expanded to include more single-handed and two-handed signals. The approach demonstrated in this study can be used to other languages such as American Sign Language (ASL) or British Sign Language (BSL), as long as the dataset meets the system's requirements [17]

3 Proposed Methodology

The Training script first loads the dataset image by image from the defined directory and appends them to a list. The images are resized to a base dimension of 32 × 32. In the preprocessing stage, the pixel values are normalized (Fig. 1).

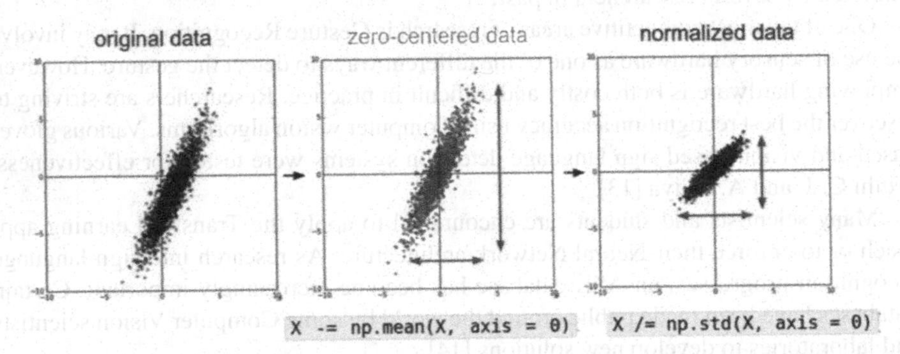

Fig. 1. Normalizing the Pixel values (stack overflow)

The images are then reshaped and a channel is added to each of the preprocessed images. Further, the process entails image augmentation to make the dataset generic for the model to fit.

The parameters mentioned are **width shift range** which can be a Float, a 1-D array-like, or an integer value. If the value passed is float and it is less than 1, it will take a percentage of the total width as the range and shift the image, if the width shift parameter is greater than 1, then it will count the pixel as the range. The width shift range is specified via a 1-D array-like parameter. The interval of pixels from (−width shift range, +width shift range) shall be defined by an integer. With width shift range = 2, the values [−1, 0, +1] are all available. Alternatively, we can directly state the same thing as with width shift range = [−1, 0, +1], whereas with width shift range = 1.0, the interval's possible values are defined as [−1.0, +1.0]. **height shift range** which can be a Float, a 1-D array-like, or an integer value. If the value passed is float and it is less than 1, it will take a percentage of the total height as the range and shift the image, if the height shift parameter is greater than 1, then it will count the pixel as the range. A 1-D array-like argument will specify the range for the height shift. The interval of pixels is defined by an integer (−height_shift_range, +height_shift_range).

$[-1, 0, +1]$ are the available values when height shift range $= 2$. or we can explicitly mention the same as with height shift range in the form of an array $[-1, 0, +1]$, whereas with height shift range $= 1.0$, the interval's potential values are defined as $[-1.0, +1.0)$. The zoom range can be specified as a Float value or as an interval such as [LWR(lower), UPR(Upper)]. Range. If a float value is passed then the range is given by, [LWR, UPR] = [1-zoom_range, 1 + zoom_range], **shear range** is a Float value that represents the Shear Intensity (Shear angle in counter-clockwise direction in degrees) **rotation range** another Integer value that defines the range of degree for random rotations [18].

to_categorical() of the keras.util package, on training, test, and validation datasets, is often used to transform a class vector (integers) to a binary class matrix (Fig. 2).

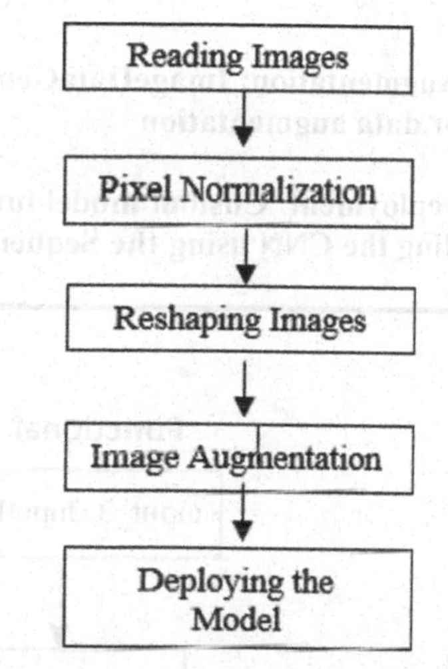

Fig. 2. The Proposed Methodology for model deployment

The model uses sequential API that groups a linear stack of layers sequentially (Table 1).

Whenever the model had several inputs/outputs, or when the layers have many inputs/outputs and layer sharing is required, and when a non-linear topology is required, Sequential model API is not recommended (skip connections or a multi-branch model) (Fig. 3).

Table 1. The algorithm for the deployment of ASL detection system

Algorithm: Proposed Methodology
1. Input: Dataset (Reading the images file by file) **For loop 1: till the Len (Number of classes)** **Nested For loop 2:** **Use cv2 object for reading and resizing**
2. Pixel Normalization: Use the custom pre-processing function **cv2 object for Grayscale using equalizeHist() Method**
3. Reshaping Images: Convert the Train, Test and Validation images using reshape () method to add a channel
4. Image Augmentation: ImageDataGenerator() Method for data augmentation
5.Model Deployment: Custom model function() For modeling the CNN using the Sequential API

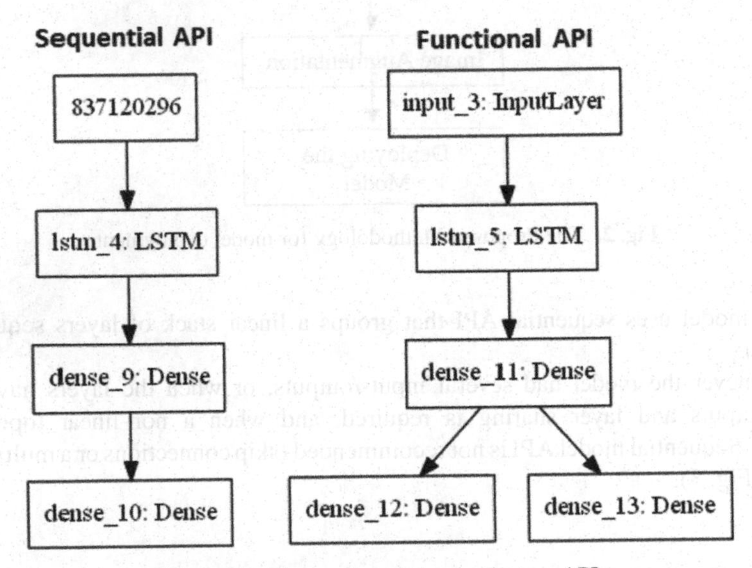

Fig. 3. The Comparison between the two APIs

4 Dataset Description

The set of data used in the project was downloaded from Kaggle containing approximately 87,000 images and about 1600 images per Alphabet. The images used in the project are **split** by an 80:20 **ratio** in both the Training and **Validation** Set. The Training Set is one in which - hot encoded to convert categorical values into numeric to make the data set more nuanced and generic [19] (Fig. 4).

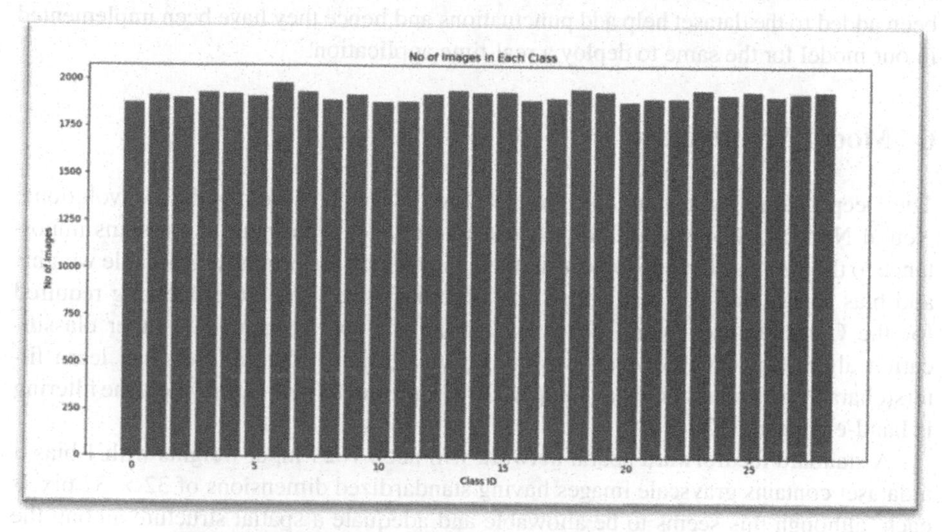

Fig. 4. Plot representing the total number of images per class

The images below represent the gestures that the dataset entails for the 26 alphabets in addition to special characters (Fig. 5).

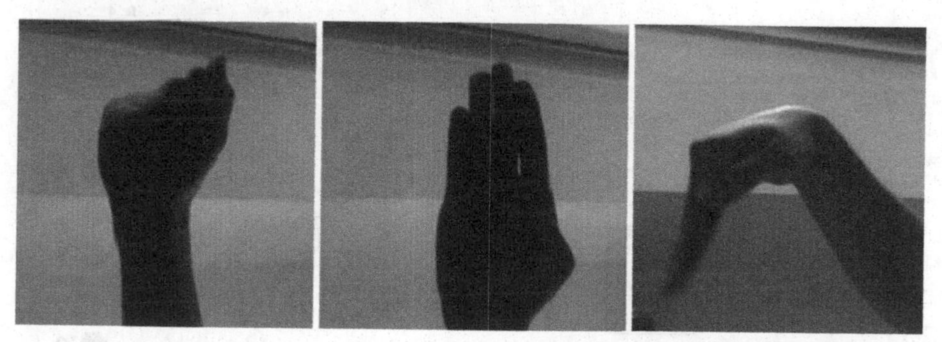

Fig. 5. Representing the Character 'A' and 'B' of ASL along with a Custom Character 'del'

5 Experimental Setup

The images were downloaded from Kaggle. The data set is a compendium of images for the American Sign Language Alphabets and it is separated into 29 folders that represent the different classes. The training data set has 87,000 images which are in the format of 200x200 pixels. A total of 29 classes have been mentioned and described out of which 26 folders/classes are for the ASL letters A-Z and the rest of the 3 classes represent the special characters for *SPACE*, *DELETE*, and *NOTHING*. The 3 extra classes that have been added to the dataset help add punctuations and hence they have been implemented in our model for the same to deploy a real-time application.

6 Model Architecture

The Deep Learning algorithm used in the Drive Through System is the Convolutional Neural Network (ConvNet / CNN). It takes an image as an input and assigns importance to the various elements/objects in the image which are primarily learnable weights and bias so that you can distinguish one from the other. The pre-processing required for the Convolutional Neural Network is relatively low compared to other classification algorithms. With adequate training Convolutional, Neural Nets can learn filters(characteristics) implicitly as compared to the primitive/traditional where the filtering is hand-engineered [20].

A standard feedforward neural network will need 1024 input weights with 1 bias if a dataset contains grayscale images having standardized dimensions of 32×32 pixels each. although this seems to be allowable and adequate a spatial structure among the image is lost because image matrix of pixels has been flattened into a protracted vector of pixel values. Unless all the images are properly resized, neural network will have trouble spatially reconstructing the pixel matrix into an absurdly lengthy linear vector of pixel values.

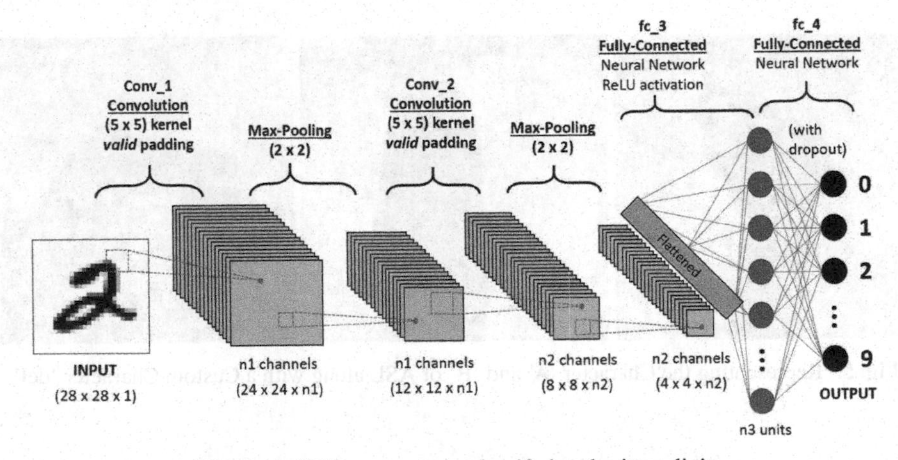

Fig. 6. A CNN sequence to classify handwritten digits

Convolutional Neural Networks are used because they can retain spatial relationships among pixels via learning internal feature representations from small squares of input data [21] (Fig. 6).

The model proposed in this paper has the following Architecture (Fig. 7):

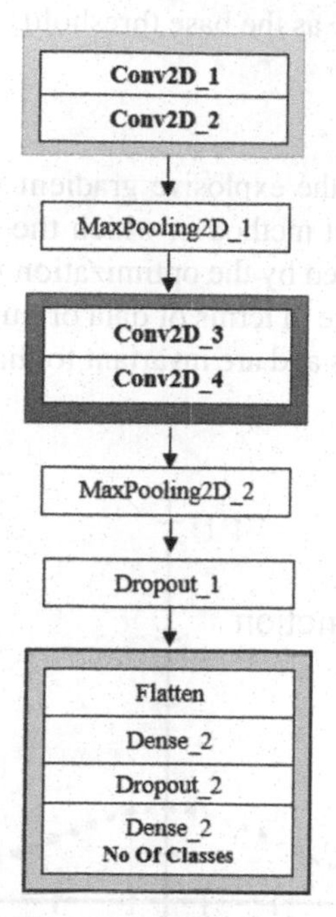

Fig. 7. Representation of the CNN Architecture

The CNN model has 4 convolution layers and two layers of max-pooling are added in between that act as a noise suppressant by taking the maximum of all the values in a matrix of image pixels. The drop-out layer was added with a set rate of frequency

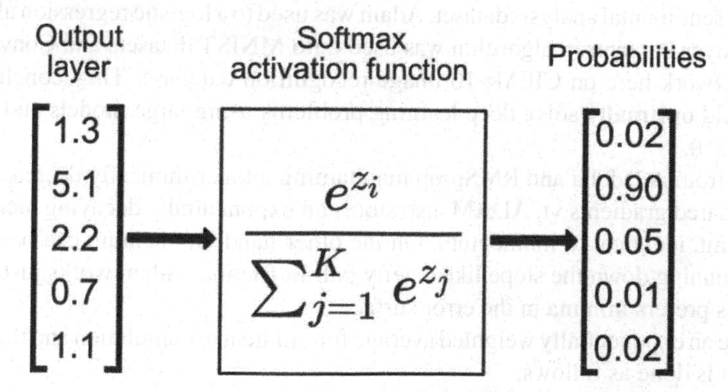

Fig. 8. SoftMax function with the output probabilities

to avoid overfitting on the validation set along with two dense layers with one of them having SoftMax as the activation function to produce the probabilities (Fig. 8).

The SoftMax is used in the final output (Dense) layer of the Neural Network that predicts a multinomial probability distribution. This way we get a probability for each of the class labels based on the threshold mentioned for the classification. The algorithm proposed in this paper uses 60% as the base threshold.

7 Model Optimization

The problem of vanishing and the explosive gradient were solved using ADAM. It is a stochastic descending gradient method in which the first order and the second-order moments are adaptively measured by the optimization strategy i.e. Adam. This method works for problems that are large in terms of data or number of parameters, which have very little memory requirements and are invariant to diagonal rescaling of the gradients [22] (Fig. 9).

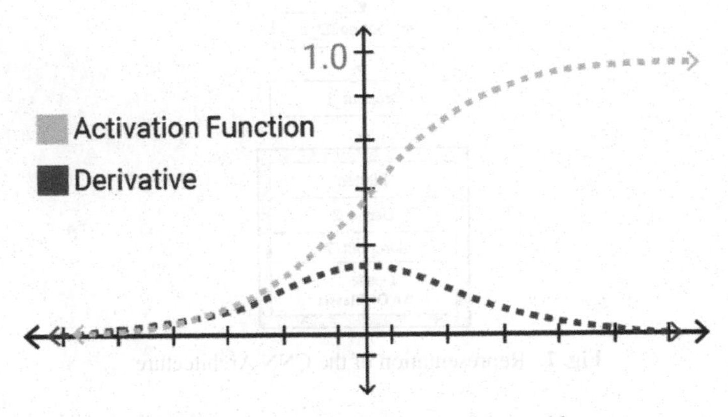

Fig. 9. Illustrating vanishing gradient problem

In the **original** paper, Adam's **empirical demonstration** showed that the **convergence** met the theoretical analysis expectations. On the MNIST digit recognition and the IMDB sentimental analysis dataset, Adam was used to a logistic regression algorithm, the Multilayer Perceptron algorithm was used onto MNIST datasets and Convolutional Neural Network here on CIFAR-10 image recognition database. They concluded that Adam could **optimally** solve deep learning problems using large models and data sets [23] (Fig. 10).

Apart from Adadelta and RMSprop maintaining a logarithmically decreasing mean of past squared gradients vt, ADAM also stores an exponentially decaying mean of past gradients mt, identical to momentum. On the other hand, momentum can be imagined as a ball running down the slope like a hefty ball in friction, Adam works just like that, which thus prefers minima in the error surface.

We use an exponentially weighted average for gradient accumulation and the moment estimation is done as follows:

$$m_t = \beta_1 m_{t-1} + (1 - \beta_1) g_t \tag{1}$$

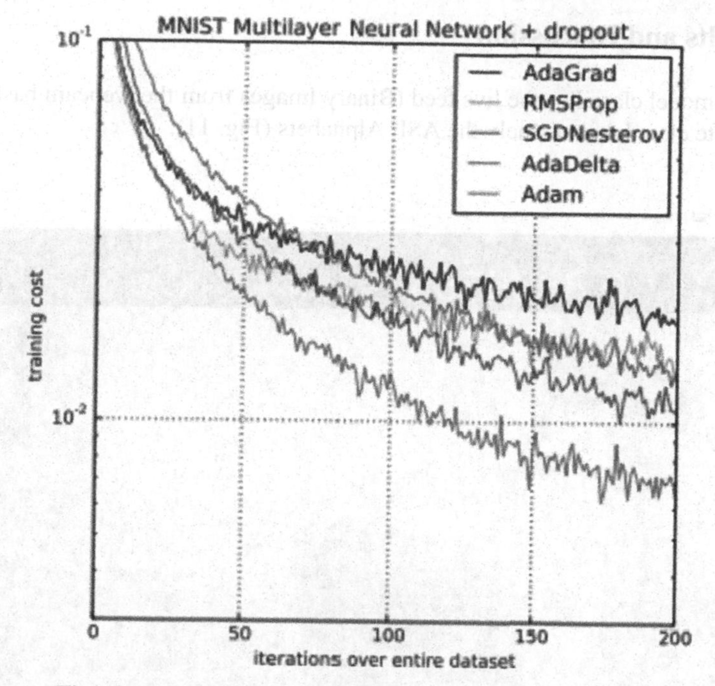

Fig. 10. Comparing Adam with other optimization algorithms

$$v_t = \beta_2 v_{t-1} + (1 - \beta_2) g_t^2 \tag{2}$$

M_t and v_t are the first moments and the second-moment estimates of the gradients respectively. Because mt and vt are both initialised as 0's, they are skewed towards zero in Adam, especially during the first time steps and while decay rates are low. (i.e., β1 and β2 are close to 1).

To perform bias corrections on the two running average variables, we use the following equations:

$$\hat{m}_t = \frac{m_t}{1 - \beta_1^t} \tag{3}$$

$$\hat{v}_t = \frac{v_t}{1 - \beta_2^t} \tag{4}$$

where 't' is the number of current iterations (we perform this before updating the weights). Adam is the combination of RMSprop added with the Momentum factor. Hence, the Adam update rule works well in practice and it is robust to hyperparameters:

$$\theta_{t+1} = \theta_t - \frac{\eta}{\sqrt{\hat{v}_t + \varepsilon}} \hat{m}_t \tag{5}$$

The default values are 0.9 for β1, 0.999 for β2, and 10–8 for ϵ are preferred. It has experientially been proven that Adam optimization works well in practice as compared to other adaptive learning-method algorithms [24].

8 Results and Discussion

The CNN model classifies the live feed (Binary Image) from the webcam based on the Multivariate class labels namely the ASL Alphabets (Fig. 11).

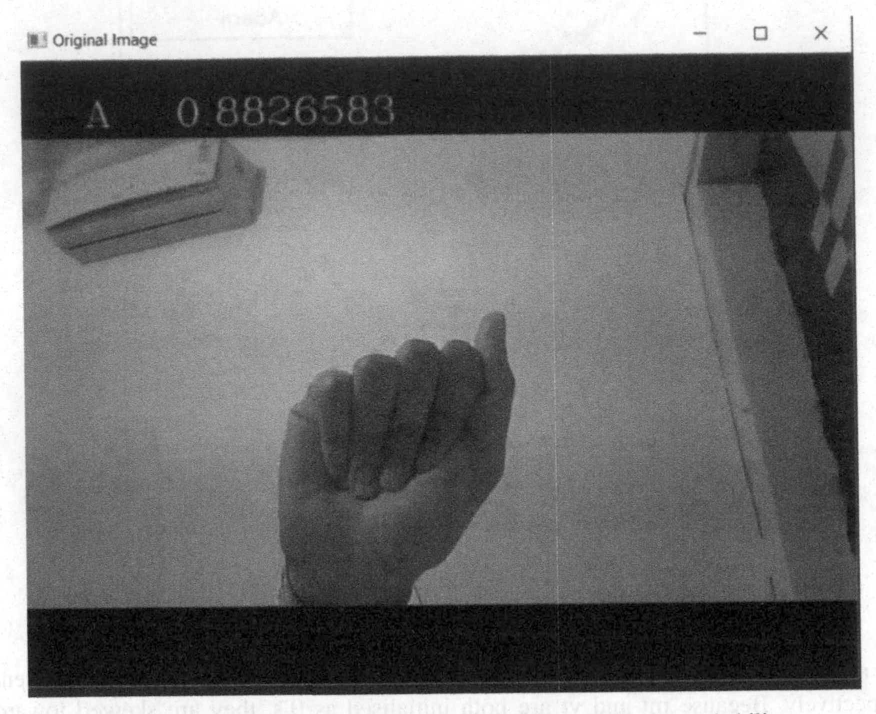

Fig. 11. Model classifying the ASL gesture along with the probability

8.1 Loss and Accuracy

The model ran 21 Epochs with a batch size of 50 sample images. The came out to be 0.89 (89%, rounded to the first two decimals) (Fig. 12).

One of the primary concerns during the modeling stage was the depth of the neural network (Fig. 13).

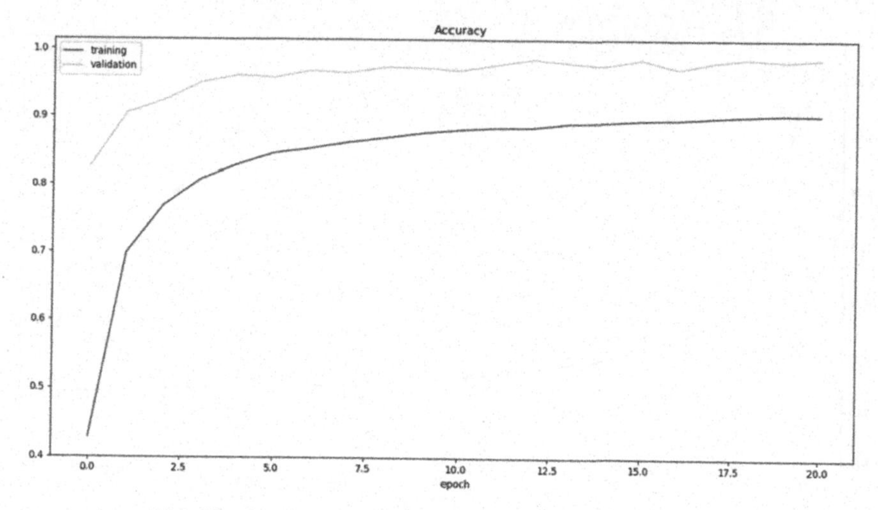

Fig. 12. Graph representing accuracy plotted against epochs

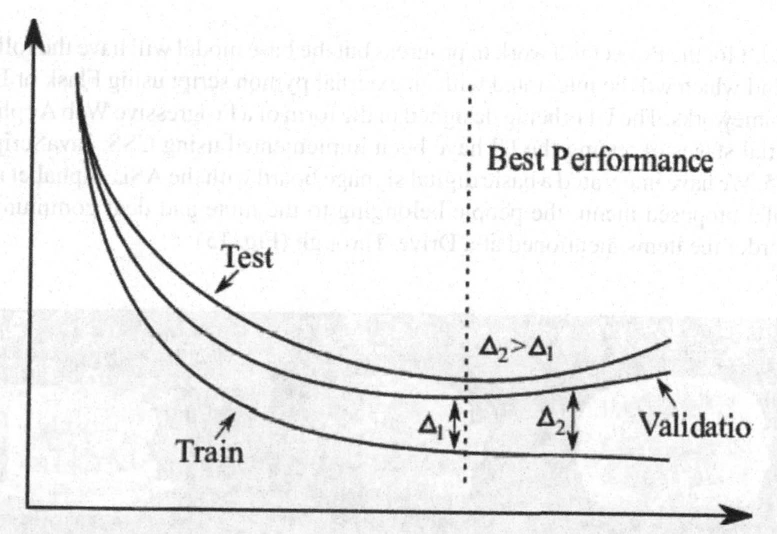

Fig. 13. Graph representing early Stopping for the optimal accuracy

The validation loss decreased with the increase in epochs. Thus, representing that the CNN model is not overfitting on the validation set making it robust to outliers (Fig. 14).

9 Conclusion

The model was able to classify the images with an accuracy of 89% however different backgrounds affected the model's performance and will be dealt with by using the inverse square law to highlight the pixels closer to the camera.

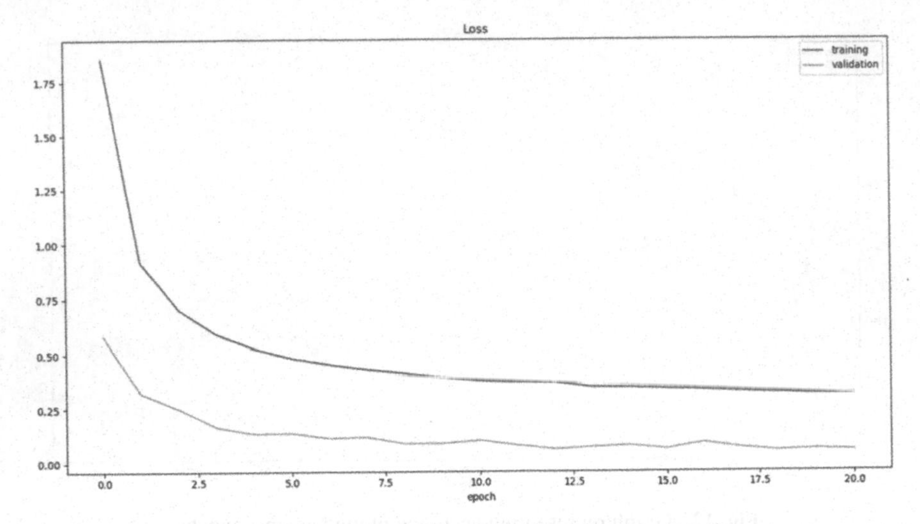

Fig. 14. Graph representing loss plotted against epochs

The UI for the Project is a work in progress but the base model will have the following Front End which will be integrated with an external python script using Flask or Django Web Frameworks. The UI is being designed in the form of a Progressive Web Application The initial stages of testing the UI have been implemented using CSS, JavaScript, and HTML5. We have integrated a basic digital signage board with the ASL Alphabet matrix. Using the proposed menu, the people belonging to the mute and deaf community can easily order the items mentioned at a Drive-Through (Fig. 15).

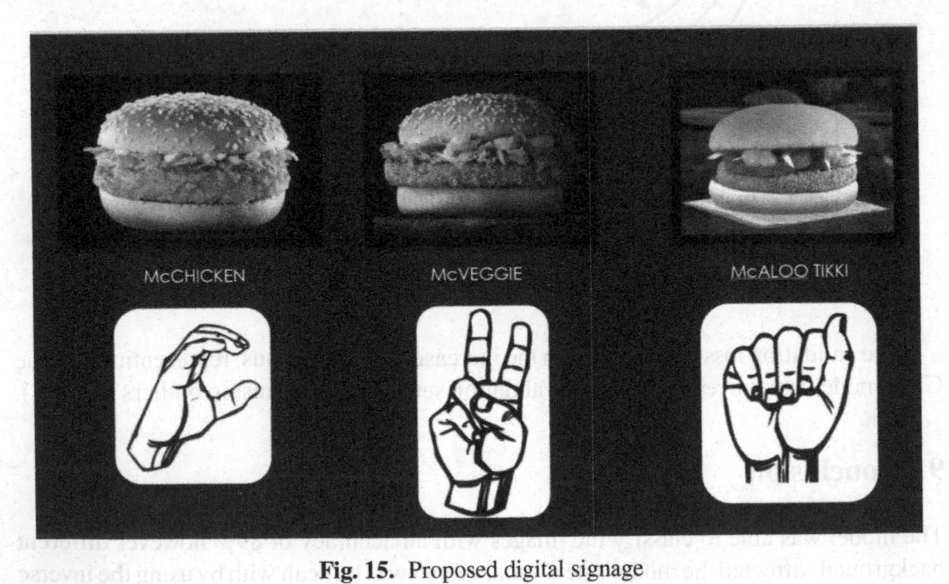

Fig. 15. Proposed digital signage

10 Future Works

We will also be implementing Transfer learning in the future to compare the performances of different neural networks.

The main reason for the need to implement transfer learning in our real-time application is to integrate the ability of pre-trained Neural Networks with the SOTA structure (ImageNet and VGG-16) to achieve better classification and results along with the innate need to retain the reuse of previously learned knowledge of neural networks. It was addressed at the NIPS-95 seminar "Learning to Learn," which concentrated on the ever-present need for machine learning approaches that preserve and reuse previously acquired knowledge [25].

Along with an in-depth understanding of the working of CNNs using Feature Maps and GradCAMS (*class activation mappings*).

To build trust in intelligent systems and move towards their meaningful integration into our daily lives, it is also important to build trust in intelligent systems and integrate them into our daily lives. Therefore, we should create 'obvious' models that explain why they are predicting what they are predicting [26].

References

1. Machine Learning methods for Sign Language recognition: A critical review and analysis By: I.A.Adeyanju | O.O.Bello and M.A.Adegboye
2. Analysis of Recent Trends in Continuous Sign Language Recognition using NLP | Vijayshri Khedkar*, Sonali Kothari, Aarohi Prasad, Arunima Mishra, Varun Saha and Vinay Kumar
3. A Review Paper on Sign Language Recognition for The Deaf and Dumb | Reddygari Sandhya Rani, R Rumana, R. Prema (2021)
4. https://keras.io/api/layers/regularization_layers/dropout/
5. Wang, Z., Wu, S., Liu, C., Wu, S.: The regression of MNIST dataset based on convolutional neural network. In: Proceedings of the 2020 International Conference on Advanced Machine Learning Technologies and Applications, Jaipur, India (2020)
6. Thakur, A., Saini, D.S.: Correlation processor-based sidelobe suppression for polyphase codes in radar systems. Wirel. Personal Commun. **115**, 377–389 (2020)
7. Zhao, S., Tan, W., Wen, S., Liu, Y.: An Improved algorithm of hand gesture recognition under intrikate pond. International Conference on Intelligent Robotics and Applications, pp. 786–787 (2008)
8. Sancho Rieger, J., Waldert, S., Pistohl, T., et al.: Vision based gesture recognition using neural networks approaches: a review. Int. J. Hum. Comput. Inter. **32**(1), 480–494 (2008)
9. Hagen, M., Ayadi, A.E., Wang, J., Vasiloglou, Afshar, E.: Optimizing training data for image classifiers. In: Proceedings of the KDD 2019 Workshop on Data Collection, Curation, and Labeling for Mining and Learning (2019)
10. Shubankar, B., Chowdhary, M., Priyaadharshini, M.: IoT device for disabled people IoT device for disabled people. Procedia Comput. Sci. **165**, 189–195 (2020)
11. Gu, J., Wang, Z., Kuen, J., et al.: Recent advances in convolutional neural networks. Pattern Recogn. **77**, 354–377 (2006)
12. Sharma, N., Jain, V., Mishra, A.: An analysis of convolutional neural networks for image classification. Procedia Comput. Sci. **132**, 377–384 (2018)

13. Sruthi, C.J., Lijiya, A.: Signet: a deep learning based indian sign language recognition system. In: 2019 International Conference on Communication and Signal Processing (ICCSP), pp. 0596–0600. IEEE (2019)
14. 'ASL-100-RGBD Dataset, Media Lab, The City College, City University of New York'. http://medialab.ccny.cuny.edu/wordpress/datecode/#Dataset
15. Lee, K.B., Cheon, S., Kim, C.O.: A convolutional neural network for fault classification and diagnosis in semiconductor manufacturing processes. IEEE Trans. Semicond. Manuf. **30**(2), 135–142 (2017)
16. Pasi, K.G., Naik, S.R.: Effect of parameter variations on accuracy of Convolutional Neural Network. In: 2016 International Conference on Computing, Analytics and Security Trends (CAST), pp. 398–403. IEEE (2016)
17. Patil, M., et al.: Indian sign language recognition. Int. J. Sci. Res. Eng. Trends (2020)
18. https://www.tensorflow.org/api_docs/python/tf/keras/preprocessing/image/ImageDataGenerator
19. https://www.kaggle.com/grassknoted/asl-alphabet
20. A Comprehensive Guide to Convolutional Neural Networks — the ELI5 way by: Sumit Shah
21. https://machinelearningmastery.com/crash-course-convolutional-neural-networks/
22. Kingma, D.P., Ba, J.L.: Adam: a method for stochastic optimization. In: International Conference on Learning Representations, pp. 1–13 (2015)
23. Show, Attend and Tell: Neural Image Caption Generation with Visual Attention | DRAW: A Recurrent Neural Network for Image Generation
24. An overview of gradient descent optimization algorithms - Sebastian Ruder
25. A Survey on Transfer Learning Sinno Jialin Pan and Qiang Yang Fellow, IEEE
26. Grad-CAM: Visual Explanations from Deep Networks via Gradient-based Localization Ramprasaath R. Selvaraju1 ∗ Michael Cogswell1 Abhishek Das1 Ramakrishna Vedantam1 ∗ Devi Parikh1,2 Dhruv Batra1,2 1Georgia Institute of Technology 2Facebook AI Research

Performance Analysis of Optimized ACO-AOMDV Routing Protocol with AODV and AOMDV in MANET

Veepin Kumar[1(✉)] and Sanjay Singla[2]

[1] IK Gujral Punjab Technical University, Jalandhar, Punjab, India
kumarvipi@gmail.com
[2] GGS College of Modern Technology Kharar, Mohali, Punjab, India

Abstract. MANET is made up of nodes that uses routing protocols to transfer a packet from source to sink using store and forward method. These packets are sent through each intermediate node as a result, routing the packets is an extremely expensive procedure. Traditional protocols also ignore energy usage in nodes when choosing path from source to destination. These protocols choose a path based on the number of nodes with the least amount of hop and if a single node fails due to a loss of battery power, then entire path will go down and the routing procedure will have to be repeated which results in a delay. As a result, node power consumption and excessive delay tends to the network performance degradation. So, to deal with these problems, a suitable routing protocol for MANETs must be selected. This research work will discuss the routing protocols that are used for the selection of data transfer. The need for an optimum communication path among nodes has prompted the use of optimization techniques like Ant-Colony Optimization (ACO), Particle Swarm Optimization etc. This research paper discusses the concept of ACO in detail and analyze the performance of AODV, AOMDV, and ACO-AOMDV protocols. The performance of a network is analyzed using several metrics such as E2E delay, energy consumption (EC), packet delivery ratio (PDR), and throughput. We use NS 2.35 simulator to conduct this research and Simulation findings reveals that when the nodes size increases gradually, ACO-AOMDV outperformed as compared to other routing protocols in terms of throughput, delay, energy consumption and PDR.

Keywords: ACO · Energy consumption · MANET · NS 2.35 · Optimization · PDR · PSO and routing protocols

1 Introduction

MANET is a network made up of wireless hosts that creates an adhoc network without the requirement for a separate infrastructure or centralized control [1]. Because the nodes in the network are mobile, they self-organize and configure themselves. Nodes act as both hosts and routers, routing data to and from other network nodes. There are two types of communication strategies for wireless mobile nodes that are infrastructure oriented and infrastructure-less. In an infrastructure-oriented structure, devices

M. Singh et al. (Eds.): ICACDS 2022, CCIS 1613, pp. 415–425, 2022.
https://doi.org/10.1007/978-3-031-12638-3_34

connected with a base station or other access point through fixed infrastructure, but in an infrastructure-less structure, nodes communicate without relying on any pre-existing network infrastructure.

Since every node in a MANET is free to travel in any route, it often changes its links with other devices, therefore routing protocols must be efficient enough to meet the network's requirements. As devices in MANET are battery operated and a lot of power gets exhausted when nodes move from one location to another which tends to the reduction of network lifespan as the network dies when all the nodes energy gets exhausted [2]. Therefore, most of the today's research is focused on lowering the amount of energy required by the adhoc network to extend its life. The draining of the batteries on which the adhoc node's function has been the key source of worry in terms of energy conservation. Therefore, route selection for transmitting data from source to sink is the vital concern in MANET.

Several MANET routing methods for route estimation have been described in the past, which can be broadly split into two types: proactive and reactive routing protocol [3]. In Proactive routing protocol, every mobile node has its own routing table, which keeps track of all possible routes to all destinations [4]. Because the MANET topology is dynamic, these routing tables are updated often as the network's topology changes. Its drawback is that it does not scale effectively in large networks since the routing table entries grow large as route information for all reachable nodes is stored. Proactive protocols include the Destination Sequenced Distance Vector Routing Protocol (DSDV) and Global State Routing (GSR) protocol.

In DSDV, a destination sequence number (DSN) is assigned in the routing table to each routing entry of node [5]. Only if the entry has a new updated route to a destination with a higher sequence number will incorporate the new update in the database. This protocol uses the concept of Dijkstra algorithm and extends the wired network's link state routing. Because each node sends link status routing information to the whole network, global flooding can occur which causes congestion. As a result, a solution called the Global State Routing Protocol was created. GSR does not swap the network with link state routing packets. In GSR, each forwarding node converts its neighbor positions into graph nodes and selects the next node with the shortest path to the sink, after which the packet is sent to the next hop, bringing the data closer to the sink [6].

In reactive protocol, the route is only found when it is required. Route discovery is accomplished through the transmission of route request packets throughout the network. The Dynamic Source Routing (DSR), Ad-Hoc On Demand Vector (AODV), and Ad-hoc On-Demand Multipath Distance Vector (AOMDV) are all reactive protocols [7]. DSR is source-initiated protocol used with mobile nodes in multi-hop networks and there is no requirement for network infrastructure or management [8]. As a result, the network can entirely self-organize and configure itself. It is divided into two phases: route discovery and route maintenance. In former phase, the most optimal path for data packet transmission between the source and sink is determined whereas, latter phase maintains the network as the topology of a MANET is dynamic in nature, therefore there is a connection breakage which results in network failure between mobile nodes [9]. AODV is designed for both unicast and multicast routing. When the source node transmits the data to sink in DSR, then whole path is included in the header. The entire path lengthens as the network grows, decreasing the network's overall performance. The AODV protocol was designed to address the issue [10]. DSR stores route information in the header of the data packet, whereas AODV saves it in the routing table. It uses DSN to discover the best relevant route, and the primary distinction between AODV and DSR is that DSR employs source routing, in which a data packet contains the entire path to be travelled whereas, the next-hop information relating to each flow for data packet transfer is held by the source and intermediate nodes in AODV [11]. It does not add to the amount of communication traffic and the time it takes to establish a connection is shorter than DSR.

AOMDV is a multipath enhancement to the AODV protocol [12]. In AOMDV, multiple paths between the source and sink are established. It uses alternate routes in the case of a route breakdown. A new route discovery is required when all routes fail. Multipath routing is an improvement over unipath routing, with the added benefit of handling network traffic, minimizing congestion, and enhancing reliability [13]. Because it is a multipath routing protocol, the sink responds to numerous route request packets which results in lengthier overhead packets in response to a single RREQ packet, which may result in significant control overhead.

The advantages of on-demand and table-driven routing protocols are integrated in this protocol [10]. This protocol can modify the zone and position of mobile nodes between source and sink since they are adaptive in nature. One of the commonly used hybrid protocol is Zone Routing Protocol in which the entire network is divided into zones, and the source and sink locations are tracked. Table-driven protocol is used to transfer packets between source and sink when they are in the same zone whereas, on-demand routing is applied when they are in different zones. Figure 1 illustrate the taxonomy of routing protocols used in MANET.

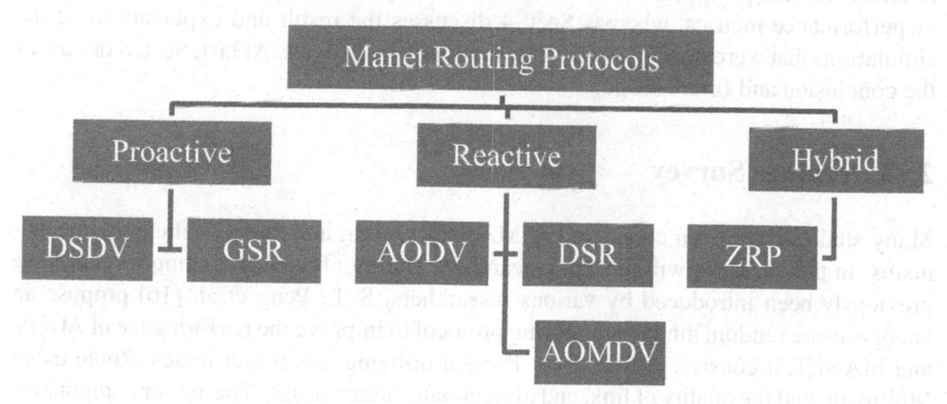

Fig. 1. Routing protocol taxonomy

Despite their widespread use, traditional routing protocols are frequently failed to meet network performance expectations. As a result, the researchers have concentrated on optimization strategies that employ natural animal behavior to tackle complex issues. The node's energy efficiency is the most important consideration in MANET because data transmission consumes most of the energy. As a result, to address this issue of energy efficiency, optimization techniques have proven to be the most effective solution, resulting in a longer network lifetime.

Because of the network's dynamic nature, designing and implementing protocols in MANET is particularly complex. MANET's resources, such as power and infrastructure, are limited. The use of SI-based techniques such as ACO and Particle Swarm Optimization has been encouraged by the necessity for an optimal communication path among nodes. These methods are based on evolutionary process, in which different agents interact with one another to discover the best or optimal path. The basic idea underlying these approaches is to calculate the fitness function over an initial population and then iteratively calculate the fitness value on a new population, which is usually acquired by performing operations on populations. In this research work, we explored the concept of ACO and apply its optimized concept to the routing protocols (AODV and AOMDV) and then performed a comparison analysis to validate its performance.

ACO is a meta-heuristic approach which is based on the concept of population that is inspired by the foraging behavior of various ant species and can be used to resolve a variety of MANET problems, including routing of packets across the network [14]. The chemical compounds named as Pheromone is created by ants used to identify a suitable path for the other members of the colony to follow. Mobile nodes can exhibit the same type of behavior when determining the best path for their route discovery. Every node includes an artificial pheromone that indicates the route discovery from source sink has used the edge between nodes. The ACO algorithm begins with parameter setup and pheromone networks. This is repeated recursively on the cycle until the first ant's path is constructed, and then the ant path is formed continually using local search. The pheromone network is finally updated [15].

The remaining paper is structured as follow: Sect. 2 explores the literature survey related to routing protocols and optimization techniques. Section 3 covers the evaluation or performance metrics, whereas Sect. 4 discusses the result and explanations of the simulations that were done to validate the comparison analysis. At last, Sect. 5 describes the conclusion and future scope.

2 Literature Survey

Many studies have been conducted in MANET routing and energy efficiency mechanisms. In this area, we will present research on energy efficiency techniques that have previously been introduced by various researchers. S. L. Peng et al. [16] propose an energy-aware random multi-path routing protocol to improve the performance of AODV and MANET. It consists of four parts: Prevent utilizing less power nodes, Route using Multipath, find the quality of link, and disseminate energy usage. These four components have a role in route selection at various stages. Researchers avoid utilizing low battery nodes in route finding to minimize network partitioning. To disperse power consumption, researcher employ the concept of multipath routing. Furthermore, in the packet forwarding phase, the author develops a probabilistic packet forwarding technique based on node leftover energy.

N. Muthukumaran [17] examined the throughput capacity of MANET with reduced packet loss and uses the MAC 802.11 protocol to discover the nodes. It can automatically discover neighbor nodes and uses the Flooding method to select the next node. Each node in this technique attempts to send all information to each of its nearest source nodes before receiving acknowledgement from the destination nodes. The shortest path between the source and sink is calculated using the K- Nearest Neighbor method. It analyses the distance between nodes and sorts the closest neighbors according to the shortest distance. The DSR algorithm is used to perform the routing, which permits every controller node to choose and manage the routes used for transferring the packets. When throughput of nodes is considered, packet loss is reduced. M. Saxena et al. [18] designed a clustering-based algorithm for MANET larger lifespan. To improve the network longevity, this technique divides the network into small, self-managing groups. When determining cluster structure, the proposed method considers both scalability and the energy measure. The Maxheap concept is used to select the cluster head. Clusters are formed using max heap based on energy level, with the cluster head being the node in the cluster with the highest energy.

S. Gangwar et al. [19] MANET employs a variety of soft computing techniques such as Neural Networks, Genetic Algorithms (GA), and Fuzzy Logic (FL). Because the sensors in MANET are deployed at random in the field, there is a problem with coverage for all or some of the nodes. The network's overall throughput may be impacted by the coverage issue. The solutions to such difficulties are provided by neural networks, which uses proactive and reactive protocols. The proactive protocol keeps track of routing table information, while the reactive protocol establishes a connection between source and sink. The best path between the source and destination nodes is found using GA. FL is used when there is low communication overhead and storage need. M. Kumar et al. [20] suggested a Maximum (MAX) energy multipath routing strategy with the goal of improving energy efficiency and network connection. MAX always chooses the neighbor nodes with more energy out of all the neighbors. The receiving and transmitting of data in a complex network quickly deplete the energy of intermediary nodes. MAX made use of the energy of these nodes by picking them based on their MAX energy level and establishing the multiple paths between source and sink node. The sender has received the responses to each request and has picked the MAX energy node to transmit data. In comparison to existing shortest multipath routing, the suggested energy-efficient strategy enhances network performance.

K. Majumder et al. [21] examines the performance of numerous reactive protocols in terms of two performance measures: average E2E delay and PDR, using identical loads and environment conditions. The simulation findings show that as the number of nodes rises, the average End-to-End delay increases and when the pause period increases, the mean E2E latency decreases, whereas the number of nodes increases the mean loop detection time. U. S. et al. [22] proposed a new Ant colony optimization with delay aware energy efficient algorithm for choosing the best path while minimizing network delay time. The main goal of this research work is to keep the network's optimal path while sending data efficiently. This method is used to build an energy efficient layer routing protocol within the network for military applications, and the simulation is done with the use of a novel cross layer strategy to design for enhancing the network's reliability and lifetime.

D. Rupérez Cañas et. al. [23] suggested a unique routing protocol named as HACOR (Hybrid Ant Colony Optimization Routing) which is a hybrid bioinspired protocol that employs the concept of Swarm Intelligence. It works in a multipath mode because it creates many paths for sending data to the sink node. It is also adaptable because it adjusts to changing traffic and network circumstances. HACOR does not use the evaporation process during the route exploration phase and while creating a route with all visited nodes, HACOR employ the free-loop method. In HACOR there is no need to set or change the initial pheromone value for each one-hop neighbor, to accomplish this, researcher use the message HELLO.

3 Evaluation Metrices

For the performance analysis of AODV, AOMDV, and ACO-AOMDV, the following measures are used [24].

E2E Delay: It is the average time taken for packets to be successfully transmitted from source to sink. It can be computed as:

$$End\ to\ End\ delay = \sum\nolimits_{j=1}^{n} ((Dj - Sj))/n$$

where D stands for destination and S stands for source.

Throughput: It refers to how much data the destination receives in a certain length of time. It is expressed in kbps.

$$Throughput = (number\ of\ bytes\ received\ * 8 \\ / \ the\ time\ taken\ for\ simulation) * 1000\ kbps.$$

PDR: It is the proportion of total packets received by the sink node (P_R) to total packets sent by the sender node (P_S) [25].

$$PDR = (P_R/P_S) * 100$$

Energy Consumption (EC): It is the quantity of energy utilized by nodes during the simulation time. This is determined by computing the energy level of each node at the end of the simulation.

$$Energy\ Consumption = \sum\nolimits_{j=1}^{n} (init(j) - ener(j))$$

4 Result and Discussion

For simulation, the NS-2 network simulator tool is utilized because it provides accurate implementations of several network protocols 12. It is used to compare and examine the results of AODV, AOMDV and ACO-AOMDV protocols. With a node count from 20 to 100 and deployment area of 1000 * 1000 m^2 the routing techniques are implemented. Table 1 below, lists the simulation parameters.

Table 1. Simulation parameters

Parameters	Value
Type of channel	Wireless
Radio propagation model	Two-way Ground
Mac type	Mac/802_11
Length of queue	50
Node size	20, 30, 40, 50, 60, 70, 80, 90, 100
Nodes speed	20
Area	1000 m * 1000 m
Type of traffic	CBR
Simulation time	80 s
Routing protocol	AODV, AOMDV
Speed type	Uniform
Pause time	5 s
Pause type	Uniform

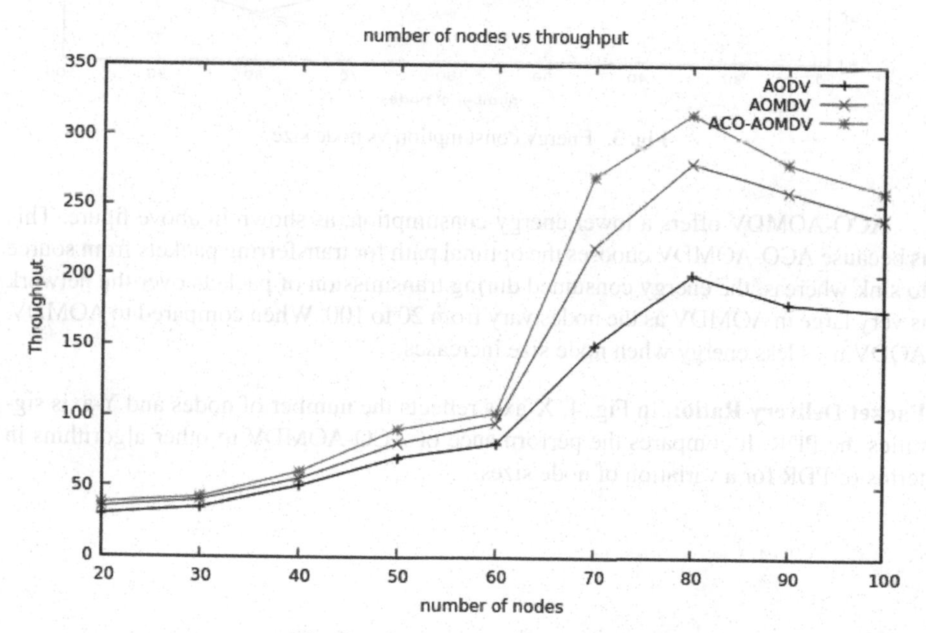

Fig. 2. Throughput vs no. of nodes

Throughput: The most critical performance metric of any routing technique is throughput. Figures 2 exhibits throughput versus number of nodes in which X-axis indicates the node size and Y-axis depicts the throughput.

Findings of the simulation reveal that as the network size gradually increases from 20 to 100, the throughput of ACO-AOMDV stays higher than that of other alternative techniques. Among the other methodologies in the comparison, AOMDV has the second greatest throughput, whereas the AODV has least throughput among all protocols.

Energy Consumption: Figure 3 represents the plots for energy EC versus node size respectively.

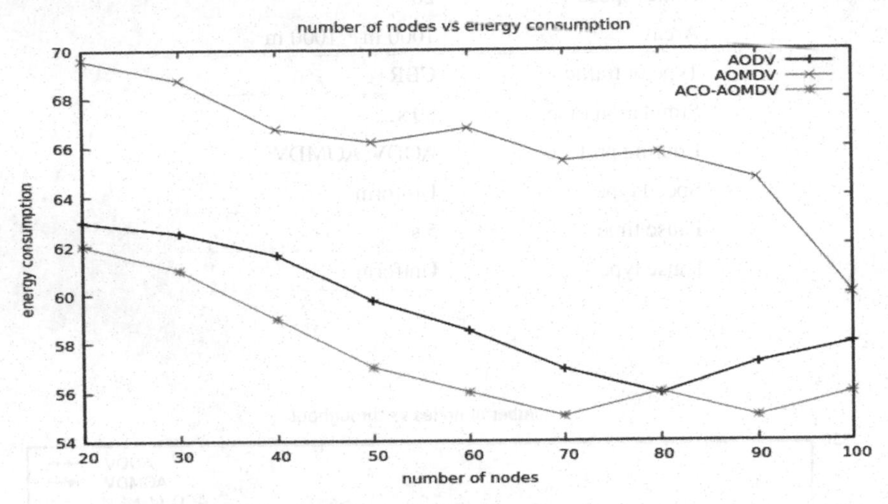

Fig. 3. Energy consumption vs node size

ACO-AOMDV offers a lower energy consumption, as shown in above figure. This is because ACO-AOMDV chooses the optimal path for transferring packets from source to sink whereas, the energy consumed during transmission of packets over the network is very large in AOMDV as the nodes vary from 20 to 100. When compared to AOMDV, AODV uses less energy when node size increases.

Packet Delivery Ratio: In Fig. 4, X-axis reflects the number of nodes and Y-axis signifies the PDR. It compares the performance of ACO-AOMDV to other algorithms in terms of PDR for a variation of node sizes.

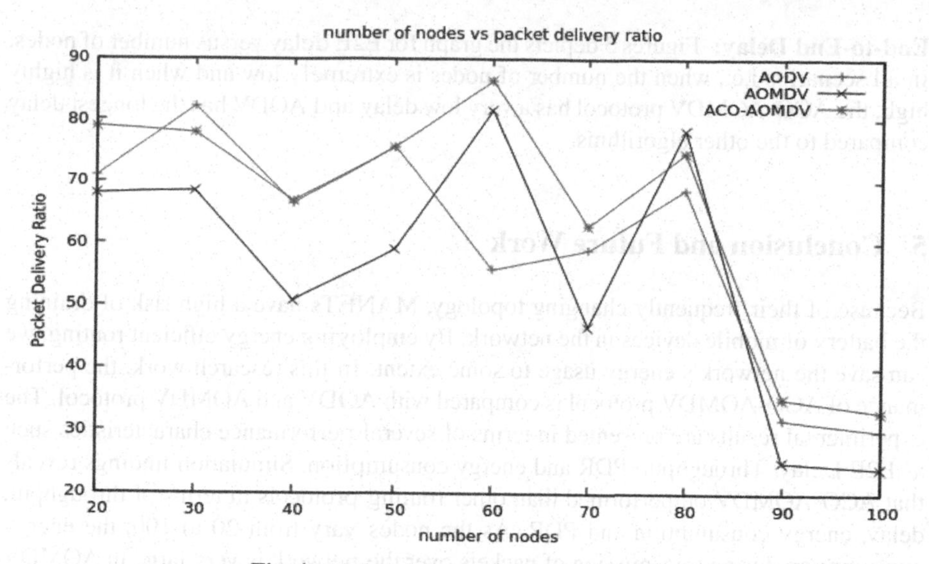

Fig. 4. Packet delivery ratio vs no. of nodes

When the nodes count is not very large, the performance of ACO-AOMDV is better than other protocols, but as the number of nodes increases, the performance of ACO-AOMDV steadily diminishes. AODV has the second highest PDR among the other protocols in the consideration.

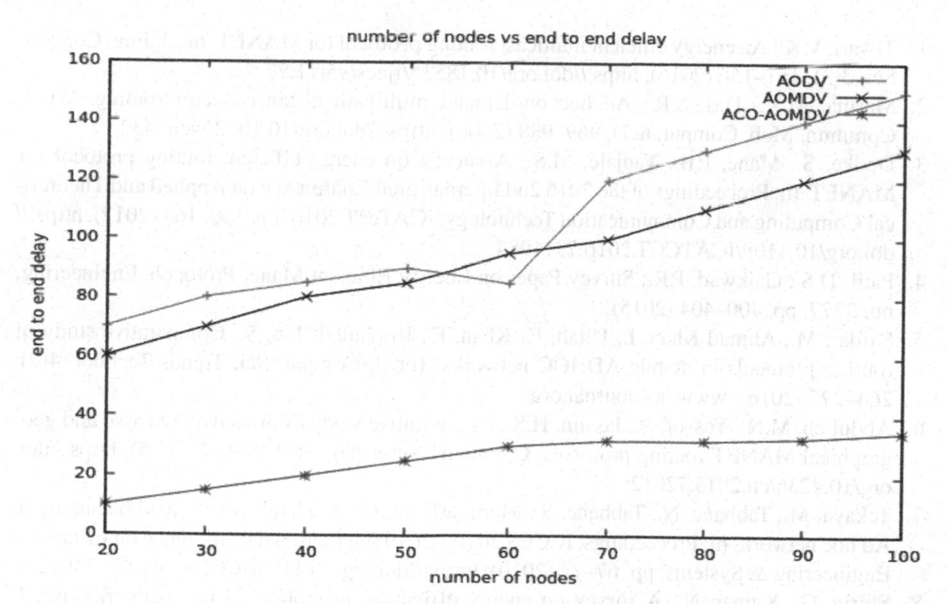

Fig. 5. E2E delay vs number of nodes

End-to-End Delay: Figures 5 depicts the graph for E2E delay versus number of nodes. In all scenarios i.e., when the number of nodes is extremely low and when it is highly high, the ACO-AOMDV protocol has a very low delay and AODV has the longest delay compared to the other algorithms.

5 Conclusion and Future Work

Because of their frequently changing topology, MANETs have a high risk of draining the battery of mobile devices in the network. By employing energy efficient routing, we can save the network's energy usage to some extent. In this research work, the performance of ACO-AOMDV protocol is compared with AODV and AOMDV protocol. The experimental results are presented in terms of several performance characteristics such as E2E Delay, Throughput, PDR and energy consumption. Simulation findings reveals that ACO-AOMDV outperformed than other routing protocols in terms of throughput, delay, energy consumption and PDR. As the nodes vary from 20 to 100, the energy consumption during transmission of packets over the network is very large in AOMDV whereas, AODV uses less energy when compared to AOMDV. Results shows that AODV method has second highest PDR and longest delay compared to the other algorithms. In future, the performance analysis of different metaheuristic methodologies could be studied to validate the use of Swarm Intelligence techniques for finding optimal MANET routing paths.

References

1. Tiwari, V.K.: An energy efficient multicast routing protocol for MANET. Int. J. Eng. Comput. Sci. **3**(3), 151–156 (2016). https://doi.org/10.18535/ijecs/v5i11.77
2. Marina, M.K., Das, S.R.: Ad hoc on-demand multipath distance vector routing. Wirel. Commun. Mob. Comput. **6**(7), 969–988 (2006). https://doi.org/10.1002/wcm.432
3. Dodke, S., Mane, P.B., Vanjale, M.S.: A survey on energy efficient routing protocol for MANET. In: Proceedings of the 2016 2nd International Conference on Applied and Theoretical Computing and Communication Technology, iCATccT 2016, pp. 160–164 (2017). https://doi.org/10.1109/ICATCCT.2016.7911984
4. Patil, D.S., Gaikwad, P.R.: Survey Paper on Energy Efficient Manet Protocols Engineering, no. 2277, pp. 400–404 (2015)
5. Safdar, M., Ahmad Khan, I., Ullah, F., Khan, F., Roohullah Jan, S.: Comparative study of routing protocols in mobile ADHOC networks. Int. J. Comput. Sci. Trends Technol. **4**(2), 264–275 (2016). www.ijcstjournal.org
6. Abdulleh, M.N., Yussof, S., Jassim, H.S.: Comparative study of proactive, reactive and geographical MANET routing protocols. Commun. Netw. **07**(02), 125–137 (2015). https://doi.org/10.4236/cn.2015.72012
7. Tekaya, M., Tabbane, N., Tabbane, S.: Multipath routing mechanism with load balancing in Ad hoc network. In: Proceedings, ICCES 2010 - 2010 International Conference on Computer Engineering & Systems, pp. 67–72 (2010). https://doi.org/10.1109/ICCES.2010.5674892
8. Shirly, G., Kumar, N.: A survey on energy efficiency in mobile ad hoc networks. Int. J. Eng. Technol. **7**(2.21 Special Issue 21), 382–385 (2018). https://doi.org/10.14419/ijet.v7i2.21.12447

9. Salem Jeyaseelan, W.R., Hariharan, S.: Comparative study on MANET routing protocols. Asian J. Inf. Technol. **15**(9), 1411–1415 (2016). https://doi.org/10.3923/ajit.2016.1411.1415
10. Hinds, A., Ngulube, M., Zhu, S., Al-Aqrabi, H.: A review of routing protocols for Mobile Ad-Hoc NETworks (MANET). Int. J. Inf. Educ. Technol. **3**(1), 1–5 (2013). https://doi.org/10.7763/ijiet.2013.v3.223
11. Arya, A., Singh, J.: Comparative study of AODV, DSDV and DSR routing protocols in wireless sensor network using NS-2 simulator, vol. 5, no. 4, pp. 5053–5056 (2014)
12. Agrawal, J., Singhal, A., Yadav, R.N.: Multipath routing in mobile Ad-hoc network using meta-heuristic approach. In: 2017 International Conference on Advances in Computing, Communications and Informatics, ICACCI 2017, January 2017, pp. 1399–1406 (2017). https://doi.org/10.1109/ICACCI.2017.8126036
13. Aggarwal, P., Garg, E.P.: AOMDV protocols in MANETS : a review. Int. J. Adv. Res. Comput. Sci. Technol. (IJARCST) **4**(2), 32–34 (2016)
14. Kaur, K., Pawar, L.: Review of various optimization techniques in MANET routing protocols. Int. J. Sci. Eng. Technol. Res. **4**(8), 2830–2833 (2015)
15. Anibrika, B.S.K., Asante, M., Hayfron-Acquah, B., Ghann, P.: A Survey of Modern Ant Colony Optimization Algorithms for MANET?: Routing Challenges, Perspectives and Paradigms. Int. J. Eng. Res. Technol. (IJERT) **9**(05), 952–959 (2020)
16. Peng, S.L., Chen, Y.H., Chang, R.S., Chang, J.M.: An energy-aware random multi-path routing protocol for MANETs. In: Proceedings - 2015 IEEE International Conference on Smart City, SmartCity 2015, Held Jointly with 8th IEEE International Conference on Social Computing and Networking, SocialCom 2015, 5th IEEE International Conference on Sustainable Computing and Communic, pp. 1092–1097 (2015). https://doi.org/10.1109/SmartCity.2015.214
17. Muthukumaran, N.: Analyzing throughput of MANET with reduced packet loss. Wireless Pers. Commun. **97**(1), 565–578 (2017). https://doi.org/10.1007/s11277-017-4520-9
18. Saxena, M., Phate, N., Mathai, K.J., Rizvi, M.A.: Clustering based energy efficient algorithm using max-heap tree for MANET. In: Proceedings - 2014 4th International Conference on Communication Systems and Network Technologies, CSNT 2014, pp. 123–127 (2014). https://doi.org/10.1109/CSNT.2014.33
19. Gangwar, S., Kumar, P.: Introduction of various soft computing techniques in mobile ad-hoc network. Int. J. Sci. Eng. Res. **8**(10), 419–423 (2017)
20. Kumar, M., Dubey, G. P.: Energy Efficient Multipath Routing with Selection of Maximum Energy and Minimum Mobility in MANET. In: International Conference on ICT in Business Industry & Government (ICTBIG). IEEE (2016)
21. Majumder, K., Sarkar, S.K.: Performance analysis of AODV and DSR routing protocols in hybrid network scenario. In: Proceedings of the INDICON 2009 - An IEEE India Council Conference, vol. 2, no. 2 (2009). https://doi.org/10.1109/INDCON.2009.5409480
22. Srilakshmi, U., et. al.: Modified energy efficient with ACO routing protocol for MANET. Turkish J. Comput. Math. Educ. **12**(2), 1739–1745 (2021). https://doi.org/10.17762/turcomat.v12i2.1510
23. Rupérez Cañas, D., Sandoval Orozco, A.L., García Villalba, L.J., Hong, P.S.: Hybrid ACO routing protocol for mobile Ad hoc networks. Int. J. Distrib. Sens. Netw. **2013** (2013). https://doi.org/10.1155/2013/265485
24. Singh, S.K., Prakash, J.: Energy efficiency and load balancing in MANET: a survey. In: 2020 6th International Conference on Advanced Computing and Communication Systems. ICACCS 2020, pp. 832–837 (2020). https://doi.org/10.1109/ICACCS48705.2020.9074398
25. Ramesh, P., Devapriya, M.: An optimized energy efficient route selection algorithm for mobile ad hoc networks based on LOA. Int. J. Eng. Adv. Technol. **8**(2), 298–304 (2018)

Author Index

Printed in the United States
by Baker & Taylor Publisher Services